AF885493

Cover Page Designed by:

Ms. Gowri Vemuri

First Edition

May, 2019

Author:

MATH-KNOTS

Index

Preface

The word “Science”brings to mind many different pictures: a fat textbook, bunch of equations, space shuttle, medicines and microscopes etc., simply discovering and learning about the nature around us. There are many branches of science and physics the most basic area of sciences, which deals with energy, matter, space, time and their interactions. The word physics is derived from the Latin word *physica*, which means "natural thing."

Physics as a fundamental science helps to uncover the laws of nature. The language of expression is mathematics. During ancient times, humans studied Stars and celestial bodies that lead to the development of disciplines like astronomy and mathematics. Based on size and speed, Physics as a subject is categorized several categories, which you can find from the following picture.

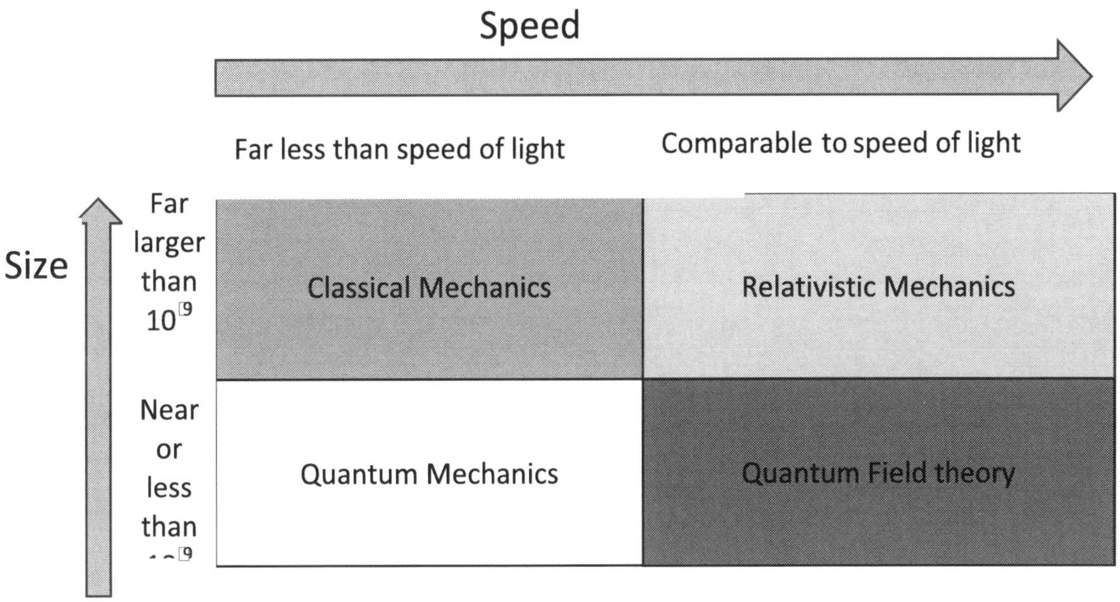

There are many ways to learn physics, but often students face difficulty in solving a problem by the concept they have learnt in the class. Our book is designed to help students in this area. Each chapter in the book, provides a summary of a concept, along with many practice problems.

Chapter - 1

UNITS AND MEASUREMENTS

MEASUREMENTS OF PHYSICAL QUANTITIES

SYNOPSIS

I. ERRORS

1) Error = measured value - true value
2) Error can be either positive or negative.
3) The true value of a measurement is the arith-metic mean of the several repeated measurements.

II. Accuracy and Precision :

1) Accuracy is a measure of closeness of the measured value to the true value. Smaller the uncertainty in the measured value, greater is its accuracy.
2) Precision refers to the agreement among a group of measured values. It does not imply anything about their relation to the true value.
3) A precise value doesn't necessarily mean an accurate value.
4) No measurement can be made with complete 'accuracy.' For this , it must be shown by a value with infinite number of decimal places.

5) True value of a measured physical quantity is something that is never known.

III. Depending on their origin, errors may be classified as follows :

Systematic errors : These are due to definite cause.Systematic errors are also called as determinate errors.Systematic errors are unidirectional w.r.t. the true value.Systematic errors in instruments can be avoided by proper design of the instrument.

Examples :

1) **Least count error :** This is due to resolution of instrumental error.
2) **Zero error :** It creeps in when zero division doesn't show zero.
3) **Theoritical error :** It is due to approximations made in derivations.
4) **Parallax error :** This personal error becomes systematic when a scale is viewed always in a particular direction.

Random errors : These errors arise from uncontrollable causes.Random errors are also called indeterminate errors.Random errors cannot be eliminated.Random errors are the ultimate limitations on the measurement.
Increasing the number of observations minimize random errors because positive and negative random errors occur with equal probability.

Examples :

1) Line voltage changes can cause random errors
2) Backlash errors : Appears in the instruments that work on the principle of screws. It is due to allowance between the screw and nut.

Gross errors : Arise due to faulty adjustment or improper usage of the instrument.

a) Increasing the number of observations has no effect on systematic and personal errors.
b) **Personal errors** can be minimized, if the experimenter is more careful and skilled.
c) The total random errors are determined by the statistical methods.

IV. Some definitions in statistical method :

1) **Variation :** When a particular physical quantity is measured a number of times, in general these values are not identical but may be nearer to one another. This is called Variation.
2) **Range :** The difference between lowest and highest readings (Measured values) is called range.
3) **Average :** If the distribution shows a central tendency about a value in the range, this value is called the average.

All the other values lie around this average value.

4) **Averages are of three types :**

i) **Median :** When the given data is written either in ascending order or in descending order, the middle value in the data is called the median. This is also called position average.

ii) **Mode :** If the same value appears a number of times in the data, this value is called 'mode' or 'most probable average'

iii) **Arithmetic mean :** If $X_1, X_2 \ldots\ldots\ldots X_N$ are the results of an experiment repeated N times, the arithmetic mean $\overline{x}$ is given by

$$\overline{x} = \frac{x_1 + x_2 + \ldots\ldots\ldots\ldots + x_n}{n} = \frac{\sum x_i}{n}$$

5) Arithmetical mean is the best value of x that can be obtained from the N measurements. This is 'true' when N is very large.
6) **Absolute error or deviation :** It is the magnitude of the difference between the arithmetical mean value and an individual measured value.

$$d_1 = |\overline{x} - x_1|, d_2 = |\overline{x} - x_2|, \ldots\ldots\ldots\ldots \text{etc.}$$

7) **Mean deviation (or final absolute error).** It is the arithmetical mean of all the deviations.

$$d = \frac{\sum d_i}{N} = \frac{d_1 + d_2 + \ldots\ldots\ldots\ldots + d_N}{N}$$

8) **Standard deviation (σ) :** It is given by the square root of the arithmetic mean of the squares of the deviations.

$$(\sigma) = \sqrt{\frac{d_1^2 + d_2^2 + \ldots\ldots\ldots + d_N^2}{N}}$$

(σ) is positive

9) In case of lesser number of observations Bessel's formula is used to determine (σ)

$$(\sigma) = \sqrt{\frac{d_1^2 + d_2^2 + \ldots\ldots\ldots + d_N^2}{\sqrt{N(N-1)}}}$$

10) **Probable error :** It is given by

$r = 0.6745\ (\sigma)$

11) Strictly speaking the true value 'μ' of a physical quantity assumed to be in between $(\bar{x} - r)$ and $(\bar{x} + r)$. It is also expressed as $\mu = \bar{x} \pm r$

V. Evaluation of Errors :

1) **Relative error :** The ratio of error in a measurement to the true value (or) arithmetical mean is called relative error.

Relative error : $\frac{\Delta x}{\mu}$ or $\frac{\Delta x}{x}$

2) **Percentage error** (ε) = (Relative error × 100%)

3) In an experiment 'probable error (r)' is calculated from the data of measurements and the result is obtained after correction.

4) If a quantity is measured as $x \pm \Delta x$, maximum deviation becomes $2\Delta x$. In the experimental data of large number of observations the values which have the variation greater than $2\Delta x$ from 'x' are omitted.

5) **Error of a sum or a difference :**

If $Z = x \pm y$, then $\Delta Z = \Delta x + \Delta y$,

Note : All terms must be positive

6) In the above case, percentage of errors are calculated as follows.

a) If $z = x + y$

$$\varepsilon_z = \left(\frac{\Delta x + \Delta y}{x + y}\right) \times 100$$

b) If $z = x - y$

$$\varepsilon_z = \left(\frac{\Delta x + \Delta y}{x - y}\right) \times 100$$

Note : i) Consider all the errors as positive

ii) $\Delta x + \Delta y$ is called maximum permissible error and $\Delta x, \Delta y$ will be taken with positive sign.

7) **In subtraction**, the percentage of error is large. So in an experiment the differences should be measured with a very high degree of accuracy.

8) **Error of a product or quotient**

if $z = k(xy)$ (or) $z = K\left(\frac{x}{y}\right)$

$$\frac{\Delta z}{z} = \frac{\Delta x}{x} + \frac{\Delta y}{y}$$

Note : K is a Constant

9) **Error due to power of a measured quantity:**

If $z = Kx^n$, then $\frac{\Delta z}{z} = n\frac{\Delta x}{x}$

Note : 'n' can be any real number.

10) If $P = x^\alpha y^\beta z^\gamma$, the relative error in P is given by

$$\frac{\Delta p}{p} = \alpha\frac{\Delta x}{x} + \beta\frac{\Delta y}{y} + \gamma\frac{\Delta z}{z}$$

Note : All the terms must be taken as positive even if the powers are negative.

11) If Resistors R_1 and R_2 are in series, then their equivalent resistance, $R = R_1 + R_2$

Maximum error in R, is, $\Delta R = \Delta R_1 + \Delta R_2$

12) If Resistors R_1 and R_2 are in parallel, their equivalent resistance is 'R' such that, $R = \frac{R_1 R_2}{R_1 + R_2}$

If ΔR = maximum error in R

$$\frac{\Delta R}{R} = \frac{\Delta R_1}{R_1} + \frac{\Delta R_2}{R_2} + \frac{\Delta R_1 + \Delta R_2}{R_1 + R_2}$$

13) Relative error or percentage error is the true index of degree of accuracy. Error doesn't give an idea of degree of accuracy.

14) The knowledge of relative or percentage errors, besides estimating the error in the experimental result, can be used for knowing the required accuracy of different quantities measured in the experiment.

15) In experiments, some physical quantities should be measured with greater accuracy as they contribute more percentage of errors if measured with same accuracy as the remaining quantities.

Example :

In the determination of coefficient of Viscosity by $\eta = \frac{\pi Pr^4}{8vl}$ more degree of accuracy is required in the measurement of 'r' because any small error in 'r' is magnified 4 times.

EXERCISE - I

1. **Zero error of an instrument introduces**
   1) Systematic error 2) Random error
   3) 1 and 2 Both 4) Gross error
2. **Which of the following is systematic error**
   1) least count error 2) parallox error
   3) theoritical error 4) all
3. **In a measurement both positive and negative errors are found to occur with equal probability. The type of errors responsible for this is**
   1) proportional errors 2) ramdom errors
   3) determinate errors 4) systematic errors
4. **By repeating the same measurement several times, the errors that can be reduced are**
   1) determinate errors 2) instrumental errors
   3) random errors 4) systematic errors
5. **In an experiment if the measured values of a physical quantity are highly concurrent, these measurements are said to be**
   1) accurate 2) precise
   3) both precise and accurate
   4) neither accurate nor precise
6. **In an experiment if the measured values of a physical quantity have a high degree of closeness to the ture value. These measurements are said to be**
   1) both precise and accurate
   2) precise 3) accurate
   4) neither accurate nor precise
7. **The measured value of physical quantity expressed to infinite number of decimal places is called**
   1) practical value 2) ideal value
   3) absolute value 4) real value
8. **The arithmetic mean of several measurements is called**
   1) practical value 2) imaginary value
   3) true value 4) ideal value
9. **Zero errors of measuring instruments are called**
   1) Indeterminate errors 2) Random errors
   3) Disproportional errors 4) Instrumental errors
10. **The type of errors that can never be completely eliminated are**
    1) determinate error 2) instrumental errors
    3) proportional errors 4) random errors
11. **Which of the following error is not systematic error?**
    1) Least count error 2) Zero error
    3) Backlash error
    4) Throretical error due to approximate
12. **Round off 20.96 to 3 significant figures**
    1) 20.9 2) 20 3) 21.0 4) 21
13. **Of the following the dimensionless error is**
    1) Systematic error 2) Gross error
    3) Random error 4) Relative error
14. **An experimenter measured the diameter of a wire without noting zero error. This error is**
    1) Random error 2) Systematic error
    3) Gross error 4) Personal error
15. **Among the following the error that can be eliminated is**
    1) Systematic errors 2) Random errors
    3) Gross errors 4) None
16. **In determing viscosity (n) by poiseuill's method the formula used $\eta = \frac{\pi pr^4}{8vl}$ which of the quantities in the formula must be measured more accurately**
    1) p 2) r 3) v 4) l
17. **a) All zeros to the right of the last non-zero digit after the decimal point are significant.**
    b)If the number is less than one, all the zeros to the right of the decimal point but to the left of the first non-zero digit are not significant.
    1) only a is correct 2) a, b are correct
    3) only b is correct 4) a, b are wrong
18. **If the immediate insignificant digit to be dropped is 5 then there will be two different cases**
    a) If the preceding digit is even, it is to be unchanged and 5 is dropped.
    b) If the preceding digit is odd, it is to be raised by 1.
    1) only a is correct 2) a, b are correct
    3) only b is correct 4) a, b are wrong

19. a) In multiplication or division, the final result should retain only that many significant figures as are there in the original number with the least number of significant figures.

b) In addition or subtraction the final result should retain only that many decimal places as are there in the number with the least number of decimal places.

1) only a is correct 2) a, b are correct
3) only b is correct 4) a, b are wrong

20. (A) : Increasing the number of observations minimises random errors

(R) : positive and negative random errors occur with equal probability.

1) both (A) and (R) are true and (R) is the correct explanation of (A)
2) both (A) and (R) are true and (R) is not the correct explanation of (A)
3) (A) is true but (R) is false
4) (A) is false but (R) is true

21. The number of significant figures in 0.007 is
1) 4 2) 2 3) 3 4) 1

22. The number of significant figures in 6.032 is
1) 4 2) 2 3) 3 4) 1

23. The number of significant figures in 0.2370 is
1) 5 2) 4 3) 3 4) 2

ANSWERS

1) 1	2) 4	3) 2	4) 4	5) 2
6) 3	7) 2	8) 3	9) 4	10) 4
11) 3	12) 3	13) 4	14) 2	15) 1
16) 2	17) 2	18) 2	19) 2	20) 1
21) 4	22) 1	23) 2		

EXERCISE - II(A)

ERRORS :

1. The radius of a sphere is measured as $(10 \pm 0.02\%)$ cm. The error in the measurement of its volume is
1) 25.1cc 2) 25.12cc
3) 2.51cc 4) 251.2cc

2. If length and breadth of a plate are (40 ± 0.2) and (30 ± 0.1) cm, the absolute error in measurement of area is
1) 10 cm^2 2) 8 cm^2 3) 9 cm^2 4) 7 cm^2

3. When 10 observations are taken the random error is x when 100 observations are taken the random error becomes
1) x/10 2) x^2 3) 10x 4) $\sqrt{x}$

4. The diameter of a wire as measured by a screw gauge was found to be 1.002cm, 1.000cm, 1.006cm. The absolute error in the second reading is
1) 0.002 cm 2) 0.004 cm
3) 0.006 cm 4) 0.003 cm

5. The percentage error in the measurement of mass and speed are 2% and 3% respectively. Maximum estimate of percentage error of K.E
1) 11% 2) 8% 3) 5% 4) 1%

6. The error in the measurement oflength of a simple pendulum is 0.1% and error in the time period is 2%. The possible maximum error in the quantity having dimensional formula LT^{-2} is
1) 1.1% 2) 2.1% 3) 4.1% 4) 6.1%

7. The length and breath of a rectangular object are 25 cm and 10cm respectively and were measured to an acuracy of 0.1 cm. The percentage error in area is
1) 2.8 % 2) 0.5% 3) 1.4% 4) 3%

8. If the length of a cylinder is measured to be 4.28 cm with an error of 0.01 cm. The percentage error in the measured length is nearly
1) 0.4% 2) 0.5% 3) 0.2% 4) 0.1%

9. The side of a cube , as measured with a vernier calipers of least count 0.01cm is 3.00 cm. The maximum possible error in the measurement of volume is
1) $\pm$ 0.01cm^3 2) $\pm$ 0.06cm^3
3) $\pm$ 0.0cm^3 4) $\pm$ 0.27cm^3

10. If $L = (20 \pm 0.01)$ m and $B = (10 \pm 0.02)$ m then L/B is
1) (2 ± 0.03)m 2) (2 ± 0.015)m
3) (2 ± 0.01)m 4) (2 ± 0.005)m

SIGNIFICANT FIGURES AND ROUNDING OFF

11. In a given number 600 if only 6 is significant we may write the number as
1) 6.00 2) 6×10^2
3) 600 4) 6.00×10^{12}

12. **The number of significant figures in 0.0006032 is**
1) 7 2) 4 3) 5 4) 2

13. **The number of significant figures in 6.320 is**
1) 4 2) 2 3) 3 4) 1

14. **The radius of a disc is 1.2cm. Its area according to the idea of significant figures is**
1) 4.5216 cm^2 2) 4.521 cm^2
3) 4.52 cm^2 4) 4.5 cm^2

15. **The mass of a block of wood is 87.2g and its volume is $25cm^3$. Its density with due regard to significant figures is**
1) 3.488g cm^{-3} 2) 3.5 g cm^{-3}
3) 3.48 g cm^{-3} 4) 3.4 g cm^{-3}

16. **The length of a rod is measured by different instruments. Of the following the accurate result is**
1) 500 mm 2) 500.00 mm
3) 500.0 mm 4) 0.5 m

17. **2.34 is obtained by rounding of the number**
1) 2.346 2) 2.355 3) 2.335 4) 2.334

18. **When 0.0003125 is reduced to 3 significant figures its value is**
1) 0.00312 2) 3.125
3) 0.000312 4) 0.000313

19. **The value of 2.2+4.08+3.125+6.3755 with due regard to significant places is**
1) 15.78 cm 2) 15.7805 cm
3) 15.780 cm 4) 15.8 cm

20. **$9765 \div 14.39 = 678.59625$. This quotient on rounding off to two significant figures is**
1) 680 2) 70 3) 678.6 4) 679

21. **The value of 124.2-52.487 with due regard to significant places is**
1) 71.7 2) 71.71 3) 72 4) 71

22. **The value of 42×0.041 with due regard to significant figures is**
1) 1.722 2) 1.72 3) 1.7 4) 1.8

23. $\sqrt{58.97} =$
1) 7.679 2) 7.68 3) 7.6 4) 7.7

ANSWERS

1) 3	2) 1	3) 1	4) 4	5) 2
6) 3	7) 3	8) 3	9) 4	10) 4
11) 2	12) 2	13) 1	14) 4	15) 2
16) 2	17) 3	18) 3	19) 4	20) 1
21) 1	22) 3	23) 2		

EXERCISE - II(B)

ERRORS

1. **The sides of a rectangle are $(5.5\pm\ 0.1)$cm and $(3.5\pm\ 0.2)$cm. Its perimeter is**
1) $(18.0\pm\ 0.1)$cm 2) $(18.0\pm\ 0.2)$cm
3) $(18.0\pm\ 0.3)$cm 4) $(18.0\pm\ 0.6)$cm

2. **The length and breadth of a plate are (6 ± 0.1) cm and (4 ± 0.2)cm respectively. The area of the plate is**
1) $(24\pm 1.6)cm^2$ 2) 24.4 cm^2
3) 23.6 cm^2 4) $(24\pm 0.02)cm^2$

3. $S=\frac{1}{2}at^2$ **where a is a constant. If relative error in measurement of t is y, relative error in measurement of S will be**
1) y^2 2) $\sqrt{y}$ 3) 2y 4) y

4. **Two objects A and B are of lengths 5cm and 7cm are determined with errors 0.1cm and 0.2cm respectively. Error in their total length and difference of their lengths are**
1) 0.1cm&0.3cm 2) 0.3cm&0.3cm
3)0.3cm&0.1cm 4) 0.4cm&0.4cm

5. **The error in the measurement of length and radius of a cylinder are 1% and 2% respectively. The maximum percentage of error in its volume is**
1) 3% 2) 5% 3) 6% 4) 10%

6. **The error in the measurement of radius of a sphere is 0.4%. Thepercentage error in its volume is**
1) 0.16% 2) 0.12% 3) 0.24% 4) 1.2%

7. **If $L_1=(2.02\pm 0.01)$m and $L_2=(1.02\pm 0.01)$m then L_1+2L_2 is (in m)**
1) 4.06 ± 0.02 2) 4.06 ± 0.03
3) 4.06 ± 0.005 4) 4.06 ± 0.01

8. **If the length of a cylinder is measured to be 50 cm with an error of 0.25 cm. The percentage error in the measured length is**
1) 0.4% 2) 0.5% 3) 0.2% 4) 0.1%

9. **The length of a cylinder is measured as 5cm using a vernier calipers of least count 0.1 mm. The percentage error is**
1) 0.5 % 2) 2 % 3) 20 % 4) 0.2 %

10. A body travels uniformly a distance of (20.0 ± 0.2) m in time (4.0 ± 0.04) s. The velocity of the body is

1) $(5.0 \pm 0.4)\ ms^{-1}$ 2) $(5.0 \pm 0.2)\ ms^{-1}$

3) $(5.0 \pm 0.6)\ ms^{-1}$ 4) $(5.0 \pm 0.1)\ ms^{-1}$

SIGNIFICANT FIGURES AND ROUNDING OFF :

11. The radius of earth is 6347km. Its rounded off value up to two significant figures is

1) 6400 km 2) 6.4×10^3 km
3) 6.3×10^3 km 4) 6.35×10^3 km

12. 3400 has significant figures

1) 4 2) 2 3) 1 4) 3

13. The number of significant figures in 0.0160kg is

1) 2 2) 3 3) 4 4) 5

14. The number of significant figures in an area $2.64 \times 10^{24} m^2$ is

1) 1 2) 2 3) 4 4) 3

15. The distance travelled by a body is 107.25 m in time 1.5s. The velocity with correct significant figures is

1) 71.5 2) 71 3) 71.50 4) 72.0

16. Less accurate of the four options given below

1) 9.27 2) 41
3) 1.01 4) 9.00×10^0

17. 9.27/41 = (consider significant digits)

1) 0.226 2) 0.23 3) 0.2 4) 0.2261

18. When 57.986 is rounded off to 4 significant figures, then it becomes

1) 58 2) 57.00 3) 57.90 4) 57.99

19. The value of 44.8 – 21.235 is

1) 23.565 2) 23.56 3) 23.57 4) 23.6

20. The length and breadth of a rectangular object are 25.2cm and 16.8cm respectively and have been measured to an accuracy of 0.1cm. Relative error and percentage error in the area of the object are

1) 0.01 & 1% 2) 0.02 & 2%
3) 0.03 & 3% 4) 0.04 & 4%

21. The value of 124.2 + 52.487 with due regard to significant places is

1) 176.7 2) 176.69 3) 176 4) 177

22. The value of 117.3×0.0024 with due regard to significant figures is

1) 0.2808 2) 0.281 3) 0.28 4) 0.29

23. The value of $(17.5)^2$ is

1) 306 2) 306.25
3) 306.2 4) 306.3

ANSWERS

1) 4	2) 1	3) 3	4) 2	5) 2
6) 4	7) 2	8) 2	9) 4	10) 4
11) 1	12) 2	13) 2	14) 4	15) 2
16) 2	17) 2	18) 4	19) 4	20) 2
21) 1	22) 3	23) 1		

EXERCISE - III

ERRORS :

1. The length of a rectangular plate is measured as 10 cm by a vernier scale of least count 0.01 cm and its breadth as 5 cm by the same scale. The percentage error in area is

1) 0.1 % 2) 0.3% 3) 0.01% 4) 0.05%

2. The length of a metallic sheet is measured as 10.0 cm using a metre scale of L.C. 0.1cm and its breadth is measured as 1.00 cm using a vernier callipers of L.C. 0.01cm, the error in area is

1) $\pm 0.01\ cm^2$ 2) $\pm 0.1\ cm^2$

3) $\pm 0.11\ cm^2$ 4) $\pm 0.2\ cm^2$

3. The initial and final temperatures are recorded by using a thermometer as $(30.4 \pm 0.2)^0 C$ and $(50.6 \pm 0.3)^0 C$ The rise in temperature is

1) 20.2^0C 2) $(20.2 \pm 0.1)^0 C$

3) $(20.2 \pm 0.5)^0 C$ 4) $(20.2 \pm 0.3)^0 C$

4. The mass of a beaker is (10.1 ± 0.1)g when empty and (17.3 ± 0.1)g when filled with a liquid. The mass of the liquid with possible limits of accuracy is

1) (7.2 ± 0.2)g 2) (7.2 ± 0.1)g

3) (7.1 ± 0.2)g 4) (7.2 ± 0.3)g

5. Force and area are measured as 20 N and 5 m^2 with errors 0.05 N and 0.0125 m^2. The maximum error in pressure is (SI units)

1) 4 ± 0.0625 2) 4 ± 0.05
3) 4 ± 0.125 4) 4 ± 0.02

6. The measured mass and volume of a body are 22.42g and 4.7cm^3 respectively with errors 0.01g and 0.1cc. The maximum error in density is

1) 0.2% 2) 2% 3) 5% 4) 2.17%

7. When two resistors of (600 ± 3) ohm and (300 ± 6) ohm are connected in parallel, value of the equivalent resistance is

1) $200\text{ohm} \pm 3.5\%$ 2) $200\text{ohm} \pm 9\%$

3) $200\text{ohm} \pm 9\%$ 4) 200ohm

8. The equivalent resistance of two resistors are $R_1 = (50 \pm 2)\Omega$, $R_2 = (150 \pm 3)\Omega$ and $R_3 = (10 \pm 1)\Omega$ The equivalent resistance when they are connected in series is

1) $(210 \pm 6)\Omega$ 2) $(100 \pm 5)\Omega$

3) $(200 \pm 12)\Omega$ 4) $(100 \pm 1)\Omega$

9. Using a screw gauge the diameter of a wire is found to be, 1.51 mm, 1.53mm, 1.48 mm, 1.49 mm, 1.51mm, 1.54 mm. The true value of the diameter of the wire is (in mm)

1) 1.49 2) 1.52 3) 1.51 4) 0.151

10. The length of a rod is measured with a meter rod having least count 0.1cm. Its diameter is measured with vernier calipers of least count 0.01 cm. If length is 5.0 cm and radius 2.00 cm, then the percentage error in volume is

1) 4% 2) 3% 3) 2% 4) 1%

11. Two resistors of 10KΩ and 20KΩ are connected in series. If tolerance of each resistor is 10% then tolerance of the combination will be

1) 5% 2) 10% 3) 15% 4) 20%

12. The equivalent resistance of two resistors are $R_1 = (200 \pm 3)\Omega$, $R_2 = (300 \pm 6)\Omega$. The equivalent resistance when they are connected in parallel is

1) $(120 \pm 5.3)\ \Omega$ 2) $(120 \pm 6)\ \Omega$

3) $(500 \pm 8)\ \Omega$ 4) $(120 \pm 4.6)\Omega$

13. A student performs experiment with a simple pendulum and measure time period for 20 vibrations. If he measures time for 100 vibrations the error in the measurement of time period will be reduced by a factor of

1) 10 2) 20 3) 5 4) 80

14. A physicist performs an experiment and takes 200 readings. He repeats the same experiment and now takes 800 readings. By doing so the probable error

1) remains the same 2) is four times

3) is halved

4) is reduced by a factor of 1/4

15. The pressure on a square plate is measured by measuring the force on the plate and the length of the sides of the plate. If the maximum error in the measurement of force and length are respectively 4% and 2%, the maximum error in the measurement of pressure is

1) 1% 2) 2% 3) 6% 4) 8%

16. In measuring 'g' using a simple pendulum a student makes a positive error of 1% in length of pendulum and a negative error of 3% in the value of time period. Percentage error in measurement of value of g is

1) 2% 2) 4% 3) 7% 4) 10%

17. The resistance R of a wire is found by determining its length l and radius r. The percentage errors in measurement of l and r are respectively 1% and 2%. The percentage error in measurement of R is

1) 4% 2) 5% 3) 3% 4) 1.5%

18. The perecentage error in surface area of a sphere is 6% then find the percentage error in volume

1) 2% 2) 9% 3) 3% 4) 12%

19. In the measurement of a physical quantity(L), the formula used is L = k(m × n), where k is a constant and m and n are the quantities to be measured. If the % errors in m and n are respectively 3% and 5%. The % error in the measurement of L is

1) 2% 2) 5% 3) 8% 4) 3%

20. In the measurement of volume of solid sphere using the formula $V = \frac{4}{3}\pi r^3$ if the error committed in the measurement of the radius r is 1.5%, the percentage error in the volume measurement is

1) 3% 2) 1.5% 3) 4.5% 4) 0.75%

21. The diameter of a wire is measured as 0.25 cm using a screw gauge of least count 0.005 cm. The percentage error is

1) 1% 2) 2.5% 3) 5% 4) 2%

22. The external and internal diameters of a hollow cylinder are determined with vernier calliperse and the results are recorded as 4.23 ± 0.01 cm and 3.89 ± 0.01 cm. The thick-ness of the cylinder wall within the limits of error is

1) 0.34 ± 0.01cm 2) 0.34 ± 0.02cm

3) 0.34 ± 0.04cm 4) 0.17 ± 0.01cm

23. The re sults after calculating the error in the length, breadth and thickness of a block of wood are, $b = 10.15 \pm 0.01$ cm and $h = 5.28 \pm 0.01$ cm. The percentage error in the volume of the block is

1) 3% 2) 0.3% 3) 0.35% 4) 0.42%

24. In an experiment the values of refractive index of glass were found to be 1.54, 1.53, 1.44, 1.54, 1.56, 1.45. The relative error and percentage error are

1) 0.04, 4% 2) 0.03, 3%

3) 0.02, 2% 4) 0.01, 1%

25. A physical quantity is represented by $X = M^a L^b T^{-c}$. If the percentage error in the measurement of M, L and T are $2\alpha\%$, $\beta\%$, $3\gamma\%$, respectively then maximum percentage error in X is

1) $(a\alpha + b\beta - c\gamma)\%$ 2) $(2a\alpha + b\beta + 3c\gamma)\%$

3) $(a\alpha - b\beta + c\gamma)\%$ 4) $(a\alpha - b\beta - c\gamma)\%$

26. The perecentage error in the measurement of current is 4% and in the measurement of voltage is 2% then find the percentage error in the measurement of power is

1) 4 2) 8 3) 2 4) 6

27. A physical quantity Q is related to measurable quantities L,M and N as $Q = K L^{-2} M^2 N^3$, where k is constant. If the % errors in the measurements of L, M and N are respectively 1%, 2% and 1%, then the % error in the measurement of Q is

1) 9% 2) 4% 3) 2% 4) 1%

28. If $x = \frac{A^2 B^3}{C\sqrt{D}}$ and percentage errors in A, B, C and D are 1%, 2%, 3% and 4% respectively, the percentage error in x is

1) 10% 2) 12% 3) 15% 4) 13%

*29. In a simple pendulum experiment, length is measured as 31.4 cm with an accuracy of 1mm. The time for 100 oscillations of pendulum is 112.0s with an acurancy of 0.01s. The percentage accuracy in g is

1) 1 2) 2 3) 1.3 4) 2.3

30. Two capacitors of capacities $(3 \pm 0.09)\mu F$ and $(6 \pm 0.09)\mu F$ are connected in parallel. Their equivalent capacity is

1) $(2 \pm 0.13)\mu F$ 2) $(9 \pm 0.18)\mu F$

3) $(2 \pm 0.09)\mu F$ 4) $(9 \pm 0.09)\mu F$

SIGNIFICANT FIGURES AND ROUNDING OFF

31. The radius of a sphere is 6.01 cm. Its volume to four significant figures is

1) 909.7 cm^3 2) 909 cm^3

3) 916 cm^3 4) 909.67 cm^3

32. The volume of a sphere is 1.76 cm^3. The total volume of 24 such spheres with due regard to the significant places is

1) 0.42×10^2 cm^3 2) 43 cm^3

3) 42.42 cm^3 4) 42.2 cm^3

33. The length of a rod is 22.4 m out of it 2.543m is cut out. The remaining length of the pencil according to the idea of significant figures is

1) 19.8 m 2) 19.9m 3) 20.0 m 4) 19.86 m

34. Three pieces of silver have masses 2.3 kg, 41.15g and 30.19g. The total mass of correct significant figures is

1) 2.37032 kg 2) 2.370 kg

3) 2.37 kg 4) 2.4 kg

35. The value of 117.4×0.0025 is

1) 0.2935 2) 0.294 3) 0.3 4) 0.29

36. The sum of the given two numbers with regard to significant figures is $(5.0 \times 10^{-8}) + (4.5 \times 10^{-6}) =$

1) 4.55×10^{-6} 2) 4.5×10^{-6}

3) 4.6×10^{-6} 4) 4×10^{-6}

37. The value of $\pi/53.2$ is

1) 0.059 2) 0.05906

3) 0.0591 4) 0.06

38. The product of 1.2, 2.54 and 3.257 is

1) 9.934 2) 9.93 3) 9.9 4) 9.9346

39. The mass of a body is measured as 2.00×10^3Kg. Its significant figures is

1) 6 2) 3 3) 1 4) 2

40. Two rods with lengths 12.321cm and 10.3 cm and placed side by side. The difference in their lengths is

1) 2.02 cm 2) 2.0 cm 3) 2 cm 4) 2.025 cm

41. The length of a rectangular sheet is 1.5cm and breadth 1.203 cm. The area of the sheet is

1) 1.8045 cm^2 2) 1.804 cm^2
3) 1.805 cm^2 4) 1.8 cm^2

42. The dimensions of a wooden block are 1.1m × 2.36 m × 3.1 m. The number of significant figures in its volume should be

1) 1 2) 2 3) 3 4) 4

ANSWERS

1) 2	2) 4	3) 3	4) 1	5) 4
6) 4	7) 1	8) 1	9) 3	10) 2
11) 2	12) 1	13) 3	14) 4	15) 4
16) 3	17) 2	18) 2	19) 3	20) 3
21) 4	22) 4	23) 3	24) 2	25) 2
26) 4	27) 1	28) 4	29) 4	30) 2
31) 1	32) 1	33) 2	34) 4	35) 4
36) 3	37) 3	38) 3	39) 2	40) 2
41) 4	42) 2			

COMPREHENSION QUESTIONS

Comprehension - I (1 - 5) :

A capacitor of capacitance $2.0 \pm 0.1\,\mu F$ is charged to a voltage of 20 ± 0.5 V. Then,

1. The relative error in the measurement of charge is

1) $\frac{1}{3}$ 2) $\frac{3}{40}$ 3) $\frac{1}{6}$ 4) $\frac{3}{5}$

2. The percentage error in the measurement of charge is

1) 2.5% 2) 2% 3) 5%4) 7.5%

3. The charge in the capacitor is

1) $(22 \pm 0.6)\mu C$ 2) $(40 \pm 0.6)\mu C$

3) $(40 \pm 3)\mu C$ 4) $(10 \pm 3)\mu C$

4. The percentage error in the measurement of energy stored in the capacitor is

1) 2% 2) 1% 3) 5%4) 10%

5. The energy stored in the condenser is

1) $(400 \pm 40) \times 10^{-6}$ J 2) $(400 \pm 4) \times 10^{-6}$ J

3) $(40 \pm 4) \times 10^{-6}$ J 4) $(800 \pm 20) \times 10^{-6}$ J

Comprehension - II (6 - 8):

The successive measurement of time period of a simple pendulum in an experiment are found to be 2.63 s, 2.56 s, 2.71 s and 2.80 s. Then,

1) 2.3404 kg 2) 2.3145 kg
3) 2.3 kg 4) 2.34 kg

6. The mean time period of oscillation of the pendulum is

1) 2.6 s 2) 2.63 s 3) 2.68 s 4) 3 s

7. The arithematic mean of all the absolute errors is

1) 0.08 s 2) 0.10 s 3) 0.1 s 4) 0.109 s

8. The percentage error in arithmatic mean is

1) 3% 2) 2% 3) 4% 4) 5%

Comprehension - III (9 - 12):

A student measures the resistances of two wires as $(4 \pm 0.5)\Omega$ and $(12 \pm 0.5)\Omega$. Then,

9. The percentage error is the measurement of resistance in series combination is

1) 6% 2) 10% 3) 12.5% 4) 6.25%

10. The percentage error is the measurement of resistance in parallel combination is

1) 23% 2) 1% 3) 3% 4) 50%

11. The equivalent resistance in series combination is

1) $(16 \pm 1)\Omega$ 2) 17Ω 3) 15Ω 4) $(16 \perp 6.25)\Omega$

12. The equivalent resistance in parallel combination is

1) $(3 \pm 0.1)\Omega$ 2) $(3 \pm 23)\Omega$

3) $(3 \pm 0.69)\Omega$ 4) $(3 \pm 1)\Omega$

Comprehension - IV (13 - 15) :

In an experiment the weight of a body in air is found to be (5.00 ± 0.05)N and its weight in water is found to be (4.00 ± 0.05)N. Then,

13. The error in measuring loss of weight in water is

1) 1 N 2) 0 N 3) 0.10 N 4) 1.10 N

14. The relative error in finding the relative density of the body is

1) 0.01 2) 0.11 3) 0.1 4) 0.2

15. The relative density of the body is

1) $5 \pm 11\%$ 2) $5 \pm 1\%$ 3) $5 \pm 10\%$ 4) $5 \pm 20\%$

ANSWERS

1) 2	2) 4	3) 3	4) 4	5) 1	6) 3
7) 1	8) 3	9) 4	10) 1	11) 1	12) 3
13) 3	14) 2	15) 1			

Chapter - 2

ELEMENTS OF VECTORS

ELEMENTS OF VECTORS

SYNOPSIS

1. Physical quantities having only magnitude are called Scalars.

   Ex : Distance, Speed, Area, Volume, Mass, Density, Work, Power, Energy, Time, Frequency, Temperature, Electric Charge, Electric Current, Potential, Resistance, Capacity, Velocity of Light, Intensity of Sound etc,.

2. Physical quantities having both magnitude and direction and that obey vector laws of addition are vectors.

   Ex : Displacement, Velocity, Acceleration, Momentum, Force, Impulse, Angular Displacement, Angular Velocity, Angular Acceleration, Moment of a Force, Torque, Magnetic Moment, Magnetic Induction Field, Intensity of Electric Field.etc,....

3. Electric current and velocity of light have both direction and magnitude but they do not obey laws of vector addition. Hence they are scalars.

4. Displacement, Velocity, Acceleration, Force etc., which have natural direction are called polar or real vectors.

5. The direction of polar vector does not change even though the coordinate system in which it is defined, changes

   Ex. : Force, Velocity, Momentum.

6. The direction of Pseudo vecotr changes with the change of coordinate system from right handed to left handed system in which it is defined. *Ex. :* All cross products

7. Torque, angular velocity, angular momentum etc., to which direction is assigned by convention are called axial or pseudo vectors.

8. Unit vector in the direction of $\vec{A}$ is given by

   $$\hat{A} = \frac{\vec{A}}{|\vec{A}|} = \frac{A_x\hat{i} + A_y\hat{j} + A_z\hat{k}}{\sqrt{A_x^2 + A_y^2 + A_z^2}}$$

9. The magnitude of null vector is zero and direction is indeterminate.

10. Equal Vectors : Vectors are equal if they have same magnitude and same direction

11. Negative vector of a vector is the vector of same magnitude but opposite direction. When a vector is added to its negative vector, the result is a Null Vector.

12. If $\vec{P}$ and $\vec{Q}$ are two vectors their resultant vector is given by $\vec{R} = \vec{P} + \vec{Q}$.

13. The magnitude of resultant $\vec{R}$ is

    $$R = \sqrt{P^2 + Q^2 + 2PQ\ \cos\theta}$$

14. If $\vec{R}$ makes an angle α with $\vec{P}$, then

    $$\tan\alpha = \frac{Q\sin\theta}{P + Q\cos\theta}$$

15. If $\vec{R}$ makes an angle β with $\vec{Q}$, then

    $$\tan\beta = \frac{P\sin\theta}{Q + P\cos\theta}$$

16. Resultant of two vectors always lies in the plane containing the vectors closer to vector of larger magnitude.

17. Special cases

    (i) $\theta = 0^0 \Rightarrow R(max) = P + Q$

    (ii) $\theta = 180^0 \Rightarrow R(min) = P - Q$

    (iii) $\theta = 90^0 \Rightarrow R = \sqrt{P^2 + Q^2}$ and $\tan\alpha = \frac{Q}{P}$

    (iv) $P{=}Q \Rightarrow R = 2P\cos\left(\frac{\theta}{2}\right)$ & $\alpha = \frac{\theta}{2}$

18. Vector addition follows.

    a) Commutative law : $\vec{P} + \vec{Q} = \vec{Q} + \vec{P}$

    b) Associative law : $\vec{P} + (\vec{Q} + \vec{R}) = (\vec{P} + \vec{Q}) + \vec{R}$

    c) Distributive law : $k(\vec{P} + \vec{Q}) = k\vec{P} + k\vec{Q}$, here 'k' is a scalar

19. The unit vector parallel to the resultant of $\vec{P}$ and $\vec{Q}$ is given by $\hat{n} = \frac{\vec{P} + \vec{Q}}{|\vec{P} + \vec{Q}|}$

20. If $\vec{P}$ and $\vec{Q}$ are two vectors then substraction of $\vec{Q}$ from $\vec{P}$ is given by

    $\Rightarrow \vec{S} = \vec{P} + (-\vec{Q}) = \vec{P} - \vec{Q}$

21. The magnitude of $\vec{P} - \vec{Q}$ is given by $S = \sqrt{P^2 + Q^2 - 2PQ\cos\theta}$

22. **Vector Substraction.**

    a) does not obey commutative law

    $\vec{P} - \vec{Q} \neq \vec{Q} - \vec{P}$

    b) does not obey associative law

    $\vec{P} - (\vec{Q} - \vec{R}) \neq (\vec{P} - \vec{Q}) - \vec{R}$

c) obeys distributive law

$$k(\vec{P}-\vec{Q}) = k\vec{P} - k\vec{Q}$$

23. If $\vec{A}$ is a vector making an angle θ with horizontal, then Horizontal component of $\vec{A}$, is $A_x = A\cos\theta$ and Vertical component of a vector $\vec{A}$ is $A_y = A\sin\theta$.

24. The components of a vector are scalars. But a vector has vector components also.

25. If $\hat{i}$ and $\hat{j}$ are unit vectors along x and y axes respectively then $\vec{A} = A_x\hat{i} + A_y\hat{j}$ and $\theta = \tan^{-1}\left(\frac{A_y}{A_x}\right)$ or $\theta = \cos^{-1}\left(\frac{A_x}{A}\right)$

26. If $\vec{A}$ is resolved into three mutually perpendicular components in a space then $\vec{A} = A_x\hat{i} + A_y\hat{j} + A_z\hat{k}$

a) $A = \sqrt{A_x^2 + A_y^2 + A_z^2}$

If angles made by $\vec{A}$ with x, y and z axis are α, β and γ, then $\cos\alpha = \frac{A_x}{|A|}$, $\cos\beta = \frac{A_y}{|A|}$, $\cos\gamma = \frac{A_z}{|A|}$ here $\cos\alpha, \cos\beta, \cos\gamma$ are known as direction cosines.

27. $\cos^2\alpha + \cos^2\beta + \cos^2\gamma = 1$.

28. $\sin^2\alpha + \sin^2\beta + \sin^2\gamma = 2$.

29. If three coplanar forces vector $\vec{F}_1, \vec{F}_2, \vec{F}_3$ acting simultaneously at a point 'O' keep it in equilibrium as shown, then $\frac{F_1}{\sin\alpha} = \frac{F_2}{\sin\beta} = \frac{F_3}{\sin\gamma}$

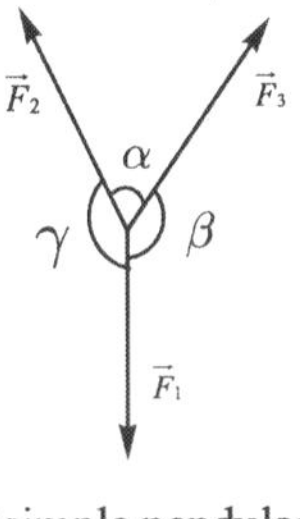

30. If the bob of simple pendulam is held atrest by appling a horizontal force F as shown in fig.

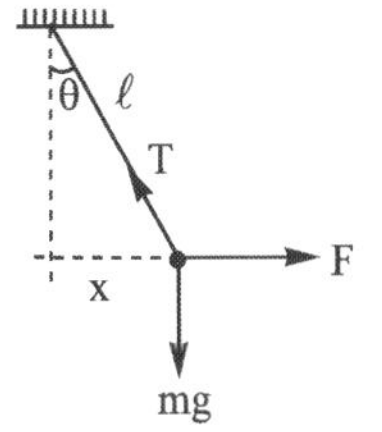

a) $T\sin\theta = F$ b) $T\cos\theta = mg$

c) $F = mg\tan\theta$ d) $\sqrt{F^2 + (mg)^2} = T$

e) $\frac{x}{F} = \frac{\ell}{T} = \frac{\sqrt{\ell^2 - x^2}}{mg}$

31. **Relative Velocity**

a) Suppose two bodies A and B are moving with velocities $\vec{V}_A$ and $\vec{V}_B$

Velocity of A relative to B is given by

$$\vec{V}_{AB} = \vec{V}_A - \vec{V}_B$$

b) When A and B are moving in perpendicular directions, θ is the angle made by $\vec{V}_{AB}$ with $\vec{V}_A$ then $|\vec{V}_{AB}| = \sqrt{V_A^2 + V_B^2}$ $\tan\theta = \frac{V_B}{V_A}$

32. **Motion of a Boat in a river :**

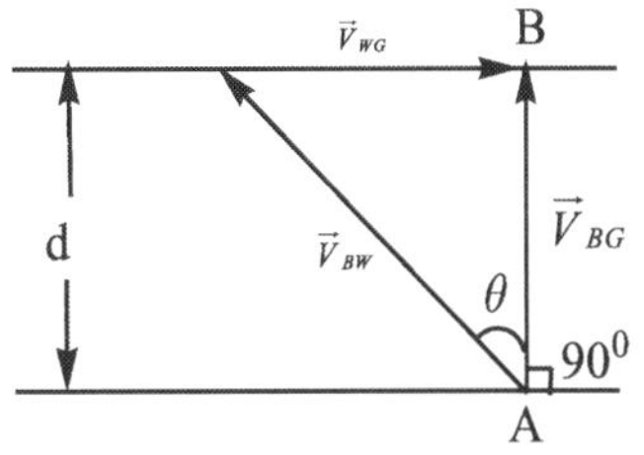

(a) To cross a river by a boat in the shortest path (or from a point on one bank to an exactly opposite point on the other bank). The angle made by the boat with normal to the flow of water is $\sin\theta = \frac{V_{WG}}{V_{BW}}$.

(b) The angle between the direction of motion of boat and the direction of river flow = 90 + θ

(c) Resultant velocity $V = \sqrt{V_{BW}^2 - V_{WG}^2}$

(d) Time taken to cross the river $t = \frac{d}{\sqrt{V_{BW}^2 - V_{WG}^2}}$ where d is the width of the river (or) $\frac{d}{V_{BW}\cos\theta}$.

(e) To cross the river in the shortest time : The boat is to be rowed in a direction perpendicular to the direction of river flow with velocity V_{BW}.

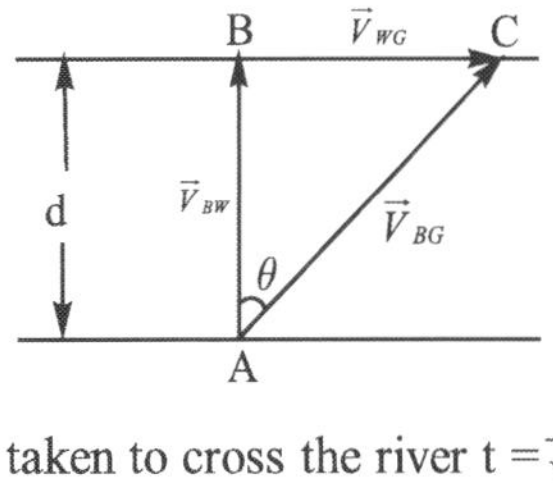

(f) Time taken to cross the river $t = \frac{d}{V_{BW}}$.

(g) Resultant velocity $V_{BG} = \sqrt{V_{BW}^2 + V_{WG}^2}$

(h) The boat reaches the other bank at an angle α, $\tan\alpha = \frac{V_{WG}}{V_{BW}}$.

(i) The drift (x) = $V_{WG}.t = V_{WG}\left(\frac{d}{V_{BW}}\right)$

(j) The boat is rowed always in a direction making an angle θ with the perpendicular drawn to the flow of river along PQ as shown. To find the drift of the boat as it reaches the opposite bank.

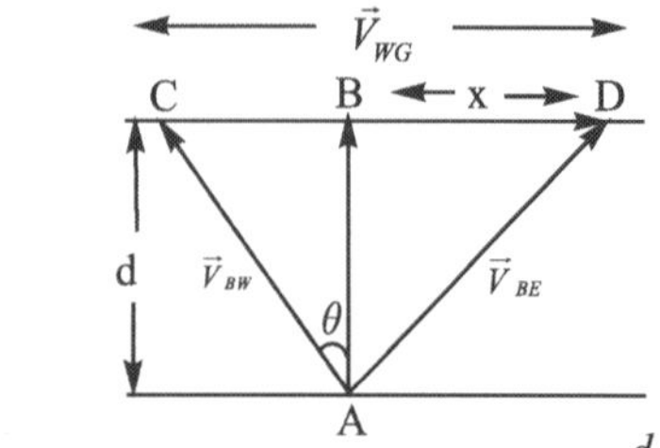

k) Time taken to cross the river $t = \frac{d}{V_{BW}\cos\theta}$

l) Resultant velocity along flow=$V_{WG} - V_{BW}\sin\theta$

m) Drift (x) = $(V_{WG} - V_{BW}\sin\theta)t$.

33. If rain drops are falling vertically with a velocity '$\vec{V}_R$' and a person is walking horizontally with a velocity $\vec{V}_M$, he should hold his umbrella at an angle θ with the vertical given by $\tan\theta = \frac{V_M}{V_R}$

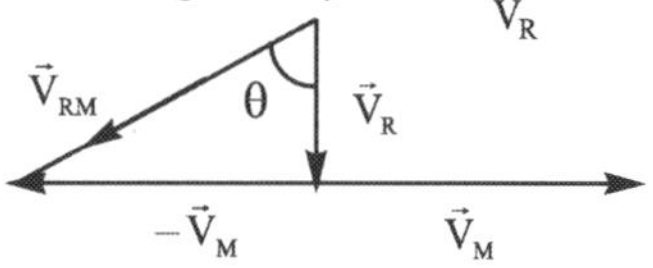

The person holds the umbrella along the direction which is the direction of rain relative to the person.

34. The dot product of two vectors $\vec{P}$ and $\vec{Q}$ is given by $\vec{P}.\vec{Q} = PQ\cos\theta$.

a) It is a Scalar

b) $\vec{P}.\vec{Q} = 0$ when $\theta = 90^0$ (Perpendicular Vectors)

c) $\vec{P}.\vec{Q} = PQ$ when $\theta = 0^0$ (Parallel Vectors)

d) Dot product is negative if $\theta > 90^0$ and $< 180^0$

e) $\vec{P}.\vec{Q} = \vec{P}.\vec{Q}$ (Commutative law)

f) $\vec{P}.\left(\vec{Q} + \vec{R}\right) = \vec{P}.\vec{Q} + \vec{P}.\vec{R}$ (Distributive law)

g) It doesnot obey associative law

h) $\vec{P}.\vec{P} = P^2$

i) Angle between two vectors $\vec{P}$ and $\vec{Q}$ is given by the relation $\cos\theta = \frac{\vec{P}.\vec{Q}}{PQ}$.

j) The component of $\vec{P}$ along $\vec{Q}$ = $P\cos\theta = \frac{\vec{P}.\vec{Q}}{|\vec{Q}|}$

k) The component of $\vec{Q}$ along $\vec{P}$ = $Q\cos\theta = \frac{\vec{P}.\vec{Q}}{|\vec{P}|}$

l) Vector component of $\vec{P}$ along $\vec{Q}$ is $\left(\vec{P}.\hat{Q}\right)\hat{Q}$.

m) Vector component of $\vec{Q}$ along $\vec{P}$ is $\left(\vec{Q}.\hat{P}\right)\hat{P}$.

n) The component of $\vec{P}$ perpendicular to $\vec{Q}$ in the same plane is $\vec{C} = \vec{P} - \left(\vec{P}.\hat{Q}\right)\hat{Q}$

o) The component of $\vec{P}$ perpendicular to $\vec{Q}$ in the same plane is $\vec{D} = \vec{Q} - \left(\vec{Q}.\hat{P}\right)\hat{P}$

p) $\hat{i}.\hat{i} = \hat{j}.\hat{j} = \hat{k}.\hat{k} = 1$, $\hat{i}.\hat{j} = \hat{j}.\hat{k} = \hat{k}.\hat{i} = 0$.

q) If $\vec{P} = P_x\hat{i} + P_y\hat{j} + P_z\hat{k}$ and $\vec{Q} = Q_x\hat{i} + Q_y\hat{j} + Q_z\hat{k}$, then $\vec{P}.\vec{Q} = P_xQ_x + P_yQ_y + P_zQ_z$

r) Two vectors $\vec{P}$ and $\vec{Q}$ are perpendicular if $\vec{P}.\vec{Q} = 0$ (or) $P_XQ_X + P_YQ_Y + P_ZQ_Z = 0$

s) Examples of dot product :

Work W = $\vec{F}.\vec{S}$

Power P = $\vec{F}.\vec{V}$

Magnetic Flux $\phi = \vec{B}.\vec{A}$

35. The cross product of two vectors $\vec{P}$ and $\vec{Q}$ is given by $\vec{P} \times \vec{Q} = PQ\sin\theta\hat{n}$

Where $\hat{n}$ is a unit vector perpendicular to the plane containing $\vec{P}$ and $\vec{Q}$.

a) It is a vector.

b) Direction is given by right hand screw rule.

c) $\vec{P} \times \vec{Q} = \vec{O}$ when $\theta = 0^0$ (Parallel Vectors)

d) $\left|\vec{P} \times \vec{Q}\right|$ = PQ when $\theta = 90^0$ (Perpendicular Vectors)

e) $\vec{P} \times \vec{Q} = -\vec{Q} \times \vec{P}$.

$\vec{P} \times \vec{Q} \neq \vec{Q} \times \vec{P}$

(commutative law is not obeyed)

f) $\vec{P} \times \left(\vec{Q} \times \vec{R}\right) \neq \left(\vec{Q} \times \vec{P}\right) \times \vec{R}$

(Associative law is not obeyed)

g) $\vec{P} \times \left(\vec{Q} + \vec{R}\right) = \vec{P} \times \vec{Q} + \vec{P} \times \vec{R}$

(Distributive law is obeyed)

h) $\hat{i} \times \hat{i} = \hat{j} \times \hat{j} = \hat{k} \times \hat{k} = 0$;

$\hat{i} \times \hat{j} = \hat{k}, \hat{j} \times \hat{k} = \hat{i}, \hat{k} \times \hat{i} = \hat{j}$

i) If $\vec{P} = P_x\hat{i} + P_y\hat{j} + P_z\hat{k}$

$\vec{Q} = Q_x\hat{i} + Q_y\hat{j} + Q_z\hat{k}$

then $\vec{P}\times\vec{Q} = \begin{vmatrix} \hat{i} & \hat{j} & \hat{k} \\ P_x & P_y & P_z \\ Q_x & Q_y & Q_z \end{vmatrix}$

$$= \left(P_yQ_z - P_zQ_y\right)\hat{i} - \left(P_xQ_z - P_zQ_x\right)\hat{j} + \left(P_xQ_y - P_yQ_x\right)\hat{k}$$

j) Unit vector normal to both $\vec{P}$ and $\vec{Q}$ is

$\hat{n} = \dfrac{\vec{P}\times\vec{Q}}{\left|\vec{P}\times\vec{Q}\right|}$.

k) If two vectors $\vec{P}$ and $\vec{Q}$ are parallel

$\dfrac{P_x}{Q_X} = \dfrac{P_y}{Q_y} = \dfrac{P_z}{Q_z} = \text{constant}$ or $\vec{P}\times\vec{Q} = 0$

l) If $\vec{P}$ *and* $\vec{Q}$ represents the sides of a parallelogram then the area of the parallelogram is $\left|\vec{P}\times\vec{Q}\right|$

m) If $\vec{d_1}$ *and* $\vec{d_2}$ represents the diagonals of a parallelogram then the area of the parallelogram is $\frac{1}{2}\left|\vec{d_1}\times\vec{d_2}\right|$

n) If $\vec{A}$ *and* $\vec{B}$ represents the sides of a triangle then the area of the triangle is $\frac{1}{2}\left|\vec{A}\times\vec{B}\right|$

o) $\vec{A}, \vec{B}$ and $\vec{C}$ are coplanar, if $\vec{A}\left(\vec{B}\times\vec{C}\right) = 0$.

36. Examples of cross product :

i) Angular momentum $\vec{L} = \vec{r}\times\vec{P}$

ii) Linear velocity $\vec{V} = \vec{\omega}\times\vec{r}$

iii) Torque $\vec{\tau} = \vec{r}\times\vec{F}$

iv) Torque on a magnet $\vec{\tau} = \vec{M}\times\vec{B}$

v) Force on a conductor $\vec{F} = i\left(\vec{l}\times\vec{B}\right)$

vi) Force on a charge $\vec{F} = q\left(\vec{V}\times\vec{B}\right)$

vii) Torque on a coil $\vec{\tau} = i\left(\vec{A}\times\vec{B}\right)$

viii) A force $\vec{F}$ acts at P and $\vec{\tau}$ is torque produced about Q. If position vector of P is $\vec{r_1}$ and position vector of Q is $\vec{r_2}$ then $\vec{\tau} = \vec{r}\times\vec{F} = \left(\vec{r_2} - \vec{r_1}\right)\times\vec{F}$.

EXERCISE - I

VECTORS & SCALARS

1. Of the following the vector quantity is

1) time 2) Electric Current
3) Velocity of light 4) gravitational force

2. Of the following the scalar quantity is

1) Temparature 2) Moment of force
3) Moment of couple 4) Magnetic moment

3. Choose the correct statement

1) Temperature is a scalar but temperature gradient is a vector
2) Velocity of a body is a vector but velocity of light is a scalar
3) Electric intensity and Electric current density are vectors 4) all the above

4. Choose the false statement :

1) Electric current is a vector because it has both magnitude and direction
2) Time is a vector which has direction always in the forward direction
3) All quantities having magnitude and direction are vector quantities
4) all the above

5. The pair containing a scalar quantity and vector quantity is

1) Impulse and Angular momentum
2) work and Frequency
3) Electromotive force and force
4) Electric power and Energy

6. The set containing only scalar quantities is

1) Temperature gradient, specific heat and latent heat
2) Electric intensity, Electric potential and Electric capacity
3) Polestrength, permeability and permittivity
4) Torque, Angular acceleration and linear momentum

7. The set containing only vector quantities is

1) Thermal capacity, Magnetic susceptibility and Electric charge
2) Magnetic moment, Electric intensity and Torque
3) Magnetic flux, Electric potential and Force
4) Magnetic induction, Electric capacity and Impulse

8. **Which of the following units could be associated with a vector quantity ?**
   1) newton/metre 2) newton metre / second
   3) kg m^2 s^{-2} 4) newton second
9. **A vector is not changed if**
   1) it is rotated through an arbitary angle
   2) it is multiptied by an arbitary scalar
   3) it is cross multiptied by a unit vector
   4) it is slid parallel to itself
10. **Which of the following is meaningful?**
   1) Vector/Vector 2) $\frac{\text{Scalar}}{\text{vector}}$
   3) Scalar + vector 4) vector / scalar
11. **Choose the correct statement.**
   1) Scalar + vector = scalar/vector
   2) $\frac{\text{vector}}{\text{vector}}$ = scalar
   3) scalar/vector = scalar (or) vector
   4) vector - vector = vector.
12. **If angle between $\vec{a}$ and $\vec{b}$ is $\frac{\pi}{3}$, then angle between $2\vec{a}$ and $-3\vec{b}$ is**
   1) $\frac{\pi}{3}$ 2) $\frac{2\pi}{3}$ 3) $\frac{\pi}{6}$ 4) $\frac{5\pi}{3}$
13. **Which one of the following is a null vector ?**
   1) Net displacement of a particle moving once around, a circle
   2) velocity of a body projected vertically up, when the body is at the highest point
   3) acceleration of a particle executing S.H.M. at the mean position
   4) all the above

ADDITION & SUBTRACTIONS OF VECTORS

14. **If $|\vec{P}+\vec{Q}| = |\vec{P}-\vec{Q}|$, then the vectors $\vec{p}$ and $\vec{Q}$ are**
   1) parallel to each other
   2) inclined at 45^0 to each other
   3) perpendicular to each other
   4) inclined at 60^0 to each other
15. **The resultant of two forces cannot exceed**
   1) average of the forces
   2) algebraic sum of the two forces
   3) difference of the two forces 4) none
16. **If the Resultant of $\vec{p}$ and $\vec{Q}$ makes an angle θ_1 with $\vec{p}$ and θ_2 with $\vec{Q}$. Then**
   1) $\theta_1 < \theta_2$ if $P \geq Q$ 2) $\theta_1 < \theta_2$ if $P > Q$
   3) $\theta_1 < \theta_2$ if $P < Q$ 4) $\theta_1 < \theta_2$ if $P = Q$
17. **The addition of two vectors will be maximum, if they are**
   1) Non coplanar vectors 2) parallel vectors
   3) orthogonal vectors 4) coplanar vectors
18. **Subtraction of vectors obeys**
   1) commutative law 2) Associative law
   3) Distributive law 4) All the above
19. **Associative law is obeyed by**
   1) Addition of vectors 2) subtraction of vectors
   3) both 4) none
20. **Choose the correct statement.**
   1) If $\vec{A}+\vec{B}=\vec{A}-\vec{B}$ then $\vec{B}$ is a null vector
   2) If $|\vec{A}+\vec{B}|=|\vec{A}-\vec{B}|$, then $\vec{A}$ and $\vec{B}$ are perpendicular vectors
   3) both of the above 4) none of the above
21. **The maximum value of magnitude of $(\vec{A}-\vec{B})$ is**
   1) A-B 2) A+B 3) A^2+B^2 4) A^2-B^2
22. **If $\vec{A}+\vec{B}=\vec{C}$ and the angle between $\vec{A}$ and $\vec{B}$ is 120^0, then the magnitude of $\vec{C}$**
   1) must be equal to $|\vec{A}-\vec{B}|$
   2) must be less than $|\vec{A}-\vec{B}|$
   3) must be greater than $|\vec{A}-\vec{B}|$
   4) may be equal to $|\vec{A}-\vec{B}|$

23. If $\vec{P}+\vec{Q}=\vec{R}$

1) R is always greater than P
2) R is always equal to P+Q
3) R is never equal to P+Q
4) R may be less than P or Q

24. When two vectors $\vec{A}$ and $\vec{B}$ of magnitudes'a' and 'b' respectively are added, the magnitude of resultant vector is always

1) Equal to (a+b) 2) Less than (a+b)
3) Greater tha (a+b) 4) Not greater than (a+b)

25. If $\vec{C}=\vec{A}+\vec{B}$ then

1) $\vec{C}$ is always greater than $|\vec{A}|$
2) C is always equal to A+B
3) C is never equal to A+B
4) It is possible to have $|\vec{C}|<|\vec{A}|$ and $|\vec{C}|<|\vec{B}|$

26. Three forces start acting simultaeously on a particle moving with velocity $\vec{V}$. The forces are represented in magnitude and direction by the three sides of a triangle ABC (as shown). The particle will now move with velocity

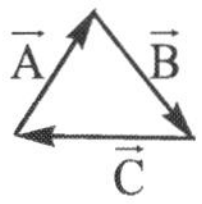

1) less than $\vec{V}$
2) greater than $\vec{V}$
3) $|\vec{V}|$ in the direction of largest force
4) $\vec{V}$ remaining unchanged

LAMI'S THEOREM

27. The minimum number of forces of equal magnitude in a plane that can keep a particle in equilibrium is

1) 4 2) 2 3) 3 4) 5

28. The minimum number of unequal forces in a plane that can keep a particle in equilibrium is

1) 4 2) 2 3) 3 4) 6

29. The minimum number of non coplanar forces that can keep a particle in equilibrium is

1) 1 2) 2 3) 3 4) 4

RESOLUTION OF VECTORS

30. If component of one vector in the direction of another vector is zero, then those two vectors

1) are parallel to each other
2) are perpendicular to each other
3) are opposite to each other
4) are coplanar vectors

31. The component of a vector is

1) always less than its magnitude
2) always greater than its magnitude
3) always equal to its magnitude
4) Less than or equal to its magnitude

32. The horizontal component of the weight of a body of mass m is

1) mg 2) $\frac{mg}{2}$ 3) Zero 4) Infinity

33. What are the maximum number of rectangular components of a vector can be split in space and in plane respectively.

1) 3,2 2) 3,3 3) 2,2 4) ∞, ∞

34. The maximum number of components a vector can be split are ?

1) 2 2) 3 3) 4 4) infnite

RELATIVE VELOCITY

35. A train is moving due east and a car is moving due north with equal speeds. A passenger in the train finds that the car is moving towards

1) North - East 2) North - West
3) South - West 4) South - East

36. A bus moves over a straight level road with a constant acceleration a. A boy in the bus drops a ball out side. The acceleration of the ball w.r.t the bus and the earth are respectively

1) $\sqrt{a^2+g^2}, g$ 2) $g, \sqrt{a^2+g^2}$
3) a, g 4) g, a

37. A particle P moves with speed 'v' along AB and BC, sides of a square ABCD. Another particle Q also starts at A and moves with the same speed but along AD and DC of the same square ABCD. Then their respective changes in velocities are

1) equal in magnitude but different in directions
2) different in magnitude but same in directions
3) different both in magnitude and direction
4) same both in magnitude and direction

38. If A and B persons are moving with V_A and V_B velocities in opposite directions. Magnitude of relative velocity of B w.r.t. A is x and magnitude of relative velocity of A w.r.t B is y. Then

1) $x > y$ 2) $x = y$ 3) $x = 2y$ 4) $2x = y$

39. Wind is blowing to east along two parallel railway tracks. Two trains moving with the same speed in opposite direction have the steam track of one double that of the other. The speed of each train is

1) Equal to that of the wind
2) Three times that of the wind
3) Double that of the wind
4) Half that of the wind

40. A river is flowing from west to east at a speed of 5 m/s. A man on the south bank of the river capable of swimming at 10 m/s in a still water wants to swim, across the river in a shortest time. He should swim in a direction

1) Due north 2) 30^0 east of west
3) 30^0 west of north 4) 60^0 east of north

DOT & CROSS PRODUCT

41. Dot product of two vectors is a

1) Vector 2) Scalar 3) Tensor 4) None

42. How much work you do on an object as you lift it to a height 'h' from the ground ?

1) mgh 2) mgh/2 3) 2mgh 4) zero

43. If ℓ_1, m_1, n_1 and ℓ_2, m_2, n_2 are the direction co-sines of two vectors and θ is the angle between them, then the value of cos θ is

1) $\ell_1\ell_2 + m_1m_2 + n_1n_2$ 2) $\ell_1m_1 + m_1n_1 + n_1\ell_1$
3) $\ell_2m_2 + m_2n_2 + n_2l_2$ 4) $m_1\ell_2 + \ell_2m_2 + n_1m_2$

44. If $\vec{A} + \vec{B} = \vec{C}$, then magnitude of $\vec{B}$ is

1) $\vec{C} - \vec{A}$ 2) $C - A$
3) $\sqrt{\vec{C}\cdot\vec{B} - \vec{A}\cdot\vec{B}}$ 4) $\sqrt{\vec{C}\cdot\vec{A} - \vec{B}\cdot\vec{A}}$

45. If $\vec{a} = m\vec{b} + \vec{c}$. The scalar m is

1) $\frac{\vec{a}.\vec{b} - \vec{b}.\vec{c}}{b^2}$ 2) $\frac{\vec{c}.\vec{b} - \vec{a}.\vec{c}}{a^2}$
3) $\frac{\vec{c}.\vec{a} - \vec{b}.\vec{c}}{c^2}$ 4) $\frac{\vec{a}.\vec{b} - \vec{b}.\vec{c}}{a^2}$

46. Cross product of vectors obeys

1) Commutative law 2) Associative law
3) Distributive law 4) All the above

47. Distributive law is obeyed by

1) scalar product 2) vector product
3) both 4) none

48. Choose the false statement

1) Scalar product and vector product obey commutative law
2) Scalar product does not obey distributive law where as vector product obeys commutative law
3) Scalar product and vector product obey associative law
4) all the above

49. Three vectors satisfy the relation $\vec{A}.\vec{B} = 0$ and $\vec{A}.\vec{C} = 0$, then $\vec{A}$ is parallel to

1) $\vec{C}$ 2) $\vec{B}$ 3) $\vec{B}\times\vec{C}$ 4) $\vec{B}.\vec{C}$

50. Let $\vec{F}$ be the force acting on a particle having position vector $\vec{r}$ and $\vec{\tau}$ be the torque of this foce about the origin. Then (AIEEE-2003)

1) $\vec{r}.\vec{F} = 0$ and $\vec{r}.\vec{\tau} \neq 0$
2) $\vec{r}.\vec{\tau} \neq 0$ and $\vec{F}.\vec{\tau} = 0$
3) $\vec{r}.\vec{\tau} \neq 0$ and $\vec{F}.\vec{\tau} \neq 0$ 4) $\vec{r}.\vec{\tau} = 0$ and $\vec{F}.\vec{\tau} = 0$

51. $(\vec{A}\times\vec{B}) + (\vec{B}\times\vec{A})$ is equal to

1) 2 AB 2) $A^2 B^2$ 3) zero 4) null vector

52. If $\vec{C} = \vec{A}\times\vec{B}$, then $\vec{C}$ is

1) parallel to $\vec{A}$ 2) parallel to $\vec{B}$
3) perpendicular to $\vec{A}$ and parallel to $\vec{B}$
4) perpendicular to both $\vec{A}$ and $\vec{B}$

53. If $\vec{A}$ and $\vec{B}$ are two vectors, then which of the following is wrong ?

1) $\vec{A}.\vec{B} = \vec{B}.\vec{A}$ 2) $\vec{A} + \vec{B} = \vec{B} + \vec{A}$
3) $\vec{A}\times\vec{B} = \vec{B}\times\vec{A}$ 4) $\vec{A}\times\vec{B} = -\vec{B}\times\vec{A}$

54. **The angle between $(\vec{A}\times\vec{B})$ and $(\vec{B}\times\vec{A})$ is (in radian)**

1) $\pi/2$ 2) π 3) $\pi/4$ 4) zero

55. **If none of the vectors $\vec{A},\vec{B}$ and $\vec{C}$ are zero and if $\vec{A}\times\vec{B}=0$ and $\vec{B}\times\vec{C}=0$ the value of $\vec{A}\times\vec{C}$ is**

1) Unity 2) Zero 3) B^2 4) AC cos θ

56. **Choose the false statement**

1) A vector having zero magnitude can have a direction
2) If $\vec{A}\times\vec{B}=\vec{0}$, then either $\vec{A}$ or $\vec{B}$ or both must have zero magnitude
3) The component of a vector is a vector
4) all the above

57. **If $\vec{A}$, $\vec{B}$ and $\vec{C}$ are coplanar vectors, then**

1) $(\vec{A}.\vec{B})\times\vec{C}=0$ 2) $(\vec{A}\times\vec{B}).\vec{C}=0$

3) $(\vec{A}.\vec{B}).\vec{C}=0$ 4) all the above are true

58. **If $\vec{A}$ along North and $\vec{B}$ along vertically upward the direction of $\vec{A}\times\vec{B}$ is along**

1) west 2) south 3) east 4) vertically downwards

59. **The magnitude of the vector product of two vectors $\vec{P}$ and $\vec{Q}$ may be**

1) equal to PQ 2) less than PQ
3) equal to zero 4) all the above

60. **i x (jxk) is**

1) unit vector 2) null vector
3) i+j+k 4) i+j-k

61. **If '$\hat{i}$' denotes a unit vector along incident ray, $\hat{r}$ a unit vector along refracted ray into a medium of refractive index μ and $\hat{n}$ is unit vector normal to boundary of medium directed towards the incident medium, the law of refraction is**

1) $i.\hat{n}=\mu(\hat{r}.\hat{n})$ 2) $i\times\hat{n}=\mu(\hat{n}\times\hat{r})$

3) $\hat{i}\times\hat{n}=\mu(\hat{r}\times\hat{n})$ 4) $\mu(\hat{r}\times\hat{n})=\hat{n}\times\hat{i}$

62. **The angle between $(\vec{A}+\vec{B})$ & $(\vec{A}\times\vec{B})$**

1) 0, 2) $\pi/4$ 3) $\pi/2$ 4) π

63. **The position vector $\vec{r}$ and linear momentum $\vec{p}$ are $\vec{r}=\hat{i}$ and $\vec{p}=4\hat{j}$ the angular momentum vector is perpendicular to**

1) x-axis 2) y-axis 3) z-axis 4) xy-plane

64. **The vector area of triangle whose sides are $\vec{a},\vec{b},\vec{c}$ is**

1) $\frac{1}{6}|\vec{b}\times\vec{c}+\vec{c}\times\vec{a}+\vec{a}\times\vec{b}|$ 2) $\frac{1}{2}|\vec{b}\times\vec{c}+\vec{c}\times\vec{a}+\vec{a}\times\vec{b}|$

3) $\frac{1}{3}|\vec{b}\times\vec{c}+\vec{c}\times\vec{a}+\vec{a}\times\vec{b}|$ 4) $\frac{1}{2}|-\vec{b}\times\vec{c}+\vec{c}\times\vec{a}+\vec{a}\times\vec{b}|$

MORE THAN ONE OPTION IS CORRECT

VECTORS & SCALAR

65. **Two vectors of same physical quantity are unequal if**

a) They have the same magnitude and same direction

b) They have different magnitudes but same direction

c) They have same magnitude but different directions

d) They have different magnitudes and different directions

1) a & b are true 2) b, c & d are true

3) only c and d are true 4) a, b, c & d are true

66. **Which of the following is a null vector**

a) velocity vector of a body moving in a circle with a uniform speed

b) velocity vector of a body moving in a straight line with a uniform speed

c) position vector of the origin of the rectangular co-ordinate system

d) displacement vector of a stationary object

1) both a & b 2) both b & c

3) a, b & c 4) both c & d

RESOLUTION OF VECTORS & RELATIVE VELOCITY

67. Consider the following statements A and B given below and identify the correct answer.

A) Vectors $\hat{i}+3\hat{j}+5\hat{k}$ and $2\hat{i}+6\hat{j}+10\hat{k}$ are parallel

B) $\hat{i}+\hat{j}+\hat{k}$ represents a unit vector

1) Both A and B false 2) A is true B is false

3) B is true but A is false 4) Both A and B are false

68. A boat moves relative to water with a velocity which is 'n' times the river flow

a) If n < 1 boat can not cross the river

b) If n = 1 boat can not cross the river without drifting

c) If n > 1 boat can cross the river along shortest path

d) Boat can cross the river what ever is the value of n excluding zero value

1) only a is correct 2) a, b are correct

3) c, d are correct 4) b, c & d are correct

DOT & CROSS PRODUCTS

69. Which of the following vector identities is/are false?

a) $\vec{A}.\vec{B}=\vec{B}.\vec{A}$ **b)** $\vec{A}.\vec{B}=-\vec{B}.\vec{A}$

c) $\vec{A}\times\vec{B}=\vec{B}\times\vec{A}$ **d)** $\vec{A}\times\vec{B}=-\vec{B}\times\vec{A}$

1) both b & c 2) only a

3) both c & d 4) both a & b

70. Which of the following physical quantities can be obtained from dot product of two vectors

a) work **b) torque**

c) power **d) magnetic flux**

1) a, b 2) a, c & d 3) only b 4) a & d

71. If $\vec{P}x\vec{Q}=\vec{R}$; $\vec{Q}x\vec{R}=\vec{P}$ and $\vec{R}x\vec{P}=\vec{Q}$ then

a) $\vec{P},\vec{Q}$ & $\vec{R}$ are coplanar

b) angle between $\vec{P}$ & $\vec{Q}$ may be less than 90°

c) $\vec{P}+\vec{Q}+\vec{R}$ can not be equal to zero

d) $\vec{P},\vec{Q}$ and $\vec{R}$ are mutually perpendicular

1) c & d are correct 2) a, b & c are correct

3) only a is correct 4) only d is correct

ASSERTION & REASONING QUESTIONS

VECTORS AND SCALARS

These Questions consist of two statements each printed as Assertion and Reason. While answering these questions choose any one of the following four responses.

1) If both (A) and (R) are true and (R) is the correct explanation of (A)

2) If both (A) and (R) are true and (R) is not the correct explanation of (A)

3) If (A) is true but (R) is false

4) If (A) is false but (R) is true

72. (A) : Electric current density is a vector

(R) : A physical quantity having magnitude and direction should be always a vector

73. (A) : The direction of velocity vector remains unchanged though the coordinate system is changed

(R) : The direction of real vector is independent of coordinate system

74. (A) : Every component of a null vector is zero

(R) : Origin and terminus of null vector are at the same point

75. (A) : $\frac{1}{\sqrt{2}}\hat{i}+\frac{1}{\sqrt{2}}\hat{j}$ is a unit vector

(R) : The component vectors of a unit vector need not be unit vectors

76. (A) : Torque is a pseudo vector

(R) : The cross product of two polar vectors is a pseudo vector

77. (A) : The division of a vector by another vector is not defined.

(R) : The division of a vector by a direction is not possible.

ADDITION AND SUBTRACTION OF VECTORS

78. (A) : Two forces 7N and 5N are acting at a point and their resultant can be 3N

(R) : If two vectors $\vec{P}$ and $\vec{Q}$ are acting at a point their resultant can have any value between $P-Q$ to $P+Q$.

79. (A) : The resultant of $\vec{P}$ and $\vec{Q}$ makes the angles α and β with $\vec{P}$ and $\vec{Q}$ respectively. If $|\vec{P}|>|\vec{Q}|$ then $\beta>\alpha$

(R) : Resultant is always closer to the vector of larger magnitude

80. (A) : $\hat{i}+\hat{j}$ is not a unit vector

(R) : The addition of unit vectors never gives unit vector

81. (A) : If $\vec{A}+\vec{B}+\vec{C}=0$ then $\vec{A}\times\vec{B}=\vec{B}\times\vec{C}=\vec{C}\times\vec{A}$.

(R) : The vector sum of three vectors can never be zero.

RESOLUTION OF VECTORS

82. (A) : When a vector is rotated, its magnitude remains constant.

(R) : The magnitude of a vector is independent of the coordinate system.

83. (A) : The projection $3\hat{j}-4\hat{k}$ along the y–axis is 3 units.

(R) : The projection of $\vec{A}$ along y–axis is A_y.

84. (A) : Vector sum of three coplanar forces can be zero

(R) : Minimum number of unequal coplanar vectors required to keep a point in equilibrium is three

RELATIVE VELOCITY

85. (A) : Relative velocity is zero when two bodies are moving opposite to each other with same velocity

(R) : Relative velocity of a body depend on direction of motion.

86. (A) : A man is on the northern bank of a river. To cross the river in a shortest time he should swim due south.

(R) : Man should move opposite to the river flow to cross in shortest time.

DOT AND CROSS PRODUCT

87. (A) : $\vec{A}\times\vec{B}=\vec{C}$ where $\vec{C}$ is a vector normal to both $\vec{A}$ and $\vec{B}$

(R) : Direction of cross product obeys right hand thumb rule

88. (A) : $\vec{A}.(\vec{B}+\vec{C})$ is a scalar

(R) : The dot product of two vectors is a scalar

89. (A) : If $\vec{\mathbf{A}}\times\vec{\mathbf{B}}=\vec{\mathbf{C}}\times\vec{\mathbf{B}}$, then $\vec{\mathbf{C}}$ must be equal to $\vec{\mathbf{A}}$.

(R) : The magnitude of the cross product of two vectors depends upon the angle between them.

MATCHING TYPE QUESTIONS

90. Match the following :

List - 1	List - 2
a) Work	e) $\vec{r}\times\vec{F}$
b) Power	f) $\vec{F}.\vec{S}$
c) Linear velocity	g) $\vec{F}.\vec{V}$
d) Torque	h) $\vec{\omega}\times\vec{r}$

1) a-g, b-h, c-f, d-e 2) a-f, b-g, c-e, d-h

3) a-f, b-g, c-h, d-e 4) a-g, b-f, c-e, d-h

91. Match the following :

List - 1	List - 2
(Angle between $\vec{A}$ & $\vec{B}$)	
a) $\vec{A}-\vec{B}=\vec{C}$ and $A-B=C$	e) π
b) $\vec{A}+\vec{B}=\vec{C}$ and $A-B=C$	f) $2\pi/3$
c) $\vec{A}-\vec{B}=\vec{C}$ and $A^2+B^2=C^2$	g) 0

d) $\vec{A}+\vec{B}=\vec{C}$ and A = B = C   h) $\pi/2$

1) a-g, b-h, c-f, d-e   2) a-h, b-e, c-f, d-g

3) a-f, b-e, c-h, d-g   4) a-g, b-e, c-h, d-f

92. If $\vec{a}$ and $\vec{b}$ are the two vectors $\vec{R}$ resultant of $\vec{a}, \vec{b}$ as shown

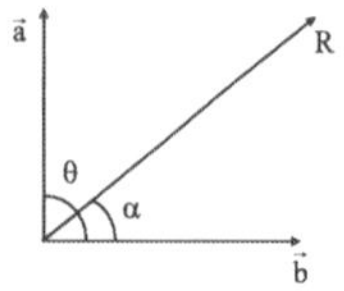

LIST - I	LIST - II
A : Component of $\vec{a}$ along $\vec{b}$	1) $a^2 + b^2$
B : Component of $\vec{b}$ along $\vec{a}$	2) 0
C : Component of R along $\vec{a}$	3) R sin α
D : Component of R along $\vec{b}$	4) R cos α

1) a–1, b–2, c–3, d–4   2) a–2, b–2, c–3, d–4
3) c–2, d–4, a–2, d–2   4) a–2, b–1, c–4, d–3

93. If $\vec{A}, \vec{B}, \vec{C}$ are the three vectors the angle between $\vec{A}, \vec{B}$ is α, $\vec{B}$ and $\vec{C}$ is β and $\vec{C}$ and $\vec{A}$ is γ, and k is constant then match the following.

LIST-I	LIST-II
a) A	1) k sin α
b) B	2) k sin γ
c) C	3) k sin β

1) a – 1, b – 2, c – 3   2) a – 1, b – 3, c – 2
3) a – 3, b – 2, c – 1   4) a – 2, b – 3, c – 1

<u>ASCENDING & DESCENDING ORDER</u>

94. Set the following vectors in the increasing order of their magnitude.

a) 3i + 4j   b) 2i + 4j + 6k   c) 2i + 2j + 2k

1) a, b, c   2) c, a, b   3) a, c, b   4) b, c, a

95. Arrange the vectors additions so that their magnitudes are in the increasing order.

a) two vector $\vec{A}$ and $\vec{B}$ are parallel

b) Two vectors $\vec{A}$ and $\vec{B}$ are antiparallel

c) Two vectors $\vec{A}$ and $\vec{B}$ making an angle 60^0

d) Two vectors A and $\vec{B}$ making 120^0.

1) b, d, c, a   2) b, c, d, a
3) a, c, d, b   4) c, d, a, b

96. Arrange the vectors subtractions so that their magnitudes are in decreasing order. If the two vectors $\vec{A}$ and $\vec{B}$ are acting at an angle $(|\vec{A}|>|\vec{B}|)$.

a) 60^0 b) 90^0 c) 180^0 d) 120^0

1) d,c,b,a   2) a,b,d,c   3) c,d,b,a   4) c,d,a,b

97. Set the drifts suffered by boat in increasing order.

a) d = 1000 m $V_R = 2$ m/s $V_b = 4$ m/s

b) d = 500 m $V_R = 1$ m/s $V_b = 6$ m/s

c) d = 1000 m $V_R = 6$ m/s $V_b = 6$ m/s

(d → width of river V_R → velocity of river V_b → velocity of Boat). The boat moves perpendicular to width of the river

1) a, b, c 2) b, c, a   3) b, a, c   4) c, b, a

98. Set the angles made by following vectors with x-axis in the increasing order.

a) 3i + 4j   b) 4i + 3j   c) i + j

1) a, b, c 2) c, b, a   3) b, c, a   4) a, c, b

99. Arrange the dot products in increasing order

a) $\vec{A}$ and $\vec{B}$ are parallel

b) $\vec{A}$ and $\vec{B}$ are making an angle 60^0

c) $\vec{A}$ and $\vec{B}$ making, angle 180^0

1) c, b, a   2) a, b, c
3) b, c, a   4) c, a, b

100. Arrange the magnitude of cross products in the decreasing order.

a) $\vec{A}$ and $\vec{B}$ making angle Zero

b) $\vec{A}$ and $\vec{B}$ making angle 30^0

c) $\vec{A}$ and $\vec{B}$ making angle 120^0

1) a, b, c   2) b, c, a   3) c, a, b   4) c, b, a

ANSWERS

1) 4	2) 1	3) 4	4) 4	5) 3
6) 3	7) 2	8) 4	9) 4	10) 4
11) 4	12) 2	13) 4	14) 3	15) 2
16) 2	17) 2	18) 3	19) 1	20) 3
21) 2	22) 3	23) 4	24) 4	25) 4
26) 4	27) 2	28) 3	29) 4	30) 2
31) 4	32) 3	33) 1	34) 4	35) 2
36) 1	37) 1	38) 2	39) 2	40) 1
41) 2	42) 1	43) 1	44) 3	45) 1
46) 3	47) 3	48) 4	49) 3	50) 4
51) 4	52) 4	53) 3	54) 2	55) 2
56) 4	57) 2	58) 3	59) 4	60) 2
61) 3	62) 3	63) 4	64) 1	65) 2
66) 4	67) 2	68) 4	69) 1	70) 2
71) 1	72) 3	73) 1	74) 1	75) 2
76) 1	77) 1	78) 1	79) 1	80) 3
81) 3	82) 1	83) 2	84) 1	85) 4
86) 3	87) 1	88) 1	89) 4	90) 3
91) 4	92) 2	93) 3	94) 2	95) 1
96) 3	97) 3	98) 3	99) 1	100) 4

EXERCISE - II(A)

ADDITION AND SUBTRACTION

1. **If A = 3i – 4j and B =– i – 4j, calculate $\vec{A}+\vec{B}$ and $|\vec{A}+\vec{B}|$.**

1) 2i–8j, 8.1 2) 2i–4j, 4.2
3) 2i, 3 4) 4j, 6

2. **If $\vec{A}$ = 3i – 4j and $\vec{B}$ = –i – 4j, calculate the direction of $\vec{A}+\vec{B}$.**

1) $\tan^{-1}(4)$ with + x– axis in clock wise
2) $\tan^{-1}(4)$ with – x– axis in clock wise
3) $\tan^{-1}(4)$ with + x– axis in anticlock wise
4) $\tan^{-1}(4)$ with – x– axis in anticlock wise

3. **A person runs along a circular path of radius 5 m. If he completes half of the circle find the magnitude of the displacement vetor, How far the person ran ?**

1) 10m, 5π m 2) 5π m
3) 5π m, 19m 4) 14m, 10π m

4. **A car makes a displacement of 100 m towards east and then 200 m towards north. Find the magnitude and direction of the resultant.**

1) 223.7m, $\tan^{-1}(2)$, N of E
2) 223.7m, $\tan^{-1}(2)$, E of N
3) 300m, $\tan^{-1}(2)$, N of E
4) 100m, $\tan^{-1}(2)$, N of E

5. **If the angle between two vectors of equal magnitude P is θ, the magnitude of the resultant vector is**

1) $2P\cos\frac{\theta}{2}$ 2) $2P\sin\frac{\theta}{2}$
3) $2P\cos\theta$ 4) $2P\sin\theta$

6. **Maximum and minimum magnitudes of the resultant of two vectors of magnitudes P and Q are found to be in the ratio 3 : 1. Which of the following relations is true ?**

1) P = Q 2) P =2Q
3) P = 4Q 4) P = Q/3

7. **The resultant of two equal forces is 141.4N when they are mutually perpendicular. When they are inclined at an angle 120°, then the resultant force will be**

1) 100 N 2) 141.4 N 3) 196 N 4) Zero

8. **The magnitudes of two vectors P and Q differ by 1. The magnitude of their resultant makes an angle of $\tan^{-1}(3/4)$ with P. The angle between P and Q is**

1) 45^0 2) 0^0 3) 180^0 4) 90^0

9. **If $\vec{P}-\vec{Q}=\vec{R}$ and P=Q=R, the ratio of angles between $\vec{P}\&\vec{R}$ to $\vec{P}\&\vec{Q}$ is**

1) 1 : 2 2) 2 : 1 3) 1 : 1 4) 1 : 3

10. **The resultant of two forces 2P and $\sqrt{2}$ P is $\sqrt{10}$ P. The angle between the forces is**

1) 30^0 2) 60^0 3) 45^0 4) 90^0

11. **Eleven forces each equal to 5N act on a particle simultaneously. If each force makes an angle 30° with the next one, the resultant of all forces is**

1) 15N 2) 55 N 3) 5 N 4) zero

12. Which of the following sets of forces acting simultaneously on a particle keep it in equilibrium?

1) 3N, 5N, 10N 2) 4N, 7N, 12N

3) 2N, 6N, 5N 4) 5N, 8N, 1N

13. The resultant of two forces 1 and P is perpendicular to '1' and equal to 1. What is the value of 'P' and angle between the forces

1) $\sqrt{2}$ N, 135^0 2) $\sqrt{2}$ N, 150^0

3) 2 N, 120^0 4) 2 N, 150^0

14. If the sum of two unit vectors is also a vector of unit magnitude, the magnitude of the difference of the two unit vectors is

1) 1 unit 2) 2 units 3) $\sqrt{3}$ units 4) Zero

15. If $\vec{P}+\vec{Q}=\vec{R}$ and $\vec{P}-\vec{Q}=\vec{S}$, then R^2+S^2 is equal to

1) P^2+Q^2 2) $2(P^2-Q^2)$ 3) $2(P^2+Q^2)$ 4) 4 PQ

16. The greater and least resultant of two forces are 7N and 3N respectively. If each of the force is increased by 3N and applied at 60°. The magnitude of the resultant is

1) 7N 2) 3N 3) 10 N 4) $\sqrt{129}$

17. Two forces are such that the sum of their magnitudes is 18N, the resultant is $\sqrt{228}$ when they are at 120^0. Then the magnitude of the forces are

1) 12N, 6N 2) 13N, 5N

3) 10N, 9N 4) 16N, 2N

18. Two vectors are given by $\vec{a}$=-2i+j-3k and $\vec{b}$ = 5i+3j-2k. If $3\vec{a}+2\vec{b}-\vec{c}=0$ then third vector $\vec{c}$ is

1) $4\hat{i}+9\hat{j}-13\hat{k}$ 2) $-4\hat{i}-9\hat{j}+13\hat{k}$

3) $4\hat{i}-9\hat{j}-13\hat{k}$ 4) $2\hat{i}-3\hat{j}+13\hat{k}$

***19. A body starts with a velocity 2i – 3j + 11k m/s and moves with an accelaration of 10i+10j+10k ms^{-2}. What is its velocity after 0.25 s ?**

1) $4.5\hat{i}+0.5\hat{j}-13.5\hat{k}$ 2) $4.5\hat{i}-0.5\hat{j}-13.5\hat{k}$

3) $4.5\hat{i}-0.5\hat{j}+13.5\hat{k}$ 4) $4.5\hat{i}+0.5\hat{j}+13.5\hat{k}$

20. Figure shows three vectors $\vec{a},\vec{b}$ and $\vec{c}$, where R is the midpoint of PQ. Then which of the following relations is correct?

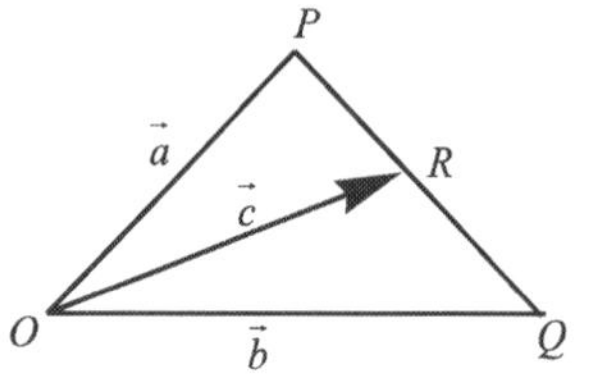

1) $\vec{a}+\vec{b}=2\vec{c}$

2) $\vec{a}+\vec{b}=\vec{c}$

3) $\vec{a}-\vec{b}=2\vec{c}$

4) $\vec{a}-\vec{b}=\vec{c}$

21. The resultant of two vectors $\vec{P}$ and $\vec{Q}$ is $\vec{R}$. If the magnitude of $\vec{Q}$ is doubled, the new resultant becomes perpendicular to $\vec{P}$, then the magnitude of $\vec{R}$ is

1) $\frac{P^2-Q}{2PQ}$ 2) $\frac{P+Q}{P-Q}$ 3) Q 4) $\frac{P}{Q}$

<u>POSITION VECTOR, COMPONENTS OF VECTORS, UNIT VECTORS</u>

22. One of the rectangular components of a velocity of 20 ms^{-1} is 10 ms^{-1} . Find the other component.

1) 10 $\sqrt{3}$ ms^{-1} 2) 20 $\sqrt{3}$ ms^{-1}

3) 5 $\sqrt{3}$ ms^{-1} 4) 35 $\sqrt{3}$ ms^{-1}

23. A car weighing 100kg is on a slope that makes an angle 30° with the horizontal. The component of car's weight parallel to the slope is (g=10ms^{-2})

1) 500N 2) 1000N 3) 15,000 N 4) 20,000N

24. The force acting on a particle makes an angle of 60^0 with the positive x-axis. If the magnitude of the force is 10 N what are its x and y-components ?

1) 5N, 5$\sqrt{3}$ N 2) 2$\sqrt{3}$ N, 2N

3) 4$\sqrt{3}$ N, 5N 4) 3$\sqrt{3}$ N, 3N

25. To go from town A to town B a plane must fly about 1780 km at an angle of 30^0 West of north. How far north of A is B ?

1) 1542km 2) 1452 km

3) 1254 km 4) 11 km

26. **Displacement vectors $\vec{A}$ = (3i+3j)m, $\vec{B}$ = (i – 4j) m and $\vec{C}$ = (–2i + 5j) m. Use the components method to determine the magnitude and direction of $\vec{E}=-\vec{A}-\vec{B}+\vec{C}$.**

1) $5\sqrt{2}$, 60^0 clockwise with + ve x – axis

2) $6\sqrt{2}$, 45^0 clockwise with + ve x – axis

3) $5\sqrt{2}$, 60^0 clockwise with – ve x – axis

4) $6\sqrt{2}$, 45^0 clockwise with – ve x – axis

27. **An ant moves on the floor of a room making two displacements. One of magnitude 3 m and another of magnitude 4 m . If the angle between the two displacements is 120^0 what will be the magnitude of the resultant displacement ?**

1) 6.082m 2) 5.2m

3) 4.8 m 4) 3.606m

28. **If A = 6.00i – 8.00j units, B = 8.00i + 3.00j units and C = 26.0i + 19.0j units, determine a and b when a A + bB+C = 0**

1) a = 0.9, b = –3.93 2) a = 0.9, b = 3.93

3) a = 0.6, b = –3 4) a = 0.4, b = 0.8

29. **Two vectors a and b have equal magnitudes of 12 units. These vectors are making angles 30^0 and 120^0 with the x axis respectively. Their sum is $\vec{r}$. Find the x and y components of $\vec{r}$.**

1) $(6\sqrt{3}-6),(6+6\sqrt{3})$ 2) $(6\sqrt{3}+6),(6+6\sqrt{3})$

3) $(6\sqrt{3}-6),(6-6\sqrt{3})$ 4) $(6\sqrt{3}+6),(6-6\sqrt{3})$

30. **The displacement of the point of a wheel initially in contact with the ground when the wheel rolls forward half a revolution where radius of the wheel is 1m, is (Assume the forward direction as x - axis)**

1) $\sqrt{2}\text{m}, \tan^{-1}(1/\pi)$ with x - axis

2) $\left(\sqrt{\pi^2+4}\right)\text{m}, \tan^{-1}(2/\pi)$ with x - axis

3) $\left(\sqrt{\pi^2+4}\right)\text{m}, \tan^{-1}(1/2)$ with x - axis

4) $\left(\sqrt{\pi^2+4}\right)\text{m}, \tan^{-1}(1/2)$ with x - axis

31. **Angle (in rad) made by the vector $\sqrt{3}\hat{i}+\hat{j}$ with the x-axis is (2005 M)**

1) $\pi/6$ 2) $\pi/4$ 3) $\pi/3$ 4) $\pi/4$

32. **A bird moves in such a way that it has a displacement of 12 m towards east, 5 m towards north and 9 m vertically upwards. Find the magnitude of its displacement.**

1) $5\sqrt{2}$ m 2) $5\sqrt{10}$ m 3) $5\sqrt{5}$ m 4) 5m

33. **The position vector of a moving particle at 't' sec is given by $\vec{r}=3\hat{i}+4t^2\hat{j}-t^3\hat{k}$. Its displacement during an interval of t = 1s to 3 sec is**

1) $\hat{j}-\hat{k}$ 2) $3\hat{i}+4\hat{j}-\hat{k}$

3) $9\hat{i}+36\hat{j}-27\hat{k}$ 4) $32\hat{j}-26\hat{k}$

34. **The direction cosines of a vector $\vec{A}$ are $\cos\alpha=\frac{4}{5\sqrt{2}}$, $\cos\beta=\frac{1}{\sqrt{2}}$ and $\cos\gamma=\frac{3}{5\sqrt{2}}$, then the vector A is**

1) $4\hat{i}+\hat{j}+3\hat{k}$ 2) $4\hat{i}+5\hat{j}+3\hat{k}$

3) $4\hat{i}-5\hat{j}-3\hat{k}$ 4) $\hat{i}+\hat{j}-\hat{k}$

35. **If $\vec{p}$ = i+2j+6k, its direction cosines are**

1) $\frac{1}{41},\frac{2}{41}$ and $\frac{6}{41}$ 2) $\frac{1}{\sqrt{41}},\frac{2}{\sqrt{41}}$ and $\frac{6}{\sqrt{41}}$

3) $\frac{3}{\sqrt{41}},\frac{8}{\sqrt{41}},\frac{7}{\sqrt{41}}$ 4) 1,2 and 6

36. **If $\vec{A}$ =2i-3j+4k, its components in yz plane and zx plane are respectively**

1) $\sqrt{13}$ and 5 2) 5 and $2\sqrt{5}$

3) $2\sqrt{5}$ and $\sqrt{13}$ 4) $\sqrt{13}$ and $\sqrt{29}$

37. **If vectors $\vec{A}$ and $\vec{B}$ are 3i – 4j + 5k and 2i + 3j – 4k respectively find the unit vector parallel to $\vec{A}+\vec{B}$.**

1) $\frac{(5i-j+k)}{\sqrt{27}}$ 2) $\frac{(5i+j+k)}{\sqrt{27}}$

3) $\frac{(5i+j+k)}{27}$ 4) $\frac{(5i-j-k)}{27}$

38. A vector $3\hat{i}+4\hat{j}$ rotates about its tail through an angle 37^0 in anticlockwise direction then the new vector is

1) $-3\hat{i}+4\hat{j}$ 2) $3\hat{i}-4\hat{j}$

3) $5\hat{j}$ 4) $5\hat{i}$

LAMI'S THEOREM

39. If 'O' is at equilibrium then the values of the tension T_1 and T_2 are x, y, if 20 N is vertically down. Then x, y are

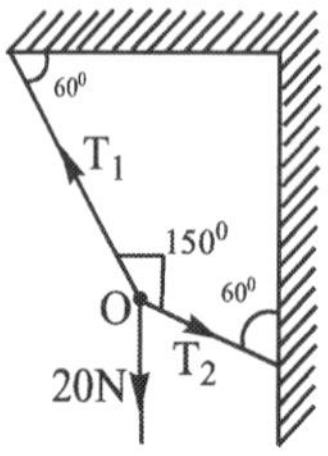

1) 20 N 30 N

2) $20\sqrt{3}$N, 20 N

3) $20\sqrt{3}$N , $20\sqrt{3}$N

4) 10 N, 30 N

40. A body of mass $\sqrt{3}$ kg is suspended by a string to a rigid support. The body is pulled horizontally by a force F until the string makes an angle of 30° with the vertical. The value of F and tension in the string are.

1) 9.8 N, 9.8 N 2) 9.8 N, 19.6 N

3) 19.6 N, 19.6 N 4) 19.6 N, 9.8 N

41. Two light strings of length 4cm and 3cm are tied to a bob of weight 500gm. The free ends of the strings are tied to pegs in the same horizontal line and separated by 5cm. The ratio of tension in the longer string to that in the shorter string is

1) 4:3 2) 3:4 3) 4:5 4) 5:4

42. A boy is hanging from a horizontal branch of a tree. The tension in the arms will be maximum when the angle between the arms is

1) 0^0 2) 30^0 3) 60^0 4) 120^0

RELATIVE VELOCITY &

RAIN UMBRELLA PROBLEMS

43. An aeroplane is flying with the velocity of V_1 = 800 kmph relative to the air towards south. A wind with velocity of V_2 =15 ms^{-1} is blowing from West to East. What is the velocity of the aeroplane with respect to the earth.

1) 221.7ms^{-1} 2) 150ms^{-1}

3) 82ms^{-1} 4) 40ms^{-1}

44. The wind is blowing from south at 10ms^{-1}, but to a cyclist it appears blowing from the east at 10ms^{-1}. The velocity of cyclist is

1) $10\sqrt{2}\ ms^{-1}$ towards S – W

2) $10\sqrt{2}\ ms^{-1}$ towards N – W

3) $10\sqrt{2}\ ms^{-1}$ towards S – E

4) $10\sqrt{2}\ ms^{-1}$ towards N – E

45. A particle is moving east wards with a velocity of 5 ms^{-1}. In 10 S, the velocity changes to 5 ms^{-1} north wards. The average acceleration in this time is (AIEEE-2005)

1) $1/\sqrt{2}\ ms^{-2}$ towards north east

2) $1/2\ ms^{-2}$ towards north

3) Zero 4) $\frac{1}{\sqrt{2}}\ ms^{-2}$ towards north-west.

46. A Ship 'A' steams down north at a speed of 8 kmph. and ship B due west at a speed of 6kmph. The velocity of A wrt. B is.

1) 10 kmph, $Tan^{-1}\left(\frac{4}{3}\right)$ N of E

2) 10kmph, $tan^{-1}\left(\frac{4}{3}\right)$ E of N

3) 10 kmph NE 4) 2kmph, $Tan^{-1}\left(\frac{4}{3}\right)$ N of E

47. Rain drops are falling down ward vertically at 4 kmph. For a person moving forward at 3 kmph feels the rain at

1) 7 kmph 2) 1 kmph

3) 5 kmph 4) 25 kmph

48. A man is travelling at 10.8 kmph in a topless car on a rainy day. He holds an umberella at an angle of 37^0 with the vertical so that he does not wet. If rain drops falls vertically downwards, what is the rain velocity ?

1) 1ms^{-1} 2) 2ms^{-1} 3) 3ms^{-1} 4) 4ms^{-1}

49. When a man is standing, rain drops appears to him falling at 60^0 from the horizontal from his front side. When he is travelling at 5 kmph on a horizontal road they appear to him falling at 30^0 from the horizontal from his front side. The actual speed of the rain is (in kmph)

1) 3 2) 4 3) 5 4) $5\sqrt{3}$

RIVER–BOAT PROBLEMS

50. A man can swim in still water at a speed of 6 kmph and he has to cross the river and reach just opposite point on the other bank. If the river is flowing at a speed of 3 kmph, he has to swim in the direction

1) 30° with the river flow 2) 60° with river flow
3) 135° with the river flow 4) 120° with the river flow

51. A person can swim in still water at 5 m/s. He moves in a river of velocity 3 m/s. First down the stream next same distance up the steam the ratio of times taken are

1) 1 : 1 2) 1 : 2 3) 1 : 4 4) 4 : 1

52. A man can swim in still water at a speed of 6 kmph and he has to cross the river and reach just opposite point on the other bank. If the river is flowing at a speed of 3 kmph, and the width of the river is 2km, the time taken to cross the river is (in hours)

1) $\frac{2}{27}$ 2) $\frac{2}{\sqrt{27}}$ 3) $\frac{2}{3}$ 4) $\frac{2}{\sqrt{45}}$

53. The velocity of water in a river is 2 kmph, while width is 400 m. A boat is rowed from a point rowing always aiming opposite point at 8 kmph of still water velocity. On reaching the opposite bank the drift obtained is

1) 93 m 2) 100. 8m 3) 112.4 m 4) 100 m

54. A person swims at 135^0 to current of river, to meet target on reaching opposite point. The ratio of person's velocity to river water velocity is

1) $\sqrt{3}:1$ 2) $\sqrt{2}:1$ 3) $1:\sqrt{2}$ 4) $1:\sqrt{3}$

55. A boat moves perpendicualar to the bank with a velocity of 7.2 km/h. The current carries it 150 m downstream, find the velocity of the current. (The width of the river is 0.5 km).

1) $0.4ms^{-1}$ 2) $1.2ms^{-1}$ 3) $0.5ms^{-1}$ 4) $0.6ms^{-1}$

56. A swimmer is capable of swimming $1.65\ ms^{-1}$ in still water. If she swims directly across a 180 m wide river whose current is 0.85 m/s, how far downstream (from a point opposite her starting point) will she reach ?

1) 92.7 m 2) 40 m 3) 48 m 4) 20 m

DOT PRODUCT

57. If P = 2i +3j – 4k and Q = 5i +2j + 4k find the angle between the two vectors.

1) 30^0 2) 45^0 3) 60^0 4) 90^0

58. A force $\vec{F}$ = 5i – 3j + 2k moves a particle from $\vec{r}_1$ = 2i +7j+4k to $\vec{r}_2$ = 5i+2j+8k. Calculate the workdone.

1) 38 units 2) 20 units 3) 30 units 4) 15 units

59. A vector perpendicular to the vector (i + 2j) and having magnitude $3\sqrt{5}$ units is

1) $3\hat{i}+6\hat{j}$ 2) $6\hat{i}-3\hat{j}$
3) $4\hat{i}-2\hat{j}$ 4) $\hat{i}-2\hat{j}$

60. If the vectors $\vec{P}=a\hat{i}+a\hat{j}+3\hat{k}$ and $\vec{Q}=a\hat{i}-2\hat{j}-\hat{k}$ are perpendicular to each other then the positive value of 'a' is

1) Zero 2) 1 3) 2 4) 3

61. A force of 2i + 3j + 2k N acts a body for 4 s and produces a displacement of 3i + 4j + 5k m calculate the power ?

1) 5 W 2) 6 W 3) 7 W 4) 9 W

62. If $\vec{A}=2i+3j$ and $\vec{B}=2j+3k$ the component of $\vec{B}$ along $\vec{A}$ is

1) 6 2) $\frac{1}{6}$ 3) $\frac{6}{13}$ 4) $\frac{6}{\sqrt{13}}$

63. The component of $\vec{A}$ along $\vec{B}$ is $\sqrt{3}$ times that of the component of $\vec{B}$ along $\vec{A}$. Then A:B is

1) $1:\sqrt{3}$ 2) $\sqrt{3}:1$ 3) $2:\sqrt{3}$ 4) $\sqrt{3}:2$

64. When a force (8i + 4j) newton displaces a particle through (3i - 3j) metre, the power is 0.6W. The time of action of the force is

1) 20s 2) 7.2s 3) 72s 4) 2s

65. If θ is the angle between unit vectors $\vec{A}$ and $\vec{B}$, then $\dfrac{(1-\vec{A}.\vec{B})}{(1+\vec{A}.\vec{B})}$ is equal to

1) $\tan^2(\theta/2)$ 2) $\sin^2(\theta/2)$
3) $\cot^2(\theta/2)$ 4) $\cos^2(\theta/2)$

CROSS PRODUCT

66. The cross product of the vectors (2i -3j+ 4k) and (i + 4j - 5k) is

1) $-\hat{i} - 14\hat{j} + 11\hat{k}$ 2) $\hat{i} + 14\hat{j} + 11\hat{k}$
3) $-\hat{i} + 14\hat{j} + 11\hat{k}$ 4) $-\hat{i} + 14\hat{j} + 5\hat{k}$

67. A particle of mass 0.2kg is moving with linear velocity (i-j+2k). If the radius vector r = 4i+j-k, the angular momentum of the particle is

1) 2.14 units 2) $\sqrt{4.28}$ units
3) $\sqrt{107}$ units 4) 5 units

68. The magnitude of scalar and vector products of two vectors are $48\sqrt{3}$ and 144 respectively. What is the angle between the two vectors ?
1) 30^0 2) 45^0 3) 60^0 4) 90^0

69. The torque of force $\vec{F} = 2\vec{i} - 3\vec{j} + 4\vec{k}$ newton acting at the point $\vec{r} = 3\vec{i} + 2\vec{j} + 3\vec{k}$ metre about the origin is

1) $6\vec{i} - 6\vec{j} + 12\vec{k}$ 2) $17\vec{i} - 6\vec{j} - 13\vec{k}$
3) $-6\vec{i} + 6\vec{j} - 12\vec{k}$ 4) $-17\vec{i} + 6\vec{j} + 13\vec{k}$

70. The adjacent sides of a parallelogram are $\vec{P}$ = 2i – 3j + k and $\vec{Q}$ = – 2i +4j – k. What is the area of the parallelogram ?

1) 4 units 2) 7 units 3) $\sqrt{5}$ units 4) $\sqrt{8}$ units

71. What is the condition for the vectors 2i+3j – 4k and 3i – aj + bk to be parallel ?
1) a = –9/2, b = – 6 2) a = –6, b = –9/2
3) a = 4, b = 5 4) a = 8, b = 2

72. a, b are the magnitudes of vectors $\vec{a}$ & $\vec{b}$. If $\vec{a} \times \vec{b} = 0$ the value of $\vec{a}.\vec{b}$ is

1) 0 2) $\sqrt{ab}$ 3) ab 4) a/b

73. If $\vec{A} = 4N, \vec{B} = 3N$ the value of $|\vec{A} \times \vec{B}|^2 + |\vec{A}.\vec{B}|^2$ then
1) 5 N 2) 25 N 3) 144 N 4) 169 N

74. If $\vec{A} = 2\hat{i} + 3\hat{j} + 6\hat{k}$ and $\vec{B} = 3\hat{i} - 6\hat{j} + 2\hat{k}$ then vector perpendicular to both $\vec{A}$ and $\vec{B}$ has magnitude K times that of $6\hat{i} + 2\hat{j} - 3\hat{k}$. Then K =

1) 1 2) 3 3) 7 4) 9

75. Vertices of a triangle are given by $\hat{i} + 3\hat{j} + 2\hat{k}$, $2\hat{i} - \hat{j} + \hat{k}$ and $-\hat{i} + 2\hat{j} + 3\hat{k}$, then area of triangle is (in units)

1) $\sqrt{107}/2$ 2) $\sqrt{107}/6$
3) $\sqrt{\dfrac{107}{2}}$ 4) $\sqrt{207}/3$

ANSWERS

1) 1	2) 1	3) 1	4) 1	5) 1
6) 2	7) 1	8) 4	9) 3	10) 3
11) 3	12) 3	13) 1	14) 3	15) 3
16) 4	17) 4	18) 1	19) 3	20) 1
21) 3	22) 1	23) 1	24) 1	25) 1
26) 4	27) 4	28) 1	29) 1	30) 2
31) 1	32) 2	33) 4	34) 2	35) 2
36) 2	37) 1	38) 3	39) 2	40) 2
41) 2	42) 4	43) 1	44) 4	45) 4
46) 1	47) 3	48) 4	49) 3	50) 4
51) 3	52) 2	53) 4	54) 2	55) 4
56) 1	57) 4	58) 1	59) 2	60) 4
61) 3	62) 4	63) 2	64) 1	65) 1
66) 3	67) 2	68) 3	69) 2	70) 3
71) 1	72) 3	73) 3	74) 3	75) 1

EXERCISE - II(B)

ADDITION AND SUBTRACTION

1. If $\vec{A} = 3i - 4j$ and $\vec{B} = -i - 4j$, calculate $\vec{A} - \vec{B}$ and $|\vec{A} - \vec{B}|$.

   1) 2i–8j, 8.1 2) 2i–4j, 4.2
   3) 2i, 3 4) 4i, 4

2. If $\vec{A} = 3i - 4j$ and $\vec{B} = -i - 4j$, calculate the direction of $\vec{A} - \vec{B}$.

   1) along positive x–axis 2) along negative x–axis
   3) along positive y–axis 4) along negative y–axis

3. A person runs along a circular path of radius 5 m. find the magnitude of the displacement if he runs all the way around the circle ?

   1) 10m 2) 20m 3) 3m 4) Zero

4. A car is moving 40m due east, turns towards north moves 30m. then turns 45^0 east of north moves $20\sqrt{2}$ m. The net displacement of car is (East is taken positive x – axis ,North as positive y – axis)

   1) $50\hat{i} + 60\hat{j}$ 2) $60\hat{i} + 50\hat{j}$
   3) $30\hat{i} + 40\hat{j}$ 4) $40\hat{i} + 30\hat{j}$

5. If the angle between two vectors of equal magnitude P is θ, the magnitude of the difference of the vectors is

   1) $2P\cos\frac{\theta}{2}$ 2) $2P\sin\frac{\theta}{2}$ 3) $P\cos\frac{\theta}{2}$ 4) $P\sin\frac{\theta}{2}$

6. Two forces are in the ratio of 5 : 2. The maximum and minimum of their resultants are in the ratio of

   1) 5 : 2 2) 2 : 5 3) 7 : 3 4) 3 : 7

7. Two equal forces of maguitude 'p' each are angled first at 60° later at 120°. The ratio of magnitude of their resultants is

   1) $1:\sqrt{3}$ 2) $\sqrt{3}:1$ 3) 1 : 1 4) 1 : 2

8. Three vectors $\vec{P}, \vec{Q}, \vec{R}$ obey $\vec{P} + \vec{Q} = \vec{R}$ and $P^2 + Q^2 = R^2$ the angle between $\vec{P}$ & $\vec{Q}$. is

   1) 0° 2) 30° 3) 60° 4) 90°

9. $\vec{P}, \vec{Q}, \vec{R}, \vec{S}$ are vector of equal magnitude. If $\vec{P} + \vec{Q} - \vec{R} = O$ angle between $\vec{P}$ and $\vec{Q}$ is θ_1. If $\vec{P} + \vec{Q} - \vec{S} = O$ angle between $\vec{P}$ and $\vec{S}$ is θ_2. The ratio of θ_1 to θ_2 is

   1) 1 : 2 2) 2 : 1 3) 1 : 1 4) $1:\sqrt{3}$

10. The resultant of two forces 2N and 3N is $\sqrt{19}$ N. The angle between the forces is

    1) 30° 2) 45° 3) 60° 4) 90°

11. Twelve forces each of 5N act on a body simultaneously. If each force makes an angle of 30° with other their resultant is

    1) 5N 2) 60 N 3) 5N 4) Zero

12. Which of the following sets of forces acting simultaneously on a particle keep it in equilibrium?(in newton)

    1)2,3,6 2) 7,8,19 3) 3,4,5 4)1,3,7

13. The resultant of two forces, one double the other in magnitude is perpendicular to the smaller of the two forces. The angle between the two forces is

    1) 150° 2) 90° 3) 60° 4) 120°

14. If the difference of two unit vectors is also a vector of unit magnitude, the magnitude of the sum of the two unit vectors is

    1)1 2)$1/\sqrt{3}$ 3) $\sqrt{3}$ 4) 2

15. The resultant of two vectors $\vec{A}$ & $\vec{B}$ is $\vec{C}$. If direction of $\vec{B}$ is reversed the resultant is $\vec{D}$. Then which of the following is true ?

    1) $C^2 + D^2 = A^2 + B^2$ 2) $A^2 + C^2 = B^2 + D^2$
    3) $C^2 + D^2 = 2(A^2 + B^2)$ 4) $C^2 - D^2 = A^2 + B^2$

16. The greater and least resultant of two forces are 9N and 5N respectively. If they are applied at 60°. The magnitude of the resultant is

    1) 100N 2) 119N 3) $\sqrt{119}$ N 4) $\sqrt{67}$ N

17. The resultant of two forces at right angles is 13N. The minimum resultant of the two forces is 7N. The forces are

    1) 20N, 6N 2) 10N, 20N
    3) 5N, 12N 4) 8 N, 15 N

18. The resultant of the forces $\vec{F_1} = 4\hat{i} - 3\hat{j}$ and $\vec{F_2} = 6\hat{i} + 8\hat{j}$ is

1) $5\sqrt{5}$ 2) $10\hat{i} - 5\hat{j}$ 3) 125 4) $-2\hat{i} - 3\hat{j}$

19. A particle has an initial velocity $(6\bar{i} + 8\bar{j})\,ms^{-1}$ and an acceleration of $(0.8\bar{i} + 0.6\bar{j})\,ms^{-2}$. Its speed after 10s is

1) $20\ ms^{-1}$ 2) $7\sqrt{2}\ ms^{-1}$

3) $10\ ms^{-1}$ 4) $14\sqrt{2}\ ms^{-1}$

20. In an equilateral triangle ABC, AL,BM and CN are medians. Forces along BC and BA represented by them will have a resultant represented by

1) $2\overrightarrow{AL}$ 2) $2\overrightarrow{BM}$ 3) $2\overrightarrow{CN}$ 4) $\overrightarrow{AC}$

21. Given that $\vec{A} + \vec{B} + \vec{C} = 0$, out of three vectors two are equal in magnitude and the magnitude of third vector is $\sqrt{2}$ times that of either of two having equal magnitude. Then angle between vectors are given by

1) $30^0, 60^0, 90^0$ 2) $45^0, 135^0, 150^0$

3) $90^0, 135^0, 150^0$ 4) $90^0, 135^0 135^0$

POSITION VECTOR, COMPONENTS OF VECTORS, UNIT VECTORS

22. The horizontal and vertical components of a force are 8N and 15N respectively. The force is

1) 23 N 2) 20N 3) 17 N 4) 7 N

23. An aeroplane is heading north east at a speed of $141.4 ms^{-1}$. The northward component of its velocity is

1) $141.4\ ms^{-1}$ 2) $100\ ms^{-1}$

3) zero 4) $50\ ms^{-1}$

24. The horizontal and vertical components of a vector are 20 and 30 units respectively. The angle made by the vector with the horizontal is

1) $\tan^{-1}\frac{3}{2}$ 2) $\cos^{-1}\frac{2}{\sqrt{13}}$

3) $\sin^{-1}\frac{3}{\sqrt{13}}$ 4) all the above

25. To go from town A to town B a plane must fly about 1780 km at an angle of 30^0 West of north. How far West of A is B ?

1) 1542km 2) 1452 km

3) 1254 km 4) 890 km

26. Displacement vectors $\vec{A}$ = (3i+3j)m, $\vec{B}$ = (i – 4j)m and $\vec{C}$ = (–2i + 5j)m. Use the components method to determine a) the magnitude and direction of the vector $\vec{D} = \vec{A} + \vec{B} + \vec{C}$.

1) $2\sqrt{5}, \tan^{-1}(2)$ 2) $2\sqrt{5}, 45^0$

3) $2\sqrt{5}, \tan^{-1}(1/2)$ 4) $2\sqrt{5}, 60^0$

27. An ant moves on the floor of a room making two displacements. One of magnitude 3 m and another of magnitude 4 m . If the angle between the two displacements is 60^0 what will be the magnitude of the resultant displacement ?

1) 6.082m 2) 5.2m 3) 4.8 m 4) 6.8m

28. Three forces $\vec{F}_1$= a(i–j+k) $\vec{F}_2$= 2i-3j+4k and $\vec{F_3}$= 8i-7j+6k act simultaneously on a particle. If the particle is in equilibrium, the value of a is

1) 10 2) -10 3) 8 4) 2

29. Two vectors a and b have equal magnitudes of 12 units. These vectors are making angles 30^0 and 120^0 with the x axis respectively. Their sum is r. Find the magnitude and the angle r makes with the +x axis.

1) $12\sqrt{2}, 75^0$ clockwise

2) $12\sqrt{2}, 75^0$ anticlockwise

3) 12, 45^0 clockwise 4) 12, 45^0 anticlockwise

30. A wheel of perimeter 4π m is rolling on a horizontal surface. The displacement of the point of contanct of wheel and ground when the wheel completes one quarter of revolution is

1) $\sqrt{(\pi+2)^2+4}$ along $\tan^{-1}\frac{2}{\pi}$ with x-axis

2) $\sqrt{(\pi-2)^2+4}$ along $\tan^{-1}\frac{2}{\pi}$ with x-axis

3) $\sqrt{(\pi-2)^2+4}$ along $\tan^{-1}\frac{2}{\pi-2}$ with x-axis

4) $\sqrt{(\pi+2)^2+4}$ along $\tan^{-1}\frac{2}{\pi-2}$ with x-axis

31. Given two vectors $\vec{A} = i - 2j - 3k$ and $\vec{B} = 4i - 2j + 6k$. The angle made by $(\vec{A} + \vec{B})$ with the X - axis is (2007 M)

1) 30^0 2) 45^0 3) 60^0 4) 90^0

32. A room has dimensions 3mx4mx5m. A fly starting at one corner ends up at the diametrically opposite corner. The magnitude of the displacement of the fly is

1) 12m 2) 60 m 3) $2\sqrt{5}$ m 4) $5\sqrt{2}$ m

33. The position vector of a particle is given by $\vec{r} = 3t^2 i + 4t^2 j + 7k$ m at a given time 't'. The net displacement of the particle after 10s is.

1) 500 m 2) 400 m 3) 300 m 4) 700 m

34. If the two directional cosines of a vectors are $\frac{1}{\sqrt{2}}$ and $\frac{1}{\sqrt{3}}$ then the value of third directional cosine is

1) $\frac{1}{\sqrt{6}}$ 2) $\frac{1}{\sqrt{5}}$ 3) $\frac{1}{\sqrt{7}}$ 4) $\frac{1}{\sqrt{10}}$

35. If $\vec{P}$ = i+j+k, its direction cosines are

1) 1,1,1 2) $1/\sqrt{3}, 1/\sqrt{3}, 1/\sqrt{3}$

3) $\sqrt{3}, \sqrt{3}, \sqrt{3}$ 4) 0,0,0

36. If $\vec{A}$ =2i-3j+4k, its component in xy plane is

1) 4 2) $\sqrt{13}$ 3) $\sqrt{29}$ 4) 1

37. The unit vector parallel to the resultant of the vectors $\vec{A} = 4\hat{i} + 3\hat{j} + 6\hat{k}$ and $\vec{B} = -\hat{i} + 3\hat{j} - 8\hat{k}$ is (2000 M)

1) $\frac{1}{7}\left(3\hat{i} + 6\hat{j} - 2\hat{k}\right)$ 2) $\frac{1}{7}\left(3\hat{i} + 6\hat{j} + 2\hat{k}\right)$

3) $\frac{1}{49}\left(3\hat{i} + 6\hat{j} - 2\hat{k}\right)$ 4) $\frac{1}{49}\left(3\hat{i} - 6\hat{j} + 2\hat{k}\right)$

38. A vector $\hat{i} + \sqrt{3}\hat{j}$ rotates about its tail through an angle 60^0 in clockwise direction then the new vector is

1) $\hat{i} + \sqrt{3}\hat{j}$ 2) $3\bar{i} - 4\bar{j}$ 3) $2\hat{j}$ 4) $2\hat{i}$

LAMI'S THEOREM

39. A 10 kg wt is suspended as shown below then tension T_1 & T_2 are

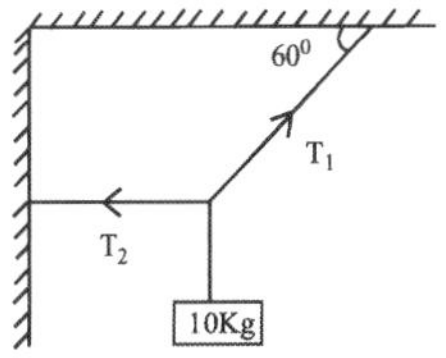

1) 20kg wt, 20kg wt

2) $\frac{20}{\sqrt{3}}$ kgwt, $\frac{20}{\sqrt{3}}$ kgwt

3) $\frac{20}{\sqrt{3}}$ kgwt $\frac{10}{\sqrt{3}}$ kgwt

4) $\frac{20}{\sqrt{3}}$ kgwt $10\sqrt{3}$ kgwt

40. A 10 kg body is suspended by a rope is pulled by means of a horizontal force to make 60^0 by rope to vertical. The horizantal force is

1) 10 kgwt 2) 30 kgwt

3) $10\sqrt{3}$ kgwt 4) $30\sqrt{3}$ kgwt

41. 'A' and 'B' are the two pegs seperated by 13 cm. A body of 169 Kgwt is suspended by thread of 17 cm connecting to A & B, such that the two segments of strings are perpendicular. Then tensions in shorter and longer parts of string having are

1) 100 kgwt, 69 kgwt 2) 65 kgwt, 156 kgwt

3) 156 kgwt, 65 kgwt 4) 69 kgwt, 100 kgwt

42. If 'P' is in equilibrium then T_1 / T_2 is

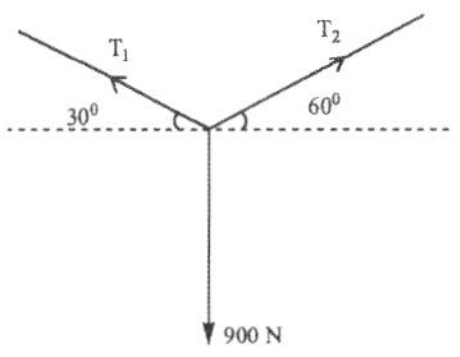

1) $\sqrt{3}$

2) 2

3) $1/\sqrt{3}$

4) 1/2

RELATIVE VELOCITY & RAIN UMBRELLA PROBLEMS

43. A person 'P' in train moving due south with 80 kmph and his friend Q moving due west at 60 kmph Q finds that P is travelling with a velocity of

1) 100 kmph, $\text{Tan}^{-1}\left(\frac{4}{3}\right)$ S of E

2) 100 kmph, $\text{Tan}^{-1}\left(\frac{4}{3}\right)$ E of S

3) 100 kmph S of E 4) 100 kmph N of E

44. A person crossing a road with a certain velocity due north sees a car moving towards east. The relative velocity of the car w.r.t the person is $\sqrt{2}$ times that of the velocity of the person. The angle made by the relative velocity with the east is

1) 30^0 2) 45^0 3) 60^0 4) 90^0

45. A particle is moving eastwards with a velocity $15ms^{-1}$. Suddenly it moves towards north and moves with the same speed in a time 10 sec. The average acceleration during this time is

1) $3/\sqrt{2}$ NE 2) $3\sqrt{2}$ NE **(1997)**
3) $3/\sqrt{2}$ NW 4) $3\sqrt{2}$ NW

46. A ship 'A' steams down to North at 16 kmph, and ship 'B' due west at 12 kmph. Relative velocity of B of w.r.t. A is

1) 10 kmph 2) 25 kmph
3) 6 kmph 4) 20 kmph

47. A person is walking in rain feels the velocity of rain as twice to his velocity. At which angle he should hold the umbrella with vertical if he moves forward, if it is raining vertically downwards

1) 30° 2) 45° 3) 60° 4) 90°

48. When it is raining vertically down, to a man walking on road the velocity of rain appears to be 1.5 times his velocity. To protect himself from rain he should hold the umbrella at an angle θ to vertical. Then $\tan\theta$ =

1) $\frac{2}{\sqrt{5}}$ 2) $\frac{\sqrt{5}}{2}$ 3) $\frac{2}{3}$ 4) $\frac{3}{2}$

49. A shower of rain appears to fall vertically downwards with a velocity of 12 kmph on a person walking west wards with a velocity of 5 kmph. The actual velocity and direction of the rain are

1) 7.5 kmph, clockwise to vertical
2) 13 kmph, anticlockwise to vertical
3) 13 kmph, clockwise to vertical
4) 17 kmph, clockwise to vertical

RIVER – BOAT PROBLEMS

50. A man can swim in still water at a speed of 4kmph. He desires to cross a river flowing at a speed of 3 kmph in the shortest time interval. The direction in which he has to swim is

1) making an angle $\tan^{-1} 3/4$ with the normal to the river flow
2) making an angle $\sin^{-1} 3/4$ with the normal to the river flow
3) making an angle $\text{Cos}^{-1} 3/4$ with the normal to the river flow
4) making an angle 90° with the river flow

51. A boat takes 4 hr upstream and 2hr down the stream for covering the same distance. The ratio of velocity of boat to the water in river is

1) 1 : 3 2) 3 : 1 3) $1:\sqrt{3}$ 4) $\sqrt{3}:1$

52. A man can swim in still water at a speed of 4kmph. He desires to cross a river flowing at a speed of 3 kmph in the shortest time interval. if the width of the river is 3km time taken to cross the river (in hours) and the horizontal distance travelled (in km) are respectively

1) $\frac{3}{4},\frac{9}{4}$ 2) $\frac{3}{5},3$ 3) $\frac{1}{4},\frac{15}{4}$ 4) $\frac{3}{\sqrt{7}},7$

53. The width of a river is $2\sqrt{3}$ km. A boat. is rowed in direction perpendicular to the banks of river. If the drift of the boat due to flow is 2km, the diplacement of the boat is

1) 3 km 2) 6 km 3) 5 km 4) 4 km

54. Person aiming to reach the exactly opposite point on the bank of a stream is swimming with a speed of $0.5\ ms^{-1}$ at an angle of 120^0 with the direction of flow of water.The speed of water in the stream is

1) $1\ ms^{-1}$ 2) $0.25\ ms^{-1}$ 3) $0.67\ ms^{-1}$ 4) $3\ ms^{-1}$

55. A boat moves perpendicualar to the bank with a velocity of 7.2 km/h. The current carries it 150 m downstream, the time required to cross the river. (The width of the river is 0.5 km).

1) 4.16min 2) 2.15min 3) 3.11min 4) 5.11min

56. A swimmer is capable of swimming $1.65\ ms^{-1}$ in still water. If she swims directly across a 180 m wide river whose current is 0.85 m/s, How long will it take her to reach the other side ?

1) 1.82 min 2) 20 min 3) 4 min 4) 4.8 min

DOT PRODUCT

57. Find the angle between $\vec{A}$ = i +2j – k and $\vec{B}$ = –i+j – 2k.
1) 30^0 2) 45^0 3) 60^0 4) 90^0

58. If $\vec{F}$ =3i + 4j +5k and $\vec{S}$ = 6i +2j+5k, find the work done by the force.
1) 51 units 2) 20 units
3) 30 units 4) 41 units

59. A vector perpendicular to the vector(3i+5j) is
1) $5\hat{i}-3\hat{k}$ 2) $5\hat{i}+3\hat{j}$ 3) $3\hat{i}-5\hat{j}$ 4) $5\hat{k}$

60. If $\vec{A}$ = 7i+2j-3k and $\vec{B}$= i+j+2ck are perpendicular vectors, the value of c is
1) -1.5 2) 1.5 3) 3 4) zero

61. If $\vec{F}$ = i +2j+k and $\vec{V}$ =4i – j + 7k find $\vec{F}.\vec{V}$
1) 6W 2) 9W 3) 13W 4) 12W

62. If $\vec{A}=5\hat{i}-2\hat{j}+3\hat{k}$ and $\vec{B}=2\hat{i}+\hat{j}+2\hat{k}$, component of $\vec{B}$ along $\vec{A}$ is
1) $\frac{\sqrt{14}}{38}$ 2) $\frac{28}{\sqrt{38}}$ 3) $\frac{\sqrt{28}}{38}$ 4) $\frac{14}{\sqrt{38}}$

63. The magnitudes of two vectors are 3 and 4 units and their scalar product is 6 units. The angle between the vectors is
1) $\frac{\pi}{3}$rad 2) $\frac{\pi}{6}$rad 3) $\frac{\pi}{2}$rad 4) $\frac{\pi}{4}$rad

64. The work done by a force (2i - j + 5k) when it displaces the body from a point (3,4,-6) to a point (7,2,5) is
1) 65 units 2) 61 units
3) 45 units 4) 35 units

65. If $\vec{p}$ and $\vec{q}$ are two unit vectors and the angle between them is 600 then $\frac{(1+\vec{p}.\vec{q})}{(1-\vec{p}.\vec{q})}$ is
1) 2 2) 3 3) 0 4) 1/2

CROSS PRODUCT

66. (i+j) x (i-j) =
1) -2k 2) 2k 3) zero 4) 2i

67. The radius vector $\vec{r}$ =2i+5k and the angular velocity of a particle is $\vec{w}=3i-4k$. Then the linear velocity of the particle is
1) $15\hat{i}-8\hat{j}+6\hat{k}$ 2) $2\hat{i}+3\hat{j}+\hat{k}$
3) $15\hat{i}+8\hat{j}-6\hat{k}$ 4) $-23\hat{j}$

68. If the dot product of two vectors is equal to the magnitude of the cross product of the same vectors, then the angle between the vectors is
1) $\frac{\pi}{2}$rad 2) $\frac{\pi}{4}$rad 3) $\frac{\pi}{3}$rad 4) π rad

69. If F = 2i + 3j – k and r = i – j + 6k find r x F
1) – 17i + 13j + 5k 2) –17i – 13j – 5k
3) 3i + 4j – 5k 4) – 3i – 4j + 5k

70. Area of a parallelogram formed by vectors $(3\hat{i}-2\hat{j}+\hat{k})$m and $(\hat{i}+2\hat{j}+3\hat{k})$ m is
1) $3\sqrt{8}\,m^2$ 2) 24 m^2 3) $8\sqrt{3}m^2$ 4) $4\sqrt{3}m^2$

71. Find the values of x and y for which vectors $\vec{A}=6\hat{i}+x\hat{j}-2\hat{k}$ and $\vec{B}=5\hat{i}-6\hat{j}-y\hat{k}$ may be parellel
1) $x=0, y=\frac{2}{3}$ 2) $x=\frac{-36}{5}, y=\frac{5}{3}$
3) $x=\frac{-15}{3}, y=\frac{23}{5}$ 4) $x=\frac{36}{3}, y=\frac{15}{14}$

72. If a,b are the magnitudes of vectors $\vec{a}\,\&\,\vec{b}$ and $\vec{a}.\vec{b}=0$ the value of $|\vec{a}\,x\,\vec{b}|$ is
1) 0 2) ab 3) $\sqrt{ab}$ 4) a/b

73. $|A \times B|^2+(A\,.\,B)^2$ is equal to
1) $A^2 + B^2$ 2) A^2B^2 3) $2A^2B^2$ 4) Zero

74. If $\vec{P}=i+j+2k$ and $\vec{Q}=3i-2j+k$, the unit vector perpendicular to both $\vec{P}$ and $\vec{Q}$ is
1) $\frac{\hat{i}+\hat{j}-\hat{k}}{\sqrt{3}}$ 2) $5(\hat{i}+\hat{j}-\hat{k})$
3) $\frac{1}{\sqrt{3}}(\hat{i}-\hat{j}+\hat{k})$ 4) $\frac{1}{25}(\hat{i}+\hat{j}-\hat{k})$

75. Two sides of a triangle are given by $\hat{i}+\hat{j}+\hat{k}$ and $-\hat{i}+2\hat{j}+3\hat{k}$, then area of triangle is

1) $\sqrt{26}$ 2) $\sqrt{26}/2$ 3) $\sqrt{46}$ 4) 26

ANSWERS

1) 4	2) 1	3) 4	4) 2	5) 2
6) 3	7) 2	8) 4	9) 2	10) 3
11) 4	12) 3	13) 4	14) 3	15) 3
16) 4	17) 3	18) 1	19) 4	20) 2
21) 4	22) 3	23) 2	24) 4	25) 4
26) 1	27) 1	28) 2	29) 2	30) 3
31) 2	32) 4	33) 1	34) 1	35) 2
36) 2	37) 1	38) 4	39) 3	40) 3
41) 3	42) 3	43) 1	44) 2	45) 3
46) 4	47) 1	48) 1	49) 3	50) 4
51) 2	52) 1	53) 4	54) 2	55) 1
56) 1	57) 3	58) 1	59) 4	60) 2
61) 2	62) 4	63) 1	64) 1	65) 2
66) 1	67) 4	68) 2	69) 1	70) 3
71) 2	72) 2	73) 2	74) 1	75) 2

EXERCISE – III

ADDITION AND SUBTRACTION

1. Two vectors of equal magnitude P are inclined at some angle such that the difference in magnitude of resultant and magnitude of either of the vectors is 0.732 times either of the magnitude of vectors. If the angle between them is increased by half of its initial value then find the magnitude of difference of the vectors

1) 2 P 2) $\sqrt{2}P$ 3) 3 P 4) $\sqrt{3}P$

2. Resultant of two vectors of magnitudes P and Q is of magnitude 'Q'. If the magnitude of $\vec{Q}$ is doubled now the angle made by new resultant with $\vec{p}$ is

1) 30^0 2) 90^0 3) 60^0 4) 120^0

3. The two forces $2\sqrt{2}N$ and xN are acting at a point their resultant is perpendicular to $\hat{x}N$ and having magnitude of $\sqrt{6}$ N. The angle between the two forces and magnitude of x are

1) $\theta=120^0, x=\sqrt{2}N$ 2) $\theta=30^0, x=\sqrt{2}N$

3) $\theta=150^0, x=\sqrt{3}N$ 4) $\theta=150^0, x=\sqrt{2}N$

4. Two forces F_1 and F_2 are acting at a point, having resultant as F. If F_2 is doubled F is also doubled. If F_2 is reversed then also F is doubled. Then $F_1 : F_2 : F$ is

1) $\sqrt{2}:\sqrt{2}:\sqrt{3}$ 2) $\sqrt{3}:\sqrt{3}:\sqrt{2}$

3) $\sqrt{3}:\sqrt{2}:\sqrt{3}$ 4) $\sqrt{2}:\sqrt{3}:\sqrt{2}$

5. When forces F_1, F_2, F_3 are acting on a particle of mass m such that F_2 and F_3 are mutually perpendicular, then the particle remains stationary, If the force F_1 is now removed then acceleration of the particle is

1) F_1/m 2) F_1F_3/mF_1

3) $(F_2-F_3)/m$ 4) F_2/m

6. The square of the resultant of two forces 4N and 3N exceeds the square of the resultant of the two forces by 12 when they are mutually perpendicular. The angle between the vectors is

1) 30° 2) 60° 3) 90° 4) 120°

7. The resultant of three vectors 1,2 and 3 units whose directions are those of the sides of an equilateral triangle is at an angle of

1) 30^0 with the first vector
2) 15^0 with the first vector
3) 100^0 with the first vector
4) 150^0 with the first vector

8. A person moving on a motor cycle in a ground takes a turn through 60^0 on his left after every 50 m. Then find the magnitude of displacement suffered by him after 9th turn

1) 100 m
2) 50 m
3) $50\sqrt{3}$ m
4) 20 m

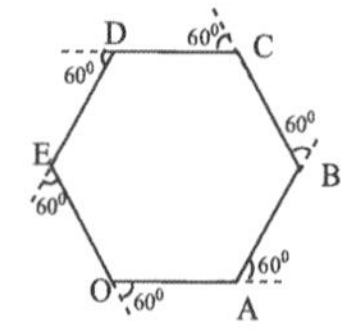

9. At what angle two vectors of magnitudes A + B and A − B act, so that their resultant is $\sqrt{3A^2+B^2}$?

1) 30^0 2) 45^0 3) 60^0 4) 90^0

POSITION VECTOR, COMPONENTS OF VECTORS, UNIT VECTORS & LAMI'S THEOREM

10. A vector $\vec{Q}$ which has a magnitude of 8 is added to the vector $\vec{P}$ which lies along the X-axis. The resultant of these two vectors is a third vector $\vec{R}$ which lies along the Y-axis and has a magnitude twice that of $\vec{P}$. The magnitude of $\vec{P}$ is (2004 M)

1) $\frac{6}{\sqrt{5}}$ 2) $\frac{8}{\sqrt{5}}$ 3) $\frac{12}{\sqrt{5}}$ 4) $\frac{16}{\sqrt{5}}$

11. If a vector $\vec{A}$ makes angles 45^0 and 60^0 with x and y axis respectively then the angle made by it with z - axis is

1) 30^0 2) 60^0 3) 90^0 4) 120^0

12. The rectangular components of a vector lying in xy plane are (n+1) and 1. If co.ordinate system is turned by 60^0. They are n & 3 respectively the value of 'n'

1) 2 2) 3 3) 2.5 4) 3.5

13. Two particles having position vectors $\vec{r}_1=(3\vec{i}+5\vec{j})$m and $\vec{r}_2=(-5\vec{i}+3\vec{j})$m are moving with velocities $\vec{V}_1=(4\hat{i}-4\hat{j})ms^{-1}$ and $\vec{V}_2=(a\hat{i}-3\hat{j})ms^{-1}$. If they collide after 2 seconds, the value of 'a' is (2003 E)

1) 2 2) 4 3) 6 4) 8

14. If four forces act at a point 'O' as shown in the figure and if O is in equilibrium then the value of 'θ' & 'P' are

1) 15^0, $10\sqrt{2}$N
2) 45^0, 10 N
3) 75^0, $10\sqrt{2}$ N
4) 90^0 , 20 N

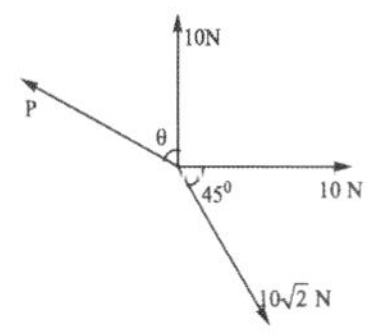

15. A uniform rod of 3kg and length 1m is suspended from a fixed point by means of two strings of length 0.6m and 0.8m which are attached to the free ends of rod, such that rod is horizan-tal Find the tensions in the string in kgwt are

1) 1.8, 2.4 2) 1.2, 1.6 3) 3, 1.8 4) 2.4, 3.6

RELATIVE VELOCITY & RAIN UMBRELLA PROBLEMS

16. A ship is moving due east with a velocity of 12 m/ sec, a truck is moving across on the ship with velocity4m/sec. A monkey is climbing the vertical pole mounted on the truck with a velocity of 3m/sec. Find the velocity of the monkey as observed by the man on the shore

1) 10 m/sec 2) 15 m/sec
3) 13 m/sec 4) 20 m/sec

17. Two stones are projected from the top of a tower in opposite directions, with the same velocity V but at 30^0 & 60^0 with horizontal respectively. The relative velocity of first stone relative to second stone is

1) 2 V 2) $\sqrt{2}$V 3) $\frac{2V}{\sqrt{3}}$ 4) $\frac{V}{\sqrt{2}}$

18. A train is travelling relative to the earth at velocity of $15ms^{-1}$. A person in the train is walking relative to train at a velocity of $1ms^{-1}$ in the direction of motion of the train. The velocity of the person relative to the earth is

1) 14 ms^{-1} 2) 16 ms^{-1} 3) 8 ms^{-1} 4) 7 ms^{-1}

19. A car 'A' is moving in N-E direction at 80 kmph another car 'B' in the S–E direction with 60 kmph. The direction of velocity A relative to B makes angle θ to the north. Then θ is

1) Tan^{-1} 1/7 2) Tan^{-1} 3/4
3) Tan^{-1} 7 4) Tan^{-1} 4/3

20. A motor car A is travelling with a velocity of 20 m/s in the north–west direction and another motor car B is travelling with a velocity of 15 m/s in the north–east direction. The magnitude of relative velocity of B with respect to A is

1) 25 m/s 2) 15 m/s
3) 20 m/s 4) 35 m/s

21. **A monkey is climbing a vetical pillar with a velocity of 5 ms^{-1} and a boy is running towards the pillar with a velocity of $5\sqrt{3}ms^{-1}$. Velocity of boy relative to monkey is**
1) 10 ms^{-1} at 30^0 with the horizontal
2) 10 ms^{-1} at 60^0 with the horizontal
3) $8\sqrt{3}$ ms^{-1} at 80^0 with the horizontal
4) $8\sqrt{3}$ ms^{-1} at 60^0 with the horizontal

22. **When two bodies approach each other with the different speeds, the distance between then decreases by 120m for every one minute. If they are moving in direction, the distance between them increases by 90m for every one minute. The speeds of the bodies are**
1) $2ms^{-1}$ and 0.5 ms^{-1} 2) $3ms^{-1}$ and $2ms^{-1}$
3) $1.75ms^{-1}$ and $0.25ms^{-1}$ 4) $2.5ms^{-1}$ and $0.5ms^{-1}$

23. **A boat going down stream crosses a float at a point A. 't_1' minutes later the boat reverses its direction and in the next 't_2' minutes it crosses the float at a distance L from the point A. The velocity of the river is**
1) $2L/t_1$ 2) $L/(t_1+2t_2)$ 3) $2L/(2t_1+t_2)$ 4) $L/2t_1$

24. **Two persons P and Q are located in x – y plane at points (0, 0) and (0, 10) respectively. At t = 0 they start moving simultaneously $\overrightarrow{V_P} = 4\hat{j}$ and $\overrightarrow{V_Q} = 4\hat{i}$ m/sec. The time after which P and Q are at closest distance is**
1) 1.25sec 2) 2.5sec 3) $5\sqrt{2}$ sec 4) $2\sqrt{5}$ sec

25. **It is raining at a speed of 5 ms^{-1} at an angle 37^0 to vertical, towards east. A man is moving to west with a velocity of 5 ms^{-1}. The angle with the vertical at which he has to hold the umbrella to protect himself from rain is**
1) Tan^{-1} (2) to west 2) Tan^{-1} (2) to east
3) Tan^{-1} (1/2) to south 4) Tan^{-1} (1/2) to east

26. **To a man walking at 7 kmph due west, the wind appears to blow from the north-west but when he walks at 3 kmph due west, the wind appears to blow from north. The magnitude and actual direction of wind are**
1) 5 kmph, $\tan^{-1}\left(\frac{3}{4}\right)$ east of north
2) 5 kmph, $\tan^{-1}\left(\frac{3}{4}\right)$ north of east
3) 4kmph, $\tan^{-1}\left(\frac{3}{4}\right)$ east of north
4) $\sqrt{58}$ kmph, $\tan^{-1}\left(\frac{3}{4}\right)$ north of east

27. **Rain, pouring down at an angle α with the vertical has a speed of 10 m/s. A girl runs against the rain with a speed of 8 m/s and sees that the rain makes an angle β with the vertical, then relation between α and β is**
1) $\tan\beta = \frac{8+10\sin\alpha}{10\cos\alpha}$ 2) $\tan\beta = \frac{8+10\cos\alpha}{10\sin\alpha}$
3) $\tan\beta = \frac{10\sin\alpha}{8+10\cos\alpha}$ 4) $\tan\beta = \frac{10\cos\alpha}{8+10\sin\alpha}$

28. **To a man walking at the rate of 2 kmph the rain appears to fall vertically, when he doubles his speed, the rain appears to fall at 30^0 to the vertical. The magnitude of actual velocity of the rain is**
1) 2 kmph 2) 4 kmph 3) 6 kmph 4) 3 kmph

RIVER – BOAT PROBLEMS

29. **The velocity of a boat in still water is 10 m/s. If water flows in the river with a velocity of 6 m/s what is the difference in times taken to cross the river in the shortest path and the shortest time. The width of the river is 80 m.**
1) 1s 2) 10s 3) $\frac{\sqrt{3}}{2}$s 4) 2s

30. **A river is of width 120m which flows at a speed of 8ms^{-1}. If a man swims with a speed of 5ms^{-1} at an angle of 127^0 with the stream, his drift on reaching other bank is**
1) 50 m 2) 150 m 3) 200 m 4) 300 m

31. **A river is flowing from west to east at a speed of 5 m per minute. A man on the south bank of the river, capable of swimming at 10 m per minute in still water wants to swim across the river to a point directly opposite point. He should then swim in a direction.**
1) 30^0 W of N 2) 60^0 W of N
2) 30^0 E of N 4) 60^0 E of N

32. A boat takes 4 hrs to travel certain distance in a river in down stream and it takes 6 hrs to travel the same distance in upstream. Then the time taken by the boat to travel the same distance in still water is

1) 4.8 hrs 2) 9.8 hrs 3) 24 hrs 4) 10 hrs

DOT PRODUCT

33. If $\vec{A}=\left(2\hat{i}+3\hat{j}\right)$ and $\vec{B}=\left(\hat{i}-\hat{j}\right)$ then component of $\vec{A}$ perpendicular to vector $\vec{B}$ and in the same plane is

1) $\frac{5}{2}\left(\hat{i}+\hat{j}\right)$ 2) $\frac{5}{\sqrt{2}}\left(\hat{i}+\hat{j}\right)$

3) $\frac{\sqrt{5}}{2}\left(\hat{i}+\hat{j}\right)$ 4) $\frac{5}{\sqrt{2}}\left(\hat{i}+\hat{k}\right)$

34. If $\vec{A}+\vec{B}=\vec{R}$ and $2\vec{A}+\vec{B}$ is perpendicular to $\vec{B}$ then

1) A = R 2) B = 2R 3) B = R 4) B = A

35. Velocity and acceleration vectors of charged particle moving perpendicular to the direction of magnetic field at a given instant of time are $\vec{\vartheta}=2\hat{i}+c\hat{j}$ and $\bar{a}=3\hat{i}+4\hat{j}$ respectively. Then the value of 'c' is (2007 E)

1) 3 2) 1.5 3) – 1.5 4) –3

CROSS PRODUCT

36. A force of $\vec{F}=\hat{i}+\hat{j}+\hat{k}$ is acting at a point (–2,3,4) The moment of force about (1, 2, 3) is

1) $-4\hat{j}+4\hat{k}$ 2) $-\hat{j}+\hat{k}$ 3) $4\hat{j}-4\hat{k}$ 4) $-4\hat{k}$

37. A body of mass $\sqrt{2}$ kg is projected with a speed V at an angle 45° to the horizontal, When the body is at the highest point in it's path, angular momentum of the body about the point of projection is,

$\left(\hat{i},\hat{j}\text{ are unit vectors in vertical plane}\right)$

1) $\frac{V^3}{4g}\hat{k}$ 2) $\frac{V^3}{4g}\left(-\hat{k}\right)$

3) $\frac{V^3}{4g}\left(\hat{j}\right)$ 4) $\frac{V^3}{4g}\left(\hat{i}\right)$

38. If $\vec{A}=\frac{1}{\sqrt{2}}\cos\theta\,\hat{i}+\frac{1}{\sqrt{2}}\sin\theta\,\hat{j}$, what will be the unit vector perpendicular to $\vec{A}$

1) $\cos\theta i+\sin\theta j$ 2) $-\cos\theta i+\sin\theta j$

3) $\frac{\cos\theta i+\sin\theta j}{\sqrt{2}}$ 4) $\sin\theta i-\cos\theta j$

39. A particle of mass 80 units is moving with a uniform speed $V=4\sqrt{2}$ units In XY plane, along a line y = x + 5. The magnitude of the angular momentum of the particle about the origin is

1) 1600 units 2) $160\sqrt{2}$ units

3) $152\sqrt{2}$ units 4) $16\sqrt{2}$ units

40. Find the value of $\left(\vec{a}+\vec{b}\right)\times\left(\vec{a}-\vec{b}\right)=$

1) $\left(\vec{a}\times\vec{b}\right)$ 2) $2\left(\vec{a}\times\vec{b}\right)$

3) $-2\left(\vec{a}.\vec{b}\right)$ 4) $-2\left(\vec{a}\times\vec{b}\right)$

ANSWERS

1) 2	2) 2	3) 1	4) 4	5) 1
6) 2	7) 1	8) 1	9) 3	10) 2
11) 2	12) 4	13) 4	14) 4	15) 1
16) 3	17) 2	18) 2	19) 1	20) 1
21) 1	22) 3	23) 4	24) 1	25) 1
26) 1	27) 1	28) 2	29) 4	30) 2
31) 1	32) 1	33) 1	34) 1	35) 3
36) 3	37) 2	38) 4	39) 1	40) 4

COMPREHENSION QUESTIONS

Comprehension - I (1 - 3) :

The velocity of a boat w.r.t. water is 5 kmph. A person wants to cross the river of width 500 m in the boat. The velocity of flow of water is 4 kmph.

1. The minimum time taken by him to corss the river in minutes

1) 10 2) 8 3) 6 4) 4

2. If the person rows the boat in a direction always making an anlge 127^0 with the flow, the drift of the boat

1) 200 m 2) 100 m 3) 60 m 4) zero

3. If the person rows the boat always at an angle 127^0 with the flow, the time taken by him to cross the river in minutes

1) 10.5 2) 9 3) 7.5 4) 6

Comprehension - II (4 - 5) :

A ferry boat travels between two points A and B on the banks of a river shown in fig. always following the line AB. The distance S between points A and B is 1,200 m. The velocity of the river current v = 1.9 ms^{-1} is constant over the entire width of the river. The line AB makes an angle $\alpha = 60^0$ with the direction of the current. The angle β remains the same during the passage from A to B and from B to A.

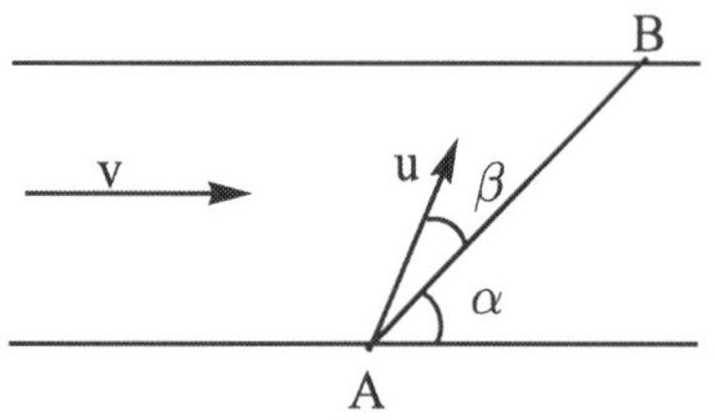

4. With what velocity u, the ferry boat move to cover the distance AB and back in a time t = 5 minutes ?

1) $8ms^{-1}$ 2) $4ms^{-1}$ 3) $6ms^{-1}$ 4) $10ms^{-1}$

5. With what angle β to the line AB should the ferry boat move to cover the distance AB and back in a time t = 5 minutes ?

1) 12^0 2) 24^0 3) 30^0 4) 60^0

Comprehension - III (6 - 8) :

A force $(2\hat{i}+\hat{j}+2\hat{k})$ N is acting on a body shifted it from (1, 2, 1) to (2, 1, 3) in the same frame of reference

6. The work doen by the force in Joule is

1) 6 2) 5 3) 4 4) 11

7. The component of the force along the displacement of the body

1) $\frac{5}{\sqrt{6}}(2\hat{i}+\hat{j}+3\hat{k})$ 2) $\frac{5}{6}(\hat{i}+2\hat{j}+\hat{k})$

3) $\frac{5}{\sqrt{10}}(\hat{i}-2\hat{j}+2\hat{k})$ 4) $\frac{5}{6}(\hat{i}-\hat{j}+2\hat{k})$

8. Angle between the displacement and force

1) $\cos^{-1}\sqrt{\frac{25}{54}}$ 2) $\cos\sqrt{\frac{48}{54}}$

3) $\sin^{-1}\sqrt{\frac{25}{54}}$ 4) $\tan\sqrt{\frac{48}{54}}$

Comprehension - IV (9 - 11) :

A force of $(\hat{i}+2\hat{j}+3\hat{k})$N is acting on a body having position vecotr $(3\hat{i}+\hat{j}+2\hat{k})$ in the same frame of reference

9. The moment of the force about the origin is

1) $\hat{i}+7\hat{j}-7\hat{k}$ 2) $7\hat{i}-11\hat{j}+5\hat{k}$

3) $\hat{i}+11\hat{j}-7\hat{k}$ 4) $7\hat{i}+10\hat{j}+7\hat{k}$

10. The direction of the moment of force is

1) $\frac{\hat{i}+11\hat{j}-7\hat{k}}{\sqrt{171}}$ 2) $\frac{7\hat{i}+10\hat{j}+7\hat{k}}{\sqrt{198}}$

3) $\frac{\hat{i}+7\hat{j}-5\hat{k}}{\sqrt{75}}$ 4) $\frac{7\hat{i}-11\hat{j}+5\hat{k}}{\sqrt{195}}$

11. The area of triangle formed by the force vector and position vector is

1) $5\sqrt{3}$ 2) $2.5\sqrt{3}$

3) $10\sqrt{3}$ 4) $1.25\sqrt{3}$

ANSWERS

1) 3	2) 2	3) 3	4) 1	5) 1
6) 2	7) 4	8) 1	9) 1	10) 3
11) 2				

Chapter - 3

MOTION IN A STRAIGHT LINE AND PLANE

MOTION IN A STRAIGHT LINE AND PLANE

KINEMATICS

SYNOPSIS

A. HORIZONTAL MOTION

1. Rest and motion are relative.
2. **Formulae related to motion :**

i) $\text{Speed} = \frac{\text{Distance}}{\text{time}}$

ii) Velocity $\vec{V} = \frac{d\vec{s}}{dt}$

iii) $\text{Average speed} = \frac{\text{Total distance}}{\text{time}}$

iv) $|\text{Average velocity}| = \frac{|\text{Net displacement}|}{\text{time}}$

v) If a body moving along a straight line suffers displacements s_1 in time t_1, $\vec{s_2}$ in time t_2........$\vec{s_n}$ in time t_n, then $<\vec{V}> = \frac{\vec{s_1} + \vec{s_2} + ... + \vec{s_n}}{t_1 + t_2 + ... + t_n}$

Acceleration :- The rate of change of velocity is known as acceleration $\vec{a} = \frac{\overrightarrow{\Delta v}}{\Delta t}$.

3. A body moving with uniform velocity has to travel along a straight line.
4. For a body moving with uniform velocity, average speed and average velocity are same.
5. For a body moving with uniform velocity instantaneous velocity is same along its motion.
6. Direction of velocity and acceleration need not be same.
7. If velocity increases with time, the body has acceleration.
8. If velocity decreases with time, the body has retardation.
9. A body can have any angle between velocity and acceleration.
10. If the angle between velocity and acceleration is accute, velocity increases.
11. If the angle between velocity and acceleration is obtuse, velocity decreases.
12. If the angle between velocity and acceleration is a right angle, velocity remain constant.
13. A body can have constant acceleration though its velocity changes both in magnitude and direction.
14. A body can have zero velocity and non zero acceleration.
15. A body can have non zero velocity and zero acceleration.
16. A body can have constant speed and changing velocity.
17. A body can not have constant velocity and changing speed.
18. **Equations of motion for a body moving with constant acceleration along a straight line.**

i) $V = u + at$ ii) $s = \left(\frac{u+v}{2}\right)t$

iii) $s = ut + \frac{1}{2}at^2$ iv) $v^2 = u^2 + 2as$

v) $s_n = u + a\left(n - \frac{1}{2}\right)$

19. For a particle starting from rest and travelling with **uniform acceleration** along a straight line.

i) $V = at$ or $V \alpha t$

ii) $S = 1/2\ at^2$ (or) $S \alpha t^2$

iii) $V^2 = 2as$ (or) $V \alpha \sqrt{s}$

iv) $s_n = a\left(n - \frac{1}{2}\right)$ or $s_n \propto (2n-1)$

20. For a particle moving with **uniform velocity**, its acceleration is zero ($a = 0$)

$\therefore S = ut$ or $S \alpha t$

21. For a particle moving with **uniform retardation** along a straight line distance travelled before coming to rest ($V=0$) is

$S = \frac{u^2}{2a}$ $\therefore S \alpha u^2$

22. If a particle starts from rest and moves with uniform acceleration 'a' such that it travels distances s_m and s_n in the mth and nth sec then

$a = \frac{s_m - s_n}{m - n}$

23. A particle starts from rest and moves along a straight line with uniform acceleration. If 's' is the distance travelled in n seconds and s_n is the distance travelled in the nth second then

$\frac{s_n}{s} = \frac{(2n-1)}{n^2}$

B. VERTICAL MOTION

If a particle is **thrown vertically downwards** with a velocity u, the equations of motion can be written as

i) $V = u + gt$ ii) $v^2 - u^2 = 2gh$

iii) $h = ut + \frac{1}{2}gt^2$ iv) $S_n = u + \frac{g}{2}(2n-1)$

C. FREELY FALLING BODY

1. If a particle is dropped from a certain height h. $a = +g$ and $u = 0$
   a) Velocity attained after falling for a time t is $V = gt$
   b) Distance fallen in a time 't' is $h = \frac{1}{2}gt^2$
   c) Time taken to fall a distance h is $t = \sqrt{\frac{2h}{g}}$
   d) Velocity attained after falling a distance h is $V = \sqrt{2gh}$
   e) Distance travelled in the nth sec, is $S_n = \frac{g}{2}(2n-1)$
   f) Ratio of distances fallen in successive equal intervals of time is 1 : 3 : 5 : 7 .... (2n-1)
   g) Ratio of distances fallen in 1s, 2s, 3s ....ns is 1 : 4 : 9 .... n^2.
   h) Ratio of the time taken to fall successive equal distances is $1:(\sqrt{2}-1):(\sqrt{3}-\sqrt{2}):(\sqrt{4}-\sqrt{3}):.....$
   i) Distances fallen in every second get increased by "g" units.
   j) Velocity is increased by g units every second.
   k) When there is no air resistance, equations of motion are independent of mass of the body.
   l) If a body falls under the influence of air resistance, as long as the velocity is increasing
      i) $a < g$
      ii) acceleration decreases with time
   m) In the above case, the body after some time has no acceleration, and subsequently falls with a constant velocity called terminal velocity.
2. A freely falling body passes through two points A and B in time intervals of t_1 and t_2 from the start, then the distance between the two points A and B is $\frac{g}{2}\left(t_2^2 - t_1^2\right)$
3. A freely falling body passes through two points A and B at distances h_1 and h_2 from the start, then the time taken by it to moves from A to B is

$$T = \sqrt{\frac{2h_2}{g}} - \sqrt{\frac{2h_1}{g}} = \sqrt{\frac{2}{g}}\left(\sqrt{h_2} - \sqrt{h_1}\right)$$

4. A stone is dropped into a well of depth 'h', the sound of splash is heard after a time of

$$t = \sqrt{\frac{2h}{g}} + \frac{h}{V_{sound}}$$

5. If two bodies are held one above the other seperated by a distance S and released simultaneously, the distance of seperation between them remains S through out their motion.

D. VERTICALLY PROJECTED BODY

1. If a particle is projected **vertically up** with a velocity u, acceleration is $a = -g$
   a) Equations of motion are

      i) $v = u - gt$ ii) $h = ut - \frac{1}{2}gt^2$

      iii) $v^2 - u^2 = -2gh$ iv) $S_n = u - \frac{g}{2}(2n-1)$

   b) i) velocity at the highest point is zero.
      ii) velocity and acceleration are in opposite directions.
      iii) Maximum Height $H_{max} = \frac{u^2}{2g}$
      iv) Time of ascent, $t_a = u/g$
      v) Time of flight, $T = 2u/g$
      vi) time of ascent = time of descent
      vii) $H_{max} = \frac{gT^2}{8}$
      viii) Speed at any point in ascent = speed at same point in descent
   c) If body rises through a height 'h' in nth second, then in $(n-1)^{th}$ sec it will rise through a height $(h+g)$ and in $(n+1)^{th}$ second it will rise through a height $(h-g)$.
   d) If velocity of body in n^{th} second is 'v' then in $(n-1)^{th}$ second it is $(v + g)$ and that in $(n + 1)$s it is $(v - g)$ while ascending,
   e) velocity of the particle one second before reaching the highest point is 9.8 m/s.
   f) The change in velocity over the complete journey is 2u (downwards)
   g) The height reached in the first second of ascent is equal to the height of fall in the last second of descent.

h) Irrespective of velocity of projection, all the bodies pass through a height g/2 in the last second of ascent.

2. A body is projected vertically up with a velocity of 'u' from ground in the presence of constant air resistance 'R'. If it reaches the ground with a velocity 'V', then

i) Height of ascent = Height of descent

ii) Time of ascent $t_a = \dfrac{mu}{mg+R}$

iii) Time of descent $t_d = \dfrac{mV}{mg-R}$

iv) $t_a < t_d$

v) $\dfrac{V}{u} = \sqrt{\dfrac{mg-R}{mg+R}}\,(V<u)$

vi) For a body projected vertically up under air resistance, retardation during motion is > g

vii) If air resistance is considered, time of ascent decreases and time of descent increases $\Rightarrow t_d > t_a$.

3. A body projected vertically upwards from the ground is at the same height h from the ground as two instants of time t_1 and t_2 (both measured from the instant of projection) Then

i) $h = \dfrac{1}{2}gt_1t_2$

ii) Velocity of projection $u = \dfrac{g}{2}(t_1+t_2)$

iii) The maximum height reached by the body

$$=\frac{1}{8}g(t_1+t_2)^2$$

4. An elevator is accelerating upwards with an acceleration a. If a person inside the elevator throws a particle vertically up with a velocity u relative to the elevator, time of flight is

$$t=\frac{2u}{g+a}$$

5. In the above case if elevator accelerates down, time of flight is $t=\dfrac{2u}{g-a}$

E. PROJECTION FROM THE TOP OF A TOWER:

1. If a body is projected vertically up from the top of a tower of height h with a velocity u and takes 't' seconds to reach the ground then height of tower

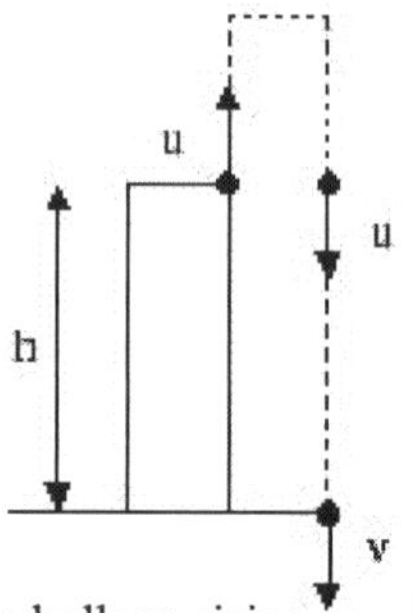

$h = -ut + \dfrac{1}{2}gt2$

2. If an object is dropped from a balloon rising up with a velocity u at a height h.

a) Equation of motion relative to earth is

$$h = -ut + \frac{1}{2}gt^2$$

b) Equation of motion relative to balloon is $h = \dfrac{1}{2}gt^2$

c) Relative to earth body goes up and then falls.

d) Relative to the balloon it falls vertically downward.

3. Distance between the object and balloon after a time of ;t;, after drop is $X=(h=ut)=\dfrac{1}{2}gt^2$

4. If a body is projected vertically up with a velocity u from a tower and it reaches the ground with a velocity nu., then the height of the tower is

$$h=\frac{u^2}{2g}(n^2-1)$$

5. A particle projected vertically up from the top of a tower takes t_1s to reach the ground. Another particle thrown downwards with the same velocity from the top of the tower takes. t_2 seconds to reach the ground.

a) In the first and second case body reaches the ground with the same velocity

b) If the particle is dropped from the top of the tower, then time taken by it to reach the ground is

$$t=\sqrt{t_1t_2}$$

c) height of the tower is $h = \dfrac{1}{2}gt_1t_2$

d) Velocity of projection is $u = \dfrac{g}{2}(t_1 - t_2)$

F. OBLIQUE PROJECTION

1. In this case, the angle of projection is $\theta < 90°$ with the horizontal

2. Projectile motion is the motion in two dimensions but possesses uniform acceleration in one dimension.

3. For the entire motion acceleration of a projectile is "g" downwards.

4. For a projectile, w.r.t stationary frame path or trajectory is parabola.

5 Path of projectile w.r.t frame of another projectile is a straight line :

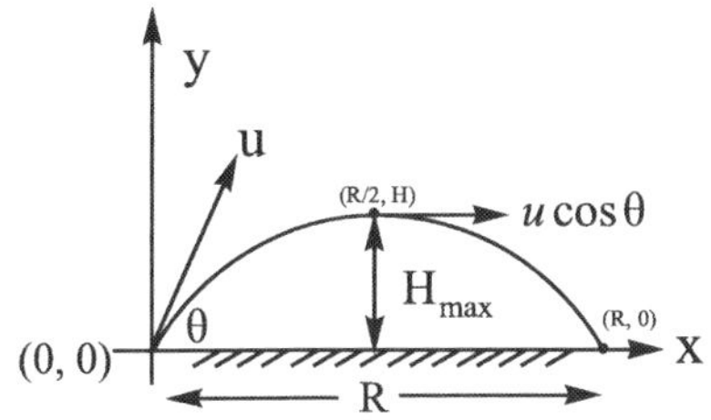

a) Horizontal component of velocity $u_x = u\cos\theta$

b) Initial vertical component of velocity $u_y = u\sin\theta$

c) Velocity of projection $u = \sqrt{u_x^2 + u_y^2}$

d) Angle of projection $\theta = \tan^{-1}\left(\frac{u_y}{u_x}\right)$

e) $\vec{u} = u_x\vec{i} + u_y\vec{j}$

6. Velocity after time "t" :

a) $a_x = 0$; $a_y = -g$

b) horizontal component of velocity through out the motion is constant, $u_x = u\cos\theta$

c) Vertical component of velocity changes with time $v_y = u\sin\theta - gt$

f) Velocity of the particle $v = \sqrt{v_x^2 + v_y^2}$

g) direction of motion w.r.t. to horizontal. $\alpha = \tan^{-1}\left(\frac{v_y}{v_x}\right)$

7) At highest point is

i) u Cos θ in the horizontal direction.

ii) Vertical component of velocity $v_y = 0$

iii) Velocity and acceleration are perpendicular to each other at the highest point.

8) If projected from level ground

i) Speed of landing and angle of landing is same in magnitude as during those in projection.

ii) velocity is maximum during projection and during landing and minimum at the highest point.

iii) Change in velocity till it reaches highest point = u sin θ

iv) Change in velocity for complete trajectory = 2 u sin θ

9) Velocity of the projectile when it moves perpendicular to its initial velocity is $u\cot\theta$.

10) time taken for the velocity to become perpendicular to the initial velocity is $\frac{u}{g\sin\theta}$

11) If V_1 and V_2 are the magnitudes of velocities at heights h_1 and h_2; $V_2^2 - V_1^2 = -2g(h_2-h_1)$

12. **Position of the projectile after time 't' :**

If x and y represent the horizontal and vertical displacements with respect to the point of projection 't' seconds after projection

i) $x = (u\cos\theta)t$ ii) $y = (u\sin\theta)t - \frac{1}{2}gt^2$

iii) Equation of trajectory is

$$Y = (\tan\theta)x - \frac{1}{2}g\frac{x^2}{u^2\cos^2\theta}$$

iv) $A = \tan\theta$; $B = \frac{g}{2u^2\cos^2\theta}$

$$H = \frac{A^2}{4B} \text{ ; Range } R = \frac{A}{B} \text{ and } \frac{H}{R} = \frac{A}{4}$$

13. i) Maximum height attained $H = \frac{u^2\sin^2\theta}{2g}$

ii) Maximum height attained is maximum when $\theta = 90°$ (body projected vertically up)

14. i) Time of ascent $t_a = \frac{U\sin\theta}{g} = \sqrt{\frac{2H}{g}}$

ii) Time of flight $t_f = \frac{2U\sin\theta}{g} = 2\sqrt{\frac{2H}{g}}$

15. i) Range $R = u_x\left(\frac{2u_y}{g}\right) = \frac{u^2\sin 2\theta}{g}$

ii) $R\tan\theta = 4H = \frac{1}{2}gT^2$, T is time of flight.

ii) If projected from level ground range is maximum when angle of projection $\theta = 45°$.

iv) $R_{max.} = \frac{u^2}{g}$; $\theta = 45°$ (with the level ground)

16. i) Time of flight $T = \frac{\sqrt{2}u}{g}$

ii) $R = \dfrac{gT^2}{2}$

17. In terms of range, equation of trajectory is
$y = x\tan\theta\left(1 - \dfrac{x}{R}\right)$

18. At maximum height,
$KE = \dfrac{1}{2}mu_x^2 = \dfrac{1}{2}mu^2\cos^2\theta$
$PE = mgH_{max} = \dfrac{1}{2}mu^2\sin^2\theta$
$T.E = \dfrac{1}{2}mu^2$
At $\theta = 45°$, KE = PE

19. Complementary angles of projection :

a) For a given velocity of projection for range is equal angles of projection are θ and $(90-\theta)$.

b) If h_1 and h_2 are the miximum heights attained for complementary angles of projection

i) $h_1 + h_2 = \dfrac{u^2}{2g}$ ii) $R = 4\sqrt{h_1 h_2}$

iii) $R_{max} = 2\,(h_1+h_2)$

c) If T_1 and T_2 are the times of flight for complementary angles of projection

Range $R = \dfrac{1}{2}gT_1T_2$

20. When range is maximum H_{max} = R/4
21. If R = H_{max} ; then $\theta = \tan^{-1}(4) = 76°$
22. If a man throws a body to maximum distance R then he can project the body to vertical height $\boxed{R/2}$
23. If a man throws a body to maximum distance R, the greatest height attained by the body during its flight is $\boxed{R/4}$
24. Velocity of a projectile relative to another projectile is constant
25. Acceleration of a projectile relative to another projectile is zero.
26. A body is projected vertically up from a topless car relative to the car which is moving horizontally relative to earth.

a) If car velocity is constant, ball will be caught by the thrower.

b) If car velocity is constant, path of ball relative to the ground is a parabola and relative to this car is straight up and then straight down.

c) If the car accelerates , ball falls back relative to the car.

d) If the car retards ball falls forward relative to the car.

27. If a gun is aimed towards a target and the bullet is fired, the moment when the target falls, the bullet will always hit the target irrespective of the velocity of the bullet.

G. HORIZONTAL PROJECTILE MOTION :

1. When a body is projected, horizontally from the top of a tower

a) path of the body is parabola relative to ground.

b) It is a vertical straight line relative to another horizontal projectile

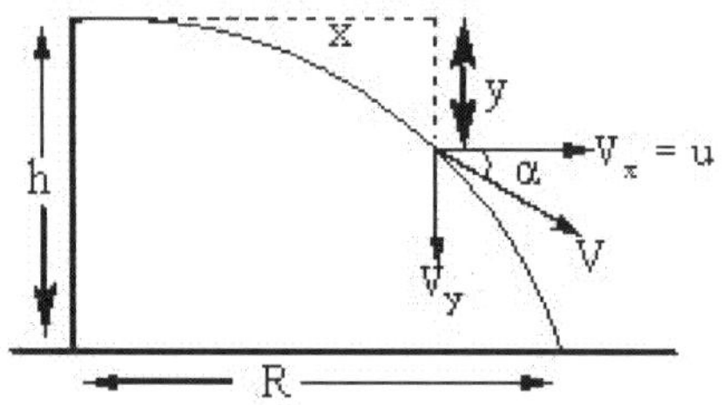

c) Velocity after time t :

i) horizontal component of velocity $v_x = u$ ($\because$ horizontal component of velocity is constant)

ii) vertical component $V_y = gt = \sqrt{2gy}$

(y = distance fallen)

iii) Net velocity $V = \sqrt{Vx^2 + Vy^2}$

iv) Direction of motion or angle made by velocity vector with the horizontal is

$\alpha = \tan^{-1}\left(\dfrac{Vy}{Vx}\right) = \tan^{-1}\left(\dfrac{gt}{u}\right)$

d) Velocity on reaching the ground is

$V = \sqrt{u^2 + 2gh}$

e) Time of flight $t = \sqrt{\dfrac{2h}{g}}$

It is independent of the velocity of projection

f) Position of the projectile after time t :

i) Horizontal displacement after time t. **x = ut**,

ii)Height fallen in time "t" $\mathbf{y = \dfrac{1}{2}gt^2}$

g) Equation of path : $\boxed{y = \dfrac{gx^2}{2u^2}}$

2. If a body is projected horizontally and another is dropped from the same height, both the bodies will take same time to reach the ground.
3. When an object is dropped from an aeroplane moving horizontally with constant velocity
   a) Path of the object relative to the earth is parabola
   b) Path of the object relative to pilot is a straight line vertically down.
4. From a certain height, if two bodies are projected horizontally with velocities u_1 and u_2 in **opposite directions**.
   a) time after which velocity vectors are perpendicular is $t = \frac{\sqrt{u_1 u_2}}{g}$
   b) Distance between the two bodies when velocity vectors are perpendicular is $\frac{\sqrt{u_1 u_2}}{g}(u_1 + u_2)$
   c) time after which displacement vectors are perpendicular is $t = \frac{2\sqrt{u_1 u_2}}{g}$
   d) Horizontal distance between the two bodies when displacement vectors are perpendicular is $\frac{2\sqrt{u_1 u_2}}{g}(u_1 + u_2)$

***Note :* Various ways of projection of body**

For (1) falling freely

$$h = \frac{1}{2}gt^2,\ V_1 = \sqrt{2gh} = gt$$

(2) $h = ut + \frac{1}{2}gt^2,\ V_2 = \sqrt{u^2 + 2gh}$

(3) $h = -ut + \frac{1}{2}gt^2,\ V_3 = \sqrt{u^2 + 2gh}$

(4) $h = -u\sin\theta t + \frac{1}{2}gt_1^2,\ V_4 = \sqrt{V_x^2 + V_y^2}$

(5) $h = \frac{1}{2}gt^2,\ V_5 = \sqrt{u^2 + (gT)^2} = \sqrt{u^2 + 2gh}$

(6) $h = u\sin\theta t + \frac{1}{2}gt^2,\ V_6 = \sqrt{v_x^2 + v_y^2}$

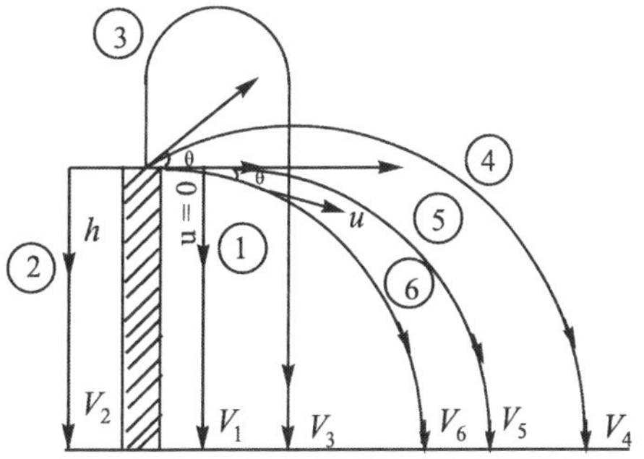

EXERCISE - I

1. The numerical ratio of displacement to the distance covered is always
1) less than one 2) equal to one
3) equal to or less than one
4) equal to or greater than one

2. The ratio of the numerical values of the average velocity and average speed of a body is always
1) unity 2) unity or less
3) unity or more 4) less than unity

3. The correct statement from the following is
1) A body having zero velocity will not necessarily have zero acceleration.
2) A body having zero velocity will necessarily have zero acceleration.
3) A body having uniform speed can have only uniform acceleration.
4) A body having non–uniform velocity will have zero acceleration.

4. Which of the following four statements is false.
1) A body can have zero velocity and still be accelerated.
2) A body can have a constant velocity and still have a varying speed.
3) A body can have a constant speed and still have a varying velocity.
4) The direction of the velocity of a body can change when its acceleration is constant.

5. The distance travelled by a particle is proportional to the square of time, then the particle travels with
1) uniform acceleration 2) uniform velocity
3) increasing acceleration
4) decreasing velocity

6. **Acceleration of a particle changes when**
   1) direction of velocity changes
   2) magnitude of velocity changes
   3) both of above 4) speed is constant
7. **What determines the nature of the path followed by the particle**
   1) speed 2) velocity
   3) acceleration 4) none of these
8. **Two bodies of different masses are dropped simultaneously from the top of a tower. If air resistance is same on both of them,**
   1) the heavier body reaches the ground earlier
   2) the lighter body reaches the ground earlier
   3) both reach the ground simultaneously
   4) cannot be decided
9. **Two bodies of different masses are dropped simultaneously from the top of a tower. If air resistance is proportional to the mass of the body,**
   1) the heavier body reaches the ground earlier
   2) the lighter body reaches the ground earlier
   3) both reach the ground simultaneously
   4) cannot be decided
10. **Choose the correct statement**
    1) If a particle is in motion average speed always equals average velocity
    2) Particle can travel with constant velocity and variable speed in a given.
    3) If acceleration is constant speed is constant in a given direction
    4) If a particle travels along a st. line average speed equals average velocity
11. **A body falls freely from a height 'h' its average velocity when it reaches earth is**
    1) $\sqrt{gh}$ 2) $\sqrt{\frac{gh}{2}}$ 3) $\sqrt{2gh}$ 4) $g\sqrt{h}$
12. **A body falls freely from a height 'h' after two seconds if accelaration due to gravity is reversed the body**
    1) continues to fall down
    2) falls down with retardation & goes up agian with accelaration
    3) falls down with uniform velocity
    4) raises up with accelaration
13. **In the above problem if we assume that gravity disappears instead of getting reversed, the body**
    1) continues to fall down
    2) falls down with accelaration
    3) falls down & floats 4) falls down with decelaration
14. **In the case of a body freely falling from small height**
    1) the changes of position are equal in equal intervals of time
    2) the changes of velocity are equal in unequal intervals of time
    3) the changes of acceleration is zero in equal or unequal intervals of time
    4) None
15. **A hydrogen ballon released on the moon from a height will**
    1) move up with acceleration 9.8ms^{-2}
    2) move down with acceleration 9.8ms^{-2}
    3) move down with acceleration $\frac{9.8}{6}$ ms^{-2}
    4) neither move up nor move down
16. **A freely falling body traveled xm in nth second distance travelled in n - 1th second is**
    1) x 2) x+g 3) x - g 4) 2x + 3g
17. **A body falls from a height 'h'.In absence of air resistance time of descent of body is**
    1) $\sqrt{\frac{2h}{g}}$ 2) $\sqrt{\frac{2h}{g-a}}$ 3) $\sqrt{\frac{2h}{g \pm a}}$ 4) $\frac{h}{g}$
18. **Three different objects of masses m_1, m_2 and m_3 are allowed to fall from rest and from the same point O along three different frictionless paths. The speeds of three objects, on reaching the ground, will be in the ratio of [AIMS 2002]**
    1) $m_1 : m_2 : m_3$ 2) $m_1 : 2m_2 : 3m_3$
    3) 1:1:1 4) $\frac{1}{m_1} : \frac{1}{m_2} : \frac{1}{m_3}$
19. **A body is projected up with a velocity 50ms^{-1} after one second if accelaration due to gravity disappears then body**
    1) floats in air
    2) continue to move up with constant velocity
    3) continue to move up with accelaration
    4) goes up and falls down

20. From the top of a tower a body A is thrown up vertically with velocity u and another body B is thrown vertically down with the same velocity u. If v_A and v_B are their velocities when they reach the ground and t_A and t_B are their times of flight, then

1) $v_A = v_B$ and $t_A = t_B$ 2) $v_A > v_B$ and $t_A > t_B$
3) $v_A = v_B$ and $t_A > t_B$ 4) $v_A < v_B$ and $t_A < t_B$

21. At the maximum height of a body thrown vertically up

1) velocity is not zero but acceleration is zero
2) acceleration is not zero but velocity is zero
3) both acceleration and velocity are zero
4) both acceleration and velocity are not zero

22. A body thrown vertically up with velocity u reaches the maximum height h after T seconds. Which of the following statements is true ?

1) At a height $\frac{h}{2}$ from the ground its velocity is $\frac{u}{2}$
2) At a time T its velocity is u
3) At a time 2T its velocity is u directed downwards
4) none of the above

23. A balloon rases up with uniform velocity 'u' . A body is dropped from ballon . The time of descent for the body is given by is

1) $\sqrt{\frac{2h}{g}}$ 2) $h = ut + \frac{1}{2}gt^2$
3) $h = -ut + \frac{1}{2}gt^2$ 4) $-h = ut + \frac{1}{2}gt^2$

24. In the above problem if body is thrown down with velocity 'u' the equation for the descent time is

1) $h = \frac{1}{2}gt^2$ 2) $h = ut + \frac{1}{2}gt^2$
3) $-h = -ut + \frac{1}{2}gt^2$ 4) $-h = -ut + \frac{1}{2}gt^2$

25. Velocity of an oblique projectile in its flight

1) Remains constant in its flight
2) First decreases, becomes zero and then increase
3) First decreases, reaches a minimum and then increases
4) First increases reaches maximum and decreases

26. A hunter aims his gun and fires a bullet directly at a monkey on a tree. At the instant the bullet leaves the gun, the monkey drops. The bullet

1) cannot hit the monkey
2) may hit the monkey it its weight is more than 30 kg.wt
3) may hit the monkey if its weight is less than 30 kg.wt
4) hits the monkey irrespective of its weight.

27. Keeping the velocity of projection constant, the angle of projection is increased from 0° to 90°. Then the horizontal range of the projectile

1) goes on increasing up to 90^0
2) decreases up to 90°
3) increases up to 45^0 and decreases afterwards
4) decreases up to 45^0 and increases afterwards

28. Keeping the velocity of projection constant, the angle of projection is increased from 0° to 90°. Then the maximum height of the projectile

1) goes on increasing upto 90^0
2) decreases upto 90^0
3) increaes upto 45^0 and decreases beyond it
4) decreases upto 45^0 and increases beyond it

29. The path of one projectile as seen from another projectile is a

1) straight line 2) parabola 3) hyperbola 4) circle

30. At the maximum height of a projectile, the velocity and acceleration are

1) parallel to each other
2) antiparallel to each other
3) perpendicular to each other
4) inclined to each other at 45°

31. A projectile has

1) minimum velocity at the point of projection and maximum at the maximum height
2) maximum at the point of projection and minimum at the maximum height
3) same velocity at any point in its path
4) zero velocity at the maximum height irrespective of the velocity of projection

32. For a projectile, the physical quantity that remains constant is

1) Vertical component of velocity and kinetic energy
2) Potential energy and kinetic energy
3) Horizontal component of velocity and acceleration
4) Potential energy and acceleration

33. Two particles are projected with same velocity but at angles of projection (45-θ) and (45+θ). Then their horizontal ranges are in the ratio of

1) 1 : 2 2) 2 : 1
3) 1 : 1 4) none of the above

34. The acceleration of a projectile relative to another projectile is

1) -g 2) g 3) 2g 4) 0

35. The direction of velocity and acceleration of a projectile at the highest point on the trajectory are [96 E]

1) parallel to each other
2) antiparallel to each other
3) perpendicular to each other
4) no specific relation

36. A number of bullets are fired horizontally with different velocities from the top of a tower they reach the ground

1) at same time with same velocity
2) at different times with different velocities
3) at same time with different velocities
4) at different times with same velocity

37. For body thrown horizontally from the top of a tower,

1) the time of flight depends both on h and v
2) the horizontal Range depends only on v but not on h
3) the time of flight and horizontal Range depend on h but not on v
4) the horizontal Range depends on both v and h

38. A stone is just dropped from the window of a train moving along a horizontal straight track with uniform speed. The path of the stone is

1) a parabola for an observer standing by the side of the track
2) a horizontal straight line for an observer inside the train
3) both of the above are true
4) none of the above is true

39. A bomb is dropped from an aeroplane flying horizontally with uniform speed. The path of the bomb is

1) a vertical straight line for a stationary observer on the ground
2) a parabola for the pilot of the aeroplane
3) a vertical straight line for the pilot and parabola for a stationary observer on the ground
4) a horizontal straight line for the pilot and parabola for a stationary observer on the ground

40. A and B are two trains moving parallel to each other. If a ball is thrown vertically up from the train A, the path of the ball is

1) parabola for an observer standing on the ground
2) vertical straight line for an observer in B when B is moving with the same speed in the same direction of A
3) a parabola for an observer in B when B is moving with same speed but in opposite direction
4) all the above are true

41. A ball is thrown from rear end to the front end of a compartment of a train which is moving at constant horizontal velocity. An observer sitting in the compartment and another observer standing on the ground draw the trajectory of the ball. They will have

1) equal horizontal and equal vertical ranges
2) equal vertical ranges but different horizontal ranges
3) different vertical ranges but equal horizontal ranges
4) different vertical and different horizontal ranges

42. From the top of a tower two bodies are projected with the same initial speed of $40ms^{-1}$, first body vertically upwards and second body vertically downwards. A third body is freely released from the top of the tower. If their respective times of flights are T_1, T_2 and T_3 identify the correct descending order of the times of flights

1) T_1, T_2, T_3 2) T_2, T_3, T_1
3) T_2, T_1, T_3 4) T_1, T_3, T_2

43. The acceleration of a moving body can be found from

1) Area under velocity–time graph
2) Area under distance–time graph
3) Slope of the velocity–time graph
4) Slope of distance–time graph

44. For a moving body at any instant of time

1) If the body is not moving, the acceleration is necessarily zero.
2) If the body is slowing, the retardation is negative
3) If the body is slowing, the distance is negative.
4) If displacement, velocity and acceleration at that instant are known, we can find the displacement at any given time in future.

45. Which of the following options is correct for having a straight line motion represented by t – s graph.

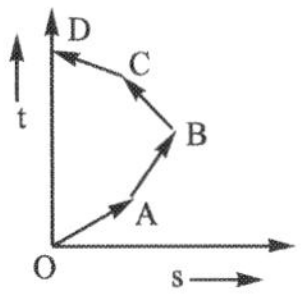

1) The object moves with constantly increasing velocity from O to A then it moves with constant velocity.
2) Velocity of the object increases uniformly.
3) Average velocity is zero.
4) The graph shown is impossible.

46. The displacement of a particle as a function of time is shown in the figure. The figure shows that

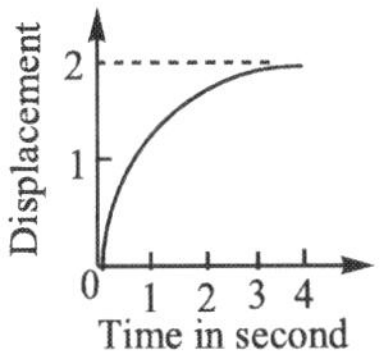

1) The particle starts with certain velocity but the motion is retarded and finally the particle stops
2) The velocity of the particle is constant throughout
3) The acceleration of the particle is constant throughout
4) The particle starts with constant velocity, then motion is accelerated and finally the particle moves with another constant velocity

47. A ball is thrown vertically upwards. Which of the following graph/graphs represent velocity–time graph of the ball during its flight (air resistance is neglected)

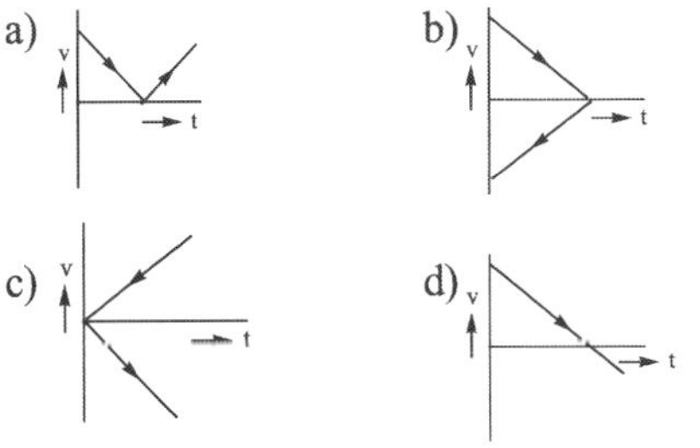

1) A 2) B 3) C 4) D

48. The graph between the displacement x and time t for a particle moving in a straight line is shown in figure. During the interval OA, AB, BC and CD, the acceleration of the particles is

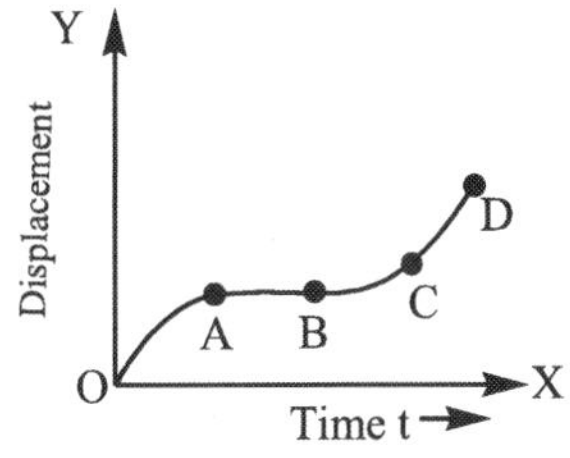

	OA	AB	BC	CD
1)	+	0	+	+
2)	–	0	+	0
3)	+	0	–	+
4)	–	0	–	0

49. The x – t graph shown in figure represents.

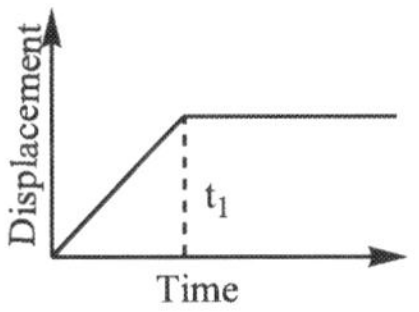

1) constant velocity
2) velocity of the body is continuously changing.
3) instantaneous velocity
4) the body travels with constant speed upto time t_1 and then stops

50. Figures (i) and (ii) below show the displacement – time graphs of two particles moving along the x – axis. We can say that

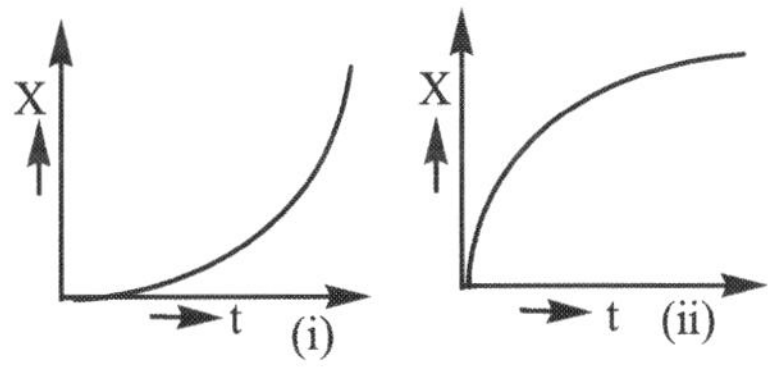

1) both the particles are having auniformly accelerated motion.
2) both the particles are having a uniformly reatrded motion.
3) particle (i) is having a uniformly accelerated motion while particle (ii) is having a uniformly retarded motion.
4) particle (i) is having a uniformly retarded motion while particle (ii) is having a uniformly accelerated motion.

51. Velocity – time curve for a body projected vertically upwards is

1) Parabola 2) Ellipse
3) Hyperbola 4) Straightline

52. **The displacement – time graph of moving particle is shown below.**

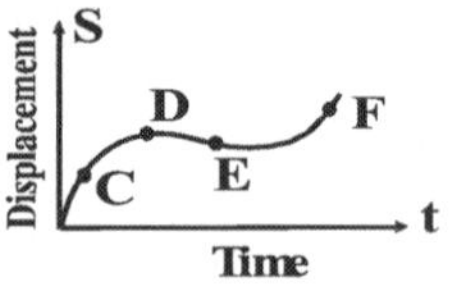

The instantaneous velocity of the particle is negative at the point.

1) D 2) F 3) C 4) E

53. **An object is moving with a uniform acceleration which is parallel to its instantaneous direction of motion. The displacement (velocity (v) graph of this object is**

1) 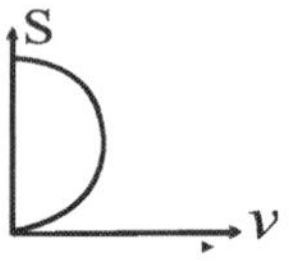

2) 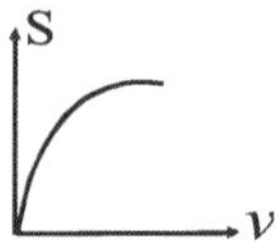

3) 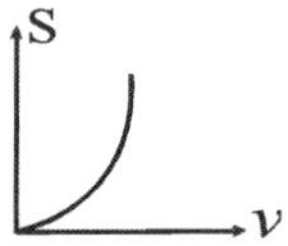

4) 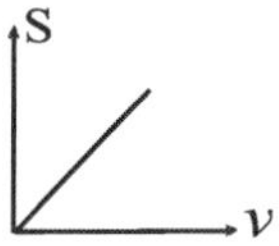

54. **Which of the following graph represents uniform motion**

1) 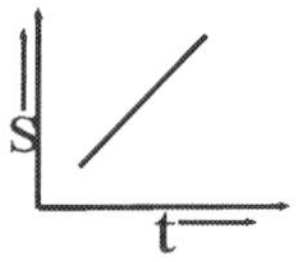

2) 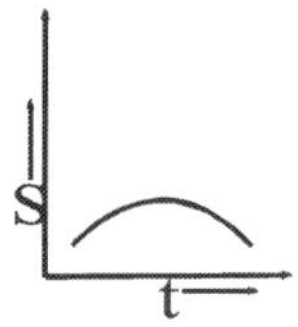

3) 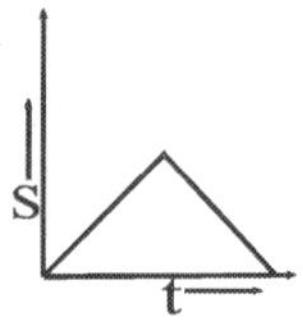

4) 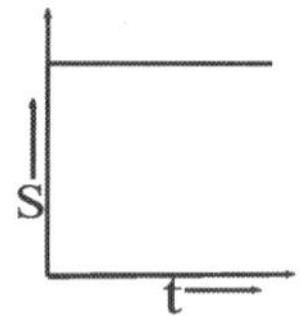

55. **The area under acceleration - time graph gives**

1) Distance travelled

2) Change in acceleration

3) Force acting 4) Change in velocity

56. **Which graph represents the uniform acceleration**

1) 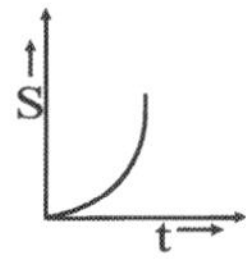

2) 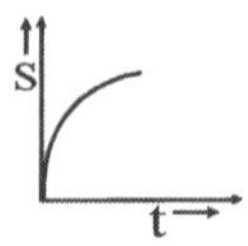

3) 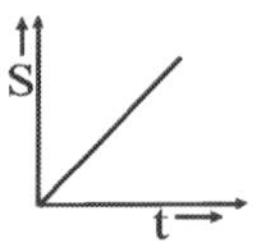

4) 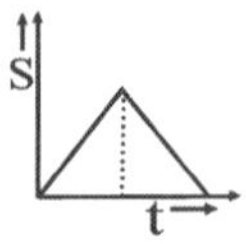

57. **Which of the following velocity-time graphs shows a realistic situation for a body in motion**

1) 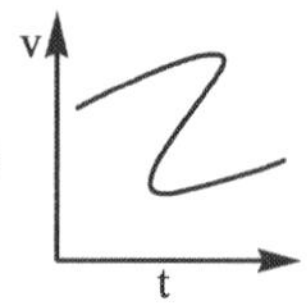

2) 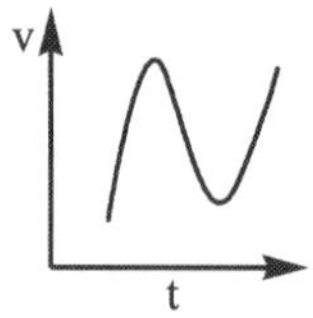

3) 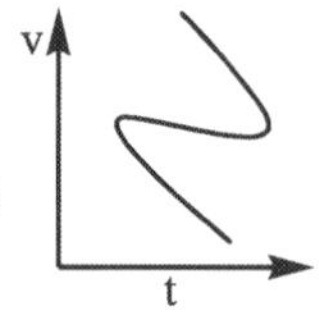

4) 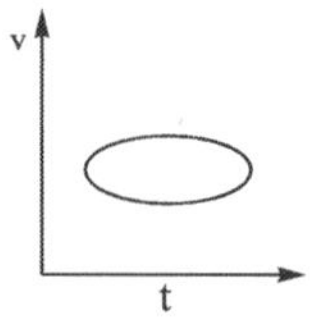

58. **Which of the following velocity-time graphs represent uniform motion**

1) 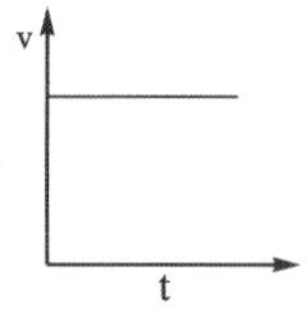

2) 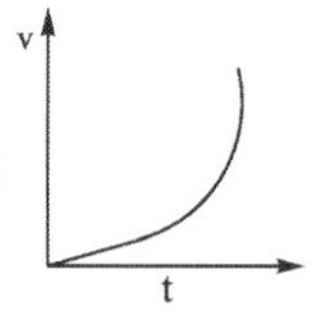

3) 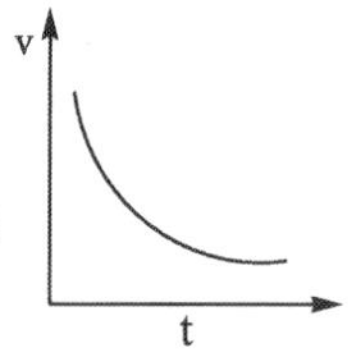

4) 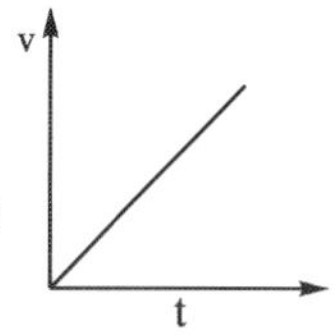

59. **A body projected vertically with a velocity 'u' from the ground. Its velocity**

a) At half of maximum height u/2

b) At $3/4^{th}$ of maximum height $\frac{u}{\sqrt{2}}$

c) At $1/3^{rd}$ of maximum height $\sqrt{\frac{2}{3}}u$

d) At $1/4^{th}$ of maximum height $\frac{\sqrt{3}}{2}u$

1) a and b correct 2) b is correct

3) c and d correct 4) a is correct

60. **Height of the body from the ground can be calculated by using the formula $h = -ut + (1/2)gt^2$ in**

a) A body projected vertically with velocity 'u' from the top of tower, reaches the ground in 't' sec.

b) A body dropped from a balloon moving up with uniform velocity, reaches the ground in 't' sec

c) A body dropped from a helicaptor moving up with uniform velocity, reaches the ground in 't' sec

d) A body projected vertically from the ground reaches the ground in 't' sec.

1) a, b and c are correct

2) a, b, c and d are correct

3) a is only correct 4) b and d are correct

61. A balloon from rest accelerates uniformly upward with 'a' ms^{-2}, for t seconds of time. A stone is released from the balloon. Now, read the following statements to pick the right ones.

a) The stone's initial velocity is zero, relative to balloon

b) The stone's initial velocity is non-zero, relative to earth

c) The time taken to reach the ground from the balloon's frame of reference is inversely propostional to $\sqrt{(a+g)}$.

d) The time take to reach the ground from earth's frame of reference is directly proprtional to $\sqrt{(a+g)}$.

1) a, b, c 2) a, c, d 3) a, b, d 4) a, c

62. A body is projected from a point with different angles of projections 20^0, 35^0, 45^0, 60^0 with the horizontal but with same initial speed. Their respective horizontal ranges are R_1, R_2, R_3 and R_4. Identify the correct order in which the horizontal ranges are arranged in increasing order

1) R_1, R_4, R_2, R_3 2) R_2, R_1, R_4, R_3

3) R_1, R_2, R_4, R_3 4) R_4, R_1, R_2, R_3

63. Two particles are projected from the same point with the same speed at different angles θ_1 and θ_2 to the horizontal. If their respective times of flights are T_1 and T_2 and horizontal ranges are same then

a) $\theta_1 + \theta_2 = 90^0$ b) $T_1 = T_2 \tan \theta_1$

c) $T_1 = T_2 \tan \theta_2$ d) $T_1 \sin \theta_2 = T_2 \sin \theta_1$

1) a, b, d are correct 2) a, c, d are correct

3) b, c, d are correct 4) a, b, c are correct

64. A body is projected with an initial speed of $100\sqrt{3}ms^{-1}$ at an angle of 60^0 above the horizontal. if $g = 10ms^{-2}$ then velocity of the projectile

a) Is perpendicular to it's acceleration at the instant t = 15 sec.

b) Is perpendicular to initial velocity of projection at t = 20 sec.

c) Is minimum at the highest point

d) Changes both in magnitude and direction, during its flight.

Mark the answer as

1) If a, b, c and d are correct

2) If a, c and d are correct

3) If b, c and d are correct

4) If a, b and d are correct

65. Two bodies are projected at angles 30^0 and 60^0 to the horizontal from the ground such that the maximum heights reached by them are equal. Then

a) Their times of flight are equal

b) Their horizontal ranges are euqal

c) The ratio of their initial speeds of projection is $\sqrt{3}:1$

d) Both take same time to reach the maximum height.

Mark the answer as

1) If a, b, c and d are correct

2) If only a, b and c are correct

3) If only a and c are correct

4) If a, c and d are correct

66. Velocity of the body on reaching the ground is same in magnetude in the following cases

a) a body projected vertically from the top of tower of height 'h' with velocity 'u'

b) a body thrown down wards with velocity 'u' from the top of tower of height 'h'

c) a body projected horizontally with a velocity 'u' from the top of tower height 'h'

d) a body dropped from the top tower of height 'h'

1) a, d, c and d are correct

2) a, b and c are correct

3) a and d are correct

4) d only correct

67. Read the following statements and choose the correct answer.

a) For a freely falling body the average velocity is proportional to square root of height of fall.

b) For a freely falling body the displacements in successive equal time intervals are in the ratio 1:4:9: .....

c) For a vertically projected body the displacement during last second of time of flight changes with velocity of projection.

d) For a body projected from the top of the tower the displacement of the body is negative when the body crosses the point of projection.

1) a, c, d are true 2) Both a,b are true
3) b & c are true 4) c and d are true

68. For an object moving with uniform acceleration, travelling 50m in 5th sec, 70m in 7th sec.

a) Its initial velocity is 5 m/s

b) Its acceleration is 20 m/s^2

c) Its travels 100m in 9th sec

d) Its average velocity during 9th sec is 90m/s

1) a is true 2) a & d are true
3) a & b are true 4) a & c are true

69. A bomber flying horizontally with constant speed releases a bomb from an aeroplane.

a) The path of bomb as seen by the observer on the ground is parabola

b) The path of the bomb as seen by a pilot is a straight line.

c) The path of the aeroplane with respect to bomb is a straight line

d) The path of the bomb as seen by pilot observed as parabola.

1) a is correct 2) a and b are correct
3) a,b and c are correct 4) only d is correct

Assertion & Reason :

These Questions consist of two statements each printed as Assertion and Reason. While answering these questions you are required to choose any one of the following four responses.

1) Both (A) and (R) are true and (R) is the correct explanation of (A)
2) Both (A) and (R) are true and (R) is not the correct explanation of (A)
3) (A) is true but (R) is false
4) (A) is false but (R) is true

70. **(A) :** A metal ball and a wooden ball of same radius are dropped from the same height in vacuum reach the ground same time.

(R) : In vacuum all the bodies dropped from same height take same time to reach the ground.

71. **(A) :** A body may be moving with uniform speed and non uniform acceleration.

(R) : A body may have uniform velocity and nonzero acceleration.

72. **(A) :** The directions of velocity and acceleration can be in any way.

(R) : The direction of acceleration depends on the direction of force, but not on the direction of velocity.

73. **(A) :** A metal ball and a wooden ball of same radius dropped from same height in vaccuum reach the ground at the same time.

(R) : In vaccuum the acceleration of the body is indipendent of the material and shape of the body moving under gravity.

74. **(A) :** A body thrown up from the top of a tower and another body thrown down from the same point strike the ground with the same velocity.

(R) : Intial velocity and acceleration are common for both.

75. **(A) :** The horizontal displacement of a projectile varies linearly with time.

(R) : Projectile motion is uniform motion along horizontal direction.

76. **(A) :** The path followed by one projectile as observed by another projectile is a straight line in uniform gravitation field.

(R) : The relative velocity between two projectiles at a given place doesnot change with time. Because their relative acceleration is zero.

77. **(A) :** If a body is projected obliquely at angle θ above horizontal with initial speed u then it's speed at the instant when it's velocity makes an angle α above the horizontal is $\left(\frac{u\cos\theta}{\cos\alpha}\right)$

(R) : Horizontal component of velocity of projectile remains constant.

78. **(A) :** If a bomb is dropped from an aeroplane moving horizontally with constant velocity then the bomb appears to move along a vertical straight line for the pilot of the plane.

(R) : Horizontal component of velocity of the bomb remains const and same as the velocity of the plane during the motion under gravity.

79. **(A) :** Time taken by the bomb to reach the ground from a moving aeroplane depends on height of aeroplane only

(R) : Horizontal component of velocity of bomb remains constant and vertical component of velocity of bomb changes due to gravity

80. **(A) :** For a body projected horizontally from the top of a tower, the velocity on reaching the ground depends both on velocity of projection and height of the tower.

(R) : For a projectile velocity varies both in horizontal and vertical directions.

81. Angle between velocity and acceleration vectors in the following cases

List - I	List - II
a) Vertically projected body	e) 90^0
b) For freely falling body	f) changes from point to point
c) For projectle	g) Zero
d) In uniform circular motion	h) 180^0

1) a – h; b – g; c – f; d – e
2) a – f; b – g; c – h; d – e
3) a – e; b – f; c – h; d – g
4) a – g; b – h; c – e; d – f

82. For a projectile 'R' is range and 'H' is maximum height

List - I	List - II
a) R = H	e) Angle of projection $tam^{-1}(1)$
b) R = 2H	f) Angle of projection $tan^{-1}(4)$
c) R = 3H	g) Angle of projection $tan^{-1}(2)$
d) R = 4H	h) Angle of projection $tan^{-1}(4/3)$

1) a – g; b – h; c – e; d – f
2) a – h; b – g; c – e; d – f
3) a – f; b – g; c – h; d – e
4) a – e; b – g; c – f; d – h

83. For a projectile relation between Range, velocity of projection and angle of projection ($g=10m/s^2$)

List - I	List - II
a) u = 20 m/s = 30^0	e) $45\sqrt{3}$m
b) u = 30 m/s = 60^0	f) 40m
c) u = 30 m/s = 45^0	g) $20\sqrt{3}$m
d) u = 20 m/s = 45^0	h) 90m

1) a – e; b – g; c – h; d – f
2) a – g; b – h; c – g; d – f
3) a – f; b – h; c – g; d – e
4) a – g; b – e; c – h; d – f

84. Match list I with list II for a projectile.

List - I	List - II
a) For two angles θ and $(90 - \theta)$ with same magnitude of velocity of projection	e) $\frac{P\bar{i}.P\bar{i}}{g}$
b) Equation of parabola of a projectile $y = Px - Qx^2$	f) Maximum height = 25% of $\frac{P^2}{Q}$
c) Radius of curvature of path of projectile projected with a velocity $(P\bar{i}+Q\bar{j})ms^{-1}$ at heighest point	g) Range = Maximum height
d) Angle of projection $\theta = \mathrm{Tan}^{-1}(4)$	h) Range is same

1) a - f, b - n, c - g, d - e
2) a - h, b - f, c - e, d - g
3) a - e, b - g, c - f, d - n
4) a - e, b - g, c - h, d - f

ANSWERS

1) 3	2) 2	3) 1	4) 2	5) 1
6) 3	7) 2	8) 1	9) 3	10) 4
11) 2	12) 2	13) 1	14) 3	15) 3
16) 3	17) 1	18) 3	19) 2	20) 3
21) 2	22) 3	23) 3	24) 2	25) 3
26) 4	27) 3	28) 1	29) 1	30) 3
31) 2	32) 3	33) 3	34) 4	35) 3
36) 3	37) 4	38) 3	39) 3	40) 4
41) 2	42) 4	43) 3	44) 4	45) 3
46) 1	47) 4	48) 2	49) 4	50) 3
51) 4	52) 4	53) 3	54) 1	55) 4
56) 1	57) 2	58) 1	59) 3	60) 1
61) 1	62) 1	63) 1	64) 1	65) 4
66) 2	67) 1	68) 2	69) 3	70) 1
71) 3	72) 1	73) 1	74) 1	75) 1
76) 1	77) 1	78) 1	79) 2	80) 3
81) 1	82) 3	83) 4	84) 2	

EXERCISE-II(A)

HORIZONTAL MOTION

1. If a cyclist takes one minute to complete half revolution on a circular path 120 m radius. What is the average velocity?

1) 1 m/s 2) 2 m/s 3) 3 m/s 4) 4 m/s

2. A person travels along a straight road for half the distance with velocity V_1 and the remaining half distance with velocity V_2 the average velocity is given by

1) V_1V_2 2) $\left(\frac{V_2}{V_1}\right)^2$ 3) $\frac{V_1+V_2}{2}$ 4) $\frac{2V_1V_2}{V_1+V_2}$

3. A person travels along a straight road for the half time with a velocity V_1 and the next half time with a velocity V_2. The mean velocity V of the man is

1) $\frac{2}{V}=\frac{1}{V_1}+\frac{1}{V_2}$ 2) $V=\frac{V_1+V_2}{2}$

3) $V=\sqrt{V_1V_2}$ 4) $V=\sqrt{\frac{V_1}{V_2}}$

4. If a car covers $2/5^{th}$ of the total distance with V_1 speed and $3/5^{th}$ distance with V_2 then average speed is

1) $\frac{1}{2}\sqrt{V_1V_2}$ 2) $\frac{V_1+V_2}{2}$

3) $\frac{2V_1V_2}{V_1+V_2}$ 4) $\frac{5V_1V_2}{3V_1+2V_2}$

5. A boy walks to his school at a distance of 6km with cons7tant speed of 2.5kmph and walks back with a constant speed of 4kmph. His average speed for round trip expressed in kmph is

1) 24/13 2) 40/13 3) 3 4) 1/2

6. If $\bar{u}=2\bar{i}-2\bar{j}+3\bar{k}$ and the final velocity is $\bar{v}=2\bar{i}-4\bar{j}+5\bar{k}$ and it is covered in a time of 10 sec, find the acceleration vector.

1) $\frac{3\bar{i}-2\bar{j}+2\bar{k}}{10}$ 2) $\frac{-3\bar{i}+\bar{j}+2\bar{k}}{10}$

3) $\frac{-3\bar{i}-2\bar{j}+2\bar{k}}{10}$ 4) $\frac{-\bar{j}+\bar{k}}{5}$

7. A train moving at a constant velocity of 54 km/hr moves east wards for 30 minutes, then due north with the same speed for 40 minutes. What is the average velocity of the train during this run? (in km/hr)

1) 30 2) 35 3) 38.6 4) 49.3

8. A particle moving with a constant acceleration describes in the last second of its motion $\frac{9}{25}$th of the whole distance. If it starts from rest, how long is the particle in motion and through what distance does it move if it describes 6 cm in the first sec.?

1) 5s; 150 cm 2) 10 s ; 150 cm

3) 15 s ; 100 cm 4) None

9. A car moving with constant acceleration covers the distance between two points 180 m apart in 6 sec. Its speed as it passes the second point is 45 m/s. What is its acceleration and its speed at the first point

1) -5 m/s^2; 15 m/s 2) -15 m/s^2 ; 5 m/s

3) -5 m/s^2 ; -15 m/s 4) 5 m/s^2 ; 15 m/s

10. A car moving with a speed of 50km/hr can be stopped by brakes after atleast 6m. If the same car is moving at a speed of 100 km/hr the minimum stopping distance is [AIEEE-2003]

1) 12m 2) 18 m 3) 24m 4) 6m

11. A body moving with a uniform acceleration had velocities of 20 m/s and 30 m/s when passing the points P and Q of its path. Find the velocity midway between P and Q (in m/s)

1) $\sqrt{450}$ 2) $\sqrt{550}$ 3) $\sqrt{650}$ 4) none

12. A particle starts moving from rest with uniform acceleration. It travels a distance x in the first 2 sec and a distance y in the next 2 sec. Then

1) y = x 2) y = 2x 3) y = 3x 4) y = 4x

13. The reaction time for an automobile driver is 0.7 sec. If the automobile can be decelarated at 5 m/s^2 calculate the total distance travelled in coming to stop from an initial velocity of 8.33 m/s after a signal is observed.

1) 12.77 m 2) 14.82 m 3) 16.83 m 4) 19.65 m

14. Two cars left place A simultaneously and reached the place B in 2 hrs. The first car travelled half the distance with a speed of 30 km/hr and the other half at a speed of 45 km/hr. The second car at the same time covered the entire distance with a constant acceleration starting from rest. Find the acceleration of the second car.

1) 36 km/hr^2 2) 18 km/hr^2

3) 72 km/hr^2 4) 144 km/hr^2

15. If the paticle is moving along a straight line given by the relation $s = 2 - 3t + 4t^3$ where s is in cms., and t in sec. its average velocity during the third sec is

1) 73 cm/s 2) 80 cm/s 3) 85 cm/s 4) 90 cm/s

16. A bullet fired into a fixed target loses half of its velocity in penetrating 15 cm. How much further it will penetrate before coming to rest?

1) 5 cm 2) 15 cm 3) 7.5 cm 4) 10 cm

17. For a body travelling with uniform acceleration, its final velocity is $v = \sqrt{180 - 7x}$, where x is the distance travelled by the body. Then the acceleration is

1) -8 m/s^2 2) -3.5 m/s^2

3) -7 m/s^2 4) 180 m/s^2

18. A bus starts from rest with a constant acceleration of $5 m/s^2$. At the same time a car travelling with a constant velocity 50 m/s overtakes and passes the bus. How fast is the bus travelling when they are side by side?

1) 10 m/s 2) 50 m/s 3) 100 m/s 4) none

19. A man walks up a stationary escalator in 90sec. When this man stands on a moving escalator he goes up in 60 sec. The time taken by the man to walk up the moving escalater is

1) 30 s 2) 45 s 3) 36 s 4) 48 s

20. A car moving on a straight road accelerates from a speed of 4.1 m/s to a speed of 6.9 m/s in 5.0s. What was its average acceleration ?

1) 0.56 m/s^2 2) 1.56 m/s^2

3) 5.6 m/s^2 4) 1.2 m/s^2

21. A bullet moving with speed 150 m/s strikes a tree and penetrates 3.5 cm before stopping. Find the magnitude of its acceleration and the time taken to stop.

1) $32.14 \times 10^5 m/s^2$, 4.67×10^4 s

2) $13.214 \times 10^5 m/s^2$, 4.67×10^4 s

3) $3.214 \times 10^5 m/s^2$, 4.67×10^4 s

4) $3.214 \times 10^5 m/s^2$, 14.67×10^4 s

22. A motor vehicle travelled the first third of a distance s at a speed of v_1 = 10 kmph, the second third at a speed of v_2 = 20 kmph and the last third at a speed of v_3 = 60 kmph. Determine the mean speed of the vehicle over the entire distance s.

1) 15 kmph 2) 12 kmph 3) 10 kmph 4) 18 kmph

23. A motorist drives north for 35.0 minutes at 85.0 km/h and then stops for 15.0 minutes. He next continues north, travelling 130 km in 2.00 hours. a) What is his total displacement

1) 85 km 2) 179.6 km 3) 20 km 4) 140 km

VERTICAL MOTION

24. A body falls from 80m. Its time of descent is $[g = 10 ms^{-2}]$

1) 3s 2) 4 s 3) 5 s 4) 6 s

25. Two bodies whose masses are in the ratio 2:1 are dropped simultaneously at two places A and B where the accelerations due to gravity are g_A and g_B respectively. If they reach the ground simultaneously, the ratio of the heights from which they are dropped is

1) $g_A : g_B$ 2) $2g_A : g_B$

3) $g_A : 2g_B$ 4) $\sqrt{g_a} : \sqrt{g_b}$

26. A body falls for 5s from rest. If the acceleration due to gravity of earth ceases to act, the distance it travels in the next 3s is

1) 73.5m 2) 294m 3) 147m 4) 49m

27. A body freely falling from a height h describes $\frac{7h}{16}$ in the last second of its fall. The height h is ($g = 10ms^{-2}$)

1) 80 m 2) 45 m 3) 160 m 4) 40 m

28. A body released from the top of a tower of height h takes T seconds to reach the ground. The position of the body at T / 4 seconds is

1) at $\frac{h}{16}$ from the ground

2) at $\frac{h}{4}$ from the top of the tower

3) at $\frac{15h}{16}$ from the ground

4) at $\frac{3h}{16}$ from the top of the tower

29. The ratio of times taken by freely falling body to cover first metre, second metre,.. is

1) $\sqrt{1}:\sqrt{2}:\sqrt{3}$ 2) $\sqrt{1}:\sqrt{2}-\sqrt{1}:\sqrt{3}-\sqrt{2}$

3) $\sqrt{2}:\sqrt{4}:\sqrt{8}$ 4) 2:3:4

30. A body is droped from a height 122.5 m. If its stopped after 3 seconds and again released the further time of descent is

1) 2 s 2) 3 s 3) 4 s 4) 5 s

31. A freely falling body travels __ of total distance in 5th second

1) 8% 2) 12% 3) 25% 4) 36%

32. If the distance travelled by a freely falling body in the last second of its journey is equal to the distance travelled in the first 2s, the time of descent of the body is

1) 5 s 2) 1.5 s 3) 2.5 s 4) 3 s

33. A ball dropped on to the floor from a height of 10m rebounds to a height of 2.5m. If the ball is in contact with the floor for 0.02s, its averge acceleration during contact is

1) 2100 ms^{-2} 2) 1050 ms^{-2}

3) 4200 ms^{-2} 4) 9.8 ms^{-2}

34. A splash is heard 3.12s after a stone is dropped into a well 45m deep. The speed of sound in air is [$g=10ms^{-2}$]

1) 330 ms^{-1} 2) 375 ms^{-1} 3) 340 ms^{-1} 4) 346 ms^{-1}

35. A body is thrown up with a velocity 29.23ms^{-1} distance travelled in last second of upward motion is

1) 2.3 m 2) 6m 3) 9.8 m 4) 4.9 m

36. A body is thrown up with a velocity 40ms^{-1}. At same time another body is droped from a height 40m. Their relative accelaration after 1.3 seconds is

1) 4g 2) g/2 3) 2g 4) zero

37. A stone is dropped into a well of 20m deep. Another stone is thrown downward with velocity v one second later. If both stones reach the water surface in the well simultaneously, v is equal to ($g=10ms^{-2}$)

1) 30 ms^{-1} 2) 15 ms^{-1} 3) 20 ms^{-1} 4) 10 ms^{-1}

38. A body is projected with a velocity 50ms^{-1}. Distance travelled in 6^{th}second is [$g=10ms^{-2}$]

1) 5m 2) 10 m 3) 15m 4) 20 m

39. In above problem ratio of distance traveled in first second of upward motion to first second of downward motion is

1) 1 : 7 2) 5 : 3 3) 9 : 1 4) 3 : 5

40. A body is projected vertically up with u. Its velocity at half its maximum height is

1) $\frac{u}{2}$ 2) $\frac{u^2}{2}$ 3) $\sqrt{2}$ u 4) $\frac{u}{\sqrt{2}}$

41. A body projected up reaches a point A in its path at the end of 4th second and reaches the ground after 5 seconds from the start. The height of A above the ground is ($g = 10m/s^2$)

1) 19.6 m 2) 30.6 m 3) 11 m 4) 20 m

42. A stone is projected vertically up from the ground with velocity 40ms^{-1}. The interval of time between the two instants at which the stone is at a height of 60m above the ground is ($g=10ms^{-2}$)

1) 4 s 2) 6 s 3) 8 s 4) 12 s

43. A ball is dropped freely while another is thrown vertically downward with an initial velocity u from the same point simultaneously. After t seconds they are separated by a distance

1) $\frac{ut}{2}$ 2) $\frac{1}{2}gt^2$ 3) ut 4) $ut+\frac{1}{2}gt^2$

44. A body is projected vertically up with velocity $98ms^{-1}$. After 2s if the acceleration due to gravity of earth disappears, the velocity of the body at the end of next 3s is

1) 49 ms^{-1} 2) 49.6 ms^{-1}
3) 78.4 ms^{-1} 4) 94.7 ms^{-1}

45. The distance travelled by a body during last second of its total flight is d when the body is projected vertically up with certain velocity. If the velocity of projection is doubled, the distance travelled by the body during last second of its total flight is

1) 2 d 2) d 3) $2d+\frac{g}{2}$ 4) $2d-\frac{g}{2}$

46. A stone is thrown vertically up from a bridge with velocity 3 ms^{-1}. If it strikes the water under the bridge after 2s, the bridge is at a height of ($g=10ms^{-2}$)

1) 26 m 2) 14 m 3) 7 m 4) 20 m

47. A bullet fired vertically up from the ground reaches a height 40m in its path from the ground and it takes further time 2 seconds to reach the same point during descent. The total time of flight is ($g=10$ ms^{-2})

1) 4 s 2) 3 s 3) 6 s 4) 8 s

48. A boy throws n balls per second at regular time intervals. When the first ball reaches the maximum height he throws the second one vertically up. The maximum height reached by each ball is

1) $\frac{g}{2(n-1)^2}$ 2) $\frac{g}{2n^2}$ 3) $\frac{g}{n^2}$ 4) $\frac{g}{n}$

49. A stone is dropped freely, while another thrown vertically downward with an initial velocity of $2ms^{-1}$ from the same point, simultaneously. The time required by them to have a distance of separation 22 m between them is [99 M]

1) 11s 2) 5.5 s 3) 44s 4) 22s

50. A body is throw up with a velocity 'u'. It reaches maximum height 'h'. If its velocity of projection is doubled the maximum height it reaches is

1) 4h 2) h 3) 2h 4) 3h

51. A person in lift which ascents up with acceleration $10ms^{-2}$ drops a stone from a height 10m. The time of decent is_ [$g=10ms^{-2}$]

1) 1 s 2) 2s 3) 15 m 4) 20 m

52. A body is projected up with velocity u. It reaches a point in its path at times $t_1$and t_2 seconds from the time of projection. Then (t_1+t_2) is

1) $\frac{2u}{g}$ 2) $\frac{u}{g}$ 3) $\sqrt{\frac{2u}{g}}$ 4) $\sqrt{\frac{u}{g}}$

53. A stone thrown vertically up with velocity v reaches three points A,B and C with velocities $\frac{v}{2}$, $\frac{v}{4}$ and $\frac{v}{8}$ respectively. Then AB:BC is

1) 1 : 1 2) 2 : 1 3) 4 : 1 4) 1 : 4

54. A stone projected vertically up with velocity v from the top of a tower reaches the ground with velocity 2v. The height of the tower is

1) $\frac{v^2}{2g}$ 2) $\frac{3v^2}{2g}$ 3) $\frac{3v^2}{g}$ 4) $\frac{v^2}{g}$

55. A stone is thrown vertically from the ground. It reaches the maximum height of 500m in 10sec. After what time it will reach the ground from the maximum height reached ? [96 M]

1) 5s 2) 10s 3) 15s 4) 20s

56. A stone is dropped from the top of a cliff. It is seen to hit the ground below after 4.2 s. How high is the cliff ?

1) 86.44 m 2) 860 m 3) 160 m 4) 180

57. How long does it take a brick to reach the ground if dropped from a height of 65 m? What will be its velocity just before it reaches the ground ?

1) 2 s, 10 m 2) 3.64 s, 35.67 m/s
3) 12s, 120 m 4) 5s, 20m

58. A stone is dropped from top of a tower 300 m high and at the same time another is projected vertically upward with a velocity of100 ms^{-1}. Find when and where the two stones meet.

1) 2 s, 200.9m 2) 3 s, 255.9m
3) 4s, 250.8m 4) 5s, 255.10m

59. A stone is thrown vertically upward with a speed of 10.0 ms^{-1} from the edge of a cliff 65m high. What will be its speed just before hitting the bottom ?

1) 3.14 m/s 2) 37.14 m/s
3) 13.71 m/s 4) 14.71 m/s

***60.** **An object reaches a maximum vertical height of 23.0m when thrown vertically upward on the earth. How high would it travel on the moon where the acceleration due to gravity is about one sixth that on the earth ? Assume that initial velocity is the same.**

1) 138 m 2) 100m 3) 10 m 4) 69 m

***61.** **A helicopter is ascending vertically with a speed of 8.0 ms^{-1}. At a height of 12 m above the earth, a package is dropped from a window. How much time does it take for the package to reach the ground ?**

1) 1.23 s 2) 3.23 s 3) 5.83 s 4) 7.83s

OBLIQUE PROJECTION

62. **A body is projected with a velocity 60 ms^{-1} at 30^0 to horizontal. Its initial velocity vector is**

1) $10\hat{i} + 10\sqrt{3}\hat{j}$ 2) $30\hat{i} + 30\sqrt{3}\hat{j}$

3) $30\sqrt{3}\hat{i} + 30\hat{j}$ 4) $30\sqrt{3}\hat{i}$

63. **In the above problem velocity after 3 seconds is __**

1) $20\hat{i} + 20\sqrt{3}\hat{j}$ 2) $30\hat{i}$

3) $10\sqrt{3}\hat{j}$ 4) $30\sqrt{3}\hat{i}$

64. **In the above problem the displacement after 2 s is __**

1) $30\sqrt{3}\hat{i} + 30\hat{j}$ 2) $60\sqrt{3}\hat{i} + 40\hat{j}$

3) $10\sqrt{3}\hat{i} + 10\hat{j}$ 4) $40\sqrt{3}\hat{j} + 40\hat{i}$

65. **A body is projected from height of 60 m with a velocity 10 ms^{-1} at angle 30^0 to horizontal. The time of flight of the body is [g =10ms^{-2}]**

1) 1 s 2) 2 s 3) 3 s 4) 4 s

66. **A body is projected with velocity u such that its horizontal range and maximum vertical heights are same. The maximum heights is**

1) $\frac{u^2}{2g}$ 2) $\frac{3u^2}{4g}$ 3) $\frac{16u^2}{17g}$ 4) $\frac{8u^2}{17g}$

67. **A stone is projected from the ground with a velocity of 14ms^{-1} one second later it clears a wall 2m high. The angle of projection is (g = 10 ms^{-2})**

1) 45° 2) 30° 3) 60° 4) 15°

68. **The height y and horizontal distance x covered by a projectile in a time t seconds are given by the equations $y = 8t - 5t^2$ and $x = 6t$. If x and y are measured in metres, the velocity of projection is**

1) 8 ms^{-1} 2) 6 ms^{-1} 3)14 ms^{-1} 4)10 ms^{-1}

69. **A body is projected at an angle 30° to the horizontal with a speed of 30 ms^{-1}. The angle made by the velocity vector with the horizontal after 1.5 s is (g=10 ms^{-2})**

1) zero 2) 60^0 3) 45^0 4) 90^0

70. **A ball is projected obliquely with a velocity 49ms^{-1} strikes the ground at a distance of 245m from the point of projection. It remained in air for**

1) 10 sec 2) $5\sqrt{2}$ sec 3) 3sec 4) 2.5sec

71. **The potential energy of a projectile at maximum height is 3/4 times kinetic energy of projection. Its angle of projection is**

1) 30^0 2) 45^0 3) 60^0 4) none

72. **Two bodies are thrown from the same point with the same velocity of 50ms^{-1}. If their angles of projection are complimentary angles and the difference of maximum heights is 30m, their maximum heights (g=10ms^{-2})**

1) 50m and 80m 2) 47.5m and 77.5m

3) 30m and 60m 4) 25m and 55m

73. **A grass hopper can jump a maximum horizontal distance of 0.3m. If it spends negligible time on ground, its horizontal component of velocity is (g=10ms^{-2})**

1) $\frac{3}{2}$ ms^{-1} 2) $\sqrt{\frac{3}{2}}$ ms^{-1} 3) $\frac{1}{2}$ ms^{-1} 4) $\sqrt{\frac{2}{3}}$ ms^{-1}

74. **A gun fires a bullet at a speed of 140ms^{-1}. If the bullet is to hit a target at the same level as the gun and at 1km distance, the angle of projection may be**

1) 60^0 or 30^0 2) 40^0 or 50^0

3) 15^0 or 75^0 4) 20^0 or 70^0

75. **A ball is thrown with velocity u making an angle q with the horizontal. Its velocity vector is normal to initial velocity (u) vector after a time interval of**

1) $\frac{u \sin\theta}{g}$ 2) $\frac{u}{g \cos\theta}$ 3) $\frac{u}{g \sin\theta}$ 4) $\frac{u \cos\theta}{g}$

76. A body is projected with velocity $24ms^{-1}$ making an angle 30° with the horizontal. The vertical component of its velocity after 2s is ($g=10ms^{-1}$)

1) $8ms^{-1}$ upward 2) $8ms^{-1}$ down ward
3) 32 ms^{-1} upward 4) $32ms^{-1}$down ward

77. A particle is thrown with a velocity u at an angle θ from the horizontal. Another particle is thrown with the same velocity at an angle a from the vertical. the ratio of times of flight of the two particles will be

1) Tan 2 θ : 1 2) Cot 2θ : 1
3) Tan θ : 1 4) Cot θ : 1

78. A body is projected with velocity u making an angle θ with the horizontal. Its velocity when it is perpendicular to the initial velocity vector is

1) u sinθ 2) u cotθ 3) u tanθ 4) u cosθ

79. The horizontal and vertical displacement of a projecticle are given by x = 12t and y = 16t - 5t², all the quantities being measured in S.I. system. The maximum height of the projectile is ($g=10ms^{-2}$)

1) 25.6 m 2) 12.8 m 3) 64 m 4) 6.4 m

80. The equation of trajectory of a projectile is $y = 10x-\left(\frac{5}{9}\right)x^2$. If we assume g = $10ms^{-2}$ the range of projectile (in meters) is [05 E]

1) 36 2) 24 3) 18 4) 9

81. The horizontal and vertical displacements x and y of projectile at a given time 't' are given by x =6t meters and y= 8t - 5t² metres. The range of the projectile in metres is [2004 E]

1) 9.6 2) 10.6 3) 19.2 4) 38.4

82. An object is projected with a velocity of $20ms^{-1}$ making an angle of 45⁰ with horizontal. The equation for the trajectory is $h=Ax-Bx^2$ where 'h' is height, x is horizontal distance, A and B are constants. The ratio A and B is ($g= 10\ ms^{-2}$) [2001 E]

1) 1:5 2) 5:1 3) 1:40 4) 40 :1

83. For a projectile the ratio of maximum height reached to the square of flight time is ($g=10ms^{-2}$) [2000 M]

1) 5:4 2) 5:2 3) 5:1 4) 10 :1

84. The speed of a projectile at its maximum height is $\frac{\sqrt{3}}{2}$ times its initial speed. If the range of the projectile is p times the maximum height attained by it, then p = [99 E]

1) 4/3 2) $2\sqrt{3}$ 3) $4\sqrt{3}$ 4) 3/4

85. The velocity at the maximum height of a projectile is half of its initial velocity of projection. The angle of projection is

1) 30^0 2) 45^0 3) 60^0 4) 76^0

86. A person throws a bottle into a dustbin at the same height as he is 2m away at an angle of 45⁰. The velocity of thrown is [98 E]

1) g 2) $\sqrt{g}$ 3) 2g 4) $\sqrt{2g}$

87. A body is projected at angle 30⁰ to horizontal with a velocity 50 ms^{-1}. Its time of flight is ($g=10\ m/s^2$)

1) 4 s 2) 5 s 3) 6 s 4) 7 s

88. In the above problem range of body is m. [$g = 10\ ms^{-2}$]

1) $125\sqrt{3}$ 2) $250\sqrt{3}$ 3) 125 4) 250

89. A body is projected at angle 30⁰ to horizontal on a planet with a velocity of 80 ms^{-1} its time of flight is 4 seconds then acceleration due to gravity on that planet is

1) $2ms^{-2}$ 2) 5 ms^{-2} 3) 10 ms^{-2} 4) 20 ms^{-2}

90. A body is thrown with velocity (4i + 3j) metre per second. Its maximum height is ($g = 10\ ms^{-2}$)

1) 2.5 m 2) 0.8 m 3) 0.9 m 4) 0.45 m

91. Two particles are projected with same velocity but at angles of projection 35^0 and 55^0. Then their horizontal ranges are in the ratio of

1) 1 : 2 2) 2 : 1 3) 1 : 1 4) 4 : 1

92. The potential energy of a projectile at its maximum height is equal to its kinetic energy there. If the velocity of projection is $20ms^{-1}$, its time of flight is ($g=10\ ms^{-2}$)

1) 2s 2) $2\sqrt{2}$ s 3) $\frac{1}{2}$s 4) $\frac{1}{\sqrt{2}}$ s

93. If the maximum vertical height and horizontal ranges of a projectile are same, the angle of projection will be

1) 45^0 2) $Tan^{-1}4$ 3) $Tan^{-1}2$ 4) 30^0

94. Two bodies are thrown from the same point with the same velocity of projection angles of projection being complimentary angles. If R_1 and R_2 are the ranges and h_1 and h_2 are maximum heights respectively, then

1) $R_1=R_2=\frac{u^2}{g}\sin 2\theta$ 2) $h_1+h_2=\frac{u^2}{2g}$

3) both 4) none

95. A body is projected with a certain speed at angles of projection of θ and $90-\theta$, The maximum height attained in the two cases are 20m and10m respectively. The maximum possible range is

1) 20m 2) 30 m 3) 60m 4) 80m

*96. Two bodies are thrown with the same initial velocity at angles α and $(90-\alpha)$ to the horizon. What is the ratio of the maximum heights reached by the bodies.

1) $\cot^2\alpha$ 2) $\tan^2\alpha$ 3) $\sec^2\alpha$ 4) $\cos^2\alpha$

Horizontal Projectile Motion

97. A body is projected horizontally from a height of 78.4 m with a velocity 10 ms^{-1}. Its velocity after 3 seconds is _ [g = 10 ms^{-2}] (Take direction of projection on $\vec{i}$ and vertically upward direction on $\vec{j}$).

1) $10\hat{i}-30\hat{j}$ 2) $10\hat{i}+30\hat{j}$

3) $20\hat{i}-30\hat{j}$ 4) $10\hat{i}+10\sqrt{3}\hat{j}$

98. In the above problem angle made by velocity vector with x axis after 4 seconds is $\tan^{-1}$

1) 3 2) 4 3) 5 4) 6

99. A stone is thrown horizontally with velocity g ms^{-1} from the top of a tower of height g metre. The velocity with which it hits the ground is (in ms^{-1})

1) g 2) 2g 3) $\sqrt{3g}$ 4) 4g

100. A body is thrown horizontally from the top of a tower. It reaches the ground after 4s at an angle 45° to the ground. The velocity of projection is

1) 9.8 ms^{-1} 2) 19.6 ms^{-1}

3) 29.4 ms^{-1} 4) 39.2 ms^{-1}

101. Two cliff of heights 120m and 100.4m are separated by a horizontal distance of 16m. If a car has to reach from the first cliff to the second, the horizontal velocity of the car should be

1) 16 ms^{-1} 2) 4 ms^{-1} 3) 2 ms^{-1} 4) 8 ms^{-1}

102. An aeroplane flying horizontally at an altitude of 490m with a speed of 180 kmph drops a bomb. The horizontal distance at which it hits the ground is

1) 500 m 2) 1000 m 3) 250 m 4) 50 m

103. A ball thrown horizontally with velocity v from the top of a tower of height h reaches the ground in t seconds. If another ball of double the mass is thrown horizontally with velocity 3v from the top of another tower of height 4h it reaches the ground in (seconds)

1) 3t/4 2) t/2 3) 4t 4) 2t

104. In the above problem if the first ball reaches the ground at a horizontal distance d, the second ball reaches the ground at a horizontal distance

1) 6d 2) 3d 3) 12 d 4) 4d

105. A body is projected horizontally from the top of a hill with a velocity of 9.8 m/s. What time elapses before the vertical velocity is twice the horizontal velocity?

1) 0.5 sec 2) 1 sec 3) 2 sec 4) 1.5 sec

106. A ball is projected horizontally from the top of a building 19.6 m high. If the line joining the point of projection to the point where it hits the ground makes an angle of 45° to the horizontal, the initial velocity of the ball is

1) 4.9 ms^{-1} 2) 9.8 ms^{-1}

3) 19.6 ms^{-1} 4) 14.7 ms^{-1}

107. A stone is projected horizontally with a velocity 9.8ms^{-1} from a tower of height 100m. Its velocity one second after projection is

1) 9.8 ms^{-1} 2) 4.9 ms^{-1}

3) $9.8\sqrt{2}$ ms^{-1} 4) $4.9\sqrt{2}$ ms^{-1}

108. A body is thrown horizontally from the top of a tower of 5m height. It touches the ground at a distance of 10m from the foot of the tower. Then initial velocity of the body is ($g=10ms^{-2}$)

1) 2.5 ms^{-1} 2) 5 ms^{-1} [2000 E]

3) 10 ms^{-1} 4) 20 ms^{-1}

109. Two thin wood screens A and B are separated by 200m. A bullet travelling horizontally at a speed of 600ms^{-1} hits the screen A, penetrates through it and finally emerges out from B making holes in A and B.If the resistance of air and wood are negligible, the difference of heights of the holes in A and B is

1) 5 m 2) $\frac{49}{90}$ m 3) $\frac{7}{\sqrt{90}}$ m 4) zero

110. The height and width of each step of a staircase are 20cm and 30cm respectively. A ball rolls off the top of a stair with horizontal velocity v and hits the fifth step. The magnitude of v is ($g=10\ ms^{-2}$)

1) $1.5\sqrt{5}\ ms^{-1}$ 2) $3\sqrt{5}\ ms^{-1}$
3) $7.5\ ms^{-1}$ 4) $1.5\ ms^{-1}$

111. An elevator is going up. The variation in the velocity of the elevator is as given in the graph. What is the height to which the elevator takes the passengers

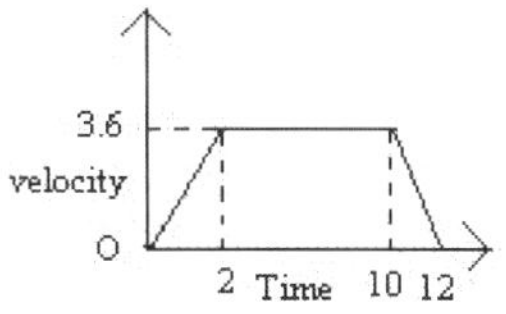

1) 3.6 m
2) 28.8m
3) 36.0m
4) 72.0m

112. The displacement time graphs of two moving particles make angles of 30^0 and 45^0 with the x-axis.The ratio of the two velocities is

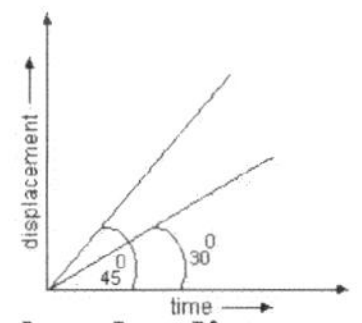

1) $\sqrt{3}:1$
2) 1 : 1
3) 1 : 2
4) $1:\sqrt{3}$

113. In the given v - t graph, the distance travelled by the body in 5 second will be

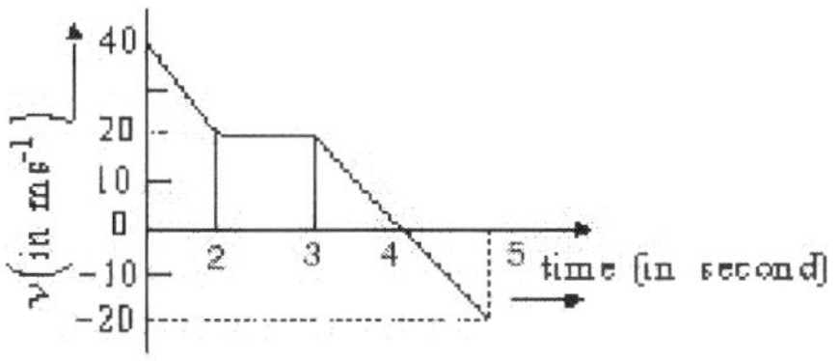

1) 20m 2) 40m 3) 80m 4) 100m

114. For the displacement-time graph shown in figure, the ratio of the magnitudes of the speeds during the first two second and the next four second is

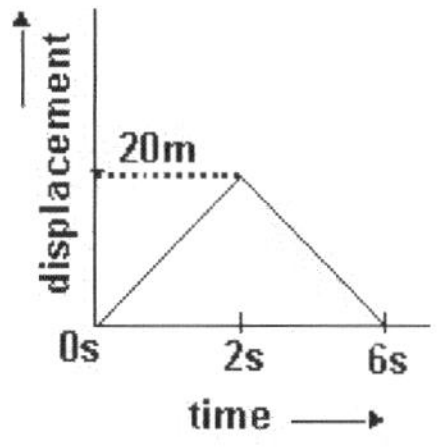

1) 1 : 1
2) 2 : 1
3) 1 : 2
4) 3 : 2

115. The displacement-time graph of a moving object is shown in figure. Which of the velocite-time graphs shown in figure could represent the motion of the same body ?

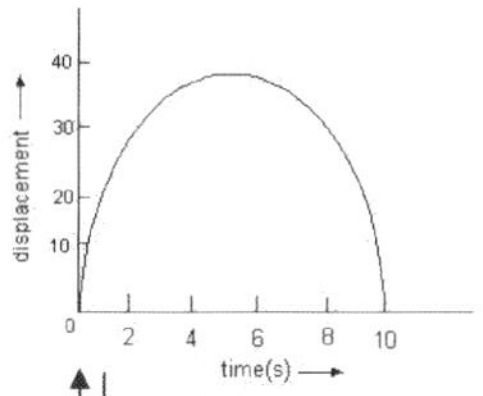

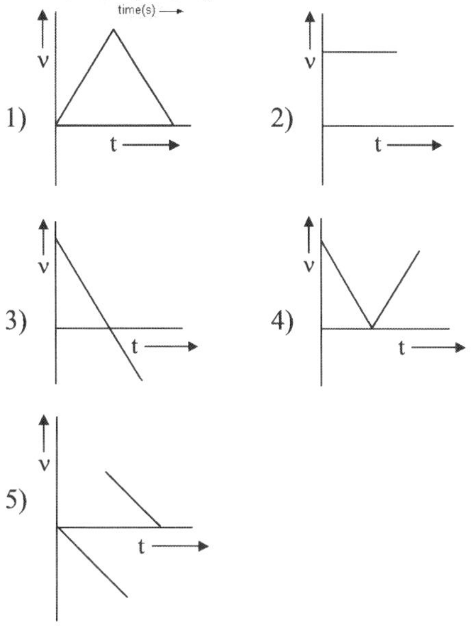

ANSWERS

1) 4	2) 4	3) 2	4) 4	5) 2
6) 4	7) 3	8) 1	9) 4	10) 3
11) 3	12) 3	13) 1	14) 1	15) 1
16) 1	17) 2	18) 3	19) 3	20) 1
21) 3	22) 4	23) 2	24) 2	25) 1
26) 3	27) 1	28) 3	29) 2	30) 3
31) 4	32) 3	33) 2	34) 2	35) 4
36) 4	37) 2	38) 1	39) 3	40) 4
41) 4	42) 1	43) 3	44) 3	45) 3
46) 2	47) 3	48) 2	49) 1	50) 1
51) 1	52) 1	53) 3	54) 2	55) 2
56) 1	57) 2	58) 2	59) 2	60) 1
61) 3	62) 3	63) 4	64) 2	65) 4
66) 4	67) 2	68) 4	69) 1	70) 2
71) 3	72) 2	73) 2	74) 3	75) 3
76) 2	77) 3	78) 2	79) 2	80) 3
81) 1	82) 4	83) 1	84) 3	85) 4
86) 4	87) 2	88) 1	89) 4	90) 4
91) 3	92) 2	93) 2	94) 3	95) 3
96) 2	97) 1	98) 2	99) 3	100) 3

101) 4	102) 1	103) 4	104) 1	105) 3
106) 2	107) 3	108) 3	109) 2	110) 1
111) 1	112) 4	113) 4	114) 2	115) 4

EXERCISE - IIB

HORIZONTAL MOTION

1. **If a cyclist takes two minutes to compete half revolution on a circular path of 120 m radius. What is the average speed?**

   1) 1 m/s 2) 3.14 m/s
   3) 2 m/s 4) 4 m/s

2. **A car travels the first half of a distance between two places a speed of 30 km/hr and the second half of the distance a 50 km/hr. The average speed of the car for the whole joumey is**

   1) 42.5 km/hr 2) 40.0 km/hr
   3) 37.5 km/hr 4) 35.0 km/hr

3. **A car moves for half of its time at 80 km/h and for rest half of time at 40 km/h. Total distance covered is 60 km. What is the average speed of the car**

   1) 60 km/h 2) 80 km/h
   3) 120 km/h 4) 180 km/h

4. **One car moving on a straight road covers one third of the distance with 20 km/hr and the rest with 60 km/hr. The average speed is**

   1) 40 km/hr 2) 80 km/hr
   3) $46\frac{2}{3}$ km/hr 4) 36 km/hr

5. **A man walks on a straight road from his home to a market 2.5 km away with a speed of 5 km/h. Finding the market closed, he instantly turns and walks back home with a speed of 7.5 km/h. The average speed of the man over the interval of time 0 to 40 min, is equal to**

   1) 5km/h 2) $\frac{25}{4}$ km/h
   3) $\frac{30}{4}$ km/h 4) $\frac{45}{8}$ km/h

6. **A body starts with a velocity $(2\hat{i}+3\hat{j}+11\hat{k})$ m/s and moves with an acceleration $(5\hat{i}+5\hat{j}-5\hat{k})$ m/s^2. What is its velocity after 0.2 sec ?**

   1) $7\hat{i}+8\hat{j}+6\hat{k}$ 2) $2\hat{i}-3\hat{j}+11\hat{k}$
   3) $3\hat{i}-4\hat{j}-10\hat{k}$ 4) $3\hat{i}+4\hat{j}+10\hat{k}$

7. **A car moving at a constant speed of 36kmph moves north wards for 20 minutes then due to west with the same speed for $8^1/_3$ minutes. What is the average velocity of the car during this run in kmph**

   1) 27.5 2) 40.5 3) 20.8 4) 32.7

8. **Moving with a uniform acceleration from the state of rest a body covers 22 m during the 6th sec of its motion. What is the distance covered by if during the first 6 sec?**

   1) 36 m 2) 48 m 3) 72 m 4) 150 m

9. **A car travelling at 108 km/hour has its speed reduced to 36 km/hour while travelling a distance of 200m. its retardation and the time taken are**

   1) 2 m/s^2; 5 sec 2) 2 m/s^2 10 sec
   3) 4 m/s^2; 5 sec 4) 4 m/s^2; 10 sec

10. **Speeds of two identical cars are U and 4U at a specific instant. The ratio of the respective distances in which the two cars are stopped from that instant is [AIEEE 2002]**

    1) 1:1 2) 1:4 3) 1:8 4) 1:16

11. **A car moving with constant acceleration crosses two points A and B with velocities 10m/s and 20 m/s in its path. Find the velocity of the car midway between A and B. (in m/s).**

    1) $\sqrt{500}$ 2) $5\sqrt{5}$ 3) $5\sqrt{10}$ 4) $\sqrt{300}$ m/s

12. **A body covers 200 cm in the first 2 sec and 220 cm in the next 4 sec under constant accelertion. Then the velocity of the body after 7 sec is**

    1) -15 cm/s 2) 115 cm/s 3) 10 cm/s 4) 20 cm/s

13. **A car A is trvelling on a straight level road with a uniform speed of 60 km/h. It is followed by another car B which is moving with a speed of 70 km/h. When the distance between them is 2.5km, the car B is given a deceleration of 20 km/h^2. After what distance and time will B catch up with A [IIT]**

    1) 32.5 km, 0.5 hr 2) 16.5 km, 1 hr
    3) 4 km, 2 hr 4) 3.25 km, 1.5 hr

14. **Two cars X and Y start off to a race on a straight path with initial velocities of 8 m/s and 5 m/s respectively. Car X moves with uniform acceleration of 1 m/s^2 and car Y moves with uniform acceleration of 1.1 m/s^2. If both the cars reach the winning post together find the length of the track.**

    1) 1000 m 2) 2000 m 3) 2500 m 4) 2280 m

15. The displacement x of a particle moving in one direction is given by $t=\sqrt{x}+3$, where x ius meter and t in sec. What is its displacement when its velocity is zero

1) 3 m/s 2) 2 m/s 3) 1 m/s 4) zero

16. If a body loses half of its velocity on penetrating 3 cm in a wooden block, then how much will it penetrate more before coming to rest [AIEEE-2002]

1) 1cm 2) 2cm 3) 3cm 4) 4cm

17. A particle moving with uniform acceleration has its final velocity $\left(\vartheta=\sqrt{150+8x}\right)$ m/s where x is the distance travelled by the body. Then the acceleration is

1) $+4m/s^2$ 2) $-4m/s^2$ 3) $+8m/s^2$ 4) $-8m/s^2$

18. A person is running at his maximum speed of 4 m/s to catch a train. When he is 6m from the door of the compartment the train starts to leave the station at a constant acceleration of 1 m/s^2. Find how long it takes him to catch up the train

1) 2 sec 2) 3 sec 3) 4 sec 4) none

19. When a man stands on a moving escalator he goes up in 50 sec. and when he walks up the moving escalator he goes up in 30 sec. Then the man walks up the stationary escalator in a time of .... sec

1) 60 2) 75 3) 90 4) 18.75

20. A vehicle moving in a straight road has its speed changed from 5 m/s to 8 m/s in 2 sec. What is its average acceleration ?

1) 1 m/s^2 2) 1.5 3) 2 m/s^2 4) zero

21. A bullet moving at 20 m/sec. strikes a wooden plank and penetrates 4 cm before coming to stop. The time taken to stop is

1) 0.08 sec 2) 0.16 sec 3) 0.04 sec 4) 0.02 sec

22. A vehicle covers first half of its total distance at a speed 20 kmph. and the other half of the distance at 40 kmph. The mean speed over the entire distance.

1) 30 kmph 2) 50 kmph

3) $\frac{40}{3}$ kmph 4) $\frac{80}{3}$ kmph

23 A vehicle moves for 10 minutes at 20 ms^{-1} due north and stops for five minutes. Then at continues due north at 40 ms^{-1} for 20 minites. Average velocity in the entire journey.

1) 2.857 ms^{-1} 2) 285.7 ms^{-1}

3) 28.57 ms^{-1} 4) 2857 ms^{-1}

<u>VERTICAL MOTION</u>

24. A body is projected down with velocity 5ms^{-1} from a height of 60 m. Its time of descents is

1) 2 s 2) 3 s 3) 4 s 4) 5 s

25. Two stones are falling at a place from heights in the ratio 2:3. Their velocities on reaching the ground are in the ratio

1) $\sqrt{3}:\sqrt{2}$ 2) $\sqrt{2}:\sqrt{3}$ 3) 2 : 3 4) 3 : 2

26. A bodyfalls for 4s from rest. If the acceleration due to gravety of earth ceases to act, then the velocity and distance it travells in the next 3sec is

1) 40 m/s, 120 m 2) 30 m/s, 400m

3) 50 m/s, 50m 4) 20 m/s, 60 m

27. A body freely falling from a height h describes $\frac{11h}{36}$ in the last second of its fall. The height h is ($g=10ms^{-2}$)

1) 125 m 2) 180 m 3) 360 m 4) 90 m

28. A body is released from the top of a tower of height h and takes 'T' sec to reach the ground. The position of the body at T/2 sec is

1) $\frac{h}{4}$ from ground 2) $\frac{3h}{4}$ from ground

3) $\frac{h}{8}$ from ground 4) $\frac{5h}{4}$ from ground

29. A particle is released from rest from a tower of height 3h. The ratio of times to fall equal heights 'h' i.e. $t_1:t_2:t_3$ is ($\sqrt{3}=1.7$ & $\sqrt{2}=1.4$)

1) $\sqrt{3}:\sqrt{2}:1$ 2) 3 : 2 : 1

3) 1 : 1.4 : 1.7 4) 10 : 4 : 3

30. A body is dropped from a height 135m. If after 3 sec acceleration due to gravity of earth cases to act then the further time of descent is ($g = 10\ m/s^2$)

1) 3 sec 2) 6sec 3) 4 sec 4) 2 sec

31. A freely falling body covers ________ of total distance in 3rd second.

1) 8% 2) 35.5%

3) 55.55% 4) 45.5%

32. A stone is dropped from the top of a tower of height 125m. The distance travelled by it during last second of its fall is ($g=10ms^{-2}$)

1) 22.5m 2) 45 m 3) 75 m 4) 90 m

33. A ball dropped on to the floor from a height of 40m and rebounds to a height of 10 m. If the ball is in contact with the floor for 0.02s, its average occeleration during contact is

1) 1050 m/s^2 2) 2100 m/s^2
3) 9.8 m/s^2 4) 210 m/s^2

34. A splash is heard 3.91 after a stone is dropped into a well of 67.6m deep. The speed of sound in air is

1) 346 m/s 2) 320m/s 3) 330 m/s 4) 338m/s

35. A body is thrown up with a velocity 29.23 ms^{-1}. The distance traveled in last but one second of upward motion is

1)14.7 m 2) 9.8 m 3) 4.9 m 4) 1 m

36. Two balls are dropped simultaneously from two points separated by a vertical height of 6m. The distance of separation between them after next 2s is

1) 9.8 m 2) 6 m 3) 12 m 4) zero

37. A ball is dropped from the top of a tower of height 78.4m Another ball is thrown down with a certain velocity 2 sec later. If both the balls reach the ground simultaneously, the velocity of the second ball is

1) 29.4 m/s 2) 26.4 m/s 3) 30 m/s 4) 41.6 m/s

38. A body is projected with a velocity 60ms^{-1} vertically upwards the distance travelled in last second of its motion is [g = 10 ms^{-1}]

1) 35 m 2) 45 m 3) 55m 4) 65 m

39. A body is thrown vertically upwards with initial velocity 'u' reaches maximum height in 6 seconds. The ratio of distances travelledby the body in the first second and seventh second is [2000 E]

1) 1:1 2) 11:1 3) 1:2 4) 1:11

40. A body is thrown vertically up with certain velocity. If h is the maximum height reached by it, its position when its velocity reduces to 1/3 of its velocity of projection is at

1) $\frac{8h}{9}$ from the ground

2) $\frac{8h}{9}$ from the top most point

3) $\frac{4h}{9}$ from the ground

4) $\frac{h}{3}$ from the top most point

41. A stone projected vertically up from the ground reaches a height y in its path at t_1 seconds and after further t_2 seconds reaches the ground. The height y is equal to

1) $\frac{1}{2}g(t_1+t_2)$ 2) $\frac{1}{2}g(t_1+t_2)^2$
3) $\frac{1}{2}g\,t_1t_2$ 4) $g\,t_1t_2$

42. A bullet fired vertically up from the ground reaches a point in its path after 3s and 13s, both times being taken from the instant of firing the bullet. The velocity of projection of the bullet is (g=10 ms^{-2})

1) 80 ms^{-1} 2) 40 ms^{-1} 3) 160 ms^{-1} 4) 20 ms^{-1}

43. A ball is dropped from a tall building. After 1s another ball is dropped from the same point. Their distance of separation after next 2s is

1) 24.5 m 2) 49 m 3) 98 m 4) 19.6 m

44. A body is projected vertically up with a velocity of 58.8 m/s. After 3s. if the acceleration due to gravity of earth disappears. The distance travelled and the velocity of the body at the end of next 5 sec is

1) 147m, 29.4 m/s 2) 29.4m, 147 m/s
3) 14.7m, 2.94 m/s 4) 19.6m, 240 m/s

45. The distance travelled by a body during last second of its upward journey is d when the body is projected with certain velocity vertically up. If the velocity of projection is doubled, the distance travelled by the body during last second of its upward journey is

1) 2d 2) 4 d 3) d/2 4) d

46. A packet is dropped from a balloon rising up with uniform velocity 9.8ms^{-1}. If the ballon is at a height of 39.2m from the ground at the time of dropping the stone, the stone reaches the ground after

1) 2 s 2) 6 s 3) 4 s 4) 8 s

47. A body is projected vertically upwards with a velocity ‘U’. It crosses a point in its journey at a height ‘h’ twice, just after 1 and 7 seconds. The value of U in ms^{-1} is [g=10ms^{-2}] [06 -AM]

1) 50 2) 40 3) 30 4) 20

48. A boy throws a ball in air in such a manner that when the ball is at its maximum height he throws another ball. If the balls are thrown with the time difference 1second, the maximum height attained by each ball is

1) 9.8 m 2) 19.6 m 3) 4.9 m 4) 2.45 m

49. A stone is dropped freely while another thrown vertically downward with an initial velocity of 4 m/s from the same point simultaneously. The distance of seperation between them will become 30m after a time of

1) 8 sec 2) 10 sec 3) 15 sec 4) 7.5 sec

50. A body thrown up with a velocity reaches a maximum height of 50m. Another body with double the mass thrown up with double the initial velocity will reach the maximum height of [96 M]

1) 100 m 2) 200 m 3) 400 m 4) 50 m

51. A person in a lift which descends down with an acceleration of 1.8 m/s^2 drops a stone from a height of 2m. The time of decent is

1) $\sqrt{2}$ sec 2) $2\sqrt{2}$ sec

3) $1/\sqrt{2}$ sec 4) $\frac{1}{2}$ sec

52. A body projected up with a velocity of 39.2 m/s reaches a point in its path after t_1 and t_2 sec from start. Then (t_1+t_2) is

1) 3 sec 2) 8 sec 3) 4 sec 4) 5 sec

53. The velocity of a body moving vertically up is $49ms^{-1}$ at half the maximum height. The height to which it could further rise is

1) 245 m 2) 122.5 m 3) 61.25 m 4) none

54. A stone is projected vertically upward from the top of a tower with a velocity u and strikes the bottom of the tower with a velocity 3u. The distance travelled by the stone is

1) $\frac{4u^2}{g}$ 2) $\frac{5u^2}{g}$ 3) $\frac{3u^2}{g}$ 4) $\frac{u^2}{g}$

55. A body is thrown vertically up reaches a maximum height of 78.4m. After what time it will reach the ground from the maximum height reached ?

1) 4 sec 2) 8 sec 3) 3 sec 4) 5 sec

56. A stone dropped from a cliff. hits the ground in 4 sec. Height of the cliff is .

1) 39.2 m 2) 29.4 m 3) 9.8 m 4) 78.4m

57. A stone is dropped from a bridge 45.1m above water. Time it takes to strike water is

1) 2 sec 2) 3 sec 3) 5 sec 4) 0.8 sec

58. A stone is dropped from a tower of height 200m and at the same time another stone is protected vertically up at 50 m/s. The height at which they meet ? [$g = 10\ m/s^2$]

1) 5 m 2) 100 m 3) 120 m 4) 125 m

59. A stone is thrown vertically up at 20 m/s from a tower 80 m height. The speed with which it hits the ground is [$g = 10\ m/s^2$].

1) 14.4 m/s 2) 28.14 m/s

3) 80 m/s 4) $20\sqrt{5}$ m/s

60. A vertically thrown up body reaches 20m at a place on the earth. The height to which it goes on the moon if projected with same velocity is [$g_{moon} = \frac{g_{earth}}{6}$]

1) 40m 2) 80 m 3) 100 m 4) 120 m

61. A balloon rises with uniform velocity of 10 ms^{-1} . When the balloon is at a height of 72.5m if astone is dropped from it the time taken by it to reach the ground is

1) 3 sec 2) 4 sec 3) 5 sec 4) 7 sec

OBLIQUE PROJECTION

62. A body is projected with an initial velocity 20 m/s at 60^0 to the horizontal. Its initial velocity vector is ___ ($g=10\ m/s^2$)

1) $10i-20j$ 2) $10\sqrt{3}i+10j$

2) $10i+10\sqrt{3}j$ 4) $5i+5\sqrt{3}j$

63. In the above problem velocity after 1 sec is

1) $10i+7.32j$ 2) $10i-7.32j$

3) $10i+\sqrt{3}j$ 4) $10i-\sqrt{3}j$

64. In the above problem the displacement after 2s is

1) $20[i+(\sqrt{3}-1)j]$ 2) $20[i-(\sqrt{3}-1)j]$

3) $10[i+(\sqrt{3}+1)j]$ 4) $10[i+(\sqrt{3}-1)j]$

65. A bullet is projected upwards from the top of a tower of height 90 m with the velocity 30 ms^{-1} making an angle 30° with the horizontal. The time taken by it to reach the ground is ($g=10\ ms^{-2}$)

1) 2 s 2) 3 s 3) 24 s 4) 6 s

66. A body is projected with velocity 'u' so that the maximum height is thrice the horizontal range. Then the maximum height is

1) $\frac{72u^2}{145g}$ 2) $\frac{6}{\sqrt{145}}\frac{u^2}{g}$

3) $\frac{u^2}{2g}$ 4) $\frac{145}{72}\frac{u^2}{g}$

67. Astone is projected from the ground with a velocity of 20 m/s at angle 30^0 with the horizontal. After one second it clears a wall then height of the wall is ($g=10\ ms^{-2}$)

1) 10m 2) 2m 3) 5m 4) 15m

68. The horizontal and vertical displacement of projectile at time 't' are x=36t and y=48-4.9t² respectively. Initial velocity of theprojectile in ms^{-1} is [2002 E]

1) 15 2) 30 3) 45 4) 60

69. A body is projected with velocity $24ms^{-1}$ making an angle 30^0 with the horizontal. The angle made by the direction of the projectile with the horizontal at 2s from start is

1) $\tan^{-1}\frac{2}{3\sqrt{3}}$ 2) $\tan^{-1}\frac{1}{3\sqrt{3}}$

3) $\tan^{-1}\frac{2}{3}$ 4) $\tan^{-1}\frac{1}{3}$

70. A body projected with velocity $30ms^{-1}$ reaches its maximum height in 1.5s. Its range is ($g=10ms^{-2}$)

1) 45m 2) 108 m 3) $45\sqrt{3}$ m 4) 54 m

71. The potential energy of a projectile at its maximum height is equal to its kinetic energy there. Its range for velocity of projection u is

1) $\frac{u^2}{2g}$ 2) $\frac{u^2}{g}$ 3) $\frac{2u^2}{g}$ 4) $\frac{2u}{g}$

72. Two bodies are thrown at angles θ and ($90-\theta$) from the same point with same velocity $25ms^{-1}$. If the difference between their maximum heights is 15m, the respective maximum heights are ($g=10ms^{-2}$)

1) $\frac{185}{8}$ m and $\frac{65}{8}$ m 2) $\frac{125}{4}$ m and $\frac{65}{4}$ m

3) 40m and 25m 4) 25 m and 40 m

73. The minimum and maximum velocities of a projectile are $10ms^{-1}$ and $20ms^{-1}$ respectively. The horizontal range and maximum height are respectively ($g=10ms^{-2}$)

1) $10\sqrt{3}$ m and 20m 2) $20\sqrt{3}$ m and 15m

3) 20m and 15m 4) $10\sqrt{3}$ m and 10m

74. A body is projected with a velocity of 30 m/s to have a horizontal range of 45m. The angle of projection may be

1) 30^0 & 60^0 2) 20^0 & 70^0

3) 15^0 & 75^0 4) 40^0 & 50^0

75. A ball is thrown with velocity $8ms^{-1}$ making an angle 60° with the horizontal. Its velocity will be perpendicular to the direction of initial velocity of projection after a time of ($g = 10ms^{-2}$)

1) $\frac{1.6}{\sqrt{3}}s$ 2) $\frac{4}{\sqrt{3}}s$ 3) 0.6 s 4) $1.6\sqrt{3}s$

76. A body is projected with an initial velocity of 58.8 m/s at angle 60° with the vertical. The vertical component of velocity after 2 sec is

1) 9.8 m/s upwards 2) 9.8 m/s downwards

3) 10.3 m/s upwards 4) 19.6m/s downwards

77. Two bodies are thrown from the same point with the same velocity of projection, angles of projection 30^0 and 60^0. If R_1 and R_2 are the ranges and h_1 and h_2 are maximum heights respectively, then

1) $R_1=R_2=\frac{u^2}{g}$ 2) $h_1+h_2=\frac{u^2}{2g}$

3) both 4) none

78. A body is projected at 60° with the horizontal with velocity of $10\sqrt{3}$ ms^{-1}. The velocity of the projectile when it moves perpendicular to its initial direction is

1) 30 ms^{-1} 2) $\frac{10}{3}ms^{-1}$ 3) 10 ms^{-1} 4) 5 ms^{-1}

79. The equations of motion of a projectile are given by x = 36t metre and 2y=96t - 9.8t² metre. The angle of projection is [2003-E]

1) $\sin^{-1}\left(\frac{4}{5}\right)$ 2) $\sin^{-1}\left(\frac{3}{5}\right)$

3) $\sin^{-1}\left(\frac{4}{3}\right)$ 4) $\sin^{-1}\left(\frac{3}{4}\right)$

80. If the equation of motion of a projectile is $y=3x-\frac{1}{8}x^2$, the range and maximum height are respectively (y and x are in metres)

1) 18 m and 24 m 2) 24 m and 18 m

3) 24 m and 6 m 4) 12 m and 9 m

81. The horizontal and vertical displacements x and y of projectile at a given time 't' are given by x = 12t m. and y = 5t – 5t²m. The range of the projectile in meteres is

1) 6 2) 12 3) 18 4) 10

82. If the equation of motion of a projectile is $y=3x-\frac{1}{8}x^2$, (y and x are in metres) the angle of projection and velocity of projection are respectively ($g=10$ ms^{-2})

1) $\tan^{-1}\left(\frac{1}{3}\right)$ and 10 ms^{-1} 2) $\tan^{-1}(3)$ and 20 ms^{-1}

3) $\tan^{-1}$ (3) and 10ms^{-1} 4) $\cos^{-1}\left(\frac{1}{3}\right)$ and 20ms^{-1}

83. The time of flight of a projectile is related to its horizontal range by the equation $gT^2 = 2R$. The angle of projection is

1) 30° 2) $\tan^{-1}(2)$ 3) $\sin^{-1}\left(\frac{1}{2}\right)$ 4) 45°

84. If the horizontal velocity of a projectile is $\sqrt{\frac{2}{5}}$ times the velocity at half the maximum height, the angle of projection is

1) 60° 2) 30°
3) 45° 4) none of the above

85. The velocity at the maximum height of a projectile is $\frac{\sqrt{3}}{2}$ times the initial velocity of projection. The angle of projection is

1) 60^0 2) 45^0 3) 30^0 4) 15^0

86. A ball is projected at an angle 30° with the horizontal with the velocity $49ms^{-1}$. The horizontal range is

1) 122.5 m 2) 245 m 3) 245 $\sqrt{3}$ m 4) 122.5 $\sqrt{3}$ m

87. A body is projected at angle 30^0 to the horizontal with a velocity 50 ms^{-1} maximum height of projectile is

1) 40 m 2) 31.25 m 3) 28 m 4) 21 m

88. A boy throws a ball with a velocity of 15 m/s at an angle of 15^0 with the horizontal. The distance at which the ball strikes the ground is

1) 5m 2) 20m 3) 10m 4) 11.25m

89. A body is projected at an angle 30° with a velocity $42ms^{-1}$. Its maximum height is

1) 22.05m 2) 32.5 m
3) 42.5 m 4) 52.5 m

90. A body is thrown with velocity (4i+3j) metre per second. Its horizontal range and time of flight are

1) 2.4m and 0.6 s 2) 3.2m and 0.8 s
3) 5m and 2s 4) 7m and 1s

91. Two particles are projected with same velocity but at angles of projection 25^0 and 65^0 with horizontal. The ratio of their horizontal ranges is

1) 1 : 2 2) 1 : 1 3) 2 : 1 4) cann't be determined

92. A body is projected with kinetic energy E such that its range is maximum. Its potential energy at the maximum height is

1) E 2) $\frac{E}{2}$ 3) 2 E 4) $\sqrt{2}$ E

93. A body is projected with a velocity u so that the horizontal range is twice the maximum height. Then the maximum height is

1) $\frac{u^2}{2g}$ 2) $\frac{u^2}{g}$ 3) $\frac{5u^2}{4g}$ 4) $\frac{2u^2}{5g}$

94. A stone projected with a velocity u at angle θ with the horizontal reaches a maximum height H_1. When it is projected with velocity u at an angle $(\pi/2-\theta)$ with the horizontal reaches maximum height H_2. The relation betwen the horizontal range R of the projectile, $H_1$and H_2 is [2000 M]

1) $R=4\sqrt{H_1H_2}$ 2) $R=4(H_1-H_2)$
3) $R=4(H_1+H_2)$ 4) $R=(H_1/H_2)^2$

95. A body is projected with a certain speed at angles of projection of θ and $90-\theta$, The maximum height attained in the two cases are 20m and10m respectively. The range for angle of projection θ is

1) $40\sqrt{2}$ m 2) $20\sqrt{2}$ m
3) $80\sqrt{2}$ m 4) $200\sqrt{2}$ m

96. Two bodies are thrown with the same inlitial velocity of 30 m/s. One at 17^0, other at 73^0 to the horizontal. The sum of the maximum heights reached by them is [g = 10 m/s^2]

1) 45 m 2) 450 m 3) 4.5 m 4) 20 m

<u>HORIZONTAL PROJECTILE MOTION</u>

97. A body is projected horizontally from the top of a tower of height 10m with a velocity 10 m/s. Its velocity after 1 second is

1) 10i - 10j 2) $5i+5\sqrt{2}j$
3) $15i+10\sqrt{2}j$ 4) $10\sqrt{2}i+5j$

98. In the above problem angle made by velocity vector with x – axis after 1/2 second is

1) $\tan^{-1}(2)$ 2) $\tan^{-1}(4)$
3) $\tan^{-1}(1/4)$ 4) $\tan^{-1}(1/2)$

99. A ball is projected horizontally with a velocity of $5ms^{-1}$ from the top of a tower of height 10m. When it reaches the ground ($g=10ms^{-2}$)

1) vertical component of its velocity is $15ms^{-1}$
2) Horizontal component of its velocity is $14ms^{-1}$
3) Its velocity is $15ms^{-1}$
4) Its velocity is $\sqrt{29}$ ms^{-1}

100. A body is projected horizontally from the top of a tower. When the body strikes the ground its velocity is V and the direction of motion makes 30° with the horizontal. The velocity of projection is

1) $\frac{\sqrt{3}v}{2}$ 2) $\frac{v}{\sqrt{3}}$ 3) $\frac{v}{2}$ 4) $\frac{2v}{\sqrt{3}}$

101. Two tall buildings are 40m apart. With what speed must a ball be thrown horizontally from a window 145 m above the ground in one building, so that it will enter a window 22.5m above the ground in other?

1) 4 m/s 2) 10 m/s 3) 8 m/s 4) 16 m/s

102. A bomb is dropped from an aircraft travelling horizontally at $150ms^{-1}$ at a height of 490m. The horizontal distance travelled by the bomb before it hits the ground is [96 E]

1) 1000 m 2) 1200 m 3) 1500 m 4) 1800 m

103. A body is thrown horizontally with a velocity of ϑ m/s from the top of a tower of height 2h reaches the ground in 't' seconds. If another body double the mass is thrown horizontally with a velocity 5ϑ from the top of another twoer of height 8h it reaches the ground in a time of

1) 2t 2) 4t 3) 6t 4) 8t

104. In the above problem if the first reaches the ground at a horizontal distance 'x' the second body reaches the ground at a horizontal distance

1) 5x 2) x/10 3) 10x 4) x/5

105. A body is projected horizontally from the top of a tower with a velocity of 20 m/s. After what time the vertical component of velocity is four times the horizontal component of velocity ? ($g = 10m/s^2$)

1) 16 sec 2) 8sec 3) 4 sec 4) 2 sec

106. A object is thrown horizontally from a point A. The ball hits the ground at a point B. the line of sight from A to B makes 60° with the horizontal. If $g = 10ms^{-2}$, the velocity of projection, if the time of flight is $\sqrt{3}$ sec. is

1) $10\ ms^{-1}$ 2) $5\ ms^{-1}$ 3) $10\sqrt{3}\ ms^{-1}$ 4) $\frac{10}{\sqrt{3}}\ ms^{-1}$

107. A body is projected horizontally from the top of a tower with a velocity of 30 m/s. the velocity of the body 4 seconds after projection is ($g = 10ms^{-2}$)

1) $40\ ms^{-1}$ 2) $20\ ms^{-1}$ 3) $50\ ms^{-1}$ 4) $100\ ms^{-1}$

108. A body thrown horizontally from the top of a tower touches the ground at a distance of 160m from the foot of the tower. If the velocity of projection is 40 m/s then the height of the towers is

1) 98 m 2) 9.6 m 3) 58.8 m 4) 78.4 m

109. Two paper screens P and Q are seperated by a distance of 100m. A bullet pierces P and Q the hole in Q is 10 cm below the hole in P. If the bullet is travelling horizontally at the time of hitting the screen P. Calculate the velocity of the bullet when it hits the screen P. (Neglect the resistance of paper and air).

1) 400 m/s 2) 500m/s 3) 700 m/s 4) 600m/s

110. A sphere rolls off the top of a stairway with a horizontal velocity of magnitude 200 cm/sec. The steps are 10cm high and 10cm wide. Which step will the ball hit first ? ($g=10m/s^2$)

1) 8 2) 2 3) 4 4) 6

111. The variation of velocity of a particle moving along a straight line is shown in the figure. The distance travelled by the particle in 4s is

1) 55m
2) 30m
3) 25m
4) 60m

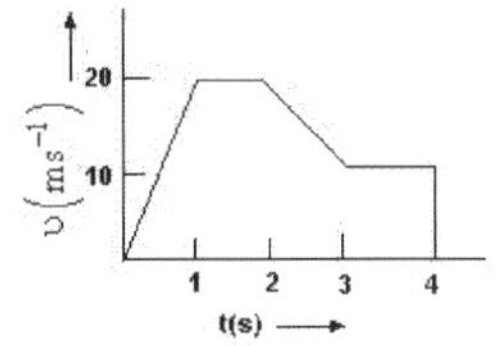

112. The graph of displacement time for two bodies travelling in a straight line are given. The ratio of velocities is

1) 3 : 1
2) 1 : 3
3) $\sqrt{3}$: 1
4) 1 : $\sqrt{3}$

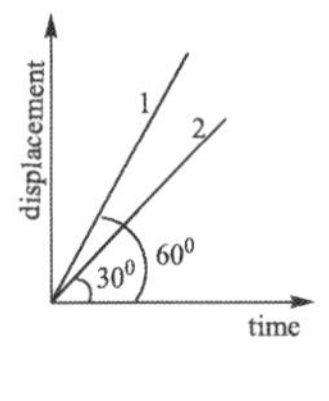

113. In the graph shown the displace-ment of the body in 5 seconds will be

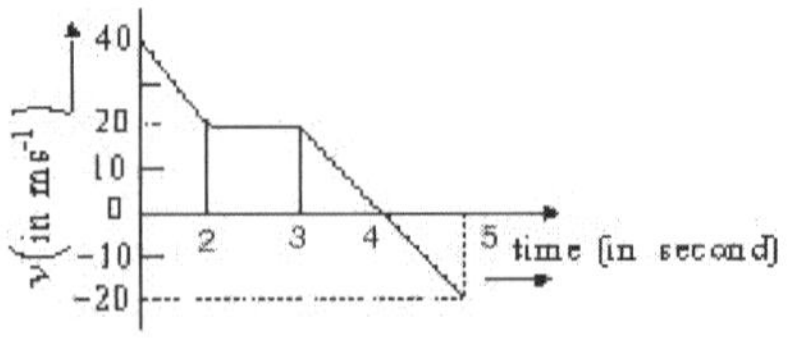

1) 60m 2) 40m
3) 80m 4) 100m

114. The velocity - time graph of a body is shown in fig. The ratio of the .....during the intervals OA and AB is ....

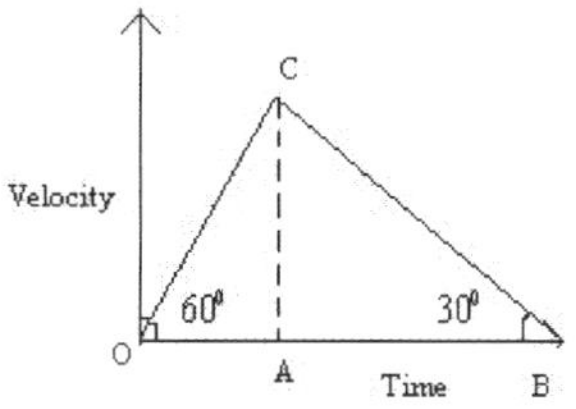

1) average velocities: 1 : 1   2) $\frac{OA}{AB}:\frac{1}{4}$

3) average accelerations, same as distances covered

4) distances covered : 1/2

115. Figure shows the acceleration - time graph of a particle. Which of the following represents the corresponding velocity-time graph ?

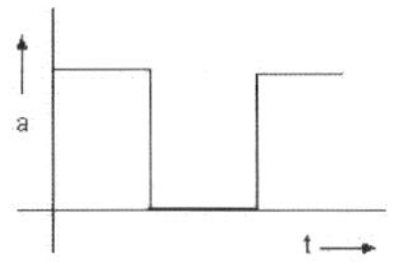

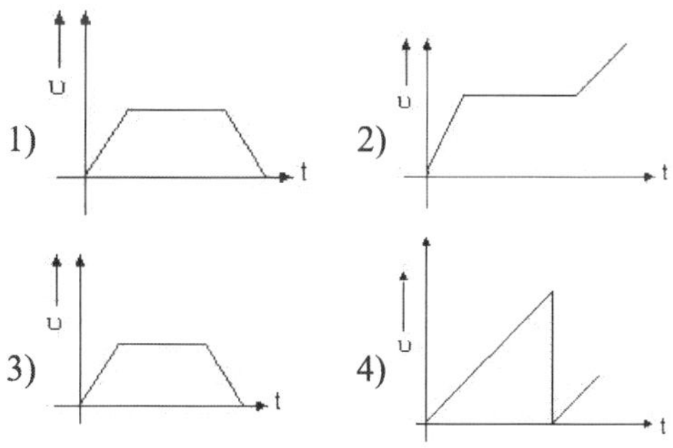

ANSWERS

1) 2	2) 3	3) 1	4) 4	5) 4
6) 4	7) 1	8) 3	9) 2	10) 4
11) 3	12) 3	13) 1	14) 4	15) 4
16) 1	17) 1	18) 1	19) 2	20) 2
21) 4	22) 4	23) 3	24) 2	25) 2
26) 1	27) 2	28) 2	29) 4	30) 1
31) 3	32) 2	33) 2	34) 4	35) 1
36) 2	37) 1	38) 3	39) 2	40) 1
41) 3	42) 1	43) 1	44) 1	45) 4
46) 3	47) 2	48) 3	49) 4	50) 2
51) 3	52) 2	53) 2	54) 2	55) 1
56) 4	57) 2	58) 3	59) 4	60) 4
61) 3	62) 2	63) 1	64) 1	65) 4
66) 1	67) 3	68) 4	69) 1	70) 3
71) 2	72) 1	73) 2	74) 3	75) 1
76) 1	77) 3	78) 3	79) 1	80) 2
81) 2	82) 2	83) 4	84) 1	85) 3
86) 4	87) 2	88) 4	89) 1	90) 1
91) 2	92) 2	93) 4	94) 1	95) 1
96) 1	97) 1	98) 4	99) 3	100) 1
101) 3	102) 3	103) 1	104) 3	105) 2
106) 2	107) 3	108) 4	109) 3	110) 1
111) 1	112) 1	113) 3	114) 1	115) 2

EXERCISE - III

HORIZONTAL MOTION

1. An express train moving at 30 m/s reduces its speed to 10 m/s in a distance of 240m. If the breaking force is increased by 12.5% in the beginning find the distance that it travels before coming to rest

1) 270 m   2) 240 m   3) 210 m   4) 195 m

2. A particle moving with uniform retardation covers distances 18m. 14m and 10m in successive seconds. It comes to rest after travelling a further distance of

1) 50m   2) 8 m   3) 12m   4) 42 m

3. A body is moving along the +ve X - axis with uniform acceleration of -4 ms^2. Its velocity at x =0 is 10 ms^{-1}. The time taken by the body to reach a point at x = 12m is

1) (2s, 3s)   2) (3s, 4s)   3) (4s, 8s)   4) (1s, 2s)

4. Four persons A, B, C and D initially at the corners of a square of side of length d. If every person starts moving with same speed v such that each one faces the other always, the person will meet after time

1) $\frac{d}{v}$ 2) $\frac{\sqrt{2}d}{v}$ 3) $\frac{d}{2v}$ 4) $\frac{d}{\sqrt{2}v}$

5. The coordinates of a moving particle at any time 't' are given by $x=\alpha t^3$ and $y=\beta t^3$. The speed of the particle at time 't' is given by [AIEEE-2003]

1) $3t\sqrt{\alpha^2+\beta^2}$ 2) $3t^2\sqrt{\alpha^2+\beta^2}$

3) $t^2\sqrt{\alpha^2+\beta^2}$ 4) $\sqrt{\alpha^2+\beta^2}$

6. Two cars 1 & 2 starting from rest are moving with speeds V_1 and V_2 m/s ($V_1 > V_2$). Car 2 is ahead of car '1' by 'S' metres when the driver of car '1' sees car '2'. What minimum retardation should be given to car '1' avoid collision [AIEEE-2002]

1) $\frac{V_1-V_2}{S}$ 2) $\frac{V_1+V_2}{S}$

3) $\frac{(V_1+V_2)^2}{2S}$ 4) $\frac{(V_1-V_2)^2}{2S}$

7. A particle moving with a velocity equal to 0.4 ms^{-1} is subjected to an acceleration of 0.15 ms^{-2} for 2 seconds in a direction at right angles to the direction of motion. The magnitude of the final velocity is

1) 0.3ms^{-1} 2) 0.4ms^{-1} 3) 0.5ms^{-1} 4) 0.7ms^{-1}

8. Two boys are standing at the ends A and B of a ground where AB = a. The boy at B starts runing in a direction perpendicular to AB with velocity v_1. The boy at A starts runing simultaneously with velocity v and catches the other boy in a time t, where t is

1) $a/\sqrt{v^2+v_1^2}$ 2) $\sqrt{a^2/\sqrt{v^2-v_1^2}}$

3) $a/(v-v_1)$ 4) $a/(v+v_1)$

9. An electron starting from rest has a velocity that increases linearly with the time that is v = kt, where k = 2m/sec^2. The distance travelled in the first 3 seconds will be

1) 9m 2) 16m 3) 27m 4) 36m

10. The x and y coordinates of a particle at any time t are given by $x=7t+4t^2$ and $y=5t$, where x and y are in metre and t in seconds. The acceleration of particle at t = 5s is

1) Zero 2) 8m/s^2 3) 20m/s^2 4) 40m/s^2

11. The position x of a particle varies with time t as $x = at^2 - bt^3$. The acceleration of the particle will be zero at time t equal to

1) $\frac{a}{b}$ 2) $\frac{2a}{3b}$ 3) $\frac{a}{3b}$ 4) Zero

12. The accelration 'a' in m/s^2 of a particle is given by $a = 3t^2 + 2t + 2$ where t is the time. If the particle starts out with a velocity u = 2m/s at t = 0, then the velocity at the end of 2 second is

1) 12m/s 2) 18m/s 3) 27 m/s 4) 36 m/s

13. The relation $3t=\sqrt{3x}+6$ describes the displacement of a particle in one direction where x is in metres and t in sec. The displacement, when velocity is zero, is

1) 24 metres 2) 12 metres 3) 5 metres 4) Zero

14. The acceleration of a particle is increasing linearly with time t as bt. The particle starts from the origin with an initial velocity v_0. The distnace travelled by the particle in time t will be

1) $v_0t+\frac{1}{3}bt^2$ 2) $v_0t+\frac{1}{2}bt^2$

3) $v_0t+\frac{1}{6}bt^3$ 4) $v_0t+\frac{1}{3}bt^3$

15. A car accelerates from rest at a constant rate α for some time, after which it decelerates at a constant rate β and comes to rest. If the total time elapsed is t, then the maximum velocity acquired by the car is

1) $\left(\frac{\alpha^2+\beta^2}{\alpha\beta}\right)t$ 2) $\left(\frac{\alpha^2-\beta^2}{\alpha\beta}\right)t$

3) $\frac{\alpha\beta t}{\alpha+\beta}$ 4) $\frac{(\alpha+\beta)t}{\alpha\beta}$

16. Velocity of a particle at time t = 0 is 2 ms^{-1}. A constant acceleration of 2 ms^{-2} acts on the particle for 1 second at an angle of 60^0 with its initial velocity. Find the magnitude of velocity at the end of 1 second.

1) $\sqrt{3}$ m/s 2) $2\sqrt{3}$ m/s

3) 4 m/s 4) 8 m/s

VERTICAL MOTION

17. A stone is dropped from a hill of height 180m. Two seconds later another stone is dropped from a point P below the top of the hill. If the two stones reach the ground simultaneously, the height of P from the ground is (g=10ms^{-2})

1) 100 m 2) 90 m 3) 80 m 4) 90 m

18. Water drops fall from the roof of a building 20m high at regular time intervals. If the first drop strikes the floor when the sixth drop begins to fall, the heights of the second and fourth drops from the ground at that instant are (g=10ms^{-2})

1) 12.8 m and 3.2 m 2) 12.8 m and 7.2 m

3) 19.2 m and 0.8 m 4) 7.2 m and 16.8 m

19. A body is dropped from a height of 16m. The body strikes the ground and losses 25% of its velocity. The body rebounds to a height of

1) 12 m 2) 9m 3) 4m 4) 8m

20. The distance travelled by a falling body in the last second of its motion, to that in the last but one second is 7 : 5, the velocity with which body strikes the ground is

1) 19.6 m/s 2) 39.2 m/s

3) 29.4 m/s 4) 49 m/s

21. A ball dropped from a point P crosses a point Q in t seconds. The time taken by it to travel from Q to R, if PQ = QR

1) t 2) $\sqrt{2}$ .t 3) 2 t 4) ($\sqrt{2}$ -1) t

22. A lead ball dropped into a lake from a diving board 5m above the water hits the water with certain velocity and then sinks to the bottom with the same constant velocity. If it reaches the bottom in 3s after it is dropped the depth of the lake is ($g=10ms^{-2}$)
1) 30 m 2) 15 m 3) 10 m 4) 20 m

23. Two balls are dropped from the same height at two different places A and B where the acceleration due to gravities are g_A and g_B. The body at 'B' takes 't' seconds less to reach the ground and strikes the ground with a velocity greater than at 'A" by υ m/s. Then the value of 'υ/t' is
1) $\frac{1}{\sqrt{g_A g_B}}$ 2) $2\sqrt{g_A g_B}$ 3) $\frac{1}{g_A g_B}$ 4) $\sqrt{g_A g_B}$

24. A body P is thrown vertically up with velocity $30\ ms^{-1}$ and another body Q is thrown up along the same vertically line with the same velocity but 1 second later from the ground. When they meet ($g=10ms^{-2}$)
1) P travels for 2.5 s 2) Q travels for 3.5s
3) P travels for 3.5 s 4) Q travels for 1s

25. A boy sees a ball go up and then down through a window 2.45m high. If the total time that ball is in sight in 1s, the height above the window the ball rises is approximately
1) 2.45 m 2) 4.9 m 3) 0.3 m 4) 0.49 m

26. From an elevated point P a stone is projected vertically upward. When it reaches a distance y below the point of projection its velocity is double the velocity when it was at a height y above p. The greatest height reached by it above P is
1) $\frac{2y}{3}$ 2) $\frac{5y}{3}$ 3) $\frac{y}{3}$ 4) 2y

27. A particle is projected vertically up and another is let fall to meet at the same instant. If they have velocities equal in magnitude when they meet, the distances travelled by them are in the ratio of
1) 1 : 1 2) 1 : 2 3) 3 : 1 4) 2 : 3

28. A ball is projected from the bottom of a tower and is found to go above the tower and is caught by the thrower at the bottom of the tower after a time interval t_1. An observer at the top of the tower sees the same ball go up above him and then come back at this level in a time interval t_2. The height of the tower is
1) $\frac{1}{2}gt_1t_2$ 2) $\frac{gt_1t_2}{8}$
3) $\frac{g}{8}(t_1^2-t_2^2)$ 4) $\frac{g}{2}(t_1-t_2)^2$

29. A stone is dropped from a height of 10cm above the top of a window 80 cm high. The time taken by the stone to cross the window is ($g = 9.8\ ms^{-2}$)
1) $\frac{1}{7}$s 2) $\frac{3}{7}$s 3) $\frac{2}{7}$s 4) $\frac{4}{7}$s

30. A vertical disc has three grooves directed along chords AB, AC and AD. Three bodies begin to slide down the respective grooves from A simultaneously. If AB>AC>AD, the respective time intervals to reach the bottoms of the respective grooves t_1,t_2 and t_3 are
1) $t_1>t_2>t_3$ 2) $t_1<t_2<t_3$
3) $t_1>t_2<t_3$ 4) $t_1=t_2=t_3$

31. If a stone dropped from the top of a tower travels half of the height of the tower during last second of its fall, the time of fall is (in seconds)
1) $3+\sqrt{2}$ 2) $4+\sqrt{2}$ 3) $2-\sqrt{2}$ 4) $2+\sqrt{2}$

32. A parachutist drops freely from an aeroplane for 8s before the parachute opens out. Then he descends with a net retardation of $2ms^{-2}$ reaching the ground with a velocity of $6ms^{-1}$. The height from which he bails out of the aeroplane is ($g = 10ms^{-2}$)
1) 1929 m 2) 1911 m 3) 2195 m 4) 2000m

33. Two balls are projected simultaneously with the same speed from the top of a tower - one upwards and the other downwards. If they reach the ground in 6s and 2s, the height of the tower is ($g=10ms^{-2}$)
1) 120 m 2) 60 m 3) 80 m 4) 30 m

34. A girl is standing on the top edge of an 18 m high building. She tosses a coin upward with a speed of $7.0\ ms^{-1}$. How long does it take for the coin to hit the ground ? How fast is the coin going just before it strikes the ground ? ($g = 10\ ms^{-2}$)
1) 2.70s, 20 m/s 2) 1.70s, 30 m/s
3) 4.70s, 50m/s 4) 10s, 50 m/s

35. Ball A is dropped from the top of a building . At the same instant another identical ball B is thrown vertically upward from the ground. When the balls collide, they are moving in opposite directions and the speed of a is twice the speed of B. At what fraction of the height of the building did the collision occur ?
1) 2/3 2) 4/3 3) 5/3 4) 1/3

36. **An object falls from a bridge that is 45m above the water. It falls directly into a small row - boat moving with constant velocity that was 12m from the point of impact when the object was released. What was the speed of the boat ? (g = 10 ms^{-2})**

1) 2 m/s 2) 3 m/s 3) 5 m/s 4) 4 m/s

37. **Two balls are dropped to the ground from different heights. One ball is ddropped 2.0s after the other, but both strike the ground at the same time 5.0 s after the 1st is dropped.**

a) What is the difference in the heights from which they were dropped ?

b) From what height was the first ball dropped ?

1) 78.4m, 122.5m 2) 80 m, 250m
3) 40m, 80m 4) 140m, 180m

OBLIQUE PROJECTION

38. **A projectile is thrown at an angle of 30° with a velocity of 10m/s. the change in velocity during the time interval in which it reaches the highest point is**

1) 10 m/s 2) 5 m/s 3) $5\sqrt{3}$ m/s 4) $10\sqrt{3}$ m/s

39. **A body projected obliquely with velocity 19.6 ms^{-1} has its kinetic energy at the maximum height equal to 3 times its potential energy there. Its position after 1second of projection from the ground is (h = maximum height)**

1) h/2 2) h/4 3) h/3 4) h

40. **A player kicks a foot ball obliquely at a speed of 20ms^{-1} so that its range is maximum. Another player at a distance of 24m away in the direction of kick starts running at that instant to catch the ball. Before the ball hits the ground to catch it, the speed with which the second player has to run is (g = 10ms^{-2})**

1) 4 ms^{-1} 2) $4\sqrt{2}$ ms^{-1} 3) $8\sqrt{2}$ ms^{-1} 4) 8 ms^{-1}

41. **It is possible to project a particle with a given speed in two possible ways so that it has the same horizontal range 'R'. The product of time taken by it in the two possible ways is [2001 M]**

1) R/g 2) 2R/g 3) 3R/g 4) 4R/g

42. **The range of a projectile, when launched at an angle of 15° with the horizontal is 1.5 km. The additional horizontal distance the projectile would cover when projected with same velocity at 45° is**

1) 3km 2) 4.5 km 3) 1.5 km 4) 2.5 km

43. **A ball A is projected from the ground such that its horizontal range is maximum. Another ball B is dropped from a height equal to the maximum range of A. The ratio of the time of flight of A to the time of fall of B is**

1) $\sqrt{2}:1$ 2) 1 : 2 3) 1 : 1 4) $1:\sqrt{2}$

44. **A particle is projected from the ground with velocity u making an angle θ with the horizontal. At half of its maximum heights,**

1) its horizontal velocity is u cos θ

2) its vertical velocity is $\frac{u\,Sin\theta}{\sqrt{2}}$

3) its velocity is $u\left(\frac{1+Cos^2\theta}{2}\right)^{\frac{1}{2}}$

4) all the above are true

45. **A projectile has the maximum range of 400m. If it is pushed up smooth inclined plane of angle 30° with the same speed, the distance travelled by it along the inclined plane will be**

1) 200 m 2) 100 m 3) 400 m 4) 50 m

46. **A hose pipe lying on the ground shoots a stream of water upward at an angle 60° to the horizontal at a speed of 20 ms^{-1}. The water strikes a wall 20m away at a height of (g=10ms^{-2})**

1) 14.64 m 2) 7.32 m 3) 29.28 m 4) 10 m

47. **A particle is projected with velocity $2\sqrt{gh}$ and at an angle 60° to the horizontal so that it just clears two walls of equal height h which are a distance 2 h from each other. The time interval for which the particle travels between this two walls is**

1) $2\sqrt{\frac{h}{g}}$ 2) $\sqrt{\frac{h}{g}}$ 3) $\sqrt{\frac{2h}{g}}$ 4) $\sqrt{\frac{h}{2g}}$

48. **A particle is aimed at a mark which is in the same horizontal plane as that of point of projection. If falls 10 m short of the target when it is projected of an angle of 75° and falls 10m ahead of the target when it is projected with an elevation of 45°. The angle of projection for which the particle exactly hits this target is (g=10ms^{-1})**

1) $\frac{1}{2}Sin^{-1}\left(\frac{3}{4}\right)$ 2) $\frac{1}{2}Sin^{-1}\left(\frac{4}{5}\right)$

3) $\frac{1}{2}tan^{-1}\left(\frac{1}{2}\right)$ 4) tan^{-1} (2)

49. From a point on the ground at a distance 2m from the foot of a vertical wall, a ball is thrown at an angle of 45° of which just clears the top, of the wall and afterward strikes the ground at a distance 4m on the otherside. The height of the wall is

1) 2/3 m 2) 3/4 m 3) 1/2 m 4) 4/3m

50. A body is projected with a velocity u at an angle of 60° to the horizontal. The time interval after which it will be moving in a direction of 30° to the horizontal is

1) $\frac{u}{\sqrt{3}g}$ 2) $\frac{\sqrt{3}u}{g}$ 3) $\frac{\sqrt{3}u}{2g}$ 4) $\frac{2u}{\sqrt{3}g}$

51. When a body is projected from a level ground the ratio of its speed in the vertical and horizontal direction is 4 : 3. If the velocity of projection is u, the time after which, the ratio of the velocities in the vertical and horizontal directions are reversed is

1) $\frac{7u}{20g}$ 2) $\frac{35u}{10g}$ 3) $\frac{9u}{g}$ 4) $\frac{10u}{g}$

52. Two balls are projected simultaneously in same verticle plane from the same point with velocities V_1 and V_2 with angles θ_1 and θ_2 respectively with the horizontal. If $V_1 \cos\theta_1 = V_2 \cos\theta_2$, the path of one ball as seen from the position of another ball is [05 M]

1) parabola 2) horizontal straight line

3) vertical straight line

4) straight line making 45° with vertical

53. A body of mass 2kg is projected from the ground with a velocity $20ms^{-1}$ at an angle 30° with the verticle. If t_1 is the time in seconds at which the body is projected and t_2 is the time in seconds at which it reaches the ground, the change in momentum in $kgms^{-1}$ during the time $(t_2 - t_1)$ is [06 -AM]

1) 40 2) $40\sqrt{3}$ 3) $50\sqrt{3}$ 4) 60

54. A projectile has initially the same horizontal velocity as it would acquire if it had moved from rest with uniform acceleration of $3ms^{-2}$ for 0.5 minutes. If the maximum height reached by it is 80m then the angle of projection is $[g = 10ms^{-2}]$ [2002 M]

1) $\tan^{-1}(3)$ 2) $\tan^{-1}\left(\frac{3}{2}\right)$

3) $\tan^{-1}\left(\frac{4}{9}\right)$ 4) $\sin^{-1}\left(\frac{4}{9}\right)$

55. A body of mass m_1 projected vertically upwards with an initial velocity 'u' reaches a maximum height 'h'. Another body of mass m_2 is projected along an inclined plane making an angle 30^0 with the horizontal and with speed 'u'. The maximum distance travelled along the incline is [2001 M]

1) 2h 2) h 3) $\frac{h}{2}$ 4) $\frac{h}{4}$

56. The maximum distance to which a man can throw a ball by projecting it horizontally from a height h is h. The maximum distance to which he can throw it vertically up is

1) h 2) 2h 3) h/2 4) h/4

57. A body projected at 45° with a velocity of 20 m/s has a range of 10m. The decrease in range due to air resistance is $(g = 10ms^{-1})$

1) 0 2) 10 m 3) 20 m 4) 30 m

58. Two seconds after projection, a projectile is moving at 30° above the horizontal. After one more second it is moving horizontally. If $g = 10ms^{-2}$, the velocity of projection is

1) $10\sqrt{3}\ ms^{-1}$ 2) 20 m/s

3) 10 m/s 4) $20\sqrt{3}$ m/s

59. A stone is projected from the top of a tower with velocity $20ms^{-1}$ making an angle 30° with the horizontal. If the total time of flight is 5s and $g=10ms^{-2}$,

1) the height of the tower is 75m

2) the maximum height of the stone from the ground is 80m

3) both of the above are true

4) none of the above is true

***60.** A body is thrown with the velocity $u = 12.0\ ms^{-1}$ at an angle of $\alpha = 45^0$ to the horizon. It touches the ground at the distance s from the point where it was thrown. From what height h should stone be thrown in a horizontal direction, with the same initial velocity u so that it falls at the same spot ?

1) 17 m/s 2) 1.7 m/s 3) 3.4 m/s 4) 7.3 m/s

***61.** A ball is tossed from an upper storey window of a building. The ball is given an initial velocity of $8.00\ ms^{-1}$ at an angle of 20.0^0 below the horizontal. It strikes the ground 3.00 s later.

a) How far horizontally from the base of the building does the ball strike the ground ?

b) Find the height from which the ball was thrown.

c) How long does it take the ball to reach a point 10.0 m below the level of launching ?

1) 22.6m, 52.3m, 1.18s

2) 2.26m, 5.23m, 0.118s

3) 5m, 6m, 0.1sec 4) 4m, 10m, 2 sec

HORIZONTAL PROJECTILE MOTION

62. A fighter plane flying horizontally at an altitude of 2 km with speed of 540 kmph passes directly over head an antiaircraft gun. If the gun can fire a bullet at the muzzle speed of 500 ms^{-1}, at what angle with the vertical the gun should fire the bullet so that the bullet hits the plane ?

1) $\cos^{-1}(3/10)$ 2) $\sin^{-1}(3/10)$

3) $\tan^{-1}(10/3)$ 4) 45°

63. At a certain height a body at rest explodes into two equal fragments with one fragment receving a horizontal velocity of 10 ms^{-1}. The time interval after the explosion for which the velocity vectors of the two fragments become perpendicular to each other is ($g=10ms^{-2}$)

1) 1s 2) 2 s 3) 1.5 s 4) 1.75 s

64. In this above problem the horizontal distance between the two fragments when their velocity vectors are perpendicular to each other is

1) 10 m 2) 20m 3) 40m 4) 50m

65. In the above problem, the time taken by the displacement vectors of the two fragments to become perpendicular to each other is

1) 1s 2) 1.5 s 3) 2s 4) 4s

66. In the above problem, this horizontal distance between the two fragments when their position vectors are perpendicular to each other is

1) 40 m 2) 20m 3) 10m 4) 5m

67. At a certain height a shell at rest explodes into two equal fragments. One of the fragments receives a horizontal velocity u. The time interval after which, the velocity vectors will be inclined at 120° to each other is

1) $\frac{u}{\sqrt{3}g}$ 2) $\frac{\sqrt{3}u}{g}$ 3) $\frac{2u}{\sqrt{3}g}$ 4) $\frac{u}{2\sqrt{3}g}$

68. A bomb at rest at the summit of a cliff breaks into two equal fragments. One of the fragments attains a horizontal velocity of $20\sqrt{3}$ ms^{-1}. The horizontal distance between the two fragments, when their displacement vectors is inclined at 60° relative to each other is ($g = 10ms^{-2}$)

1) $40\sqrt{3}$ m 2) $80\sqrt{3}$ m 3) $120\sqrt{3}$ m 4) $480\sqrt{3}$ m

69. A particle starts from rest. Its acceleration (a) versus time(t) is as shown in the figure. The maximum speed of the particle will be

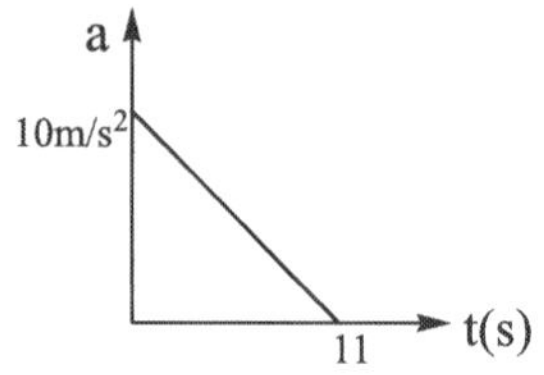

1) 110 m/s

2) 55 m/s

3) 550 m/s

4) 660 m/ss

70. The velocity time graph of a body moving in a straight line is shown in the figure. The displacement and distance travelled by the body in 6 sec are respectively.

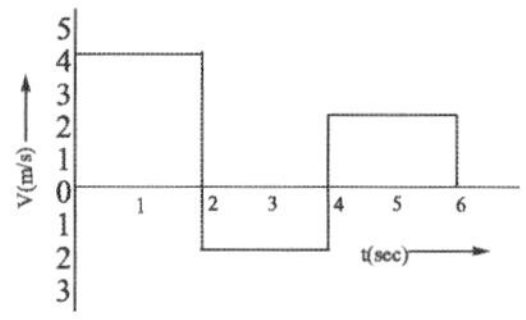

1) 8 m, 16 m

2) 16m, 8m

3) 16 m, 16m

4) 8 m, 8m

71. For the velocity - time graph shown in figure below the distance covered by the body in last two seconds of its motion is what fraction of the total distance covered by it in all the total distance covered by it in all the seven seconds

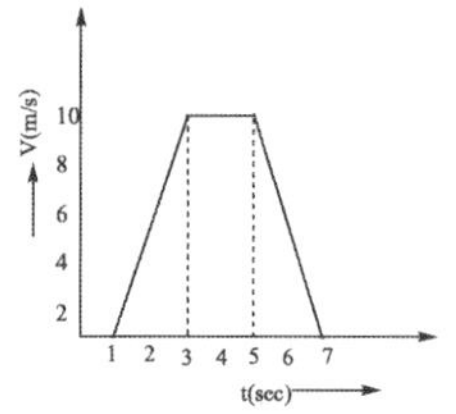

1) $\frac{1}{2}$ 2) $\frac{1}{4}$

3) $\frac{1}{3}$ 4) $\frac{2}{3}$

72. A particle starts from rest at time t=0 and moves on a straight line with acceleration as plotted in fig. The speed of the particle will be maximum after time.

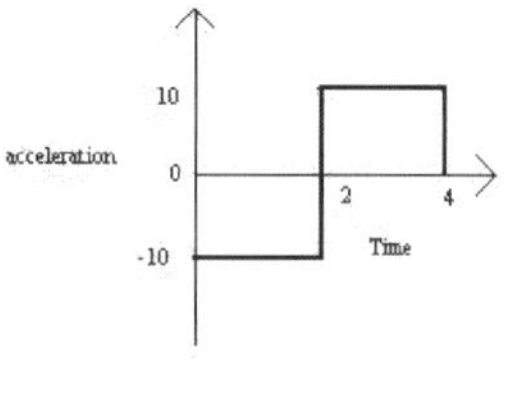

1) 1 s

2) 2 s

3) 3 s

4) 4 s

73. **Velocity - time (v-t) graph for a moving object is shown in the figure. Total displacement of the object during the time interval when there is non - zero acceleration and retardation is**

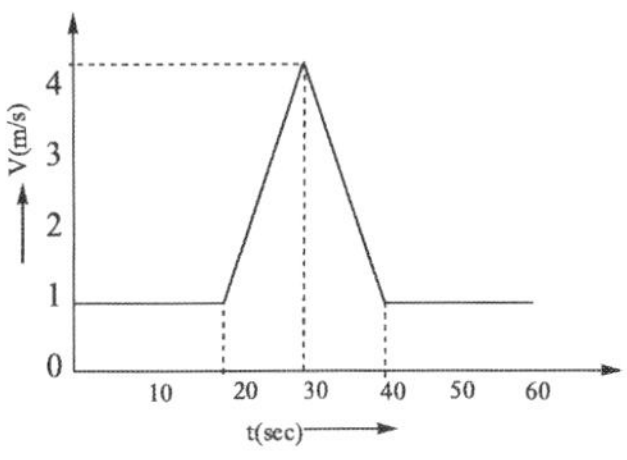

1) 60m 2) 50 m 3) 30 m 4) 40 m

74. **A body is at rest at x = 0, At t = 0, it starts moving in the positive x-direction with a constant acceleration. At the same instant another body passes through x = 0 moving in the positive x-direction with a constant speed. The position of the first body is given by $x_1(t)$ and the position of the second body is $x_2(t)$ after the same time interval. Which of the following graphs correctly describes (x_1-x_2) as a function of time 't'?** **(AIEEE2008)**

1) 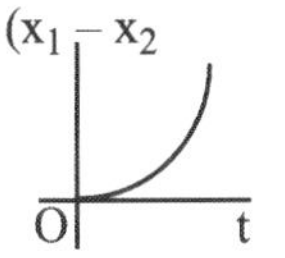

2) 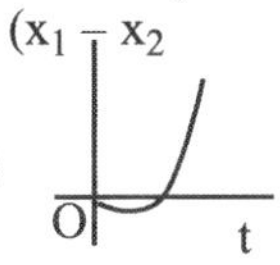

3) 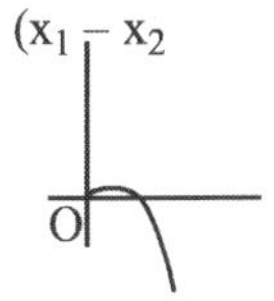

4) 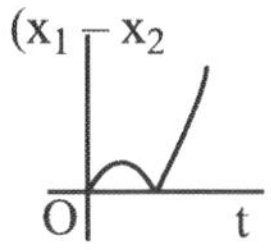

75. **A particle has an initial velocity $3\hat{i}+4\hat{j}$ and an acceleration of $0.4\hat{i}+0.3\hat{j}$. Its speed after 10s is** **(AIEEE-2009)**

1) 10 units 2) $7\sqrt{2}$ units
3) 7 units 4) 8.5 units

76. **A particle is moving with velocity $\vec{\upsilon}=K\left(y\hat{i}+x\hat{j}\right)$, where K is a constant. The general equation for its path is (AIEEE2010)**

1) $y^2=x^2+\text{constant}$ 2) $y=x^2+\text{constant}$
3) $y^2=x+\text{constant}$ 4) $xy=\text{constant}$

ANSWERS

1) 2	2) 2	3) 1	4) 1	5) 2
6) 4	7) 3	8) 2	9) 1	10) 2
11) 3	12) 2	13) 2	14) 3	15) 3
16) 2	17) 3	18) 4	19) 2	20) 2
21) 4	22) 4	23) 4	24) 3	25) 3
26) 2	27) 3	28) 3	29) 3	30) 4
31) 4	32) 2	33) 2	34) 1	35) 1
36) 4	37) 1	38) 2	39) 4	40) 2
41) 2	42) 3	43) 3	44) 4	45) 3
46) 1	47) 1	48) 1	49) 4	50) 1
51) 1	52) 3	53) 2	54) 3	55) 1
56) 4	57) 4	58) 4	59) 3	60) 4
61) 1	62) 2	63) 1	64) 2	65) 3
66) 1	67) 1	68) 4	69) 2	70) 1
71) 2	72) 2	73) 2	74) 2	75) 2
76) 1				

COMPREHENSION QUESTIONS

Comprehension - I :

A body is projected horizontally with a velocity u of a tower of height h. If the straiking angle is 60^0 with horizontal then

1. **Time of fall is**

1) $\sqrt{\frac{h}{g}}$ 2) $\sqrt{3}\frac{h}{g}$ 3) $\sqrt{\frac{h}{2g}}$ 4) $\sqrt{3}\frac{u}{g}$

2. **The velocity of the body while striking the ground is**

1) $\sqrt{2gh}$ 2) $u+\sqrt{2gh}$
3) $\sqrt{u^2-2gh}$ 4) $\sqrt{u^2+2gh}$

3. **The ratio of vertical component of velocity to resultant velocity at the time of striking is**

1) $\sqrt{3}:4$ 2) $4:\sqrt{3}$
3) $\sqrt{3}:2$ 4) $2:\sqrt{3}$

Comprehension - II :

A ball is projected vertically upwards with velocity of 24.5 m/s from the bottom of a tower. A boy, who is standing at the top of the tower is unable to catch the ball when it passes him in the upward dirction. But the ball again reaches him after 3 sec., when it is falling and then he catches it. (g=9.8 m/s^2)

4. **The maximum height reached the ball is**

1) 29.4 m 2) 30.625 m
3) 25 m 4) 19.6 m

5. **The velocity of the ball when the boy got it is**

1) 9.8 m/s 2) 14.7 m/s
3) 24.5 m/s 4) 19.6 m/s

6. The height of the tower is
   1) 9.8 m 2) 4.9 m
   3) 19.6 m 4) 29.4 m

Comprehension - III :

Two particles A and B are projected from the same point with equal angle 'θ' from the horizontal. A has the velocity of projection equal to V and B has projection velocity double that of A.

7. The time after which the velocity vectors would be parallel (or) antiparallel.
   1) $\frac{4V\sin\theta}{3g}$ 2) $\frac{4V}{4g}$
   3) $\frac{V Cos\theta}{2g}$ 4) $\frac{V Sin\theta}{4g}$

8. The horizontal separation between the two when the velocity vectors are anti parallel is
   1) $\frac{4V^2 Sin2\theta}{g}$ 2) $\frac{4V^2 Cos2\theta}{g}$
   3) $\frac{3V^2 Sin2\theta}{g}$ 4) $\frac{2V^2 Sin2\theta}{g}$

9. The ratio of the range of two particles is
   1) 2:1 2) 1:2 3) 4:1 4) 1:4

Comprehension - IV :

The trajectory of a projectile in a vertical plane is $y=Ax-Bx^2$, where A and B are constants and x and y are respectively the horizontal and vertical distance of the projectile from the point of projection.

10. The maximum height reached by the projectile is
    1) $\frac{A^2}{2B}$ 2) $\frac{A^2}{B}$ 3) $\frac{A}{B}$ 4) $\frac{A^2}{4B}$

11. The range of the projectile is
    1) $\frac{B}{A}$ 2) $\frac{B^2}{A}$ 3) $\frac{A}{B}$ 4) $\frac{B}{A^2}$

12. The time of flight of projectile is
    1) $\frac{A}{B}\sqrt{\frac{2}{g}}$ 2) $\frac{B}{A}\sqrt{\frac{2}{g}}$
    3) $A\sqrt{\frac{2}{Bg}}$ 4) $B\sqrt{\frac{2}{Ag}}$

Comprehension - V :

A ball is projected horozontally with a velocity 'u' from the top of a building of height 'h'. The ball reached the ground at a horizontal distance 'd' from the bottom of building. If the line joining the point of projection to the point where it hits the ground makes an angle of 'α' to the horizontal.

13. The time taken by the ball to reach the ground is
    1) $\frac{h}{u}$ 2) $\frac{d}{u}$ 3) $\sqrt{\frac{2h}{g}}$ 4) $\sqrt{\frac{h}{g}}$

14. The initial velocity(u) of the ball at the point of projection is
    1) $\frac{d}{h}$ 2) $d\sqrt{\frac{2h}{g}}$ 3) $d\sqrt{\frac{g}{2h}}$ 4) $d\sqrt{\frac{2g}{h}}$

15. The angle (α) made by velocity vector with the horizontal after time 't' sec is
    1) $\tan^{-1}\left(\frac{u}{gt}\right)$ 2) $\tan^{-1}\left(\frac{ug}{t}\right)$
    3) $\tan^{-1}\left(\frac{tu}{g}\right)$ 4) $\tan^{-1}\left(\frac{gt}{u}\right)$

ANSWERS

1) 4	2) 4	3) 3	4) 2	5) 2
6) 3	7) 1	8) 4	9) 4	10) 4
11) 3	12) 3	13) 3	14) 3	15) 4

Chapter - 4

ROTATORY MOTION

ROTATIONAL MECHANICS

SYNOPSIS

1. In translatory motion, every particle travels the same distance along parallel paths, which may be straight or curved. Every particle of the body has the same speed and acceleration.
2. In rotatory motion the body rotates about a fixed axis. Every particle of the body describes a circular path and centres of concentric circles lie on the axis of rotation. Every particle of the body undergoes the same angular displacement, but linear velocity of particles differ depending on their distance from axis of rotation.

UNIFORM CIRCULAR MOTION :

1) If an object moves in a circular path with constant speed, its motion is called uniform circular motion.
2) The line perpendicular to plane of motion of the body about which the body rotates is called axis of rotation.
3) The line joining the instantaneous position of the body and centre of the circle is called radius vector.
4) The angle swept by the radius vector in a given interval of time is called angular displacement θ .
   a) S.I unit of θ is radian.
   b) Finite angular displacements are scalars.
   c) Infinitesimally small angular displacements are vectors
   d) This is a pseudo vector.
   e) $\theta = 2\pi$ N rad, N-Number of rotations.
5) Angular velocity 'ω'
   a) Rate of angular displacement is called angular velocity.
   b) S.I unit is rad s^{-1}
   c) $\omega = \theta/t$ or $\omega = d\theta/dt$.
   d) If T is time period and n is frequency then
   $\omega = 2\pi n = 2\pi/T$
   e) If the particle revolving along a circle with uniform speed completes n revolutions in t sec, then
   $\omega = 2\pi n/t$
   f) If $\bar{v}$ is the linear velocity of the body
   $\bar{v} = \bar{\omega} \times \bar{r}$
   $v = r\omega \sin\theta$
   $v = r\omega$ when $\theta = 90^0$
   g) Other units of ω are revolutions per minute(rpm). revolutions per second(rps)
   1 rps = 60 rpm
   1 rpm = $\pi/30$ rad s^{-1}
   1 rps = 2π rad s^{-1}
   h) When a body is rotating about its own axis, then the particles on the axis of rotation have "zero" linear velocity
6) The angular velocity is a pseudo vector. The direction is perpendicular to the plane of rotation and given by the right hand thumb rule. The direction of angular velocity is same as that of the angular displacement.
7) Angular accelaration α :
   a) Rate of change of angular velocity is called angular acceleration $\alpha = \dfrac{d\omega}{dt}$
   b) S.I unit is rad.s^{-2}
   c) The direction of $\bar{\alpha}$ is parallel to the axis of rotation.
   d) If angular velocity of a rotating body increases at uniform rate from ω_1 to ω_2 in t sec then

   $$\omega_2 - \omega_1 = \alpha t \text{ (or) } \alpha = \frac{\omega_2 - \omega_1}{t}$$

   $$\omega_2^2 - \omega_1^2 = 2\alpha\theta \text{ (or) } \alpha = \frac{\omega_2^2 - \omega_1^2}{2\theta}$$

   $$\theta = \omega_1 t + \frac{1}{2}\alpha t^2$$

   e) It is produced only when an unbalanced torque acts on the body
   f) When angular velocity of the body is constant $\alpha = 0$
   g) If 'a' is the tangential acceleration then $a = r\alpha$. Where r is the radius of the circular path.
8) Centripetal acceleration (a_r)
   a) The accelaration possessed by a body moving in a circular path and directed towards the centre is called centripetal or radial accelaration.
   b) $a = v\omega = v^2/r = r\omega^2$, (in m/s^2) where v, ω and r are the linear velocity, angular velocity of the body and 'r' radius of the circle
   c) It arises due to change in the **direction of velocity** of the body.
   d) $\bar{a}_r = \bar{\omega} \times \bar{v}$
9) Tangential acceleration 'a_T'
   a) If a body moves in a circular path with changing speed it possesses (i) tangential accelaration(a_T) (ii) radial accelaration (a_r) and (iii) angular acceleration (α)
   b) $a_T = dv/dt$. (in ms^{-2})
   c) It arises due to **change in speed** of the object

d) A car going in a circular track with increasing speed possesses both a_r and a_T.

e) Resultant linear accelaration of the body

$$a = \sqrt{a_r^2 + a_T^2}$$

If α is the angle made by 'a' with a_r, then

$$\operatorname{Tan}\alpha = \frac{a_T}{a_r}$$

10) Centripetal force (F_N)

a. It is the force required to keep the body in circular motion. This force changes the direction of linear velocity but not its magnitude.

b. Direction of centripetal force is always perpendicualar to the direction of linear velocity.

c. Centripetal force = mass X centripetal acceleration.

d. Centripetal force

$$F_N = \frac{mv^2}{r} = mr\omega^2 = mv\omega = \frac{4\pi^2 mr}{T^2} = 4\pi^2 n^2 mr$$

e. Tangential force $F_t = mr\,\alpha$

f. For a body in non uniform circular motion,

resultant force $F = \sqrt{F_N^2 + F_t^2}$

$$= \sqrt{\left(mr\omega^2\right)^2 + \left(mr\alpha\right)^2} = mr\sqrt{\omega^4 + \alpha^2}$$

g. When an electron moves round the nucleus in a circular orbit, the electrostatic force between the electron and nucleus is the centripetal force.

h. If an electron of mass 'm' and charge 'e' moves around the nucleus of atomic number 'Z' in a circular orbit of radius 'r', centripetal force on it is

$$\frac{mv^2}{r} = \frac{1}{4\pi\epsilon_0}\frac{Ze^2}{r^2}$$

i. When the earth moves round the sun the gravitational force of attraction between them is the centripetal force.

j. When a planet of mass 'm' moves around the sun in a circular orbit of radius r,

centripetal force on it $\dfrac{mv^2}{r} = \dfrac{GMm}{r^2}$

k. Work done by centripetal force is zero.

l. When a stone is whirled round in horizontal circle by attaching it at the end of a string, tension in the string provides the centripetal force.

m. If a stone of mass m attached at the end of a string is whirled in a horizontal circle of radius 'ℓ', tension in the string is equal to centripetal force

$$T = \frac{mv^2}{\ell}$$

n. If the centripetal force acting on a body moving in a circle vanishes suddenly, it flies off along the tangent at that point.

o. If the stone tied at the end of a string is whirled and the string breaks, then the stone flies off tangentially.

p. A stone is tied at the end of a string of length 'ℓ' having breaking tension 'F'. Then the safe maximum angular velocity which prevents the breaking off the string is $\omega = \sqrt{\dfrac{F}{m\ell}}$

q. If E is the kinetic energy of a body moving in a circular path of radius 'r' with uniform speed, the necessary centripetal force provided is $F = \dfrac{2E}{r}$

11) **Centrifugal force**

a) The Pseudo force which acts radilly outwards on the body moving along a circle is called centrifugal force.

b) Even though centripetal and centrifugal forces are equal in magnitude and opposite in direction, they do not form action - reaction pair. They act on the same body in two different frames.

c) An observer can observe centrifugal force when he is also moving along with the body which is moving along a circle.

d) Magnitude of centrifugal force is equal to that of centripetal force

e) Every person on the earth is moving in a circular path along with the spinning earth.

So, every person experiences a centrifugal force.

g) Due to centrifugal force the earth is bulged at the equator and flattend at the poles.

h) Watt's governer is used to control the flow of steam in the steam engine. It works on the principal of centrifugal force.

i) Cream can be separated from milk using centrifuge. Wet clothes can be dried by the centrifuge.

MOTION IN A VERTICAL CIRCLE WITH UNIFORM SPEED :

A particle of mass m is rotated at uniform speed 'V' in a vertical circle of radius r with the help of a light string.

a) Then the sum of its KE and PE is not constant

b) If T_1, T_2, T_0 are the tensions in the string when the particle is at lowest point, highest point of the vertical circle and at an angle θ from lowest point of vertical circle respectively, then

$$T_1 = \frac{mv^2}{r} + mg \;, T_2 = \frac{mv^2}{r} - mg$$

$$T_0 = \frac{mv^2}{r} + mg\cos\theta$$

c) The difference between maximum and minimum tensions in the string $T_1 - T_2 = 2mg$

d) A vehicle of mass m is travelling with uniform speed V along a concave road and then a convex road of same radius of curvature r. If R_1 and R_2 are the normal reactions on the vehicle at the lowest point of the concave road and highest point of the convex road respectively then

$$R_1 = mg + \frac{mv^2}{r} \text{ and } R_2 = mg - \frac{mv^2}{r}$$

e) Maximum speed of the vehicle at the highest point of convex shaped bridge so that the vehicle does not leave the surface of the bridge is

$$V_{max} = \sqrt{gr}$$

f) In the following cases, the minimum velocity of the body at the highest point should be $\sqrt{gr}$

i) To rotate a can full of water, without water spilling out even at the highest position.

ii) In circus, for a motor cyclist to drive the cycle in a vertical circle inside a cage.

iii) For the pilot of an aeroplane who is not tied to his seat, not to fall down while looping a vertical circle.

MOTION IN A VERTICAL CIRCLE WITH NON - UNIFORM SPEED :

1. A particle of mass 'm' suspended by a thread is given a horizontal speed 'u', when it is at 'A', it moves in a vertical circle of radius 'r'

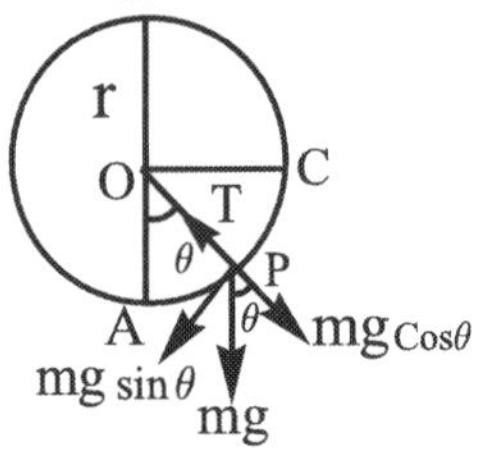

2. When the angular displacement of the particle is 'θ', ie when the particle is at p.

a) Speed of the particle

$v = \sqrt{u^2 - 2gr(1 - Cos\,\theta)}$ where u is the velocity at A, the lowest point

b) Centripetal force $mv^2/r = T - mg\cos\theta$

c) The speed of the particle continuously changes. It increase while coming down and decreases while going up

d) This is an example for non-uniform circular motion.

e) Tangential accelaration = g sin θ

f) Tangential force = mg sin θ

g) Tension in the string $T = mv^2/r + mg\,Cos\,\theta$

h) Velocity, speed, K.E, linear momentum, angular momentum, angular velocity, all are variables. Only total energy remains constant

i) if V_A and V_B are the velocities of the particle at the lowest (A) and highest (B) points of the vertical circle then

$$V_A^2 - V_B^2 = 4gr$$

3. If T_A and T_B are the tensions in the string when the particle is at lowest point and highest point of the vertical circle then

$$T_A = \frac{mv_A^2}{r} + mg \quad \text{(maximum tension)}$$

$$T_B = \frac{mv_B^2}{r} - mg \quad \text{(minimum tension)}$$

4. The difference between maximum and minimum tensions in the string

$$T_A - T_B = 6mg$$

This difference is independent of velocity of projection of the particle.

5. Conditions to be satisfied to complete the vertical circle by the particle

a) minimum tension, $T_B = 0$

b) minimum velocity at the highest point, $V_B = \sqrt{gr}$

c) minimum velocity at the lowest point, $V_A = \sqrt{5gr}$

d) minimum velocity at an angle 'θ' from lowest point, $V_o = \sqrt{3gr + 2gr\cos\theta}$

e) minimum tension at angle 'θ' from lowest point, $T_o = 3mg(1+\cos\theta)$

Note: V_{min} is independent of mass of the body and T_{min} in the string is independent of radius of vertical circle

6. When a particle of mass 'm' is rotated in a vertical circle of radius 'r' with **critical speeds** with the help of a light string, then

a) Its total energy = (5/2) mgr

b) At the lowest point, KE_{max} = (5/2) mgr

c) At the highest point, KE_{min} = (1/2) mgr

d) between lowest and highest points $\Delta KE = \Delta PE = 2mgr$

7. A simple pendulum of length 'L' is having a bob of mass 'm'. Let the bob is given a velocity 'u' along the horizontal direction.

a) If $u \geq \sqrt{5gL}$, the bob describes verticle circle of radius L.

b) If $u \le \sqrt{2gL}$ then

i) the bob oscillates about the mean position

ii) the angle with the lower vertical at which velocity becomes zero is $\cos\theta = \left(1 - \frac{u^2}{2gL}\right)$

iii) the height at which velocity becomes zero is $h = \frac{u^2}{2g}$

c) If $\sqrt{2gL} < u < \sqrt{5gL}$, then

i) the bob neither performs oscillations nor describe vertical circle but the bob travels inside the vertical circle

ii) the angle with the lower vertical at which the tension in the string becomes zero is $\cos\theta = \left(\frac{2gL - u^2}{3gL}\right)$

iii) the height at which tension in the string becomes zero is $h = \left(\frac{u^2 + gL}{3g}\right)$

8. The mass of the bob of a simple pendulum of length 'L' is 'm'. The bob is projected horizontally from its mean position with velocity 'u' and is put in vertical circular motion. Then the tension in the string at angular displacement θ is

$$T = \frac{mu^2}{L} - mg(2 - 3\cos\theta)$$

(or) $T = T_{max} - 3mg(1 - \cos\theta)$

where $T_{max} = mg + mu^2/L$

9. A ball of mass 'M' is suspended vertically by a string of length 'l'. A bullet of mass 'm' is fired horizontally with a velocity 'u' into the ball, sticks to it. For the system to complete the vertical circle, the minimum value of 'u' is given by $u = \frac{(M+m)}{m}\sqrt{5gl}$

10. A nail is fixed at a certain distance x vertically below the point of suspension of a simple pendulum of length L. The bob is released when the string makes an angle θ with vertical. The bob reaches the lowest position then describes a vertical circle whose centre coincides with the nail. then

$$x_{min} = \frac{L(3 + 2\cos\theta)}{5}$$

11. A body of mass 'm' is allowed to slide down from rest, from the top of a smooth incline of height 'h'. For the body to loop in a loop of radius 'r' on arriving at the bottom,

a) Minimum height of smooth incline $h = \left(\frac{5r}{2}\right)$

b) 'h' is independent of mass of the body

12. A small block is freely sliding down from the top of a smooth convex hemisphere of radius r, placed on a table with its flat face on the table. then

a) Normal reaction on the block is zero at the instant the block leaves the hemisphere.

b) the vertical height from the table at which the block leaves the hemisphere is $h = 2r/3$

c) If the position vector of the block with respect to the centre of curvature makes an angle θ with vertical when the block leaves the hemisphere, then $\cos\theta = 2/3$

d) velocity of the block at that instant is $V = \sqrt{\frac{2gr}{3}}$

e) If the block is given a horizontal velocity 'u' from the top of the smooth convex hemisphere then the angle θ with vertical at which the block leaves the hemisphere is

$$\cos\theta = \frac{2}{3} + \frac{u^2}{3gr}$$

13. A small block of mass m is oscillating along smooth concave hemispherical bowl of radius r with an angular amplitude θ. If R_1, R_2 are the normal reactions on the block at the mean and extreme positions repectively and V is the speed of the block at the mean position then

$V = \sqrt{2gr(1 - \cos\theta)}$

$R_1 = mg + mv^2/r = mg(3 - 2\cos\theta)$

$R_2 = mg\cos\theta$

ROTATIONAL MOTION OF A RIGID BODY:

1. If the distance between any two particles of a body remains constant in all operations, the body is called a rigid body
2. A rigid body can execute both translatory, and rotatory motions. It can also execute translational and rotational motions at the same time.

3. A rigid body executes pure rotational motion with angular acceleration when it is acted upon by an unbalanced torque.
4. **Torque (or) moment of force (τ)**
   a) It is the turning effect of a force .
   b) Magnitude of torque = force acting on the body x perpendicular distance between line of action of force and axis of rotation .
   c) It is an axial vector . Its direction is parallel to the axis of rotation (or) perpendicular to the plane of rotation $\vec{\tau} = \vec{r} \times \vec{F}$

$$\tau = Fr\sin\theta \cdot$$

   d) S.I unit is Nm
   e) Magnitude of torque is maximum when $\bar{r}$ & $\bar{F}$ are perpendicular to each other.
5. **Couple**
   a) Two equal and opposite non-collinear forces simultaneously acting on a body constitute couple.
   b) The moment of a couple or torque is equal to the product of one of the forces and normal distance between the forces.
   c) Couple always produces turning effect
      eg: turning the cap of a pen
      turning the head of a tap
      turning the key in lock
6. When a rigid body rotates with uniform angular velocity
   a) all particles undergo the same angular displacement but different linear displacements.
   b) all particles rotate with same angular velocity but with different linear velocities.
   c) Particles on the axis of rotation remain at rest.
7. **Moment of inertia or Rotational inertia**
   a) It is the property of a body due to which it opposes any change in its state of rest or uniform rotation.
   b) It is the rotational analogue of inertia in translatory motion
   c) For a particle of mass 'm' rotating at a distance r from the axis of rotation. $I = mr^2$
   d) Moment of inertia of a group or a system of particles is

$$I = m_1r_1^2 + m_2r_2^2 + ... + m_nr_n^2$$

$$I = \sum mr^2$$

   where $m_1, m_2,$ ---- m_n are masses of particles & r_1, r_2 ---- r_n are their perpendicular distances from axis of rotation.
   e) For a rigid body $I = M K^2$
   K is called radius of gyration

$$K = \sqrt{\frac{r_1^2 + r_2^2 + - - + r_n^2}{n}}$$

   f) S.I unit of moment of inertia is kg m^2
   g) Moment of inertia of a body depends on
      i) mass of the body
      ii) distribution of mass of the body
      iii) position of axis of rotation
   h) M.I. is independent of angular velocity of rotation of the body.
   i) Radius of gyration of a body depends on
      i) distribution of mass of the body
      ii) position of axis of rotation
   j) K is independent of mass of the rotating rigid body .
9. Parallel axes theorem
   a) $I = I_C + Md^2$
   I - MI of the body about any axis
   I_C - MI of the body about a parallel axis passing through centre of mass
   d - perpendicular distance between the parallel axes
   M - mass of the body
   b) $K^2 = K_c^2 + d^2$
   Where K_c = radius of gyration about the axis passing through centre of mass
   K = radius of gyration about the parallel axis
10. Perpendicular axes theorem. (Valid for lamina only)
   a) $I_z = I_x + I_y$
   I_z - MI of the body about z - axis
   I_x - MI of the body about x - axis
   I_y - MI of the body about y - axis
   b) $K_z^2 = K_x^2 + K_y^2$
   Where K_x, K_y, K_z are radii of gyration about x,y,z axes respectively

11. **Moment of inertia of some regular bodies**

S.No.	Body	Axis of rotation inertia (I)	Moment of gyration (K)	Radius of
1.	Thin circular ring of mass M & radius R	a) Passing through centre and normal to its plane	MR^2	R
		b) any diameter	$MR^2/2$	$R/\sqrt{2}$
		c) any tangent in its plane	$3MR^2/2$	$\sqrt{3}R/\sqrt{2}$
		d) any tangent perpendicular to its plane	$2MR^2$	$\sqrt{2}R$
2.	Thin circular disc of mass M and radius R	a) Passing through centre and normal to its plane	$MR^2/2$	$R/\sqrt{2}$
3.	Thin hollow cylinder of mass M, radius R and length L its centre	a) about the natural axis	MR^2	R
		b) perpendicular to the axis of cylinder and passing through	$M\left(\frac{L^2}{12}+\frac{R^2}{2}\right)$	$\sqrt{\frac{L^2}{12}+\frac{R^2}{2}}$
		c) Passing through line of contact of cylinder with floor when it is rolling	$2MR^2$	$\sqrt{2}R$
		d) perpendicular to the axis of cylinder and passing through one end	$M\left(\frac{L^2}{3}+\frac{R^2}{2}\right)$	$\sqrt{\frac{L^2}{3}+\frac{R^2}{2}}$
4.	Solid cylinder of mass M, radius R and length L	a) about the natural axis	$MR^2/2$	$R/\sqrt{2}$
5.	Thin hollow sphere of mass M, radius R	a) about the natural (or) diametrical axis	$2MR^2/3$	$\sqrt{2}R/\sqrt{3}$
		b) Passing through any tangent	$5MR^2/3$	$\sqrt{5}R/\sqrt{3}$
6.	Solid sphere of mass M and radius R	a) about the natural (or) diametrsical axis	$2MR^2/5$	$\sqrt{2}R/\sqrt{5}$
		b) About any tangent	$7\ MR^2/5$	$\sqrt{7}R/\sqrt{5}$
7.	Thin uniform rod of mass M and length L	a) Perpendicular to the length of rod and passing through its centre	$ML^2/12$	$L/2\sqrt{3}$
		b) perpendicular to the length of rod and passing through its end	$ML^2/3$	$L/\sqrt{3}$
8.	Thin uniform rectangular plate of mass M, length L and breadth b	a) perpendicular to the plane and passing through its centre	$\frac{M(L^2+b^2)}{12}$	$\sqrt{\frac{L^2+b^2}{12}}$
9.	Thin square plate of mass M and side length L.	a) perpendicular to the plane and passing through its centre	$ML^2/6$	$L/\sqrt{6}$
		b) perpendicular to the plane and passing through a corner	$2ML^2/3$	$\sqrt{2}L/\sqrt{3}$
		c) any axis passing through the center and in the plane of plate	$ML^2/12$	$L/2\sqrt{3}$

12. **Angular momentum L :**

a) Moment of momentum of a rotating body about the axis of rotation

b) $\overline{L} = \overline{r} \times \overline{p}$

c) It is an axial vector whose direction is parallel to the axis of rotation or perpendicular to the plane of rotation.

d) S.I unit is joule second

e) For a particle in circular motion $L = mvr = mr^2 \omega$

f) In case of rotational motion of rigid body

$\tau = I\alpha = I.d\omega / dt = dL / dt$

13. **Law of conservation of angular momentum :**

a) When no external torque acts, then angular mommentum of the rotating system remains conseved.

b) When there is no external torque acting on a rigid body, $L = I\omega$ = constant

$$I_1\omega_1 = I_2\omega_2 \text{ or } I_1n_1 = I_2n_2 \text{ or } \frac{I_1}{T_1} = \frac{I_2}{T_2}$$

c) If two rigid bodies with angular velocity ω_1 and ω_2 are coupled and ω is the final angular velocity of the arrangement,

$$(I_1\omega_1 + I_2\omega_2) = (I_1 + I_2)\omega$$

$$\text{or } (I_1n_1 + I_2n_2) = (I_1 + I_2)n$$

d) In above case loss in the rotational KE of the system is $\frac{1}{2}\frac{I_1I_2}{I_1+I_2}(\omega_1 - \omega_2)^2$

e) If the earth suddenly contracts, its moment of inertia decreases and angular velocity increases. As a result duration of the day would decrease.

f) If the radius of the earth shrinks to 1/n th of the present radius without any change in the mean density, duration of the new day is $(24/n^5)$hrs.

14. **Rotational KE**

$$KE_{rot} = \frac{1}{2} I \omega^2 = \frac{L^2}{2I} = \frac{1}{2} L\omega$$

15. **a) Work done by a torque :**

If a torque τ rotates a body through a small angle $d\theta$, then the work done is $dw = \tau\, d\theta$.

$\therefore$ The total work done in rotating the body from θ_1 to θ_2 is $w = \int_{\theta_1}^{\theta_2} \tau\, d\theta$

For a constant torque, $W = \tau\,\theta$

where $\theta = \theta_2 - \theta_1$

b) Work energy principle :

The work done by the constant net torque on a body is equal to the change in its rotational K.E.

$$W = \frac{1}{2} I(\omega_2^2 - \omega_1^2)$$

16. Power of a torque

The rate at which work is done by a torque

$P = dw/dt = \tau\, d\theta/dt = \tau\,\omega$

17. Rolling of a body without slipping on a horizontal surface

a) the body possesses both translational and rotational kinetic energy.

b) Total energy of the body

$E = KE_{trans} + KE_{rot}$

$$= \frac{1}{2} mV_c^2 + \frac{1}{2} I\omega^2 = \frac{1}{2} mV_c^2 \left(1 + \frac{K^2}{r^2}\right)$$

where V_c = velocity of center of mass

K = radius of gyration and r = radius

c) Fraction of energy associated with transla-tional motion $= 1/(1 + K^2/r^2)$

d) Fraction of energy associated with rotational motion $= 1/(1 + r^2/k^2)$

e) If V is the velocity of centre of mass of a rolling body, then the velocity of its highest point is 2V and that of lowest point is zero (With respect to observer out side)

18. When a body rolls down without sliding on an inclined plane with inclination θ and height 'h'

a) Acceleration of the body

$$a = \frac{g \sin\theta}{1 + K^2/r^2}$$ θ -- inclination of the plane

b) Velocity acquired by body on reaching the bottom

$$V = \sqrt{\frac{2gh}{1 + k^2/r^2}}$$ h - height of the incline.

c) Time taken by the body to roll down the inclined plane

$$t = \frac{1}{\sin\theta}\sqrt{\frac{2h}{g}\left(1 + \frac{K^2}{r^2}\right)}$$

d) frictional force $= \frac{mg \sin\theta}{(1 + r^2/K^2)}$

e) $\mu_s \geq \frac{\tan\theta}{(1 + r^2/K^2)}$

Body	K^2/r	Accelaration	Velocity at the bottom
Rolling ring (or) Hollow cylinder	1	(1/2) g sin θ	$\sqrt{gh}$
Disc (or) Solid cylinder	1/2	(2/3) g sin θ	$\sqrt{4gh/3}$
Solid sphere	2/5	(5/7)g sin θ	$\sqrt{10gh/7}$
Hollow sphere	2/3	(3/5)g sin θ	$\sqrt{6gh/5}$

19. A solid sphere, hollow sphere, disc and ring are allowed to roll down on an inclined plane simultaneously;

    a) then solid sphere reaches the bottom first, and the ring reaches last.

    b) disc and solid cylinder reach the bottom at same time.

20. When a body rolls up or down an inclined plane without slipping frictional force acts up the plane.

21. **Comparision of translatory and rotatory motions**

Translatory	Rotatory
1. Linear displacement (s)	Angular displacement (θ)
2. Linear velocity (V)	Angular velocity(ω)
3. Linear acceleration (a)	Angular acceleration(α)
4. mass (m)	moment of inertia (I)
5. Linear momentum (P = mv)	Angular momentum (L = I ω)
6. Force (F= dp/ dt)	Torque (τ = dL/dt)
7. work $(w = \int F\,ds)$	work $(w = \int \tau\,d\theta)$
8. Linear KE $(E=mv^2/2 = p^2/2m)$	Rotational KE $(E_r = I\omega^2/2 = L^2/2I)$
9. work-energy theorm $FS = \frac{1}{2}mv^2 - \frac{1}{2}mu^2$	work-energy theorm $\tau\theta = \frac{1}{2}I\omega_2^2 - \frac{1}{2}I\omega_1^2$
10. power (P=Fv)	power (P= τ ω)
11. Impulse $(I = \int F\,dt)$	Impulse $(I = \int \tau\,dt)$

Equations of motion

1. V = u + at	$\omega_2 = \omega_1 + \alpha t$
2. $S = ut + at^2/2 = (u+v)t/2$	$\theta = \omega_1 t + \alpha t^2/2 = (\omega_1 + \omega_2)t/2$
3. $V^2 - u^2 = 2as$	$\omega_2^2 - \omega_1^2 = 2\alpha\theta$
4. $S_n = u + a(n-1/2)$	$\theta_n = \omega_1 + \alpha(n - 1/2)$

EXERCISE - I

MULTIPLE CHOICE QUESTIONS :

1. **The direction of angular accelaration of a body moving in a circle in the plane of the paper is**

   1) along the tangent 2) along the radius inward

   3) along the radius outward

   4) along the perpendicular to the plane of the paper

2. **A particle of mass m is tied to a light string and rotated with a speed v along a circular path of radius r. If T is tension in the string and mg is gravitational force on the particle then, the actual forces acting on the particle are**

   1) mg, and T only

   2) mg, T and an additional force of $\frac{mv^2}{r}$ directed inwards

   3) mg.T and an additional force of $\frac{mv^2}{r}$ directed outwards

   4) only a force $\frac{mv^2}{r}$ directed outwards

3. **A motor car with a mass m moves with uniform speed V on (a) horizontal level bridge (b) convex bridge (c) concave bridge. The reaction of the bridge at it's centre, on the car is**

   1) highest in case (a) 2) highest in case (b)

   3) highest in case (c) 4) same in the three cases

4. **Suppose a disc is rotating counter clockwise in the plane of the paper then**

   1) It's angular velocity vector will be perpendicular to the page pointing up out of the page

   2) It's angular velocity vector will be perpendicular to the page pointing inwards

   3) It's angular velocity vector acts along the tangent to the disc.

   4) none of the above is correct.

5. **A particle of mass M moves in a circular path of radius r with a constant speed equal to V. Then its centripetal acceleration is**

   1) $\frac{V^2}{r}$ 2) $\frac{V^2}{r^2}$ 3) $V^2 r$ 4) zero

6. **Many great rivers flow towards the equator, what effect does the sediment they carry to sea have on the rotation of the earth?**

   1) The rotation of the earth slows down

   2) The rotation of theearth speeds up

   3) No effect on the rotation of the earth

   4) none

7. **A stone is suspended by a thread, the minimum horizontal velocity to be given to the stone at lowest position to complete the vertical circle**
   1) depends on length of thread used
   2) depends on mass of the stone
   3) independent of length of the thread used
   4) none

8. **A stone of mass 'm' tied to a string is revolved in a vertical circle of radius R. If T_1 is the tension in the string at lowest position, the net force on the stone acting in that position is**
   1) mg-T_1 2) mg + T_1 3) T_1-mg 4) T_1

9. **A stone is tied to the end of a string and whirled in a vertical circle. The tension in the string is T_1 when the stone is at the highest point and T_2 when it is at the lowest point. Then [99M]**
   1) $T_1=T_2$ 2) $T_1=-T_2$ 3) $T_1>T_2$ 4) $T_1<T_2$

10. **The minimum speed for a particle at the lowest point of a vertical circle of radius R to describe the circle is V. If the radius of the circle is reduced to one fourth its value, the corresponding minimum speed will be [99E]**
    1) V/4 2) V/2 3) 2V 4) 4V

11. **A stone tied to a string is rotated in a vertical circle. The minimum speed with which the string has to be rotated [98E]**
    1) decreases with increasing mass of the stone
    2) is independent of mass of the stone
    3) decreases with increasing length of the string
    4) is independent of length of the string

12. **Moment of inertia of a body is independent of**
    1) mass of the body
    2) distribution of mass of the body
    3) temperature of the body
    4) angular velocity of the body

13. **Moment of inertia of a body depends upon**
    1) distribution of mass of the body
    2) position of axis of rotation
    3) temperature of the body
    4) all of the above

14. **When temperature of a body made of brass increases its moment of inertia.**
    1) increases 2) decreses
    3) remains same 4) becomes zero

15. **Rotational analogue of inertia in translatory motion is**
    1) mass 2) weight
    3) moment of inertia 4) Torque

16. **Moment of inertia comes into play in**
    1) Rectilinear motion 2) curvilinear motion
    3) Rotational motion 4) translational motion

17. **The following bodies have same mass and radius. The body with large moment of inertia about geometric axis is**
    1) solid sphere 2) Hollow sphere
    3) Disc 4) Ring

18. **If I_1 I_2, and I_3 are moments of inertia of a disc about its geometric axis, diameter and a tangent in its plane, then**
    1) $I_1 > I_2 > I_3$ 2) $I_3 > I_2 > I_1$
    3) $I_3 > I_1 > I_2$ 4) $I_2 > I_1 > I_3$

19. **Moment of inertia of a ring is minimum**
    1) about its geometric axis 2) about a diameter
    3) about a tangent in its plane
    4) tangent perpendicular to its plane

20. **Moment of inertia of a solid cylinder of mass m, length l, radius r about an axis passing through its centre and perpendicular to its length is**
    1) $\frac{ml^2}{12}$ 2) $\frac{ml^2}{3}$ 3) $\frac{ml^2}{12}+\frac{mr^2}{4}$ 4) $\frac{ml^2}{12}+\frac{mr^2}{3}$

21. **Moment of inertia of a uniform plate of mass M, length 'l' and breadth 'b' about an axis passing through its centre of mass and normal to the plane is**
    1) $M(l^2+b^2)/2$ 2) $M(l^2+b^2)/12$
    3) $M(l^2+b^2)/3$ 4) $M(l^2+b^2)$

22. **The moment of inertia of a thin uniform rod of mass M and length l about an axis perpendicular to the rod, through its centre is I. The moment of inertia of the rod about an axis perpendicular to the rod through its end point is [99E]**
    1) I/4 2) I/2 3) 2I 4) 4I

23. **Moment of inertia of a thin square plate about an axis passing through its diagonal is I. Its moment of inertia about an axis passing through its centre in the plane of the plate and making an angle θ with the diagonal is**
    1) I 2) I cosθ 3) I sinθ 4) I tanθ

24. **Two identical rods are joined to form an 'X'. The smaller angle between the rods is 'θ'. The moment of inertia of the system about and axis passing through the point of intersection of the rods and perpendicular to their plane is**
    1) $\alpha\ \theta$ 2) $\alpha\ \sin^2\theta$
    3) $\alpha\ \cos^2\theta$ 4) independent of θ

25. **The correct relation of the following is**

1) $\bar{\tau}=\bar{r}.\bar{F}$ 2) $\bar{\tau}=\bar{r}\times\bar{F}$ 3) $\bar{\tau}=\frac{\bar{F}}{\bar{r}}$ 4) $\vec{\tau}=\vec{r}+\vec{F}$

26. **The rotational analogue of force in translatory motion is**

1) torque 2) rotational inertia

3) weight 4) angular impulse.

27. **When a constant torque is applied on a rigid body, then**

1) The body moves with linear accelaration

2) The body rotates with constant angular velocity

3) The body rotates with constant angular accelaration

4) The body undergoes equal angular displacement in equal intervals of time

28. **Relation between torque and angular momentum is similar to the relation between**

1) Accelaration and velocity

2) mass and moment of inertia

3) energy and displacement

4) force and linear momentum

29. **The physical quantity of the following that is not conserved in case of a planet revolving in an orbit is**

1) Angular momentum 2) Total energy

3) Areal velocity 4) Kinetic energy

30. **A body of mass M and radius of gyration k is rotating with an angular acceleration α. The torque acting on it is**

1) $Mk^2\alpha$ 2) $\frac{Mk^2\alpha^2}{2}$ 3) $Mk^2\alpha^2$ 4) $Mk\alpha$

31. **When a rigid body is rotating, every point in it describes a ciruclar path**

1) Only if the axis of rotation passes symmertrically through the centre of gravity of the body

2) Only it the axis of rotation passes symmetri-cally through the body

3) Provided the axis passes through the edge of the body

4) Whatever be the axis of roation

32. **A solid metal sphere and a solid wooden sphere are having the same mass. If they are spinning with same angular velocity, then**

1) metal sphere possesses more angular momentum

2) wooden sphere will have more angular momentum

3) both will have same angular momentum

4) none of the above

33. **Two bodies with moment of inertia I_1 and I_2 ($I_2>I_1$) are rotating with same angular momentum. If K_1 and K_2 are their kinetic energies, then**

1) $K_2 > K_1$ 2) $K_2 < K_1$

3) $K_1 = K_2$ 4) $K_2 \geq K_1$

34. **A Particle of mass M is moving in a straight line with uniform speed v parallel to x - axis in X - Y plane at a height 'h' from the X -axis. Its angular momentum about the origin is**

1) Zero

2) mvh and is directed along positive z axis

3) mvh and is directed along negative z - axis

4) mvh and is directed along positive x - axis

35. **Angular momentum of a body is equal to the product of**

1) Moment of inertia and angular velocity

2) Centripetal force and radius

3) Linear velocity and angular velocty

4) mass and angular velocity

36. **A body is moving at a distance parallel to X-axis with a constant velocity. Its angular momentum about the origin**

1) remains constant 2) increases

3) decreases 4) may increase or decrease

37. **When running boy jumps on to a rotating table, the quantity that is conserved is**

1) Linear momentum 2) Angular momentum

3) Kinetic energy 4) None of the above

38. **The principle used by a gymnast to increase the number of somersaults, is law of conservation of**

1) linear momentum 2) energy

3) mass 4) angular momentum

39. **If polar ice caps melt, the length of the day**

1) will increase 2) will decrease

3) will remain same 4) cannot be decided

40. **A boy standing on a rotating table with heavy spheres in his hands, suddenly brings his hands close to his body. The angular velocity of the table**

1) Remains unchaged 2) Becomes zero

3) decreases 4) increases

41. **When a stopper is pulled from a filled wash basin, the water drains out while circulating like a small whirl pool. The angular velocity of a fluid element about a vetical axis through the orifice appears to be greater near the orifice. The angular velocity of fluid element varies**

1) Inversely as the square of its distance from the axis through the orifice.

2) inversely as the distance from the axis through the orifice.

3) directly as the square of its distance from the axis through the orifice

4) directly as the distance from the axiz through the orifice.

42. **A man turns on rotating table with an angular speed ω. He is holding two equal masses at arms length. Without moving his arms, he just drops the two masses. How will his angular speed change?**

1) it will be less than ω

2) it will be more than ω

3) it will be equal to ω

4) it will be more than ω if the dropped mass is more than 9.8kg and it will be less than ω if the mass dropped is less than 9.8kg

43. **A solid sphere, solid cylinder and a disc are allowed to roll down from the top of an incline plane from the same height. Then**

1) Disc will reach the bottom first

2) Solid cylinder will reach the bottom first

3) Sphere will reach the bottom first

4) All will reach the bottom simultaneously

44. **Two identical hollow spheres roll down two inclined planes of same height but of different angles of inclination. Then, they reach the bottom**

1) with same speeds and in same time

2) with different speeds and in different times

3) with same speed but in different times

4) with different speeds in same time

45. **A particle moves along a horizontal circle with constant speed. If 'a' is its acceleration and 'E' is its kinetic energy**

A) a is constant **B) E is constant**

C) a is variable **D) E is variable**

1) A and B are correct

2) C and D are correct

3) A and D are correct

4) B and C are correct

46. **Identify the increasing order of angular velocities of following [05-E]**

a) Earth rotating about its own axis

b) Hour's hand of clock

c) Seconds hand of clock

d) Fly wheel of radius 2m making 300 r.p.m.

1) a, b, c, d 2) b, c, d, a

3) c, d, a, b 4) d, a, b, c

47. **When a body is whirled in a vertical circle such that its total mechanical energy is constant with critical velocities then tension in the string changes with time. Arrange the following in increasing order of tension when**

i) String is horizontal

ii) String makes an angle 0^0 with downward vertical

iii) String makes an angle of 180^0 with downward vertical

iv) String makes an angle 120^0 with downward vertical

1) iii, iv, i, ii 2) iii, i, ii, iv

3) i, ii, iii, iv 4) iv, iii, ii, i

48. **Moment of inertia of a rigid body depends on**

A) Mass of body

B) Position of axis of rotation

C) Angular velocity of the body

D) Time period of its rotation

1) A and B are correct

2) B and C are correct

3) A and C are correct

4) C and D are correct

49. **M is the mass of a rectangular lamina and L, b are its length and breadth respectively. Then**

a) M.I about AD is $\frac{ML^2}{3}$

b) M.I about F and perpendicular to the plane = $\frac{ML^2}{12}+\frac{Mb^2}{3}$

c) M.I about H and perpendicular to the plane is $\frac{ML^2}{3}+\frac{Mb^2}{3}$

d) M.I about D and perpendicular to the plane is $\frac{ML^2}{3}+\frac{Mb^2}{3}$

The correct statements are,

1) a, b, c 2) b, c, d 3) a, b, d 4) a, b, c, d

50. Identify the correct order in which the values of M.I. decreases for the following

i. M.I. of solid sphere of mass 'M' and radius 'R' about its diameter of rotation

ii. M.I. of uniform ring of mass 'M' and radius 'R' about its tangent perpendicular to its plane

iii. M.I. of uniform dise of mass 'M' and radius 'R' about its diameter

iv. M.I of a uniform solid cylinder of mass M about its own axis of rotation

1) iii, i, iv, ii 2) i, iv, iii, ii 3) i, ii, iii, iv 4) iv, iii, ii, i

51. Four objects with the same mass and radius are spinning freely about a diameter with the same angular speed. Arrange the work required to stop them in the decreasing order

a) Solid sphere b) Hollow sphere

c) Disc d) Hoop

1) d, b, c, a 2) a, b, c, d 3) b, a, d, c 4) c, a, b, d

52. A thin disc rotates about an axis passing through its centre and perpendicular to its plane with a constant angular velocity 'ω' I is the moment of inertia of that disc and 'L' is its angular momentum about the given axis. Then rotational kinetic energy of the disc 'E' is

A) $E \alpha L^2$ B) $E \alpha L^{-2}$

C) $E \alpha I$ D) $E \alpha I^{-1}$(if L constant)

1) A and C are correct 2) B and C are correct

3) B and D are correct 4) A and D are correct

53. A man is spinning in the gravity free-space changes the shape of the body by spreading his arms. By doing this he can change his

(a) moment of inertia

(b) angular momentum

(c) angular velocity

(d) rotational kinetic energy

Which one of the following are correct?

1) a, b and c 2) d, a and b

3) c, d and a 4) b, c and d

54. Match list I with list II

	List - I	List - II
a)	Ratio of angular velocities of hours hand of a clock and self rotation of the earth	e) 12 : 1
b)	Ratio of angular velocities of seconds hand to minutes hand of a clock	f) 60 : 1
c)	Ratio of angular velocities of seconds hand to hours hand of a clock	g) 2 : 1
d)	Ratio of angular velocities of minutes hand to hours hand of a clock	h) 720 : 1

1) a - g, b - f, c - e, d - h 2) a - f, b - g, c - h, d - e

3) a - g, b - f, c - h, d - e 4) a - g, b - e, c- f, d - h

55. Match list I with list II

List - I	List - II
a) Centrifugal force	e) Along the axis of rotation
b) Centripetal force	f) Towards the centre of rotation
c) Tangential force	g) Away from the centre of rotation
d) Angular velocity	h) Changes the angular velocity

1) a-h, b-g, c-f, d-e 2) a-g, b-f, c-h, d-e

3) a-f, b-g, c-h, d-e 4) a-e, b-h, c-e, d-f

56. Match list I with list II

List - I	List - II
a) Torque	e) (1/2)L ω
b) Angular momentum	f) Zero
c) Rotational K.E.	g) $I\alpha$
d) Work done by centripetal force	h) I ω

1) a - h, b - e, c - f, d - g
2) a - f, b - h, c - g, d - e
3) a - e, b - f, c - g, d - h
4) a - g, b - h, c - e, d - f

57. Match list I with list II for a particle in uniform circular motion

List - I	List - II
a) changes only in direction	e) external torque
b) zero magnitude	f) angular momentum
c) acts along axis of rotation	g) kinetic energy
d) non zero and constant (having no direction)	h) linear momentum

1) a-e, b-h, c-f, d-g
2) a-h, b-e, c-g, d-f
3) a-h, b-f, c-e, d-g
4) a-h, b-e, c-f, d-g

58. Match list I with list II

List - I	List - II
a) Angular acceleration	e) J.S
b) Moment of inertia	f) Kgm^2
c) Torque	g) rad/sec^2
d) Angular momentum	h) N.m

1) a-g, b-f, c-h, d-e
2) a-h, b-f, c-g, d-e
3) a-f, b-h, c-g, d-e
4) a-e, b-h, c-g, d-f

ASSERTION & REASON TYPE

These Questions consist of two statements each printed as Assertion and Reason. While answering these questions you are required to choose any one of the following four responses.

1) Both (A) and (R) are true and (R) is the correct explanation of (A)
2) Both (A) & (R) are true but (R) is not correct explanation of (A)
3) (A) is true and (R) is false
4) (A) is false but (R) is true

59. A) : A ball connected to a string is in circular motion on a frictionless horizontal table and is in equilibrium

R) : 'Magnitude of the centripetal force is equal to the magnitude of the tension in the string.
2008 M

60. A) : Centripetal force does no work in circular motion.

R) : Force and displacements are perpendicular to each other in circular motion.

61. A) : The speed of particle remains constant in uniform circular motion

R) : No force is acting on the particle

62. A) : There is change in direction of motion of particle in uniform circular motion due to centripetal force

R) : Force is that which produces acceleration

63. A) : A coin placed on a rotating disc flies away if the angular velocity is gradually increased

R) : Friction can not provide the sufficient centripetal force

64. A) : A particle in non uniform circular motion has both radial and tangential acceleration.

R) : In non uniform circular motion velocity changes both in magnitude and direction.

65. A) : A rectangular lanina has least moment of inertia about an axis passing through centre and lying in its plane with axis parallel to largest dimension.

R) : Any rigid body has minimum moment of Inertia about an axis passing through centre of gravity wrt any parallel axis.

66. A) : I_S and I_H are moment of inertia about the diameters of a solid and thin-walled hallow sphere respectively. If the radius and masses of above spheres are equal, $I_H > I_S$.

R) : In solid sphere, the mass is continuously and regularly distributed about the centre, where as the mass, to a large extent, is concentrated on the surface of hollow sphere [2005 - M]

67. A):Moment of momentum is angular momentum

R) : Any particle having linear momentum has same angular momentum about any point

68. A) : The dimensions of torque and work are same

R) : Torque is a vector but work is a scalar

69. A) : The angular momentum of a particle w.r.t origin moving parallel to x -axis with constant velocity is constant

R) : There is no change in the perpendicular distance of the particle from the origin when it travels prallel to x-axis.

70. A) : While a solid sphere spining about an axis, through it's centre different particles of the sphere have different speeds

R) : Different particles will be at different distances from axis of rotation of a spinning sphere.

71. A) : If body or a system has neither translational nor rotational motion it is in complete mechanical equilibrium

R) : A body or system is said to be in mechanical equilibrium when net force and net torque acting on the body are zero

72. A) : Two solid cylinders of different masses and radii take same time to roll down an inclined plane without slipping.

R) : Moment of inertia of two solid cylinders of different masses and radii about their natural axes are different.

73. A) : If polar ice cap melts, duration of day increases

R) : Moment of inertia increases and angular velocity decreases

74. A) : A gymnast curls his body while diving to perform more number of somersaults in air

R) : Curling reduces his moment of inertia and increases angular velocity

75. A) : When a boy on turn table rotating about vertical axis with dumbells in his streched hands lowers his hands angular velocity increases

R) : In non uniform circular motion velocity changes both in magnitude and direction

76. A) : A body slides down the plane and another body of same mass rolls down from the top of same inclined plane. Then the velocity of centre of mass is more for sliding body

R) : On reaching to the bottom of the plane, rolling body has both translational and rotational kinetic energies where as sliding body has only translational kinetic energy

77. A) : A soild sphere and hollow sphere when released from the top of a smooth and fixed inclined plane, reach the ground simultaneously

R) : Acceleration while sliding is independent of mass and dimension of the body

78. A) : A disc released from the top of a fixed inclined plane rolls down without slipping if the inclined plane is rough only

R) : Friction does no work in pure rolling

79. Statement - A : If a uniform metal disc is remoulded into a solid sphere, then the moment of inertia about the axis of symmetry increases than that before.

Statement - B : For a given body and for a given plane, the moment of inertia is minimum about an axis passing through the centre of mass.

1) Both A and B are wrong.

2) Both A and B are correct.

3) A is correct and B is wrong.

4) A is wrong but B is correct.

80. Consider the following two statements.

A) Linear momentum of the system remains constant.

B) Centre of mass of the system remains at rest.

1) A implies B and B implies A

2) A does not imply B and B does not imply A.

3) A implies B but B does not imply A

4) B implies A but A does not imply B

ANSWERS

1) 4	2) 1	3) 3	4) 1	5) 1
6) 1	7) 1	8) 3	9) 4	10) 2
11) 2	12) 4	13) 4	14) 1	15) 3
16) 3	17) 4	18) 3	19) 2	20) 3
21) 2	22) 4	23) 1	24) 4	25) 2
26) 1	27) 3	28) 4	29) 4	30) 1
31) 4	32) 2	33) 2	34) 3	35) 1
36) 1	37) 2	38) 4	39) 1	40) 4
41) 1	42) 3	43) 3	44) 3	45) 4
46) 1	47) 1	48) 1	49) 3	50) 1
51) 1	52) 4	53) 3	54) 3	55) 2
56) 4	57) 4	58) 1	59) 4	60) 1
61) 3	62) 1	63) 1	64) 1	65) 1
66) 1	67) 3	68) 2	69) 1	70) 1
71) 1	72) 2	73) 1	74) 1	75) 2
76) 1	77) 1	78) 2	79) 4	80) 4

EXERCISE - II(A)

Circular Motion :

1. **The angular velocity of second's hand in a wall clock is : (in rads^{-1})**

   1) $\frac{p}{30}$ 2) $\frac{p}{60}$

   3) $\frac{p}{1800}$ 4) $\frac{p}{3600}$

2. **A point on the rim of a wheel 3 m in diameter has linear velocity of 18 ms^{-1}. The angular velocity of the wheel is**

   1) 4 rads^{-1} 2) 12 rads^{-1}

   3) 6 rads^{-1} 4) 18 rads^{-1}

3. **The centripetal force required by a 1000 kg car that takes a turn of radius 50 m at a speed of 36 kmph is**

   1) 1000N 2) 3500N 3) 1600N 4) 2000N

4. **A stone of mass 0.5 kg is attached to a string of length 2 m and is whirled in a horizontal circle. If the string can with stand a tension of 9N, the maximum velocity with which the stone can be whirled is**

   1) 6 ms^{-1} 2) 8 ms^{-1} 3) 4 ms^{-1} 4) 12 ms^{-1}

5. **A car is moving with a speed of 30 ms^{-1} on a circular path of radius 500 m. If its speed is increasing at the rate of 2 ms^{-2}, the net acceleration of the car is**

   1) 3.6 ms^{-2} 2) 2.7 ms^{-2} 3) 1.8 ms^{-2} 4) 2 ms^{-2}

6. **The speed of a motor increases from 1200 rpm to 1800 rpm in 20S. How many revolutions does it make in this period of time ?**

   1) 400 2) 200 3) 500 4) 800

7. **A car is travelling at 36 kmph on a road. If m =0.5 between the tyres and the road, the minimum turning radius of the car is (g = 10ms^{-2})**

   1) 20m 2) 25m 3) 30m 4) 35m

8. **The angular displacement of a particle is given by $q = t^3 + t^2 + t + 1$ where 't' is time in seconds. Its angular velocity after 2s is**

   1) 17 rads^{-1} 2) 14 rads^{-1}

   3) 12 rads^{-1} 4) 20 rads^{-1}

9. **The angular frequency of a fan increases from 30 rpm to 60 rpm in ps. A dust particle is present at a distance of 20 cm from axis of rotation. The tangential acceleration of the particle in ps is**

   1) 0.8 ms^{-2} 2) 0.34 ms^{-2}

   3) 0.2 ms^{-2} 4) 1.2 ms^{-2}

10. **A coin is kept at distance of 10 cm from the centre of a circular turn table. If m =0.8, the frequency of rotation at which the coin just begins to slip is**

    1) 62.8 rpm 2) 84.54 rpm

    3) 54.6 rpm 4) 32.4 rpm

11. **A particle of mass 'm' is moving in a horizontal circle of radius 'r' under a centripetal force – k/r^2 where 'K' is a constant. The total energy of the particle is**

    1) – K/r 2) – K/2r 3) K/2r 4) –2K/r

12. **A car is moving in a circular horizontal track of radius 10 m with a constant speed of 10 ms^{-1}. A plumb bob is suspended from the roof of the car by a string of length 1m. The angle made by the string with vertical is (g = 10 ms^{-2})**

    1) 0^0 2) 30^0 3) 45^0 4) 60^0

Vertical Circle :

13. **A mass of 0.1 kg is rotated in a vertical circle using a string of length 20 cm. When the string makes an angle 30^0 with the vertical, the speed of the mass is 1.5 ms^{-1}. The tangential acceleration of the mass at that instant is**

    1) 4.9 ms^{-2} 2) 9.8 ms^{-2}

    3) 2.45 ms^{-2} 4) 19.6 ms^{-2}

14. A 2 kg stone is swung in a vertical circle by attaching it at the end of a string of length 2m. If the string can with stand a tension 140.6N, the maximum speed with which the stone can be rotated is
1) 22 ms^{-1} 2) 44 ms^{-1} 3) 33 ms^{-1} 4) 11 ms^{-1}

15. A bob of mass 100 g tied at the end of a string of length 50 cm is revolved in a vertical circle with a constant speed of 1 ms^{-1}. When the tension in the string is 0.7N, the angle made by the string with the vertical is (g = 10 ms^{-2})
1) 0^0 2) 90^0 3) 180^0 4) 60^0

16. A 5 kg body is rotated in a vertical circle with a constant speed of 4 ms^{-1} using a string of length 1 m, when the tension in the string is 31N, then the body will be
1) at the lowest point
2) making an angle 30^0 with vertical
3) at the highest point 4) at horizontal position

17. A simple pendulum of length 'ℓ' carries a bob of mass m. If the breaking strength of the string is 2 mg, the maximum angular amplitude from the vertical can be
1) 0^0 2) 30^0 3) 60^0 4) 90^0

18. A body is tied at the end of a string of length 'ℓ' and rotated in a vertical circle. The string is just taut when the body is at the highest point. Velocity of the body when the string is in horizontal position is
1) $3\sqrt{gr}$ 2) $\sqrt{gr}$ 3) $\sqrt{5gr}$ 4) $\sqrt{3gr}$

19. A ball of mass 'm' is rotated in a vertical circle with constant speed. The difference in tension at the top and bottom would be
1) 6 mg 2) 5 mg 3) 2 mg 4) mg

20. A 1kg ball is rotated in a vertical circle by using a string of length 0.1 m. If the tension in the string at the lowest point is 29.4N, its angular velocity at that position is
1) 7 $rads^{-1}$ 2) 14 $rads^{-1}$
3) 3.5 $rads^{-1}$ 4) 25.6 $rads^{-1}$

21. A ball of mass 'm' is rotated in a vertical circle with constant speed. The difference in tensions at the bottom and horizontal positions would be
1) 6 mg 2) 5 mg 3) 2 mg 4) mg

22. Two identical cars A and B are moving at 36 kmph. A goes on a bridge, convex upward and 'B' on concave upward. If the radius of curvature of bridge is 20 m, the ratio of normal forces exerted on the cars when they are at the middle of bridges (g = 10 ms^{-2})
1) 1 : 3 2) 1 : 2 3) 2 : 3 4) 1 : 5

23. The velocity of a body revolving in a vertical circle of radius 'r' at the lowest point $\sqrt{7gr}$. The ratio of maximum to minimum tensions in the string is
1) 8 : 1 2) 4 : 1 3) $\sqrt{7}:1$ 4) $1:\sqrt{7}$

24. The minimum speed of a bucket full of water whirled in a vertical circle of radius 10 m at the highest point so that the water may not fall is (g = 10 ms^{-2})
1) 1 ms^{-1} 2) 2 ms^{-1} 3) 3 ms^{-1} 4) 10 ms^{-1}

25. An inclined track ends in a circular loop of diameter 'D'. From what height on the track a particle should be released so that it completes that loop in the vertical plane ?
1) $\frac{5D}{2}$ 2) $\frac{2D}{5}$ 3) $\frac{5D}{4}$ 4) $\frac{4D}{5}$

26. A stone of mass 1kg is tied to one end of a string of length 0.5 m. It is whirled in a vertical circle. If the maximum tension in the string is 58.8N, the velocity at the top is
1) 1.82 ms^{-1} 2) 2.2ms^{-1}
3) 3.26 ms^{-1} 4) 2.87 ms^{-1}

Moment of inertia :

27. Two small spheres of masses 10kg and 30 kg are joined by a rod of length 0.5 m and of negligible mass. The M.I of the system about a normal axis through centre of mass of the system is
1) 1.875 kgm^2 2) 2.45 kgm^2
3) 0.75 kgm^2 4) 1.75 kgm^2

28. A ring has a mass of 0.5 kg and radius 1 m. The moment of inertia of the ring about its diameter is.
1) 0.75 kgm^2 2) 0.25 kgm^2
3) 1.25 kgm^2 4) 1.75 kgm^2

29. Four particles each of mass 1 kg are at the four corners of a square of side 1m. The M.I of the system about a normal axis through centre of square is
1) 6 kgm^2 2) 1.2 kgm^2
3) 2 kgm^2 4) 2.5 kgm^2

30. The radius of gyration of a body about an axis at a distance of 4 cm from its centre of mass is 5 cm. The radius of gyration about a parallel axis through centre of mass is
1) 2 cm 2) 5 cm 3) 4 cm 4) 3 cm

31. Two discs have the same mass and thickness. Their materials are of densities d_1 and d_2. The ratio of their M.I about their central axis is
1) $d_2 : d_1$ 2) $d_1 : d_2$
3) $d_1^2 : d_2^2$ 4) $d_2^2 : d_1^2$

32. The moment of inertia of a cylinder about its own axis is equal to its moment of inertia about an axis passing through its centre and normal to its length. The ratio of length to radius is
1) 2 : 1 2) $\sqrt{3}:1$ 3) 3 : 1 4) $\sqrt{2}:1$

33. Two circular loops of radii R and nR are made of same wire. If their M.I about their normal axis through centre are in the ratio 1 : 8, the value of n is (2001 M)
1) 6 2) 1 3) 2 4) 4

34. Three identical rings, each of mass M and radius R are placed in the same plane touching each other such that their centers form the vertices of an equilateral triangle. The M.I of the system about an axis passing through center of one of the rings and perpendicular to its plane is
1) $\frac{MR^2}{2}$ 2) MR^2 3) $\frac{5}{2}MR^2$ 4) 11 MR^2

35. The radius of gyration of a solid sphere of radius 'R' about its tangential axis is (2000E)
1) $\frac{7R}{5}$ 2) $\frac{3R}{5}$ 3) $R\sqrt{\frac{7}{5}}$ 4) $R\sqrt{\frac{3}{5}}$

36. A straight thin uniform rod of length 4L and man 4M is bent into a square. Its M.I about one side is
1) $\frac{5}{3}ML^2$ 2) $\frac{7}{6}ML^2$ 3) $\frac{ML^2}{48}$ 4) $\frac{ML^2}{3}$

37. The moment of inertia of an uniform circular disc about its central axis is 'I'. Its M.I about a tangent in its plane is equal to (2008 E)
1) 2I 2) 2.5 I 3) 1.5 I 4) $\frac{I}{2}$

38. Two circular rings of equal mass and radius are placed touching each other. The moment of inertia of the system about tangential axis in the plane of system passing through point of contact of rings is
1) 3 mr^2 2) $\frac{3}{2}mr^2$ 3) 6 mr^2 4) $\frac{5}{2}mr^2$

Conservation of Angular momentum :

39. If the mass of earth and radius suddenly become 2 times and 1/4 th of the present value, the length of the day becomes
1) 24h 2) 6h 3) 3/2h 4) 3h

40. A child is standing with folded hands at the centre of a platform rotating about its central axis. The K.E of the system is K. The child now stretches his arms so that the M.I of the system doubles. The K.E of the system now is
1) 2K 2) K/2 3) 4K 4) K/4

41. A ballet dancer spins about a vertical axis at 60 rpm with his arms closed. Now he stretches his arms such that M.I increases by 50%. The new speed of revolution is
1) 80 rpm 2) 40 rpm 3) 90 rpm 4) 30 rpm

42. A circular disc is rotating about its natural axis with angular velocity of 10 $rads^{-1}$. A second disc of same mass is joined to it coaxially. If the radius of disc is half of the radius of the first, then they together rotate with an angular velocity of
1) 2.5 $rads^{-1}$ 2) 5 $rads^{-1}$
3) 8 $rads^{-1}$ 4) 6 $rads^{-1}$

43. A constant torque acting on a uniform circular wheel changes its angular momentum from A to 4A in 4S. The torque acted on it is
1) $\frac{3A}{4}$ 2) $\frac{A}{4}$ 3) $\frac{2A}{3}$ 4) $\frac{3A}{2}$

44. Density remaining constant, if earth contracts to half of its present radius, duration of the day would be (in minutes)
1) 45 2) 80 3) 100 4) 120

45. Two wheels of M.I. 3kg m^2 and 5kg m^2 are rotating at the rate of 600 rpm and 800 rpm respectively in the same direction. If the two are coupled so as to rotate with the same axis of rotation, the resultent speed of rotation will be (in rpm)
1) 725 2) 850 3) 420 4) 630

46. A mass is whirled in a circular path with an angular momentum L.If the length of string and angular velocity, both are doubled, the new angular momentum is
1) L 2) 4L 3) 8L 4) 16L

47. A disc of mass 100g and radius 10cm has a projection on its circumference. The mass of projection is negligible. A 20g bit of putty moving tangential to the disc with a velocity of $5ms^{-1}$ strikes the projection and sticks to it. The angular velocity of disc is

1) 14.29 $rads^{-1}$ 2) 17.3 $rads^{-1}$

3) 12.4 $rads^{-1}$ 4) 9.82 $rads^{-1}$

Rotational Dynamics :

48. A constant torque of 1000 Nm turns a wheel of of M.I 200kg m^2 about an axis through centre. The angular velocity after 3s is

1) 15 rad s^{-1} 2) 22 rad s^{-1}

3) 28 rad s^{-1} 4) 60 rad s^{-1}

49. If 484 J of energy is spent in increasing the speed of a wheel from 60 rpm to 360 rpm , the M.I. of the wheel is.

1) 1.6 kg m^2 2) 0.3 kg m^2

3) 0.7 kg m^2 4) 1.2 kg m^2

50. A shaft rotating at 3000 rpm is transmitting a power of 3.14 KW. The magnitude of the driving torque is

1) 6 Nm 2) 10 Nm 3) 15 Nm 4) 22 Nm

51. The diameter of a rotating fly wheel is R. Its coefficient of linear expansion is α. If the temperature is increased by ΔT, the percentage change in its rotational KE would be.

1) $\alpha.\Delta t.100$ 2) $2.\alpha.\Delta t.\times 100$

3) $\frac{\alpha\Delta t}{2}\times 100$ 4) $\frac{2\alpha\Delta t}{5}\times 100$

52. A wheel rotating at an angular speed of 20 $rads^{-1}$, is brought to rest by a constant torque in 4s. If the M.I is 0.2 kg m^2, the work done in first 2s is

1) 50 J 2) 30 J 3) 20 J 4) 10 J

53. A thin uniform rod of length l and mass m is swinging freely about a horizontal axis passing through its end. Its maximum angular speed is ω. Its centre of mass rises to a maximum height of AIEEE - 2009

1) $\frac{1}{6}\frac{\ell\omega}{g}$ 2) $\frac{1}{2}\frac{\ell^2\omega^2}{g}$ 3) $\frac{1}{6}\frac{\ell^2\omega^2}{g}$ 4) $\frac{1}{3}\frac{\ell^2\omega^2}{g}$

54. The K.E. of a body rotating at 300 rpm is 62.8J. Its angular momentum is (2006 M)

1) 1 Js 2) 2 Js 3) 4 Js 4) 8 Js

55. If the velocity of a rotating body of mass 50 kg is given by $\overline{V} = 3\overline{i} + 4\overline{j} + 5\overline{k}$ and radius vector $\overline{r} = 2\overline{i} + 3\overline{j} + 4\overline{k}$, then its angular momentum is.

1) $50\overline{i} + 50\overline{j} + 50\overline{k}$ 2) $50\overline{i} + 100\overline{j} + 50\overline{k}$

3) $50\overline{i} + 50\overline{j} + 100\overline{k}$ 4) $-50\overline{i} + 100\overline{j} - 50\overline{k}$

56. If the Earth shrinks such that its density becomes 8 times to the present value, then the new during of the day in hours will be

1) 24 2) 12 3) 6 4) 4

57. A point sized sphere of mass 'm' is suspended from a point using a string of length 'ℓ'. It is then pulled to a side till the string is horizontal and released. As the mass passes through the portion where the string is vertical, magnitude of its angular momentum is

1) $m\ell\sqrt{g\ell}$ 2) $m\ell\sqrt{2g\ell}$

3) $m\ell\sqrt{g\ell/2}$ 4) $m\ell\sqrt{3g\ell}$

58. Two rigid bodies have same angular momentum about their axes of symmetry. If same torque is applied about their axes, the ratio of times after which they will be stopped is

1) 1 :1 2) 1 : 2 3) 2 : 1 4) 2 : 3

Rolling Motion :

59. A solid sphere rolls down without slipping from rest on a 30^0 incline. Its linear accelearation is

1) 5g/7 2) 5g/14 3) 2g/3 4) g/3

60. A hollow sphere rolls down a 30^0 incline of length 6m without slipping. The speed of centre of mass at the bottom of plane is

1) 6 ms^{-1} 2) 3 ms^{-1}

3) $6\sqrt{2}$ ms^{-1} 4) $3\sqrt{2}$ ms^{-1}

61. For a body rolling along a level surface, the translational and rotational K.E are equal. The body is

1) Solid cylinder 2) disc

3) ring 4) hallow sphere

62. A ring and a disc of same mass roll without slippling along a horizontal surface with same velocity. If the K.E of ring is 8J, then that of disc is

1) 2 J 2) 4 J 3) 6 J 4) 16 J

63. When a hollow sphere is rolling without slipping on a rough horizontal surface then the percentage of its total K.E which is Translational is

1) 72% 2) 28% 3) 60% 4) 40%

ANSWERS

1) 1	2) 2	3) 4	4) 1	5) 2
6) 3	7) 1	8) 1	9) 3	10) 2
11) 2	12) 3	13) 1	14) 4	15) 4
16) 3	17) 3	18) 4	19) 3	20) 2
21) 4	22) 1	23) 2	24) 4	25) 3
26) 2	27) 1	28) 2	29) 3	30) 4
31) 1	32) 2	33) 3	34) 4	35) 3
36) 1	37) 2	38) 1	39) 4	40) 2
41) 2	42) 3	43) 1	44) 1	45) 1
46) 3	47) 1	48) 1	49) 3	50) 2
51) 2	52) 2	53) 3	54) 3	55) 4
56) 2	57) 2	58) 1	59) 2	60) 1
61) 3	62) 3	63) 3		

EXERCISE - IIB

Circular Motion :

1. **The angular velocity of minutes hand in a wall clock is (in rad s^{-1})**

   1) $\pi/30$ 2) $\pi/60$ 3) $\pi/3600$ 4) $\pi/1800$

2. **The length of minutes hand in a pendulum clock is 10 cm. The speed of the tip of the hand is**

   1) $\pi/6000\ ms^{-1}$ 2) $\pi/18000\ ms^{-1}$

   3) $\pi/3600\ ms^{-1}$ 4) $\pi/1200\ ms^{-1}$

3. **The centripetal force required for a 1000 kg car travelling at 36 kmph to take a turn by 90^0 in travelling along an arc of length 628 m is**

   1) 250 N 2) 500 N 3) 1000 4) 125 N

4. **A string will break under a load of 5 kg. To the one end of such a 2m long string a mass of 1kg is attached. The maximum rpm in the horizontal plane so that the string does not break is ($g = 10\ ms^{-2}$)**

   1) 28.66 2) 47.73 3) 38.92 4) 54.12

5. **A car of mass 1000 kg is moving with a speed of 40 ms^{-1} on a circular path of radius 400m. If its speed is increasing at the rate of $3ms^{-1}$ the total force acting on the car is**

   1) 3000 N 2) 4000 N 3) 5000 N 4) 7000 N

6. **The speed of a motor decreases from 1200 rpm to 600 rpm in 20s. The total number of rotations it makes before coming to rest is**

   1) 400 2) 600 3) 800 4) 1000

7. **A cyclist rides round a curved path of radius 10m at a speed 18 kmph. If the cycle and the rider together have a mass of 100 kg, the firctional force the ground exerts on the wheel is ($g = 10\ ms^2$)**

   1) 250 N 2) 400 N 3) 500 N 4) 600 N

8. **The angular displacement of a particle is given by $\theta = t^3 + t^2 + t + 1$ where 't' is time in seconds. Its angular acceleration at t = 2s is**

   1) 14 rad s^{-2} 2) 17 rad s^{-2}

   3) 12 rad s^{-2} 4) 9 rad s^{-2}

9. **The angular frequency of a fan increases from 30 rpm to 60 rpm in πs. A dust particle is present on the blades at a distance of 20 cm from axis of rotation. The net acceleration of dust particle at the end of πs is**

   1) 2.8 ms^{-2} 2) 7.9 ms^{-2}

   3) 9.2 ms^{-2} 4) 6.2 ms^{-2}

10. **A coin placed on a rotating turn table just slips if it is placed at a distance of 4cm from the centre. If the angular velocity of turn table is doubled, it will just slip at a distance of**

    1) 1 cm 2) 2 cm 3) 4 cm 4) 8 cm

11. **If the centripetal force acting on a body revolving along a circular path of radius 25 m is 200 N, its KE is**

    1) 2.5 KJ 2) 4 KJ 3) 4.8 KJ 4) 6 J

12. **A bob is suspended from an ideal string of length 'ℓ'. Now it is pulled to a side through 60^0 to vertical and whirled along a horizontal circle. Then its period of revolution is**

    1) $\pi\sqrt{\ell/g}$ 2) $\pi\sqrt{\ell/2g}$

    3) $\pi\sqrt{2\ell/g}$ 4) $2\pi\sqrt{\ell/g}$

Vertical Circle :

13. **A mass of 0.1kg is rotated in a vertical circle using a string of length 1m. When the string makes an angle 30^0 with the vertical, the speed of the mass is 2 ms^{-1}. The radial acceleration of the mass at that instant is**

    1) $4ms^{-2}$ 2) $0.1ms^{-2}$ 3) $4ms^{-2}$ 4) $4ms^{-2}$

14. **A 0.5 kg ball is suspended to a string of length 1m and is rotated in a vertical plane by holding one end of the string. If it is making 30 rpm, the maximum tension in the string is $(\pi^2 = 10)$**

    1) 19.9 N 2) 9.9 N 3) 1.9 N 4) 29.9 N

15. A body of weight 10 N attached to one end of a string of length 1m is rotated in a vertical plane. At the instant when the string makes an angle of 60⁰ with the downward vertical, the speed of the bob is $\sqrt{9.8}\,ms^{-1}$. The tension in the string at that instant is equal to

1) 10N 2) 15 N 3) 20 N 4) 25 N

16. A 1kg stone at the end of a 1m long string is whirled in a vertical circle at a constant speed of 4 ms^{-1}. The tension in the string is 16N, when the stone is at (g = 10 ms^{-2})

1) The bottom 2) at the top

3) at the horizontal 4) at 60^0 to vertical

17. A simple pendulam of length 'ℓ' carries a bob of mass 'm'. The bob is released from horizontal position. The tension in the string at the lowest point would be

1) mg 2) 2 mg 3) 3 mg 4) 4 mg

18. A body tied at the end of a string of length 'ℓ' is rotated in a vertical circle. The velocity of the body at the highest point is $\sqrt{2g\ell}$. The maximum velocity of the body would be

1) $\sqrt{6g\ell}$ 2) $\sqrt{5g\ell}$ 3) $\sqrt{7g\ell}$ 4) $\sqrt{8g\ell}$

19. A ball of mass 'm' is rotated in a vertical by a string. The difference in tension at the top and bottom would be

1) 6mg 2) 1.5 mg 3) 3 mg 4) mg

20. A 0.4 kg body is rotated with a constant angular speed of 2 rps in a verticle circle of radius 1.2 m with the help of a string. The tension in the string when it is at the highest point is (g = 10 ms^{-2} and $\pi^2 = 10$)

1) 79.8 N 2) 97.8 N 3) 72.8 N 4) 85.7 N

21. A body of mass 'm' is rotating in a vertical circle of radius 'r' with critical speed at the highest point. The difference in its K.E. at the top and the bottom is

1) 2 mgr 2) 4 mgr 3) 6 mgr 4) 3 mgr

22. A bridge over a canal is in the form of a circular arc of radius 5m. The maximum speed with which a motor cycle crosses the bridge without leaving the ground at the highest point is g = 10ms^{-2},

1) 10 ms^{-1} 2) 50 ms^{-1} 3) 100 ms^{-1} 4) $5\sqrt{2}$ ms^{-1}

23. A stone tied to a string is whirled in a vertical circle with uniform speed. If the difference between the maximum and minimum tensions is 1kgwt, the mass of stone is

1) 1kg 2) 0.5 kg 3) 1.2 kg 4) 0.2 kg

24. The maximum time period of a bucket full of water whirled in a vertical circle of radius 10m so that the water may not fall is (g = 10 ms^{-2})

1) $2\pi s$ 2) πs 3) $\pi/2s$ 4) $\pi^2 s$

25. A body slides without friction from a height H = 60 cm and then loops the loop of radius R = 20 cm at the bottom of incline. The ratio of forces exerted on the body by the track at A, B and C are (g = 10 ms^{-2})

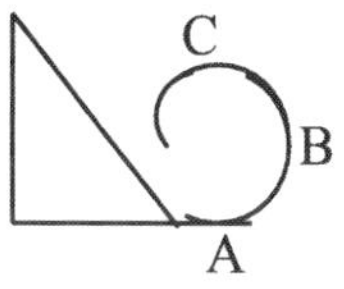

1) 1 : 4 : 7 2) 7 : 4 : 1

3) 4 : 1 : 7 4) 7 : 1 : 4

26. A stone of mass 2 kg is tied to one end of a string of length 0.5 m. It is whirled in a vertical circle. If the velocity of stone at the top of circle is 5 ms^{-1}, the tension in the string at the bottom of circle is

1) 125 N 2) 164 N 3) 198 N 4) 228 N

Moment of Inertia :

27. Two small spheres of mass 5kg and 15 kg are joined by a rod of length 0.5m and of negligible mass. The M.I. of the system about an axis passing through centre of rod and normal to it is

1) 10 kg m^2 2)1.25 kg m^2

3) 20 kg m^2 4) 8 kg m^2

28. Moment of inertia of a hoop suspended from a peg about the peg is

1) MR^2 2) $\frac{MR^2}{2}$ 3) 2 MR^2 4) $3MR^2/_2$

29. Three point masses 3kg, 2kg and 1kg are placed at A,B,C of a triangle. AB = 0.3,m AC = 0.5 m and BC = 0.4 m. The M.I of the system about an axis through A and normal to the plane of triangle is

1) 0.43 kgm^2 2) 0.25 kgm^2

3) 0.18 kgm^2 4) 0.07 kgm^2

30. The M.I of a uniform thin circular ring about an axis through its centre and normal to the plane of ring is I. Its M.I about atangent in its plane is equal to

1) 1.5 I 2) 2.5 I 3) $1/2$ 4) 2 I

31. Ratio of densities of materials of two circular discs of same mass and thickness is 5 : 6. The ratio of their M.I about natural axes is

1) 5 : 6 2) 6 : 5 3) 25 : 36 4) 1 : 1

32. M.I. of a uniform horizontal solid cylinder of mass M about an axis passing through its edge and perpendicular to the axis of cylinder when its length is 6 times of its radius R is

1) $\frac{39}{4}MR^2$ 2) $\frac{39}{2}MR^2$

3) $\frac{49}{4}MR^2$ 4) $\frac{49}{2}MR^2$

33. M.I of a solid sphere about its diameter is 64 kg m^2. If that sphere is recast into 8 identical small spheres, then M.I. of such small sphere about its diameter is

1) 8 kg m^2 2) 4kg m^2

3) 3kg m^2 4) 2 kg m^2

34. Two thin uniform circular rings each of radius 10 m and mass 0.1 kg are arranged such that they have common centre and their planes are prependicular to each other. The M.I. of the system about an axis passing through their common centre and perpendicular to the plane of either of the rings in kg m^{-2} is

1) 1.5 2) 5 3) 15 4) 18

35. The radius of gyration of a hollow sphere about a tangent is

1) $R\sqrt{7/5}$ 2) $R\sqrt{2/5}$ 3) $R\sqrt{2/3}$ 4) $R\sqrt{\frac{5}{3}}$

36. Consider a uniform square plate of side 'a' and mass 'm'. The moment of inertia of this plate about an axis perpendicular to its plane and passing through one of its corner is

1) $\frac{5}{6}ma^2$ 2) $\frac{1}{12}ma^2$ 3) $\frac{7}{12}ma^2$ 4) $\frac{2}{3}ma^2$

37. The M.I of a thin circular disc about a tangent in the same plane of disc is I. Then M.I. about one of the diameters is

1) 0.2 I 2) 0.8I 3) 1.25 I 4) 2.5 I

38. Three uniform discs each of mass 'm' and radius 'r' are kept touching each other, such that their centres form a triangle. The M.I of the system about a median is

1) $\frac{11}{4}mr^2$ 2) $\frac{5}{4}mr^2$ 3) $\frac{9}{4}mr^2$ 4) $\frac{3}{2}mr^2$

Conservation of Angular Momentum :

39. Circular disc of mass 2 kg and radius 1 metre is rotating about an axis perpendicular to its plane and passing through its centre of mass with a rotational kinetic energy of 8 Joules. The angular momentum in (J - sec) is

1) 8 2) 4 3) 2 4) 1

40. A person is spinning with his hands outstretched at the rate of 4 rad s^{-1}. When he brings his hands close to the body, he spins at the rate of 16 rad s^{-1}. The ratio of M.I. in the two cases successively is

1) 4 : 1 2) 1 : 14 3) 16 : 1 4) 2 : 1

41. A dancer spins about a vertical axis at 60 rpm with her arms folded. If she strectes her hands so that M.I. about the vertical axis increases by 25%, the new rate of revolvution is

1) 48 rpm 2) 75 rpm

3) 15 rpm 4) 24 rpm

42. Two identical discs are placed one over the other and both are rotating about a common axis through their centre and normal to their frame with angular velocity ω. If the upper disc is suddenly removed, anguler velocity of the other disc would be

1) ω 2) $\omega/2$ 3) 2ω 4) 4ω

43. $\overline{F} = a\overline{i} + 3\overline{j} + 6\overline{k}$ and $\overline{r} = 2\overline{i} - 6\overline{j} - 12\overline{k}$. The value of 'a' for which the angular momentum is conserved is

1) – 1 2) 0 3) 1 4) 2

44. If earth shrinks to $\frac{1}{64}$ of its volume with mass remaining same, duration of day will be

1) 1.5 h 2) 3h 3) 4.5 h 4) 6h

45. Two wheels of M.I. 3kg m^2 and 5 kg m^2 are rotaing at a rate of 600 rpm and 800 rpm respectively in opposite directions. If they are coupled so as to rotate with the same axis of rotation, the resultent speed of rotation (in rpm)

1) 275 2) 350 3) 210 4) 420

46. A mass is whirled in a circular path with a constant angular velocity and its angular momentum is L. If the length of string is now halved keeping the angular velocity same, the new angular momentum is

1) $L/4$ 2) $L/2$ 3) L 4) 2L

47. A square plate of moment of inertia 'I' is pivoted at its centre. It has a small projection of neglegible mass to one of its sides. A particle of mass 'm' moving with a velocity v parallel to the side having projection strikes to projection and stricks to it. If 'ℓ' is the side length, the angular velocity of plate is

1) $\frac{mv\ell}{2\left[I+\frac{m\ell^2}{4}\right]}$ 2) $\frac{mv\ell}{2\left[I+m\ell^2\right]}$

3) $\frac{mv\ell}{2I}$ 4) $\frac{mv\ell}{4I}$

Rotational Dynamics :

48. A flywheel of M.I. $6\times10^{-2}kgm^2$ is rotating with an angular velocity of 20 rad s^{-1}. The torque required to bring it to rest in 4s is

1) 1.6 Nm 2) 0.6 Nm 3) 0.8 Nm 4) 0.3 Nm

49. When 200J of work is done on a fly wheel its frequency of rotation increases from 4Hz to 9Hz. The M.I of the wheel about the axis of rotation is

1) 0.15 kg m^2 2) 0.2 kg m^2

3) 0.22 kg m^2 4) 0.3 kg m^2

50. The shaft of a motar is making 1260 rpm. The torque supplied by the motar is 100 Nm. The power of motar is

1) 100 KW 2) 21 KW

3) 13.2 KW 4) 4.8 KW

51. A solid shere is rotating with an angular velocity ω. It is of moment is inertia 'I' about its diametrical axis and is of linear expansion coefficient α. When temperature is increased by ΔT, the fractional change in its angular velocity is

1) $2\alpha\Delta T$ 2) $\alpha\Delta T$ 3) $\frac{\alpha\Delta T}{2}$ 4) $\frac{2\alpha\Delta T}{5}$

52. The moment of inertia of a wheel of radius 20 cm is 40 kg m^2. If a tangential force of 80 N applied on the wheel, its Rotational K.E. after 4s is

1) 16.2 J 2) 51.2J 3) 25.6J 4) 24.8J

53. A thin rod of mass 'm' and length 2L is made to rotate about a normal axis through centre of rod. If its angular velocity changes from 'O' to 'ω' in time 't'. The torque acting on it is

1) $mL^2\omega/12t$ 2) $mL^2\omega/3t$

3) $mL^2\omega/t$ 4) $4mL^2\omega/3t$

54. The M.I. of a body is 1.2 kg m^2. Initially the body is at rest. In order to produce a rotational kinetic energy of 1500 J an angular acceleration of 25 rad s^{-2} must be applied about the axis for a duration of

1) 2s 2) 4s 3) 8s 4) 10s

55. A uniform solid sphere of diameter 0.2m and mass 10kg is rotated about its diameter with an angular velocity of 2 rad s^{-1}. Then its angular momentum is kg m^2 s^{-1} is

1) 0.01 2) 0.02 3) 0.08 4) 0.04

56. If the K.E of a rotating body about an axis is decreased by 36%, its angular momentum about that axis

1) increases by 72% 2) decreases by 72%

3) increases by 20% 4) decreases by 20%

57. A shell of mass of mass 'm' is projected with a velocity 'v' at an angle 60^0 to horizontal. When it reaches the maximum height, its angular momentum with respect to point of projection is

1) $\sqrt{3}\frac{mv^3}{8g}$ 2) $\frac{3mv^3}{8g}$

3) $\frac{3mv^3}{16g}$ 4) $\frac{\sqrt{3}mv^3}{16g}$

58. A solid sphere and a hallow sphere of same mass and radius rotate about their diameter with same K.E. If torques of equal magnitude are applied on them, the ratio of their angular displacement before they are brought to rest is

1) 1 :1 2) 3 : 5 3) 5 : 3 4) 2 : 3

Rolling Motion :

59. A ring is allowed to roll down on an incline of 1 in 10 without slipping. The acceleration of its centre of mass is

1) 9.8 ms^{-2} 2) 4.9 ms^{-2}

3) 0.98 ms^{-2} 4) 0.49 ms^{-2}

60. A cylinder is released from rest from the top of an incline of inclination θ and length 'ℓ'. If the cylinder roles without slippling its speed at the bottom

1) $\sqrt{\frac{4g\ell\sin\theta}{3}}$ 2) $\sqrt{\frac{3g\ell\sin\theta}{2}}$

3) $\sqrt{\frac{4g\ell}{3\sin\theta}}$ 4) $\sqrt{\frac{4g\sin\theta}{3\ell}}$

61. For a body rolling along a level surface, without slipping the translational and rotational kinetic energies are in the ratio 2 : 1. The body is

1) Hollow sphere 2) Solid cylinder
3) Ring 4) Solid sphere

62. A solid sphere and a spherical shell roll down an incline from rest from same height. The ratio of times taken by them is

1) $\sqrt{21/25}$ 2) $21/25$
3) $\sqrt{25/21}$ 4) $25/21$

63. When a solid sphere is rolling along level surface the percentage of its total kinetic energy that is translational is

1) 29% 2) 71% 3) 60% 4) 40%

ANSWERS

1) 3	2) 2	3) 1	4) 2	5) 3
6) 1	7) 1	8) 1	9) 2	10) 1
11) 1	12) 3	13) 1	14) 2	15) 2
16) 3	17) 3	18) 1	19) 1	20) 3
21) 1	22) 4	23) 2	24) 1	25) 2
26) 3	27) 2	28) 3	29) 1	30) 1
31) 2	32) 3	33) 4	34) 3	35) 4
36) 4	37) 1	38) 1	39) 2	40) 1
41) 1	42) 3	43) 1	44) 1	45) 1
46) 1	47) 1	48) 4	49) 1	50) 3
51) 1	52) 2	53) 2	54) 1	55) 3
56) 4	57) 3	58) 1	59) 4	60) 1
61) 2	62) 1	63) 2		

EXERCISE - III(A)

Circular Motion in Horizontal plane :

1. A particle describes a horizontal circle on the smooth surface of an inverted cone. The plane of that circle is at a height of 9.8cm above the vertex. Then the speed of the particle is

1) 0.49 ms^{-1} 2) 0.98 ms^{-1}
3) 1.96 ms^{-1} 4) 3.92 ms^{-1}

2. Length of a simple pendulum is 2m and mass of its bob is 0.2 kg. If the tension in the string exceeds 4N, it will break. If the bob is whirled in horizontal plane, the maximum angle the string can make with vertical during rotation is

1) 30^0 2) 45^0 3) 60^0 4) 90^0

3. A chain of 100 links is 1m long and has a mass of 2kg. With the ends fastened together it is set rotating at 3000 rpm, in a horizontal plane.The centripetal force on each link is

1) 3.14 N 2) 31.4N 3) 314 N 4) 3140 N

4. A boy is sitting on a horizontal platform in the shape of a disc at a distance of 5m from its centre. The boy begins to slip when the speed of wheel exceeds 10 rpm The coefficient of friction between the boy and platform is. ($g = 10\ ms^{-2}$) (2002 M)

1) $\pi^2/6$ 2) $\pi^2/18$
3) $\pi/6$ 4) $\pi/2$

5. Three point masses each of mass 'm' are joined together using a string to form an equilateral triangle of side 'a'. The system is placed on a smooth horizontal surface and rotated with a constant angular velocity 'ω' about a vertical axis passing through the centroid. Then the tension in each string is

1) $ma\omega^2$ 2) $3ma\omega^2$
3) $\frac{ma\omega^2}{3}$ 4) $\frac{ma\omega^2}{\sqrt{3}}$

6. There is a small hole in a table. A string of length 1m passes through it. Two bodies of masses 70g and 100g are attached at its ends. The 100g mass hangs freely at a depth of 60 cm from the table. If this mass is to be in equilibrium, the other mass should rotate in a circle with a frequency equal to

1) $4\pi/140$ Hz 2) $\pi/140$ Hz
3) $\pi/\sqrt{140}$ Hz 4) $\frac{\sqrt{140}}{4\pi}$ Hz

7. A particle 'P is moving in a circle of radius 'r' with uniform speed v. AB is the diameter of circle and 'C' is the centre. The angular velocity of P about A and C are in the ratio.

1) 1 : 1 2) 1 : 2 3) 2 : 1 4) 4 : 1

8. Starting from rest a wheel rotates with uniform angular acceleration $2\pi\ rad s^{-2}$. After 4s, if the angular acceleration ceases to act, its angular displacement in the next 4s is

1) 8π rad 2) 16π rad
3) 24π rad 4) 32π rad

Vertical Circle :

9. A simple pendulum of length 'ℓ' has a bob of mass 'm'. If it is released when the string makes an angle 60^0 with the vertical, the tension in the string at the lowest position is

1) 3mg 2) $\frac{3}{2}mg$ 3) 2 mg 4) $\frac{5}{2}mg$

10. A body is revolving in a vertical circle with constant mechanical energy. The speed of the body at the highest point is $\sqrt{2rg}$. The speed of the body at the lowest point is

1) $\sqrt{7gr}$ 2) $\sqrt{6gr}$ 3) $\sqrt{8gr}$ 4) $\sqrt{9gr}$

11. A particle suspended by a thread of length 'l' is projected horizontally with a velocity $\sqrt{3g\ell}$ at the lowest point. The height from the bottom at which the tension in the string becomes zero is

1) $\frac{4\ell}{3}$ 2) $\frac{2\ell}{3}$ 3) $\frac{5\ell}{3}$ 4) $\frac{\ell}{3}$

12. A nail is located at a certain distance vertically below the point of suspension of a simple pendulum of length 1m. The pendulum is released from the horizontal position. If it rotates in a vertical circle with nail as centre, the distiance of nail from point of suspension is

1) 0.5 m 2) 0.6 m 3) 0.4 m 4) 0.8m

13. A body is moving in a vertical circle of radius 'r' by a string. If the ratio of maximum to minimum speed is $\sqrt{3}:1$, the ratio of maximum to minimum tensions in the string is

1) 3 : 1 2) 5 : 1 3) 7 : 1 4) 9 : 1

14. A test tube of mass 10g closed with a cork of mass 1g having some ether is suspended by a thread of length 4cm. On heating the tube the cork flies out. The minimum velocity of cork, so that the test tube just describes the vertical circle is

1) 14 ms^{-1} 2) 7 ms^{-1} 3) 3.5 ms^{-1} 4) 12.4 ms^{-1}

15. The simple pendulum bob is given a horizontal velocity $\sqrt{g\ell}$ at the bottom. The angle with the vertical through which the ball swings before its velocity becomes zero is

1) 120^0 2) 60^0 3) 450^0 4) 30^0

16. A coin is sliding down on a smooth hemi-spherical surface of radius R. The height from the bottom, where it looses contact with the surface is

1) $R/3$ 2) $2R/3$ 3) $3R/4$ 4) $4R/5$

17. A small glass marble of mass 'm' oscillates between the two edges, inside a hemispherical glass bowl of radius 'r'.If 'V' is the speed of marble at the lowest position, the normal reaction at that position is

1) $\frac{mv^2}{r}$ 2) $\frac{2mv^2}{r}$ 3) $\frac{3mv^2}{r}$ 4) $\frac{3mv^2}{2r}$

18. An elastic string carrying a body of mass 'm' extends by 'e'. The body rotates in a vertical circle with critical velocity. The extension in the string at the lowest position is

1) 2e 2) 4e 3) 6e 4) 8e

19. A mass 'm' is released, with a horizontal speed v from the top of a smooth and fixed, hemispherical bowl, of radius r. The angle θ w.r.t. the vertical where it leaves contact with the bowl is

1) $\sin^{-1}\left(\frac{v^2}{3rg}+\frac{2}{3}\right)$ 2) $\cos^{-1}\left(\frac{v^2}{3rg}+\frac{2}{3}\right)$

3) $\tan^{-1}\left(\frac{v^2}{3rg}+\frac{2}{3}\right)$ 4) $\cos^{-1}\left(2/3\right)$

20. A mass of 3.9 kg suspended from a string of length 0.5m is at rest. Another body of mass 0.1kg moving horizontally with a velocity 200 ms^{-1} strikes and sticks to it. The tension in the string when it makes 60^0 with the vertical is ($g = 10\ ms^{-2}$)

1) 180 N 2) 220 N 3) 80 N 4) 100 N

21. A simple pendulum of length 'ℓ' carries a bob of mass 'm'. When the bob is at its lowest position, it is given the minimum horizontal speed necessary for it to move in a vertical circle about the point of suspension. When the string is horizontal, the net force on the bob is

1) mg 2) 3mg 3) $\sqrt{10}$ mg 4) 4 mg

22. A simple pendulum is vibrating with an angular amplitude of 90^0. For what values of θ with vertical, the acceleration is directed

1) Vertically upwards 2) horizontally
3) vertically downwards

1) $0^0, \cos^{-1} 1/\sqrt{3}, 90^0$

2) $\cos^{-1}\left(1/\sqrt{3}\right), 0^0, 90^0$

3) $90^0, \cos^{-1} 1/\sqrt{3}, 0^0$ 4) $\cos^{-1}\left(1/\sqrt{3}\right), 90^0, 0^0$

Moment of Inertia :

23. Four particles each of mass 'm' are placed at the four corners of a square of side 'ℓ'. The M.I. of the system about a normal axis through centre of square is

1) $4m\ell^2$ 2) $8m\ell^2$ 3) $2m\ell^2$ 4) $m\ell^2/4$

24. Three point masses m_1, m_2 and m_3 are placed at the vertices of an equilateral triangle of side 'a'. The M.I. of the system about an altitude passing through m_1 is

1) $(m_1+m_2)\frac{a^2}{4}$ 2) $(m_3+m_1)\frac{a^2}{4}$

3) $(m_2+m_3)\frac{a^2}{4}$ 4) $(m_1+m_2+m_3)\frac{a^2}{4}$

25. Four small spheres each of radius 'r' and mass 'm' are placed with their centres at the four corners of a square of side 'L'. The M.I. of the system about any side of square is

1) $\frac{8mr^2}{5}+mL^2$ 2) $\frac{8mr^2}{5}+2mL^2$

3) $\frac{5mr^2}{8}+mL^2$ 4) $\frac{5mr^2}{8}+2mL^2$

26. In a rectangle ABCD, AB = 2ℓ and BC = ℓ. The centre of rectangle is at the origin of co-ordinate system. AB and CD are parallel to X-axis. The M.I. is least about

1) BD 2) BC 3) X -axis 4) Y-axis

27. The moment of inertia of a uniform rod of length '2ℓ' and mass 'm' about an aixs through centre and inclined at an angle θ to rod is

1) $\frac{m\ell^2}{3}\sin^2\theta$ 2) $\frac{m\ell^2}{6}\cos^2\theta$

3) $\frac{m\ell^2}{2}\cos^2\theta$ 4) $\frac{m\ell^2}{12}\sin^2\theta$

28. Three identical rods, each of length 'L' are joined to form a rigid equilateral triangle. Its radius of gyration about an axis passing through a corner and perpendicular to the plane of triangle is

1) $L/2$ 2) $\sqrt{\frac{3}{2}}L$ 3) $\frac{L}{\sqrt{2}}$ 4) $\frac{L}{\sqrt{3}}$

29. A disc of mass 'm' and radius R has a concentric hole of radius 'r'. Its M.I. about an axis through centre and normal to its plane is

1) $\frac{m}{2}(R-r)^2$ 2) $\frac{m}{2}(R^2-r^2)$

3) $\frac{m}{2}(R+r)^2$ 4) $\frac{m}{2}(R^2+r^2)$

30. Four holes of radius 'R' are cut from a thin square plate of side 4R and mass M. The M.I. of the remaining portion about a normal axis through centre of square is

1) $\frac{\pi}{12}MR^2$ 2) $\left(\frac{8}{3}-\frac{10\pi}{16}\right)MR^2$

3) $\left(\frac{4}{3}-\frac{\pi}{4}\right)MR^2$ 4) $\frac{\pi}{3}MR^2$

31. The M.I of a thin circular ring of Radius 'R' and mass M about a chord at a distance equal to half of the radius of ring is

1) $\frac{5}{4}MR^2$ 2) $\frac{MR^2}{4}$ 3) $\frac{3}{2}MR^2$ 4) $\frac{3}{4}MR^2$

32. The M.I. of a thin rod about a normal axis through its centre is I. It is bent at the centre such that, the two parts are perpendicular to each other and perpendicular to the axis. The M.I. of the system about the same axis will be

1) 2I 2) I 3) I/2 4) 4I

33. A circular disc of radius 'R' and thickness R/6 has moment of inertia 'I' about an axis passing through its centre and perpendicular to its plane. It is melted and recasted into a solid sphere. The M.I of the sphere about its diameter as axis of rotation is

1) I 2) 2I/3 3) I/5 4) I/10

34. The M.I. of a uniform circular disc about a diameter is I. Its M.I. about an axis perpendicular to its plane and passing through a point on its rim will be

1) 2I 2) 3I 3) 5I 4) 6I

35. A uniform circular disc of radicu R lies in the XY - plane with its centre at the orgin of co-oridinate system. Its moment of inertia about an axis, lying in the xy-plane, parallel to the x-axis and passing through a point on the y-axis at a distance y = 2R is I_1. Its moment of inertia about an axis lying in a plane perpendicular to xy-plane passing through a point on the x-axis at a distance x = d is I_2. If $I_1 = I_2$, the value of 'd' is (2006 M)

1) $\frac{\sqrt{19}}{2}R$ 2) $\frac{\sqrt{17}}{2}R$ 3) $\frac{\sqrt{15}}{2}R$ 4) $\frac{\sqrt{13}}{2}R$

36. **Three spheres, each of mass 'm' and radius 'R' are kept in touch with each other such that their centres from an equilateral triangle. The M.I. of the system about a median of triangle is**

1) $\frac{21}{5}MR^2$ 2) $\frac{16}{5}MR^2$ 3) $\frac{7}{2}MR^2$ 4) $\frac{4}{5}MR^2$

37. **Two uniform thin rods each of mass 'm' and length 'L' are arranged to form a cross. The moment of inertia of the sytem about an angular bisector is**

1) $\frac{mL^2}{12}$ 2) $\frac{mL^2}{6}$ 3) $\frac{mL^2}{3}$ 4) $\frac{2mL^2}{3}$

38. **A thin rod of mass 6m and length 6L is bent into regular hexagon .The M.I of the hexagon about a normal axis to its plane and through centre of system is**

1) mL^2 2) $3mL^2$ 3) $5mL^2$ 4) $11mL^2$

39. **Let 'I' be the M.I. of uniform square plate about an axis AB that passes through its centre and is parallel to two of its sides. CD is a line in the plane of the plate that passes through the centre of the plate and makes an angle θ with AB. The M.I . of the plate about the axis CD is equal to**

1) I 2) $I\sin^2\theta$ 3) $I\cos^2\theta$ 4) $I\cos^2\left(\frac{\theta}{2}\right)$

Conservation of angular momentum

40. **A person is standing on a platform rotating with an angular speed of 1 rps. with streched hands carrying weights . On drawing his hands close to the body, the M.I. of the system changes from 8 kgm² to 4kgm². The increase in his kinetic energy nearly is.**

1) 320J 2) 40J 3) 80J 4) 160 J

41. **A boy of mass 30kg is running with a velocity of 3 ms^{-1} on ground just tangentially to a merry - go -round which is at rest. It has radius R = 2m, a masss of 120 kg and its radius of gyration is 1m. If the boy suddenly jumpes on to the merry -go - round, the angular velocity acquired by the system is**

1) 1 rad s^{-1} 2) $\frac{4}{3}rad\,s^{-1}$ 3) $\frac{3}{4}rad\,s^{-1}$ 4) 0.5 rad s^{-1}

42. **A thin circular ring of mass 'm' and radius R is rotating about its axis with a constant angular velocity ω. Tho objects each of mass M are attached gently to he opposite ends of a diameter of the ring. The new angular velocity of ring is**

1) $\frac{\omega m}{M+m}$ 2) $\frac{\omega m}{m+2M}$

3) $\frac{\omega(m+2M)}{m}$ 4) $\frac{\omega(m-2M)}{m+2M}$

43. **A horizontal disc rotates freely with angular velocity 'ω' about a vertical axis through its centre. A ring, having the same mass and radius as the disc, is gently placed on the disc. The common angular velocity with which the two rotate, will be,**

1) ω 2) $\omega/2$ 3) $\omega/3$ 4) $\omega/4$

44. **A circular platform is free to rotate in a horizontal plane about a vertical axis passing through its centre. A tortoise is sitting at the edge of platform. Now the platform is given an angular velocity ω_0. When the tortoise moves along the chord of the platform with a constant velocity (with respect to the platform); The angular velocity of platform $\omega(t)$ will vary with time 't' as**

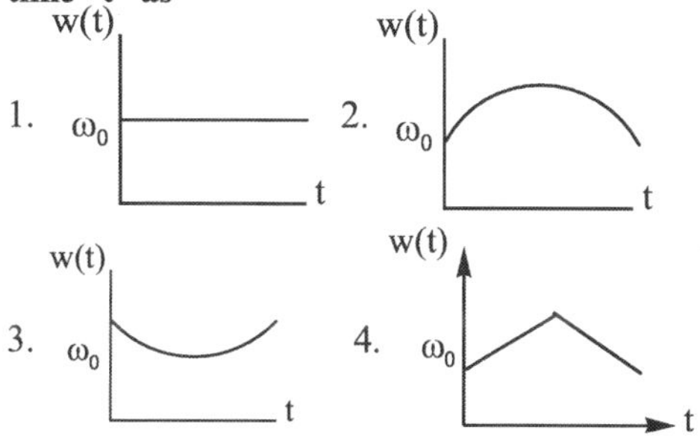

45. **A rigid uniform rod of mass M and length 'L' is resting on a smooth horizontal table. Two marbles each of mass 'm' and travelling with uniform speed 'V' collide with two ends of the rod simultaneously and inelastically as shown. The marbles get struck to the rod after the collision and continue to move with the rod. If $m = \frac{M}{6}$ and V = L mts/sec, then the time taken by the rod to rotate through $\frac{\pi}{2}$ is 2009 M**

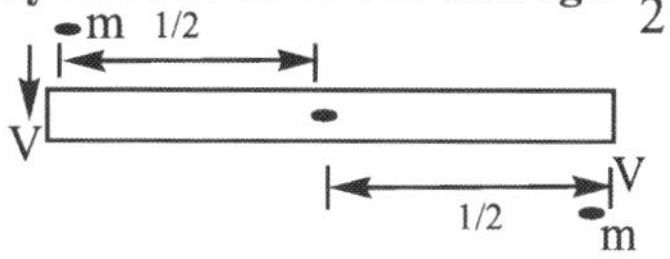

1) 1 s 2) 2π s 3) π s 4) $\frac{\pi}{2}$ s

Rotational Dynamics :

46. **A wheel at rest has M.I. $\frac{2}{\pi^2}kg\,m^2$. It is rotated by a 60 W motar for one minute. The number of rotations made by the wheel in one minute is**

1) 90 2) 450 3) 1800 4) 1200

47. **A particle performs uniform circular motion with an angular momentum L. If the angular frequency of particle is doubled and kinetic energy is halved, its angular momentum becomes (2003 M)**

1) 4L 2) 2L 3) L/L 4) L/4

48. **A heavy wheel of radius 20 cm and weight 10kg is to be dragged over a step of height 10cm , by a horizontal force F applied at the centre of the wheel. The minimum value of F is**

1) 20 kgwt 2) 1 kgwt

3) $10\sqrt{3}$kgwt 4) $10\sqrt{2}$kgwt

49. **A wheel of radius 0.4 m can rotated freely about its axis as shown in the figure. A string is wrapped over its rim and a mass of 4 kg is hung, An angular acceleration of 8 rad-s^{-2} is produced in it due to the torque. Then moment inergia of the wheel is (g = 10 ms|−2) 2009 E**

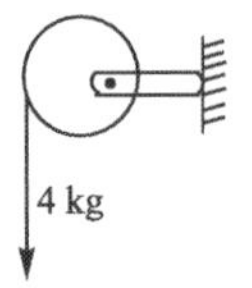

1) 2 kg - m^2

2) 1 kg - m^2

3) 1.36 kg - m^2

4) 8 kg - m^2

50. **Same torque is applied on a solid cylinder and solid sphere with same mass and radius. Cylinder rotates about its axis and sphere about its diameter. Then in a given time, the angular velocity acquired is**

1)is more for sphere 2) is more for cylinder

3) is some for both 4) none

51. **A wheel of M.I 0.1 kg m^2 is rotating at a speed of 1200 rpm. Now it is slowed down uniformly such that it comes to rest after making 314 complete revolutions. The magnitude of the retarding torque is**

1) 0.4 Nm 2) 0.2 Nm

3) 2 Nm 4) 4 Nm

52. **The M.I. of a uniform disc about an axis passing through its centre and perpendicular to its plane is 1 kg m^2. It is rotating with an angular velocity of 100 rad s^{-1}. A second disc of same mass and radius is joined to it coaxially. Now these two discs together continue to rotate about the same axis. Then the lose in kinetic energy in kilo joules is (2002 M)**

1) 2.5 2) 3 3) 3.5 4) 4

53. **A rod of lenght l is held vertically stationary with its lower end located at a point 'P', on the horizontal plane. When the rod is released to topple about 'P', the velocity of the upper end of the rod with which it hits the ground is 2009 E, M**

1) $\sqrt{\frac{g}{\ell}}$ 2) $\sqrt{3g\ell}$ 3) $3\sqrt{\frac{g}{\ell}}$ 4) $\sqrt{\frac{3g}{\ell}}$

54. **A string is wrapped around a wheel of radius 'r'. The axis of the wheel is horizontal and its M.I. about the axis is I. Weight 'mg' is tied to free end of the string which is released to fall down from rest position. The angular velocity of the wheel after it has fallen through distance 'h' will be**

1) $\left(\frac{2gh}{I+mr^2}\right)^{\frac{1}{2}}$ 2) $\left(\frac{2mgh}{I+mr^2}\right)^{\frac{1}{2}}$

3) $\frac{2mgh}{I+2mr^2}$ 4) $\sqrt{2gh}$

55. **Reel of thread in the form of solid cylinder is allowed to unroll by holding the loose end of the thread in hand. The acceleration with which the reel falls down is (g = 9.9 ms^{-2})**

1) 9.9 ms^{-2} 2) 3.3 ms^{-2}

3) 6.6 ms^{-2} 4) 1.1 ms^{-2}

56. **A particle of mass 1 kg is projected with an initial velocity 10 ms^{-1} at an angle of projection 45° with the horizontal. The average torque acting on the projectlie, between the time at which it is projected and the time at which it strikes the ground about the point of projection is**

1) 25 Nm 2) 50 Nm 3) 75 Nm 4) 100 Nm

57. **'O' is the centre of an equilateral triangle ABC. F_1, F_2 and F_3 are three forces acting along the sides AB, BC and AC as shown in figure. What should be the magnitude of F_3, so that the total torque about 'O' is zero ?**

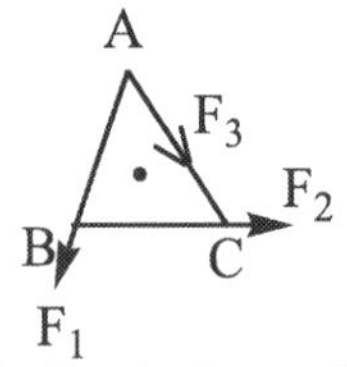

1) $\frac{F_1+F_2}{2}$ 2) $2(F_1+F_2)$

3) (F_1+F_2) 4) (F_1-F_2)

Rolling Motion :

58. **A disc at rest at the top of an inclined plane of heigh 'h' rolls down without slippling and acquires a velocity 'v' on reaching the bottom. If the same disc slides down a smooth incline and acquires the same velocity on reaching the bottom the height of smooth incline is**

1) h/3 2) h/2 3) $\frac{2h}{3}$ 4) h

59. **A uniform solid sphere of mass 0.5kg rolls on a horizontal surface without slipping. If the velocity of centre of mass is 4 cm s^{-1}, the work done in stopping sphere is**

1) 8×10^{-4} J 2) 5.6×10^{-4} J

3) 1.6×10^{-4} J 4) 2.4×10^{-4} J

60. A sphere rolls without slipping on an incline of inclination θ. The minimum coefficient of static friction to support pure rolling is to be

1) $\frac{2}{3}\text{Tan}\theta$ 2) $\frac{2}{5}\text{Tan}\theta$ 3) $\frac{2}{7}\text{Tan}\theta$ 4) $\frac{3}{7}\text{Tan}\theta$

61. A wheel of radius 'r' rolls without slipping with a speed 'v' on a horizontal road. When it is at point 'A' on the road, a small lump of mud separtes from the wheel at its highest point B and drops at point 'C' on the road. The distance AC will be

1) $v\sqrt{r/g}$ 2) $2v\sqrt{r/g}$ 3) $4v\sqrt{r/g}$ 4) $v\sqrt{3r/g}$

Equilibrium of Body & Parallel Forces :

62. The resultent of two like parallel forces is 14 kgwt and acts at distance of 30 cm and 40 cm from them. The forces are

1) 6 kgwt, 8 kgwt 2) 3 kgwt, 4 kgwt
3) 4 kgwt, 6 kgwt 4) 8 kgwt, 3 kgwt

63. A unifrom rod weighs 10kg and with a load 5kg attached to one end, it balances on a knife edge at 2m from that end. The length of rod is

1) 10 m 2) 8m 3) 6m 4) 4m

64. A boy and a man carry a uniform rod of length 'L' horizontally in such a way that the boy get 1/4 of the load. If the boy is at one end of the rod, the distance of the man from the other end is

1) $L/6$ 2) $L/5$ 3) $L/4$ 4) $L/3$

ANSWERS

1) 2	2) 3	3) 3	4) 2	5) 3
6) 4	7) 2	8) 4	9) 3	10) 2
11) 1	12) 2	13) 3	14) 1	15) 2
16) 2	17) 4	18) 3	19) 2	20) 1
21) 3	22) 1	23) 3	24) 3	25) 2
26) 3	27) 1	28) 3	29) 4	30) 2
31) 4	32) 2	33) 3	34) 4	35) 3
36) 2	37) 3	38) 3	39) 1	40) 4
41) 3	42) 2	43) 3	44) 2	45) 4
46) 3	47) 4	48) 3	49) 3	50) 1
51) 1	52) 1	53) 4	54) 2	55) 3
56) 2	57) 3	58) 3	59) 2	60) 3
61) 3	62) 1	63) 3	64) 4	

EXERCISE - III(B)

Circular Motion in Horizontal plane

1. A hemispherical bowl of radius R is set rotating about its axis of symmetry which is kept vertical. A small block kept in the bowl rotates with the bowl without slipping on its surface. If the surface of bowl is smooth, and the angle made by the radius through the block with the vertical is θ, the angular speed at which the bowl is rotating is

1) $\sqrt{\frac{g}{R\cos\theta}}$ 2) $\sqrt{\frac{g\cos\theta}{R}}$

3) $\sqrt{\frac{g\sin\theta}{R}}$ 4) $\sqrt{\frac{g}{R\sin\theta}}$

2. A particle of mass 'm'is suspended from the ceiling through a string of length L. The particle moves in a ho;rizontal circle of radius r. The speed of the particle is.

1) $\frac{rg}{\sqrt{L^2 - r^2}}$ 2) $\frac{r\sqrt{g}}{\left(L^2 - r^2\right)^{\frac{1}{4}}}$

3) $\frac{r\sqrt{g}}{\left(L^2 - r^2\right)^{\frac{1}{2}}}$ 4) $\frac{mgL}{\left(L^2 - r^2\right)^{\frac{1}{2}}}$

3. A rope of mass 'm' and radius R is placed on a smooth horizontal table and is set rotating about its own axis in such a way that each point of it moves with a speed v. The tension in the string is

1) $\frac{mv^2}{2\pi^2 R}$ 2) $\frac{mv^2}{\pi R}$ 3) $\frac{mv^2}{2\pi R}$ 4) $\frac{mv^2}{R}$

4. A person stands in contact against a wall of cylindrical drum of radius r, rotating with an angular velocity ω. If μ is the coefficient of friction between the wall and the person, the minimum rotational speed which enables the person to remain stuck to the wall is

1) $\sqrt{g/\mu r}$ 2) $g/\sqrt{\mu r}$ 3) $\sqrt{\frac{\mu r}{g}}$ 4) $\sqrt{\frac{gr}{\mu}}$

5. One end of a massless spring of spring constant 100 Nm^{-1} and natural length 0.5 m is fixed and the other end is connected to a particle of mass 0.5 kg lying on a frictionless horizontal table. If the mass rotates at an angular velocity of 2 rad s^{-1}, the elongation of the spring is approxi-mately

1) 4 cm 2) 3 cm 3) 1 cm 4) 2 cm

6. Two particles of masses m_1 and m_2 are connected to a string and the system is rotated in a horizontal plane with 'P' as center. The ratio of tension in the two parts of string is

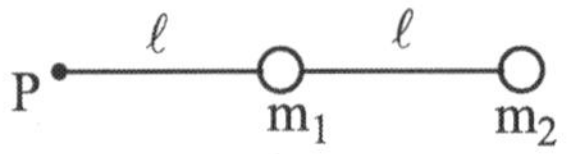

1) $\frac{m_1}{m_1+m_2}$ 2) $\frac{m_1+m_2}{m_1}$

3) $\frac{m_1+2m_2}{2m_2}$ 4) $\frac{2m_1}{m_1+m_2}$

7. A particle P is moving in a circle of diametre 'd' with uniform speed V. 'C' is the centre of circle and AB is diameter. Then angular velocity of 'P' about A and C are in the ratio.

1) 1 : 1 2) 1 : 2 3) 2 : 1 4) 4 : 1

8. A point moves along a circle of radius 20 cm with costant angential acceleration of 5 cm s^{-2}. The time after which the normal acceleration of the point will be equal to tangential acceleration is

1) 1s 2) 2s 3) 4s 4) 6s

Vertical Circle :

9. A simple pendulum of length 'ℓ' has a bob of mass m. It is released from the horizontal position. The tension in the string at the lowest position is

1) 3mg 2) $\frac{3}{2}mg$ 3) 2mg 4) $\frac{5}{2}mg$

10. A body is revolving in a verticle circle with constant mechanical energy.The speed of the body at the highest point is $\sqrt{3rg}$. The speed of the body at the horizontal position is

1) $\sqrt{rg}$ 2) $\sqrt{5rg}$ 3) $\sqrt{7rg}$ 4) $\sqrt{9rg}$

11. A simple pendulum bob is given a horizontal velocity $\sqrt{\frac{7}{2}g\ell}$ at the bottom. The string slackens after swinging through an angle θ with the vertical. The θ is

1) 30^0 2) 60^0 3) 90^0 4) 120^0

12. Anail is located at a certain distance vertically below the point of suspension of a simple pendulum. The pendulum is released at an angle 60^0 to vertical. The distance of the nail from the point of suspension such that the bob will just perform revolution with the nail as centre (length of pendulum = 1m)

1) 0.2 m 2) 0.4 m 3) 0.6 m 4) 0.8m

13. The maximum and minimum tensions in a string of length 0.1m with a stone 5g tied to it and whirled in a vertical circle are in the ratio 13 : 1. The velocity of the body at highest point is

1) 0.7 ms^{-1} 2) $0.7\sqrt{3}ms^{-1}$

3) $\sqrt{3}ms^{-1}$ 4) $7\sqrt{3}ms^{-1}$

14. A bullet of mass 'm' hits a pendulum bob of mass 2m with a velocity 'v' and comes out of the bob with a velocity V/2. The length of pendulum is 2m. The minimum value of 'v' so that the bob revolves in a vertical circle is (g = 10 ms^{-1})

1) 40 ms^{-1} 2) 2 ms^{-1} 3)15 ms^{-1} 4) 10 ms^{-1}

15. A stone tied to a string revolving in a vertical circle has critical velocity 'v' at the highest point. When the string makes an angle 120^0 with lower vertical. The speed of the stone is

1) $3V/2$ 2) $\sqrt{2}v$ 3) $\sqrt{3}v$ 4) $\sqrt{\frac{3}{2}}v$

16. A small mass lying at the top of a smooth hemispherical bowl is just pushed horizontally. The angle with the vertical where it loses contact with surface is

1) $Tan^{-1}(2/3)$ 2) $\sin^{-1}(2/3)$

3) $\cos^{-1}(2/3)$ 4) $\cot^{-1}(2/3)$

17. A body slides down from rest on a smooth incline of height of 8R. At the bottom of incline, there is a vertical groove of radius R. When the ball reaches the lowest point of the track, the normal reaction on it is

1) 17 mg 2) 12 mg 3) 8 mg 4) 10 mg

18. An elastic string carrying a body of mass 'm' extends by 'e'. If the body rotates in a vertical circle with critical velocity at highest point, the extension in the string at the horizontal position is

1) 2e 2) 4e 3) 3e 4) 5e

19. A body of mass Mkg is on the top point of a smooth hemisphere of radius 5m. It is released to slide down the surface of hemisphere. It leaves the surface when its velocity is 5 ms^{-1}. At this instant the angle made by the radius vecotor of the body with the vertical is (g = 10 ms^{-2}) (2002 E)

1) 30^0 2) 45^0 3)60^0 4) 90^0

20. A mass of 3.9 kg suspended from a string of length 0.5 m is at rest. Another body of mass 0.1 kg moving horinzontally with a velocity 200 ms^{-1} strikes and stricks to it. The maximum angle through which the system swings just before the tension in the string becomes zero is

1) 60^0 2) 90^0 3) 120^0 4) 180^0

21. A stone tied to a string of length 'L' is whirled in a vertical circle with the other end of string at the centre. At a certain instant of time, the stone is at its lowest position and has a speed u. The magnitude of the change in velocity as it reaches a point where the string is horizontal is

1) $\sqrt{u^2-2gL}$ 2) $\sqrt{2gL}$

3) $\sqrt{u^2-gL}$ 4) $\sqrt{2(u^2-gL)}$

22. The breaking strength of a string is equal to double the weight of bob it carries. If the string is released from horizontal position the angle θ with the vertical at which the string breaks is

1) 60^0 2) $\cos^{-1}\left(\frac{1}{3}\right)$ 3) $\sin^{-1}\left(\frac{2}{3}\right)$ 4) 45^0

Moment of Inertia :

23. Four particles each of mass 'm' are placed at the corners of a square of side 'L'. The radius of gyration of the system about an axis normal to the square and passing through its centre is

1) $L/\sqrt{2}$ 2) $L/2$ 3) L 4) $\sqrt{2}L$

24. Three particles each of mass 'm' are located at the vertices of an equilateral triangle of side 'L'. The radius of gyration of the system about an axis through centroid and perpendiculer to its plane

1) $\frac{L}{2\sqrt{3}}$ 2) $\frac{L}{\sqrt{3}}$ 3) $\frac{L}{2}$ 4) $\frac{L}{2\sqrt{5}}$

25. Four spheres each of mass 'm' and radius 'r' are situated at the four corners of a square of side 'R'. The M.I. of the system about an axis through centre and perpendicular to its plane is

1) $\frac{5}{2}M(4r^2+5R^2)$ 2) $\frac{2}{5}M(4r^2+5R^2)$

3) $\frac{2}{5}M(4R^2+5r^2)$ 4) $\frac{5}{2}M(4R^2+5r^2)$

26. ABC is a right angled trianglar plate of uniform thichness. I_1, I_2 and I_3 are M.I. about AB, BC and AC respectively, Then which of the following relations is correct ? [AB = 3m BC = 4m]

1) $I_1 = I_2 = I_3$ 2) $I_2 > I_1 > I_3$

3) $I_3 < I_2 < I_1$ 4) $I_3 > I_1 > I_2$

27. The moment of inertia of a uniform rod of length 'ℓ' and mass 'm' about an axis passing through one end of rod and inclined at angle θ to rod is

1) $\frac{m\ell^2}{3}\cos^2\theta$ 2) $\frac{m\ell^2}{3}\sin^2\theta$

3) $\frac{m\ell^2}{12}\cos^2\theta$ 4) $\frac{m\ell^2}{12}\sin^2\theta$

28. Four identical rods, each of mass 'm' and length 'L' are arranged in the from of a square. The radius of gyration of the system about a perpendicular axis to its plane, through centre is

1) $\frac{L}{\sqrt{3}}$ 2) $\frac{L}{2\sqrt{3}}$ 3) $\frac{L}{\sqrt{2}}$ 4) $\frac{L}{6}$

29. From a complete ring of mass 'm' and radius 'R' a 30^0 sector is removed. The M.I. of the remaing ring about an axis through centre of ring and normal to its plane is

1) $\frac{3}{4}mR^2$ 2) $\frac{11}{12}mR^2$ 3) $\frac{2}{5}mR^2$ 4) mR^2

30. From a disc of mass 'm' and radius 'r' a circular portion of radius $'r'/2$ is removed from the edges. The M.I. of the remaining system about a normal axis through centre of original disc is

1) $\frac{13}{32}mr^2$ 2) $\frac{11}{12}mr^2$ 3) $\frac{17}{24}mr^2$ 4) $\frac{8}{11}mr^2$

31. A thin wire of mass 'm' and length 'L' is bent in the form of a ring. The M.I. of the ring about a tangential axis parallel to any of its diameters is

1) $\frac{mL^2}{2\pi^2}$ 2) $\frac{3mL^2}{8\pi^2}$ 3) $\frac{2mL^2}{5\pi^2}$ 4) $\frac{mL^2}{6\pi^2}$

32. A thin rod of length 'L' and mass 'm' is bent at the centre through 60^0. The M.I of the system about an axis through centre and perpendicular to the plane of system is

1) $\frac{mL^2}{6}$ 2) $\frac{mL^2}{24}$ 3) $\frac{mL^2}{12}$ 4) $\frac{mL^2}{3}$

33. If 27 small spheres each of M.I. 'I' about their diametrical axis are combined to form a big sphere, then the M.I. of the big sphere about its diameter is

1) 243 I 2) 81 I 3) 27 I 4) 9 I

34. The moment of inertia of a uniform circular disc about a normal axis through centre is I. The M.I. about a tangential axis parallel to any of its diameters is (2008 E)

1) $\frac{7I}{2}$ 2) $\frac{3}{2}I$ 3) 6I 4) $\frac{5}{2}I$

35. A uniform circular ring of radius 'R' lies in the XY - plane with its centre at the origin. Its moment of inertia about an axis, lying the XY-plane, parallel to X- axis and passing through a point on the Y - axis at a distance y = 2R is I_1. Its M.I. about an axis lying in a plane perpendicular to XY - plane. Passing through a point on the X - axis at a distance x = d is I_2. If $I_1 = I_2$, the value of 'd' is

1) $R\sqrt{7/2}$ 2) $R\sqrt{3/2}$

3) $R\sqrt{5/2}$ 4) $R\sqrt{6/7}$

36. Three rings, each of mass 'M' and radius 'R' are kept touching each other such that their centress from equilateral triangle. The M.I. of the system about a median of the triangle is

1) 3 MR^2 2) 3.5 MR^2 3) 2.5 MR^2 4) 2 MR^2

37. Two identical rods each of moment of inertia 'I' about a normal axis through centre are arranged in the from of a cross. The M.I. of the system about an axis through centre and perpendicular to the plane of system is

1) 2I 2) I 3) 2.5 I 4) 6I

38. A quarter sector is cut from a unifrom circular disc of radius R. This sector has mass M. It is made to rotate about a line perpendicular to its plane and passing through the centre of the original disc. Its moment of inertia about the axis of rotation is (IIT - 2001)

1) $\frac{MR^2}{2}$ 2) $\frac{MR^2}{4}$ 3) $\frac{MR^2}{8}$ 4) $\sqrt{2}\,MR^2$

39. The moment of inertia of a uniform semi circular disc of mass M and radius 'r' about a line perpendicular to the plane of disc through the centre is

1) MR^2 2) $MR^2/2$ 3) $MR^2/4$ 4) $\frac{2}{5}MR^2$

Conservation of Angular Momentum :

40. A person is standing on a platform rotating with an angular speed w, with stretched hands carrying equal masses each of m. Now he releases the masses such that they fall on the platform. The new angular velocity of platform would be

1) w 2) $\frac{w}{2}$ 3) 2w 4) $2w/3$

41. A cockroach of mass 'm' is moving on the rim of a disc with a velocity 'v' in anti clockwise direction. The M.I. of the system about its own axis is I and it is rotating in the clockwise direction with angular speed w. If the cockroach stops moving, the angular speed of the disc will be

1) $\frac{Iw}{I+mR^2}$ 2) $\frac{Iw+mvr}{I+mR^2}$

3) $\frac{Iw-mvr}{I+mR^2}$ 4) $\frac{Iw-mvR}{I}$

42. A thin circular disc of mass 'm' and radius R is rotating about its axis with a constant angular velocity w. It has two masses each of mass 'm' (same mass as that of disc) at the opposite ends of a diametre of the disc. Now if the massas are removed, the new angular velocity of disc would be

1) $w/5$ 2) 5w 3) $\frac{5}{2}w$ 4) $\frac{2}{5}w$

43. A disc rotates freely with rotational kinetic energy E about a normal axis through centre. A ring having the same radius but double the mass of disc is now, gently placed on the disc. The new rotational kinetic energy of the system would be

1) $E/2$ 2) $2E/5$

3) $E/5$ 4) $E/3$

44. An equilateral triangle ABC formed from a uniform wire has two small identical beads initially located at A. The triangle is set rotating about the vertical axis AO. Then the beads are released from rest simultaneously and allowed to slide down, one along AB and other along AC as shown. Neglecting frictional effects, the quantities that are conserved as the beads slide down, are (IIT - 2000)

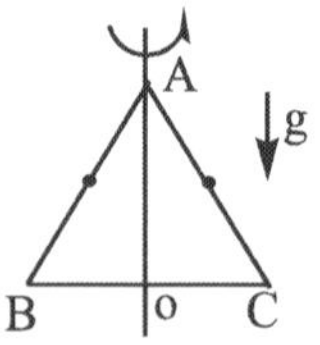

1) Angular velocity and total energy
2) Total angular momentum and total energy
3) Angular velocity and M.I. about the axis of rotation
4) Total angular momentum and M.I. about the axis of rotation

45. A rod of negligible mass but of length 'ℓ' connected with two identical masses at both ends is placed on a horizontal frictionless surface. A sudden impulse of MV is given as shown in figure. The angular velocity of rod would be

1) $\frac{4v}{\ell}$ 2) $\frac{v}{\ell}$

3) $\frac{2v}{\ell}$ 4) $\frac{3v}{\ell}$

Rotational Dynamics :

46. A constant power is supplied to a rotating disc. The relationship of angular velocity (w) of disc and number of rotations (n) made by the disc is governed by (2005 E)

1) $w \alpha n^{\frac{1}{3}}$ 2) $w \alpha n^{\frac{2}{3}}$

3) $w \alpha n^{\frac{3}{2}}$ 4) $w \alpha n^{2}$

47. A particle performing unifrom circular motion has angular momentum L. If its angular frequency is doubled and its kinetic energy is halved, then the new angular momentum is

1) $L/4$ 2) 2L 3) 4L 4) $L/2$

48. A cubical block of side 'L' rests on a rough horizontal surface with coefficient of friction μ. A horizontal force F is applied on the block as shown in figure. The minimum force required to topple the block is (IIT - 2005)

1) Infinitesimal

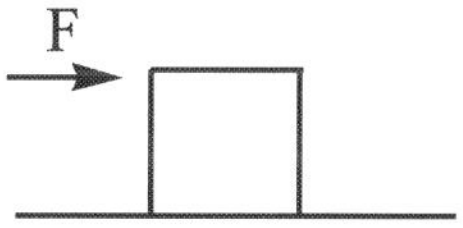

2) $mg/4$

3) $mg/2$

4) $mg(1-\mu)$

49. A rope is wound several times on a solid cylinder of radius 0.1 m and mass 10 kg . The cylinder is free to rotate about its axis. If the rope is pulled with a force of 10 N, the angular acceleration of the cylinder is

1) 10 rad s^{-2} 2) 5 rad s^{-2}

3) 20 rad s^{-2} 4) 15 rad s^{-2}

50. As shown in figure, a stick half of which is made of wood and other half of iron is pivoted as shown in fig. (a) It is pivoted at wooden end and force is applied to the other end of it at right angles to its length. (b) It is pivoted at iron end and the same force is applied to the other end at right angles to its length. Then angular acceleration produced is

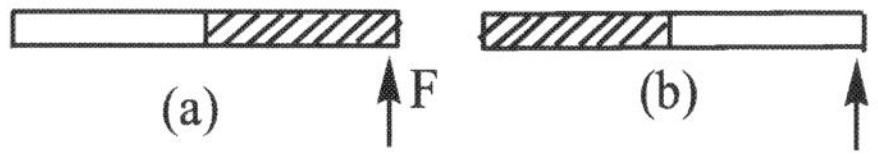

1) More in case (b) than in (a)

2) more in case (a) than in (b)

3) equal in both the cases

4) data insufficient

51. A wheel of radius 20 cm can rotate about its centre. A string is wrapped over its rim and is pulled by a force of 10N tangential to pulley. If the angular acceleration produced is 4 rad s^{-2}, then the moment of inertia of the wheel in kg m^2 is

1) 25 2) 0.5 3) 50 4) 5

52. The moment of inertia of a uniform disc about an axis passing through its centre and perpendicular to its plane is 2 kg m^2. It is rotating with an angular velocity 100 rad s^{-1}. A second identical disc rotating opposite to that of the first but with same angular velocity is joined to it coaxially. Now these two discs together continue to rotate about the same axis. Then the lose in KE in KJ is

1) 10 2) 20 3) 25 4) zero

53. A rod is free to rotate about 'O' as shown. It begins to rotate. When it has turned through an angle 'θ'. its angular velocity 'w' is given as

1) $\sqrt{\frac{6g}{\ell}}\sin\theta$ 2) $\sqrt{\frac{6g}{\ell}}\sin\left(\frac{\theta}{2}\right)$

3) $\sqrt{\frac{6g}{\ell}}\cos\frac{\theta}{2}$ 4) $\sqrt{\frac{6g}{\ell}}\cos\theta$

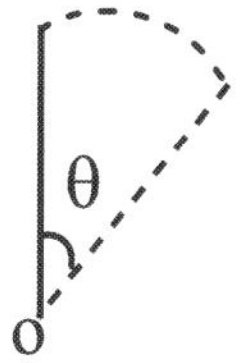

54. A block of mass 2kg hangs from the rim of a wheel of radius 0.5m. On releasing from rest, the block falls through 5m height in 2s. The M.I. of the wheel will be

1) 1 kg m^2

2) 3.2 kg m^2

3) 2.5 kg m^2

4) 1.5 kg m^2

55. A cylinder of mass M and radius R starts falling under gravity at t = 0 as shown in figure. If the mass of chord is negligible, the tension in each string is

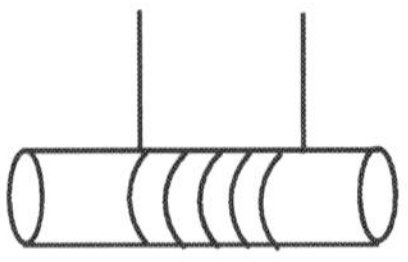

1) $Mg/6$ 2) $Mg/4$

3) $Mg/2$ 4) Mg

56. A particle of mass 'm' is projected with an initial velocity u at an angle 'θ' to horiziontal. The torque of gravity on projectile at maxi-mum height about the point of projection is

1) $\frac{mgu^2 \sin 2\theta}{2}$ 2) $mg\,u^2 \sin 2\theta$

3) $\frac{mgu^2 \sin\theta}{2}$ 4) $\frac{1}{2}mu^2 \sin 2\theta$

57. 'O' is the center of a square ABCD. F_1, F_2, F_3 and F_4 are four forces acting along the sides AB, BC, CD and AD as shown in figure. What should be the maginitude of F_4 so that, the total torque about 'O' is zero ?

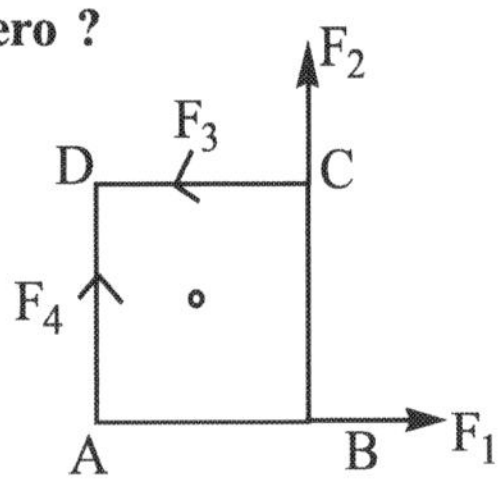

1) $F_1 + F_2 + F_3$

2) $F_1 + F_2 - F_3$

3) $F_1 - F_2 - F_3$

4) $\frac{F_1 + F_2 + F_3}{2}$

Rolling Motion :

58. A disc is at rest at the top of a rough incined plane. It rolls without slipping. At the bottom of inclined palne there is a vertical groove of radius 'R'. In order to loop the groove. the minimum height of incline required is

1) $\frac{15R}{4}$ 2) $\frac{9R}{4}$

3) $\frac{5R}{2}$ 4) $\frac{7R}{5}$

59. A solid sphere of mass 1.5 kg rolls on a frictionless horizontal surface, its centre of mass moving at a speed of 5 ms^{-1}. Then it rolls up an incline of 30^0 to horizontal. The height attained by the sphere before it stops is (g = 10 sm^{-2})

1) 5 m 2) 1.5m

3) 1.75 m 4) 3.65m

60. A solid cylinder of mass 'm' rolls without slippling down an inclined plane making an angle θ with the horizontal. The fricitional force between the cylinder and the incline is

1) mg sin θ 2) $\frac{mg\sin\theta}{3}$

3) mg cos θ 4) $\frac{2mg\sin\theta}{3}$

61. A block of mass 12 kg is attached to a string wrapped around a wheel of radius 10cm. The acceleration of the block moving down the inclined plane is 2 ms^{-2}. The tension in the string is (I of wheel = 0.117 kg m_2)

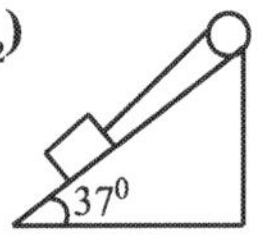

1) 24.5 N 2) 68.7N

3) 23.4N 4) 46.8N

EQUILIBRIUM OF BODY & PARALLEL FORCES

62. The resultent of two unlike parallel froces is 5N and acts at a distance of 20 cm and 40 cm from them. The forces are

1) 2N, 5N 2) 5 N, 10N 3) 6N, 10N 4) 8N, 12N

63. A uniform rod is 4m long and weighs 10kg. If it is supported on a knife edge at one metre from the end, what weight placed at that end keeps the rod horizontal ?

1) 8 kg 2) 10 kg 3) 12 kg 4) 16 kg

64. A boy and a man carry a unifrorm rod of length 'L' horizontally in such a way that the boy gets 1/3 of the load. If the boy is at one end of the rod, the distance of the man from the other end is

1) L/6 2) L/5 3) L/4 4) L/3

ANSWERS

1) 1	2) 2	3) 3	4) 1	5) 3
6) 3	7) 3	8) 2	9) 1	10) 2
11) 4	12) 4	13) 2	14) 1	15) 2
16) 3	17) 1	18) 3	19) 3	20) 4
21) 4	22) 1	23) 1	24) 2	25) 2
26) 3	27) 2	28) 1	29) 2	30) 1
31) 2	32) 3	33) 1	34) 4	35) 1
36) 2	37) 1	38) 1	39) 2	40) 1
41) 1	42) 2	43) 3	44) 2	45) 2
46) 1	47) 1	48) 3	49) 3	50) 1
51) 2	52) 2	53) 2	54) 4	55) 1
56) 4	57) 1	58) 1	59) 3	60) 2
61) 3	62) 2	63) 2	64) 3	

COMPREHENSION QUESTIONS

Comprehension - I (1-3):

Four rods of equal length l and mass m each form a square as shown in figure.

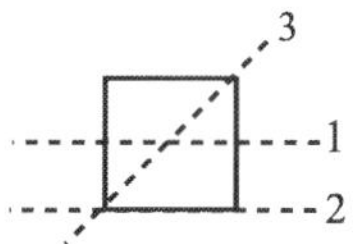

1. **The moment of inertia of the system about the axes 1 is**

   1) $\frac{2}{3}m\ell^2$ 2) $\frac{5}{3}m\ell^2$ 3) $\frac{1}{3}m\ell^2$ 4) $\frac{10}{3}m\ell^2$

2. **The moment of inertia of the system about the axes 2 is**

   1) $\frac{2}{3}m\ell^2$ 2) $\frac{5}{3}m\ell^2$ 3) $\frac{1}{3}m\ell^2$ 4) $\frac{10}{3}m\ell^2$

3. **The moment of inertia of the system about the axes 3 is**

   1) $\frac{2}{3}m\ell^2$ 2) $\frac{5}{3}m\ell^2$

   3) $\frac{1}{3}m\ell^2$ 4) $\frac{10}{3}m\ell^2$

Comprehension - II (4-6):

A particle of mass 1 kg is projected with velocity $20\sqrt{2}$ m/s at 45^0 with ground.($g = 10\ m/s^2$)

4. **When, the particle is at highest point the net torque on the particle about point of projection is**

   1) 400 Nm 2) 300 Nm

   3) 200 Nm 4) 100 Nm

5. **When, the particle is at highest point the angular momentum of the particle about point of projection is**

   1) 400 Kg m^2s^{-1} 2) 300 Kg m^2s^{-1}

   3) 200 Kg m^2s^{-1} 4) 100 Kg m^2s^{-1}

6. **When, the particle is at highest point the angular velocity of the particle about point of projection is**

   1) 0.1 rad/s 2) 0.2 rad/s

   3) 0.3 rad/s 4) 0.4 rad/s

Comprehension - III (7-9):

A disc rolls on ground without slipping. Velocity of centre of mass is v.

7. **If the point P lies on the circumference of disc at angle $\theta = 90^0$ as shown in the figure then the speed v_P at the point P is**

   1) $\sqrt{2}v$ 2) v 3) 2v 4) $\sqrt{3}v$

8. **If the point P lies on the circumference of disc at angle $\theta = 120^0$ as shown in the figure then the speed v_P at the point P is**

   1) $\sqrt{2}v$ 2) v 3) 2v 4) $\sqrt{3}v$

9. **If the point P lies on the circumference of disc at angle $\theta = 180^0$ as shown in the figure then the speed v_P at the point P is**

   1) $\sqrt{2}v$ 2) v 3) 2v 4) $\sqrt{3}v$

Comprehension - IV (10-11):

A rod of length 2m is kept vertical inside a smooth spherical shell of radius 2m. The rod starts slipping inside the shell. Mass of the rod is 4 kg.

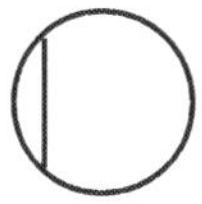

10. **Angular speed of the rod (in rad/s) in the position when it becomes horizontal is**

    1) 4.6 2) 6.8 3) 3.2 4) 7.2

11. **Velocity of centre of the rod (in m/s) at this instant is approximately.**

    1) 5.5 2) 6.2 3) 4.7 4) 10.2

Comprehension - V (12-13):

A solid sphere is rolling without slipping on rough ground as shown in figure. It collides elastically with an identical another sphere at rest. There is no friction between the two spheres. Radius of each sphere is R and mass is m.

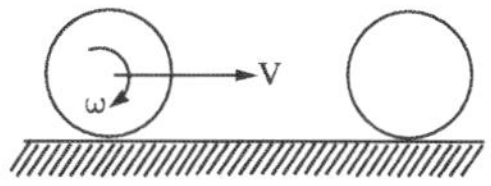

12. **Linear velocity of first sphere after it again starts rolling without slipping is:**

    1) $\frac{2}{5}\omega R$ 2) $\frac{2}{7}\omega R$ 3) $\frac{7}{10}\omega R$ 4) $\frac{7}{5}\omega R$

13. **What is the net angular impulse imparted to second sphere by the external forces?**

    1) $\frac{2}{7}mRv$ 2) $\frac{5}{7}mRv$ 3) $\frac{2}{5}mRv$ 4) $\frac{7}{10}mRv$

Comprehension - VI (14-15):

A small sphere of mass 1 kg is rolling without slipping with linear speed $v = \sqrt{\frac{200}{7}}m/s$. It leaves the inclined plane at point C.

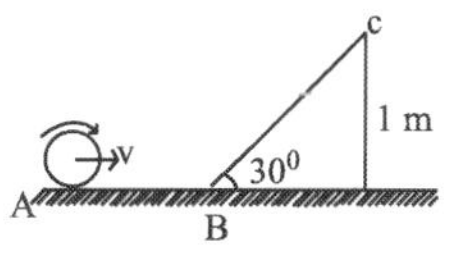

14. Its linear speed at point C

1) $\sqrt{\frac{100}{7}}m/s$ 2) $\sqrt{\frac{50}{7}}m/s$

3) $\sqrt{\frac{100}{35}}m/s$ 4) $\sqrt{\frac{200}{35}}m/s$

15. The ratio of rotational and translational kinetic energy of the sphere when it strikes the ground after leaving form point C.

1) $\frac{2}{5}$ 2) $\frac{2}{3}$ 3) $\frac{1}{6}$ 4) $\frac{1}{2}$

Comprehension - VII (16-17):

A solid sphere has linear velocity $v_0 = 4m/s$ and angular velocity $\omega_0 = 9 rad/s$ as shown . Ground on which it is moving is smooth. It collides elastically with a rough wall of coefficient of friction μ. Radius of the sphere is 1 m and mass is 2 kg.

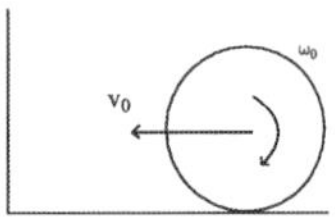

16. If the sphere after colliding with the wall rolls without slipping in opposite direction, then coefficient of friction μ is :

1) $\frac{1}{2}$ 2) $\frac{2}{3}$ 3) $\frac{1}{3}$ 4) $\frac{1}{4}$

17. What is net linear impulse imparted by the wall on the sphere during impact.

1) $\sqrt{32}N-s$ 2) $4\sqrt{14}N-s$

3) $5\sqrt{4}N-s$ 4) $15\sqrt{2}N-s$

Comprehension - VIII (18-19):

A rod AB of length 2 m and mass 2 kg is lying on smooth horizontal x-y plane with its centre at origin O as shown figure. An impulse. J of magnitude 10 N- s is applied perpendicular to AB at A.

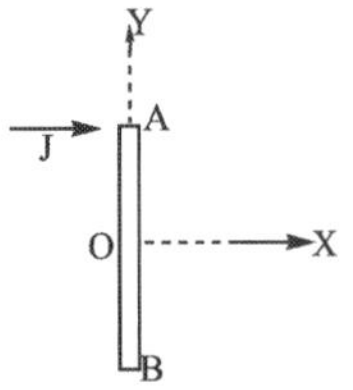

18. The distance of point P from centre of the rod which is at rest just after the impact is :

1) $\frac{2}{3}m$ 2) $\frac{1}{3}m$ 3) $\frac{1}{2}m$ 4) $\frac{1}{4}m$

19. Co-ordinates of point A of the rod after time $t = \frac{\pi}{45}s$ will be :

1) $\left[\left(\frac{\pi}{9}+\frac{\sqrt{3}}{2}\right)m, \frac{1}{2}m\right]$ 2) $\left[\left(\frac{3}{4}m, \frac{3}{2}m\right)\right]$

3) $\left[\left(\frac{\pi}{6}+\frac{1}{2}\right)m, \frac{1}{2}m\right]$ 4) $\left[\frac{1}{2}m, \frac{1}{2}m\right]$

Comprehension - IX (20-21):

Length AB in the figure shown is 5m. The body is released form A. Friction is sufficient for pure rolling to take place.

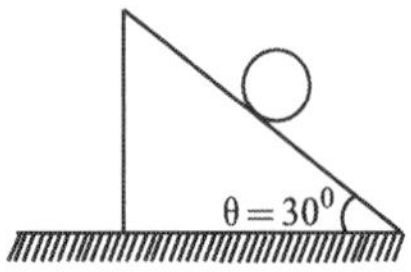

20. The maximum time which anybody (which can roll) can take to reach the bottom is :

1) 8 s 2) 6 s 3) 2 s 4) 4 s

21. In the above case suppose we have four bodies ring , disc, solid sphere and hollow sphere. The angle θ is now gradually increased. Which body will start slipping very first. All the bodies have same mass and radius, coefficient of friction is also same

1) Ring 2) Dic 3) Solid sphere 4) Hollow sphere

Comprehension - X (22-23):

Moment of inertia of a straight wire about an axis perpendicular to the wire and passing through one of its end is I.

22. This wire is now framed into a circle (a ring) of single turn. The moment of inertia of this ring about an axis passing through centre and perpendicular to its plane would be :

1) $\left(\frac{3}{\pi^2}\right)I$ 2) $\left(\frac{3}{4\pi^2}\right)I$

3) $\left(\frac{\pi^2}{3}\right)I$ 4) $\left(\frac{4\pi^2}{3}\right)I$

23. Now the same wire is bent into a ring of two turns, then the moment of inertia would be

1) $\left(\frac{\pi^2}{3}\right)I$ 2) $\left(\frac{\pi^2}{12}\right)I$

3) $\left(\frac{3}{16\pi^2}\right)I$ 4) $\left(\frac{3}{4\pi^2}\right)I$

Comprehension - XI (24-25):

In the given figure $F = 10N$, R = 1 m, mass of the body is 2 kg and moment of inertia of the body about an axis passing through O and perpendicular to plane of body is $4kg - m^2$. O is the centre of mass of the body.

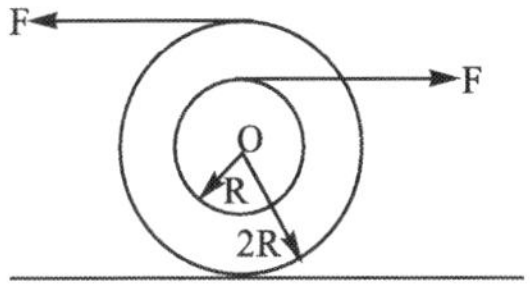

24. If the ground is smooth what is total kinetic energy of the body after 2 s :

1) 25 J 2) 50 J 3) 75 J 4) 100 J

25. If ground is sufficiently rough, what is kinetic energy of the body now in the given time interval.

1) 10.33 J 2) 25.67 J
3) 16.67 J 4) None of these

Comprehension - XII (26-27):

A rod of mass m and length l in placed on a smooth table. An another particle of same mass m strikes the rod with velocity v_0 in a distance perpendicular to the rod at distance $x(< l/2)$ form its centre. Particle sticks to the rod. Let ω be the angular speed of system after collision, then :

26. As x is increased from 0 to $l/2$, the angular speed ω :

1) Will continuosly increase
2) Will continuously decrease
3) Will first increase and then decrease
4) Will first decrease and then increase

27. Find maximum possible value of impulses (by varying x) that can be impartd to the particle during collision. Particle still sticks to the rod.

1) $\frac{mv_0}{2}$ 2) $\frac{2mv_o}{3}$
3) $\frac{3mv_0}{4}$ 4) $\frac{4mv}{5}$

Comprehension - XIII (28-29):

A disc of mass m and radius R is placed over a plank of same mass m. There is sufficient friction between disc and plank to pravent slipping. A force F is applied at the centre of the disc.

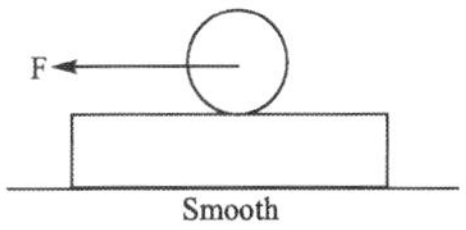

28. Acceleration of the plank is :

1) $\frac{F}{2m}$ 2) $\frac{3F}{4m}$ 3) $\frac{F}{4m}$ $\frac{3F}{2m}$

29. Force of friction between the disc and the plank is :

1) $\frac{F}{2}$ 2) $\frac{F}{4}$ 3) $\frac{F}{3}$ 4) $\frac{2F}{3}$

Comprehension - XIV (30-32):

When the force F is applied the body may topple about A or it may translate.

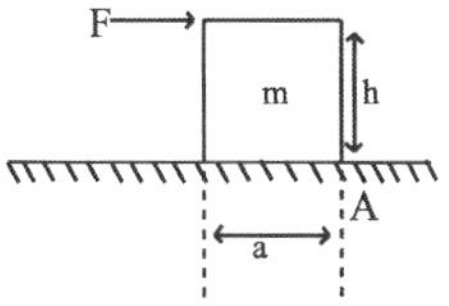

30. When the block topples about A, the normal force :

1) passing through centre mass
2) is zero
3) shifts to the right and pass through rightmost edge containing A
4) is zero if the surface is smooth

31. If the block be a cube of edge a and μ = 0.2 then

1) the body will translate
2) the body will topple
3) the body may translate or topple
4) none of the above

32. If the block is a cube of edge a and μ = 0.6 then

1) The body will translate
2) The body will topple
3) The body first translates and then topples
4) None of the above

ANSWERS

1) 1	2) 2	3) 1	4) 1	5) 1
6) 2	7) 1	8) 4	9) 3	10) 3
11) 1	12) 2	13) 1	14) 1	15) 3
16) 4	17) 2	18) 2	19) 1	20) 3
21) 1	22) 2	23) 3	24) 2	25) 3
26) 3	27) 1	28) 3	29) 2	30) 3
31) 1	32) 2			

Chapter - 5

NEWTONS LAWS OF MOTION

NEWTON'S LAWS OF MOTION

SYNOPSIS

NEWTON'S FIRST LAW :

* Every body continues to be in its state of rest or of uniform motion along a straight line unless it is acted on by an external force to change its state.
* Newton's first law of motion explains the concept of force qualitatively.
* If a net force acts on a particle it accelerates.

LINEAR MOMENTUM :

* The linear momentum of a particle is the product of its mass and velocity. $\vec{P} = m\vec{V}$
* Linear momentum is a vector, it points in the direction of the velocity of the particle.

NEWTON'S SECOND LAW :

* According to Newton II law of motion, the rate of change of linear momentum of a particle is equal to the net force on the particle.

$$\vec{F} = \frac{d\vec{p}}{dt} = \frac{d}{dt}(m\vec{\upsilon})$$

* $$\vec{F} = \frac{m\,d\vec{\upsilon}}{dt} + \vec{\upsilon}\frac{dm}{dt}$$

a) If mass is constant, $\frac{dm}{dt} = 0 \quad \vec{F} = ma$

b) If v is constant $\quad \vec{F} = \vec{v}\frac{dm}{dt}$

* S.I. unit of force is newton, C.G.S. unit is dyne. Dimensional formula is MLT^{-2}.
* $1\text{ N} = 10^5$ dyne
* If $\vec{u}$ and $\vec{v}$ are the initial and final velocities of particle of mass m and average force acting on the particle in a time interval 't' is

$$\vec{F} = \frac{\Delta\vec{P}}{t} = \frac{m\vec{v} - m\vec{u}}{t}$$

* If a particle of mass m, strikes a rigid surface normally, and rebounds with the same speed v, the change in linear momentum of the particle is 2 mV. If t is time of contact then force on particle is

$F = \frac{2mV}{t}$ (along the normal, away from wall)

* If a particle of mass m, strikes a rigid surface with a speed v, with the direction of motion making an angle θ, with the normal to the surface and rebounds with the same speed at the same angle, the magnitude of the change in momentum of the particle is 2mV Cosθ perpendicular and away from the wall. If t is time of contact then force on particle is $F = \frac{2mV\cos\theta}{t}$
* In the above case, if ϕ is the angle between direction of motion and the surface, the magnitude of change in momentum of the particle is $2mv\sin\phi$. If t is time of contact then force on particle is $F = \frac{2mV\sin\phi}{t}$
* From a nozzle of area of crosssection A, liquid of density d, issues out with a velocity 'v' horizontally and strikes a vertical wall then

a) If water does not bounce back i.e it falls dead or drops dead after striking the wall, then the force exerted on the wall is $F = Av^2d$

b) If water bounces back with the same speed then force exerted on the wall is

$F = 2Av^2d$

c) If water rebounces with velocity v^1, then the force excrted on the wall is

$F = Adv(v + v^1)$

d) In the above case if water strikes the surface at angle θ with the normal and reflects with the same speed and at the same angle, force exerted on the wall is $2dAv^2\cos\theta$.

e) If gravel is dropped on a conveyor belt at the rate of $\frac{dm}{dt}$, extra force required to keep the belt moving with velocity u is

$F = u\left(\frac{dm}{dt}\right)$

f) If a gun fires *n* bullets per second each with velocity u and m is mass of the bullet, the force required to hold the gun in its position is F= nm u.

* A metallic plate of mass 'M is kept held in mid air by firing 'n' bullets/second each of mass 'm' from below with speed u and if the bullets rebound with same speed after striking the plate, then,

Rate of change of momentum of bullets =Weight of the plate

$\therefore 2\,mnu = Mg$

* The force needed to stop a body moving with a speed u in a distance S is

$F = \frac{mu^2}{2S}$

Stopping distance $S \propto u^2$ (If F = constant)

* The force needed to stop a body of mass m travelling with a speed u in a time t second is
$F = \frac{mu}{t}$

IMPULSE :

* Forces acting for a short duration and which vary with time are impulsive forces.

Ex.: a) Force acting between the balls during collision between two billiard balls

b) Force exerted when nail driven into a wall by striking it with a hammer.

* The impulse of a force is defined as the product of the average force and the time interval for which it acts.
Impulse, $\vec{J} = \overrightarrow{F_{AV}}\,\Delta t = m\vec{v} - m\vec{u}$
Incase of variable force

$$\vec{J} = \int_{t_1}^{t_2} \vec{F}\,dt = m\vec{v} - m\vec{u}$$

* S.I. Unit of impulse is Ns,
It's dimensional formula is MLT^{-1}
* Impulse of a force is in the direction of change in momentum produced by the force.
* Force, impulse, change in momentum, change in velocity, acceleration these five quantities are always in the same direction.
* The area under F – t graph gives impulse.
* If force $\vec{F_1}$ acts on a body for time t_1, $\vec{F_2}$ for t_2, $\vec{F_3}$ for t_3, ... then total change in momentum is equal to $\vec{F_1}t_1 + \vec{F_2}t_2 + \vec{F_3}t_3 + ...$
* For a given impulse, the average force is inversely proportional to the time for which force acts,

a) A person falling on a concrete floor from a certain height receives greater injuries than when he falls from same height on loose soil.

b) The glass ware, China ware, Television sets etc. are transported from one place to another by packing them with felt, syfo foam etc.

NEWTON'S THIRD LAW :

* For every action there is always an equal and opposite reaction
* Action and reaction are equal in magnitude and opposite in direction and they act on two different bodies (So they do not cancel)
* Newton's third law leads to law of conservation of linear momentum
* A bird is in a wire cage hanging from a spring balance. When the bird starts flying in the cage, the reading of the balance decreases.
* In the above case, if the bird is in a closed cage or air-tight cage and it hovers in the cage, the reading of the spring balance does not change.
* In the above case for a closed cage if the bird accelerates upward reading of the balance is
$R = W_{bird} + ma$, where m is the mass of the bird and a is acceleration.
* A block of mass m hangs at the end of a massless string. When the block accelerates in the upward direction or decelerates in downward direction, the tension in string is $T = m(g+a)$.
* When the block is accelerated in the downward direction or decelerates in upward direction $T=m(g-a)$.
* When the block is moved up or down with uniform speed the tension $T = mg$.

a) ***If elevator accelerates up :***

Apparent weight $N = m(g + a)$

Tension in the cable $T = (M_{elevator} + M_{man})(g+a)$.

Note : Same is the case if elevator moving downward is decelerated

b) ***If elevator accelerates downward :***

Apparent weight $N = m(g-a)$

Tension in the cable $T = (M_{elevator} + M_{man})(g-a)$

Note : Same is the case if elevator moving upward is decelerated

c) ***If elevator falls freely (cable breaks) :***

N=0 i.e apparent weight of a body in a free fall=0

d) ***If elevator moves up or down with uniform velocity :***

Apparent weight = N = mg

Tension in the cable $T = (M_{elevator} + M_{man})\,g$

* **Man inside an artificial satellite**

An artificial satellite orbiting the earth in a circular orbit is like a freely falling body because its centripetal acceleration is equal to the acceleration due to gravity in that orbit.
In this case, apparent weight of the man inside the satellite is zero.

$mg^1 - N = \frac{mV_0^2}{r}$ where V_0 = orbital velocity

But $V_0 = \sqrt{g^1 r}$

$\Rightarrow$ N = 0, apparent weight of a body = 0

LAW OF CONSERVATION OF LINEAR MOMENTUM :

i) In the absence of an external force, the linear momentum of a particle or a body remains constant i.e.

if $\vec{F} = 0$ then $\vec{p} = \text{constant}$

ii) In the absence of external force, the linear momenta of individual particles can change but the total linear momentum of the whole system remains constant.

iii) The law of conservation of linear momentum is based on the Newton's third law of motion. This is the fundamental law of nature and there is no exception to it.

iv) This law is valid only for linear motion.

v) Internal forces cannot change the total momentum of the system, however they may change the momentum of each particle of the system.

vi) Motion of a rocket, firing of a bullet from a gun and explosion of a shell fired from a cannon are some examples where we can apply the law of conservation of linear momentum

RECOIL OF GUN :

i) If a stationary gun fires a bullet horizontally, the total momentum of the gun and bullet is zero before and after firing.

ii) According to conservation of linear momentum,

$$m_1\vec{v}_1 + m_2\vec{v}_2 = 0 \quad \therefore \quad \vec{v}_2 = \frac{-m_1\vec{v}_1}{m_2}$$

Negative sign shows that as bullet moves in forward direction the gun moves in backward direction i.e. the gun recoils.

iii) If '$\vec{v}$' is the muzzle velocity of bullet (velocity of bullet w.r.t. gun) the velocity of gun $\vec{V} = \frac{-m_1\vec{v}}{m_1 + m_2}$.

iv) The magnitude of momentum of the gun is equal to magnitude of momentum of the bullet.

v) $KE = \frac{P^2}{2m}$. As p = constant, the bullet has greater kinetic energy than the gun.

$$\frac{K.E.gun}{K.E.bullet} = \frac{m_1}{m_2}$$

EXPLOSIONS :

i) In an explosion, linear momentum is conserved, but kinetic energy is not conserved. The kinetic energy of the system after explosion increases. The internal energy of the system is used for the above purpose.

ii) If a stationary shell breaks into two fragments, they will move in opposite directions, with velocities in the inverse ratio of their masses.

$$\frac{V_1}{V_2} = \frac{-m_2}{m_1}$$

iii) In the above, the two fragments have the same magnitude of linear momentum.

iv) In the above case, the Kinetic energy of the two fragments is inversely propotional to their masses.

$$\frac{K.E_1}{K.E_2} = \frac{m_2}{m_1}$$

v) If a shell at rest breaks into **three fragments**, the total momentum of any two of the fragments must be equal and opposite to the momentum of the third fragment

i.e. $\vec{P}_1 + \vec{P}_2 = -\vec{P}_3$

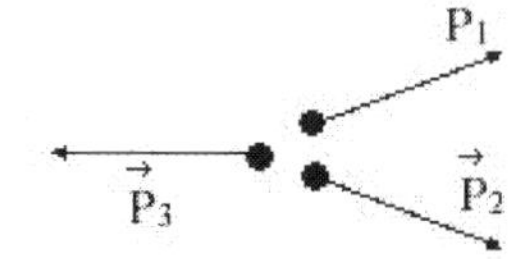

vii) When a stationary shell explodes, its total momentum is zero, before or after explosion.

viii) If a stationary body breaks into **three parts** out of which the first two parts move at right angles to each other with momenta p_1 and p_2 respectively, then the momentum of third part is determined as follows

$p_3 = \sqrt{p_1^2 + p_2^2}$ and direction of p_3 is given by

$$\tan\theta = \frac{p_2}{p_1}$$

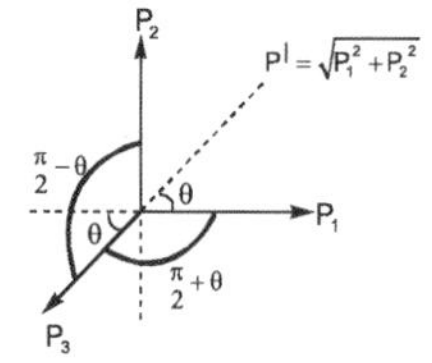

P_3 makes an angle $\frac{\pi}{2} - \theta$ with P_1

P_3 makes an angle $\frac{\pi}{2} + \theta$ with P_2

ROCKET PROPULSION :

* Velocity of a rocket at any instant of time is given by

$$v = u_0 + v_r \log_e \frac{m_0}{m}$$

m_0 = mass of the rocket at t = 0

v_0 = velocity of the rocket at t = 0

m = mass of the rocket at any instant of time

v = velocity of the rocket at any instant of time

v_r = velocity of the exhaust gases relative to the rocket

a) If the rocket is to moveup with constant speed over coming its weight then

$$\text{Thrust on the rocket} = \left(\frac{dm}{dt}\right) u = Mg$$

b) If the rocket moves up with constant acceleration 'a'. then

Thrust on the rocket $= \left(\frac{dm}{dt}\right)u$ = Mg + Ma

Applications of pseudoforce :

★ An object kept on a smooth inclined plane can be kept stationary relative to the incline by giving a horizontal acceleration of g tan θ as shown.

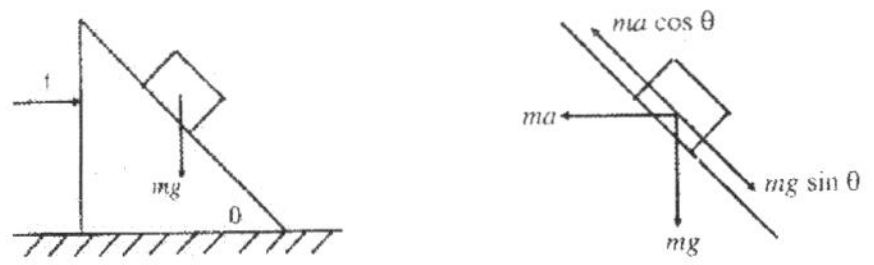

(M + m) g tan θ (M is mass of wedge) is the force to be applied on the wedge.

★ Acceleratomer (A pendulum is suspended from the roof a moving car.)

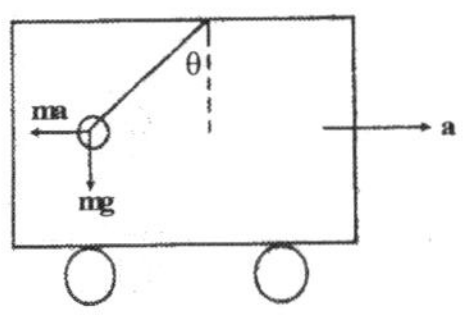

(acceleration)

a) If the car is moving with uniform velocity, the position of the bob does not change

b) If the car moves with acceleration or retardation, the bob moves in opposite direction to that of acceleration.

The forces acting on the bob are, weight mg, tension in the string T and pseudo force ma.

For the equilibrium of bob,

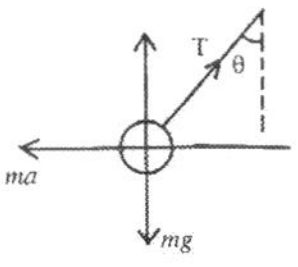

$\tan\theta = ma/mg$

$\Rightarrow$ tan θ = a/g

T = $m\sqrt{a^2+g^2}$

c) In the above case if the car is moving along a horizontal circular track, pseudo force is $mr\omega^2$ and

$\tan\theta = \frac{ma}{mg} = \frac{mr\omega^2}{mg}$ or $r\omega^2 = g\tan\theta$

$\therefore \frac{v^2}{r} = g\tan\theta$

Here $mr\omega^2$ is centrifugal force.

★ An object of mass m is kept on a wedge of mass M. The wedge is moved with an acceleration 'a' such that the object moves down like a freely falling body. (no friction)

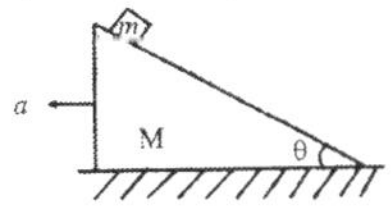

Here a = g cot θ.

MOTIONS OF BODIES IN CONTACT

If two bodies with masses m_1 and m_2 are put in contact and if a force F is applied on the mass m_1, then equal acceleration is produced in both the bodies.

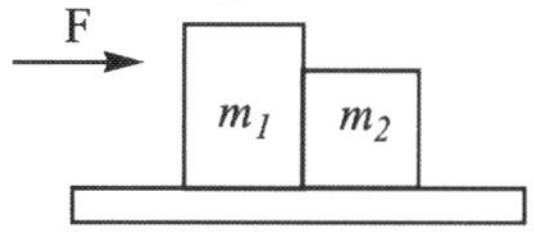

a) Acceleration produced in both bodies

$$a = \frac{F}{m_1+m_2} = \frac{Force}{Total\ mass}$$

Force required to produce an acceleration 'a' in m_1 $F_1 = \frac{m_1F}{m_1+m_2}$

The contact force on $F_2 = \frac{m_2F}{m_1+m_2}$

b) When three bodies are lying in contact on a horizontal smooth table :

when the force acts on a body of mass m_1.

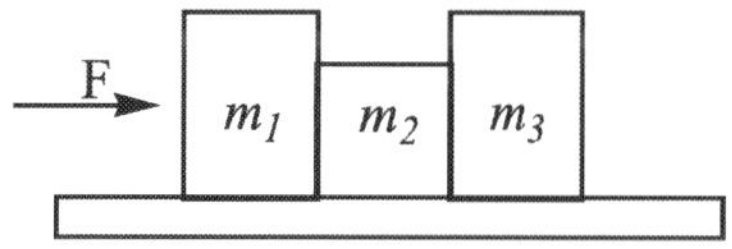

Representation of action and reaction forces.

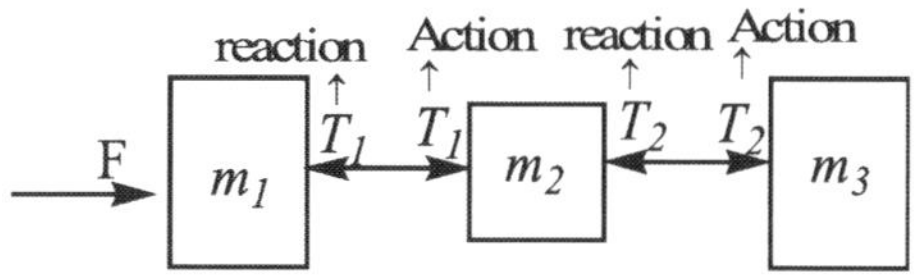

The contact force between m_1 and m_2 is T_1 and that between m_2 and m_3 is T_2.

$$a = \frac{F}{m_1+m_2+m_3}$$

$$T_1 = \frac{(m_2+m_3)F}{m_1+m_2+m_3}; \qquad T_2 = \frac{m_3F}{m_1+m_2+m_3}$$

If the force F acts on the body of mass m_3.

$$T_1 = \frac{m_1F}{m_1+m_2+m_3}; \qquad T_2 = \frac{(m_1+m_2)F}{m_1+m_2+m_3}$$

MOTION OF BODIES CONNECTED BY STRINGS

i) When two bodies of masses m_1 and m_2 are tied to the ends of a string and these kept on smooth horizontal surface are pulled by a force F.

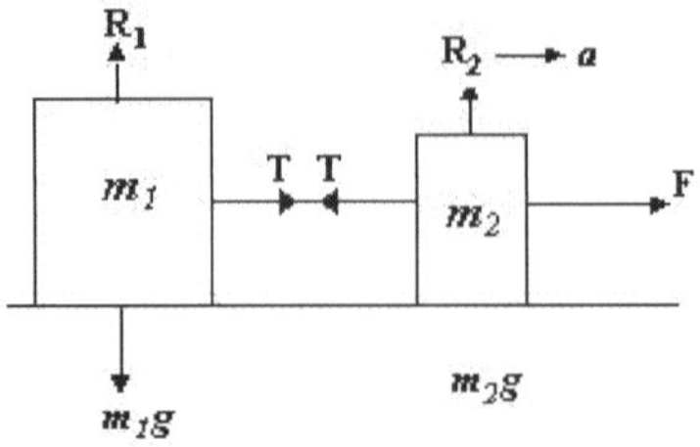

$a = \dfrac{F}{m_1 + m_2}$ and $T = \dfrac{m_1 F}{m_1 + m_2}$

ii) If in the above example three bodies are pulled with a force F.

Acceleration produced in the system.

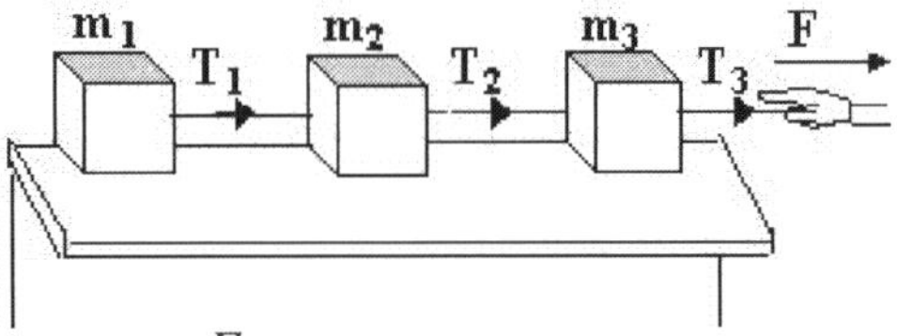

$a = \dfrac{F}{m_1 + m_2 + m_3}$

Tension produced in the first string

$T_1 = m_1 a = \dfrac{m_1 F}{m_1 + m_2 + m_3}$

Tension produced in second string

$T_2 = \dfrac{(m_1 + m_2)F}{m_1 + m_2 + m_3}$ $(\because F - T_2 = m_3 a)$

Tension produced in the third string

$T_3 = F$

ATWOOD'S MACHINE :

i) Let two bodies of masses m_1 and m_2 respectively are connected by a light inextensible string passing over a massless smooth pulley. If $m_1 > m_2$, then mass m_1 moves downwards and mass m_2 moves upwards.

$a = \dfrac{m_1 - m_2}{m_1 + m_2} g$;

and $T = \dfrac{2m_1 m_2}{m_1 + m_2} g$

The reaction at the pulley

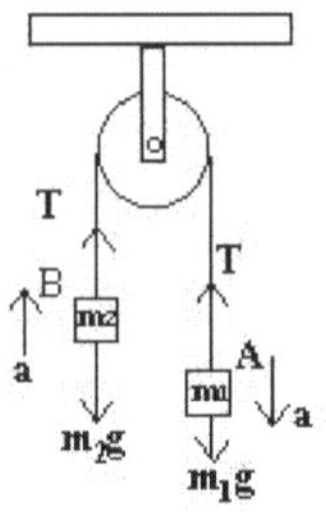

$R = 2T = \dfrac{4m_1 m_2}{m_1 + m_2} g$

ii) In the above case, if the pulley is not massless rather has finite mass M and radius R, then tensions in two segment strings are different, say T_1 and T_2. Then equations of motion.

$$a = \left(\frac{m_1 - m_2}{m_1 + m_2 + \frac{M}{2}} \right) g$$

$$T_1 = \frac{m_1 \left[2m_2 + \frac{M}{2} \right]}{m_1 + m_2 + \frac{M}{2}} g;$$

$$T_2 = \frac{m_2 \left[2m_1 + \frac{M}{2} \right]}{m_1 + m_2 + \frac{M}{2}} g$$

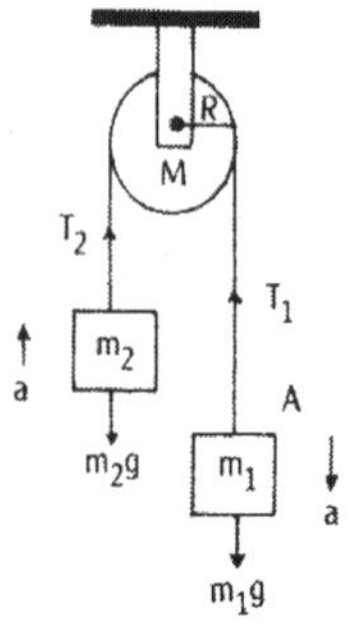

iii) Suppose a body B with mass m_2 rests on a smooth table. Mass m_1 is connected to m_2 by a light string and it hangs freely. If $m_1 > m_2$, then equation of motion

$a = \dfrac{m_1 g}{m_1 + m_2}$; and $T = \dfrac{m_1 m_2 g}{m_1 + m_2}$

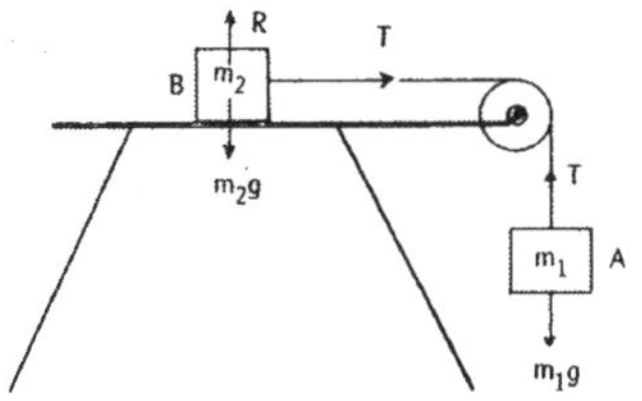

If the coefficient of kinetic friction between the table and mass is μ, then

$a = \left(\dfrac{m_1 - \mu m_2}{m_1 + m_2} \right) g$; $T = \dfrac{m_1 m_2 (1 + \mu) g}{m_1 + m_2}$

iv) Suppose a mass M rests on a friction less horizontal table. Let the mass M is connected to two masses m_1 and m_2 by two light strings passing over two light pulleys. If $m_1 > m_2$ and m_1 is moving downwards with acceleration 'a', then the equations of motion.

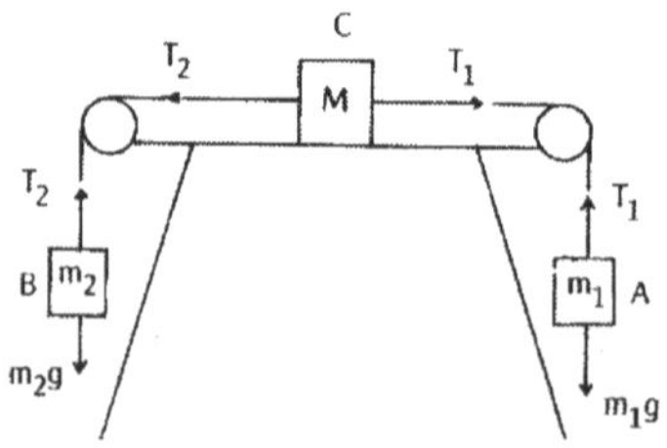

Solving, we get $\;; T_1 = \dfrac{m_1 (2m_2 + M)}{(m_1 + m_2 + M)} g$

$T_2 = \dfrac{m_2 (2m_2 + M)}{(m_1 + m_2 + M)} g$

v) When the two masses m_1 and m_2 are hanging as shown in the figure and $m_2 > m_1$.

For the mass m_1, $T - m_1 g \sin\theta = m_1 a$

For the mass m_2, $m_2 g - T = m_2 a$

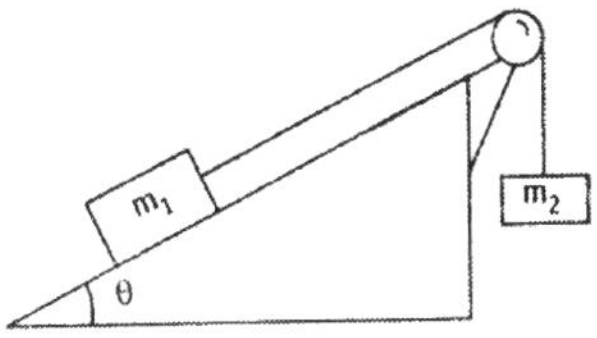

solving these equations, we get

$$a = \frac{m_2 - m_1 \sin\theta}{m_1 + m_2} g \; ; \; T = \frac{m_1 m_2 g (1 + \sin\theta)}{m_1 + m_2}$$

$$T = m_2 (g - a) \; ; \quad F = 2T \cos\frac{(90^0 - \theta)}{2}$$

vi) If in the system shown in the figure, $m_1 > m_2$, the acceleration 'a' of the system is given by

$$a = \frac{(m_1 \sin\alpha - m_2 \sin\beta)}{m_1 + m_2} g$$

and the tension in the string is given by

$$T = \frac{m_1 m_2 (\sin\alpha + \sin\beta)}{m_1 + m_2} g; \quad T = m_2 (g - a)$$

$$F = 2T \cos\frac{(90^0 - \theta)}{2}$$

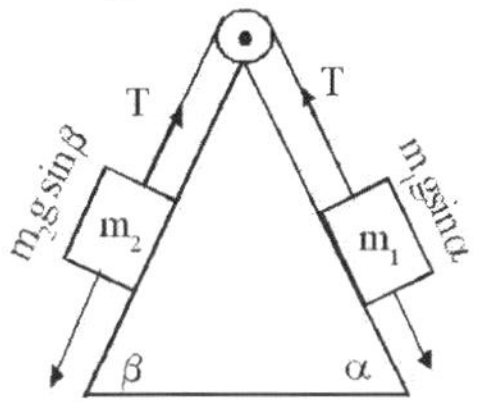

vii) a) If a body A of mass M is hanging from one end of a string passing over a pulley and a person B of same mass starts climbing up along the same string, the body and the person rises up with the same speed conserving the momentum.

b) If in the above example, two persons of equal mass start climbing up along the string and if one of them climbs faster than the other, then the two persons reach the pulley simultaneously conserving the momentum.

MASSES CONNECTED BY SPRINGS

i) When two bodies of masses m_1 and m_2 connected by a spring are pressed and released

$m_1 a_1 = m_2 a_2$. Where a_1 and a_2 are their initial accelerations after release

ii) If a force F is applied on a body of mass m_1 which connected to another body of mass m_2 and the two bodies move with accelerations a_1 and a_2 respectively, then

$F = m_1 a_1 + m_2 a_2$

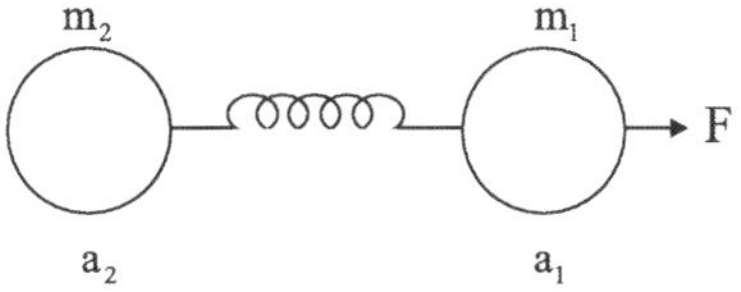

EXERCISE - I

1. **To keep a particle moving with constant velocity on a frictionless horizontal surface an external force**

   1) Should act continuously
   2) Should be a variable force
   3) Should not act
   4) Should act opposite to the direction of motion

2. **The internal forces acting on an isolated system**

   1) Can change the velocity the system
   2) Cannot change the velocity the system
   3) Can change the acceleration of the system
   4) Can change the direction of motion

3. **When we jump out of a boat standing in water, it moves**

   1) backward 2) forward
   3) upward 4) opposite to the direction of jump

4. **The apparent weight of a man in a lift is less than the real weight when the lift is going**

   1) down with uniform speed
   2) up with an acceleration
   3) up with a retardation
   4) up with an uniform speed

5. **The apparent weight of a freely falling body is**

   1) zero 2) increased
   3) decreased 4) constant

6. **A lift is ascending with a constant speed V. A passenger in the lift drops a coin. The acceleration of the coin towards the floor will be**

   1) zero 2) g
   3) less than g 4) greater than g

7. **A block can slide down an a smooth inclined plane of inclination θ, kept on a floor of a lift. When the lift is descending with a retardation a. The acceleration of the block relative to the incline is**

   1) $a - g \sin\theta$ 2) $g - a$
   3) $(g \quad a) \sin\theta$ 4) $(g + a) \sin\theta$

8. **A person in a car tends to fall back when it suddenly starts is an example for**

1) inertia of rest 2) inertia of motion
3) inertia of direction 4) both 1 & 2

9. **The area under force time curve gives**

1) work 2) power
3) displacement 4) impluse

10. **A reference frame attached to the earth with respect to an observer in space**

1) is an inertial frame because Newton's laws of motion are applicable in it
2) is an inertial frame by definition
3) cannot be an inertial frame because earth is rotating about its axis
4) can be an inertial frame because earth is revolving around the sun

11. **A stationary railway platform on earth is**

1) an inertial frame of reference for an observer on earth
2) a non inertial frame of reference for an observer on moon
3) both are true 4) both are false

12. **A rotating platform for a stationary observer outside it is**

1) inertial frame of reference
2) non inertial frame of reference 3) both
4) some times inertial (or) some times non inertial

13. **Frames moving uniformly with respect to an inertial frame are**

1) inertial frames 2) non inertial frames
3) both 4) accelerated

14. **If particle stays at rest as seen from a frame of reference**

1) the frame may be inertial and a resultant force acts on it
2) the frame may be non inertial and $F_{real} = -F_{pseudo}$
3) the frame may be non inertial but a resultant force acts on it
4) the frame must always be inertial

15. **The acceleration of a particle is found to be non zero while no force acts on the particle. This is possible if the measurement is made from**

1) inertial frame 2) non inertial frame
3) both
4) some times inertial (or) some times non inertial

16. **A train is moving with acceleration along a straight line with respect to ground. A person in the train finds that**

1) Newton's 2nd law is false but newtons 3rd law is true
2) Newton's 3rd law is false but newtons 2nd law is true
3) All the three newtons laws are false but can apply 2nd law by considering a pseudo force
4) All the three Newton's laws are true

17. **A particle is acted upon by a force of constant magnitude which is always perpendicular to the velocity of the particle. The motion of the particle takes place in a plane. It follows that**

1) the kinetic energy of the particle changes with time
2) the acceleration of the particle is constant
3) the velocity of the particle is constant
4) the speed of the particle is constant

18. **A particle is acted upon by a force of constant magnitude which is always perpendicular to the velocity of the particle. The motion of the particle takes place in a plane. It follows**

1) its velocity is constant
2) its kinetic energy is constant
3) its acceleration is constant
4) it moves in a straight line

19. **A man is standing on a spring platform Reading of spring balance is 60Kg.wt. If man jumps outside platform, then reading of the spring balance**

1) first increase and then decreases to zero
2) decreases 3) increases 4) remains same

20. **If an elevator is moving vertically upwards with an acceleration a, the force exerted on the floor by a passenger of mass M is**

1) Ma 2) Mg 3. M(g-a) 4. M(g+a)

21. **A ball with initial momentum P collides with rigid wall elastically. If P^1 be it's momentum after collision then**

1) $P^1 = P$ 2) $P^1 = -P$
3) $P^1 = 2P$ 4) $P^1 = -2P$

22. **There are three Newton's laws of motion namely first, second and third laws. We can derive**

1) Second and third laws from the first law
2) Third and first laws from the second law
3) First and second laws from the third law
4) All the laws are independent of each other

23. A force P acts on a particle so as to accelerate it from rest to certain velocity. If P is replaced then by Q which decelerates it to rest.

1) P must be equal to Q

2) P may be equal to Q

3) P must be unequal to Q 4) P $\gtreqless$ Q

24. A block of weight W is suspended from the mid-point of a rope whose ends are at the same horizontal level. The force required to straighten the rope is

1) W 2) 2W

3) W/2 4) infinitely large

25. A ball hangs from a string inside a rail road car moving along a straight track. The string is observed to be inclined towards the rear of the car making a constant small angle with the vertical. It shows that the car is

1) moving with a uniform acceleration

2) moving with a uniform velocity

3) moving with a uniform retardation

4) moving with an acceleration which is increasing uniformly

26. A body of mass m has its position x at a time t, expressed by the equation: $x=3t^{3/2}+2t-\frac{1}{2}$. The instantaneous force F on the body is proportional to

1) $t^{3/2}$ 2) t 3) $t^{-1/2}$ 4) t^0

27. A jet of water with a cross sectional area 'a' is striking against a wall at an angle θ to the horizontal and rebounds elastically. If the velocity of water jet is v and the density is ρ, the normal force acting on the wall is

1) $2av^2\rho\cos\theta$ 2) $av\rho\cos\theta$

3) $2av\rho\cos\theta$ 4) $av\cos\theta$

28. A vessel contains some water. It is moved towards right on a horizontal plane with a constant acceleration 'a'. Which of the following figures represents correct water surface ?

1) 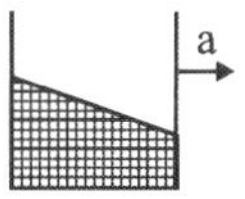

2) 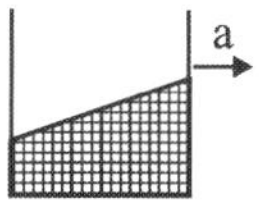

3) 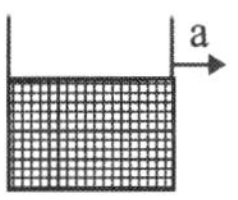

4) 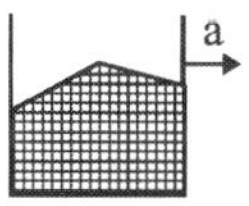

29. A cork and a metal bob are connected by a string as shown in the figure. If the beaker is given an acceleration towards left, then the cork will be thrown towards

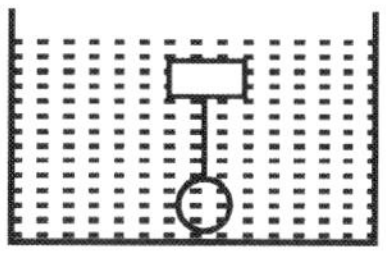

1) right 2) left

3) upwards 4) downwards

30. A block slips with constant velocity on a plane inclined at an angle 'θ'. The same block is pushed up the plane with an initial velocity v_0. The distance covered by the block before coming to rest is

1) $\frac{V_0^2}{2g\sin\theta}$ 2) $\frac{V_0^2}{4g\sin\theta}$

3) $\frac{V_0^2\sin^2\theta}{2g}$ 4) $\frac{V_0^2\sin^2\theta}{4g}$

31. A constant force(F) is applied on a stationary particle of mass 'm'. The velocity attained by the particle in a certain displacement will be proportional to

1) m 2) 1/m 3) $\sqrt{m}$ 4) $\frac{1}{\sqrt{m}}$

32. In the above problem, the velocity attained by the particle in a certain interval of time will be proportional to

1) m 2) 1/m

3) $\sqrt{m}$ 4) $\frac{1}{\sqrt{m}}$

33. A force F varies with time in accordance with the following figure. The mean force for one cycle

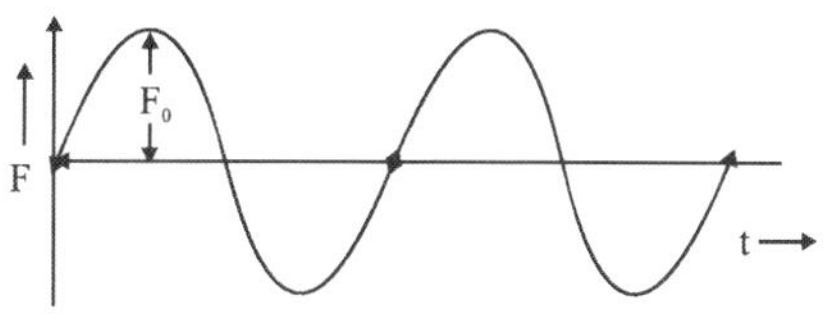

1) F_0 2) $F_0/2$ 3) $2F_0$ 4) Zero

34. You are on a frictionless horizontal plane . How can you get off if no horizontal force is exerted by pushing against the surface ?

1) by jumping
2) by spitting or sneezing
3) by rolling your body on the surface
4) by running on the plane

35. Graph of a body is shown below .It explains that

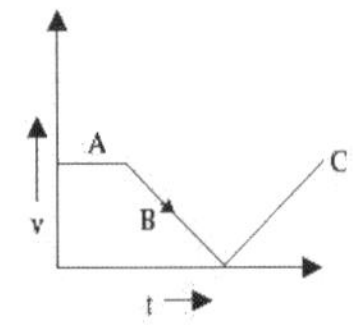

1) at B force is zero
2) at B there is a force but towards motion
3) at B there is a force which opposes motion
4) at B the force may be in any direction

36. A satellite in force-free space sweeps stationary interplanetary dust at a rate $\frac{dM}{dt}$ =av, Where M is the mass and v the speed of the satellite, and a is a constant. What is the deceleration the satellite experiences.

1) $-av^2$ 2) $-av^2/2M$
3) $-av^2/M$ 4) $-2av^2/M$

37. A body is moving with an acceleretion 'a' under the action of a force 'g'. The weight of the body is

1) g/a 2) $-g^2/a$
3) g^2/a 4) a^2/g

38. A force produces an acceleration of a_1 in a body and the same force produces an acceleration of a_2 in another body. If the two bodies are combined and the same force is applied on the combination, the acceleration produced in it is

1) $a_1 + a_2$ 2) $\frac{a_1 + a_2}{a_1a_2}$
3) $\frac{a_1a_2}{a_1 + a_2}$ 4) $\sqrt{a_1a_2}$

39. A rubber ball falls from a height h and rebounds to a height h/2. A rubber ball of double the mass falling from the same height h rebounds to a height

1) h 2) h/2 3) 3h/4 4) 2h

40. A fat hose pipe is held horizontally by a fireman. It delivers water through a constricting nozzle at 1 litre/sec. If by increasing the pressure, the water is delivered at 2 litre/sec, the fireman now has to

1) push forward twice as hard
2) push forward four times as hard
3) push forward eight times as hard
4) push backward four times as hard

41. Two trains A and B are running in the same directions on parallel tracks such that A is faster than B, packets of equal weight are exchanged between them. Then

1) A will be retarded and B will be accelerated
2) B will be retarded and A will be accelerated
3)There will no change in A but B will be retarded
4) There will no change in B but A will be retarded

42. Match list (I) with List (II)

	LIST - I	LIST - II
a)	Recoil of gun	e) Newtons first law
b)	Acceleration in one direction dimensions and motion in two dimensions	f) Newton's second law
c)	Definition of force	g) Projectile motion
d)	Measurement of force	h) Newton's third law

1) $a \to e, b \to g, c \to h$ and $d \to f$
2) $a \to h, b \to g, c \to e$ and $d \to f$
3) $a \to e, b \to h, c \to g$ and $d \to f$
4) $a \to h, b \to e, c \to g$ and $d \to f$

43. A force of constant magnitude starts acting on a moving particle when it is at some point 'P'. Depending on the orientation of the force, the particle may

a) pass through point P at some time later
b) not return to point P
c) describe a circular path
d) describe a parabolic path

1) a is correct 2) a, b, c, d are correct
3) c only correct 4) d only correct

44. The force exerted by the floor of an elevator on the foot of a person standing there is more than the weight of the person if the elevator is

a) going up and slowing down

b) going up and speeding up

c) going down and slowing down

d) going down and speeding up

1) b and c are correct

2) a and b are correct

3) b and d are correct 4) d only correct

45. A reference frame attached to the earth

a) is an inertial frame by definition

b) can not be an inertial frame because the earth is revolving around the sun

c) is an inertial frame because Newton's laws are applicable in this frame

d) cannot be an inertial frame because the earth is rotating about its axis

1) a, b, c are correct 2) b only correct

3) b and d are correct 4) All are correct

46. a) In a frame of reference S_1, though the net force is zero, the net acceleration is not zero.

b) In a frame of reference S_2, though the net force is not zero, the net acceleration is zero.

c) In a frame of reference S_3, the net acceleration is zero whenever the net force is zero.

1) S_1 and S_3 are inertial and S_2 is non-inertial

2) S_1 and S_2 are non-inertial and S_3 is inertial

3) S_1, S_2, S_3 are non-inertial

4) S_1, S_2, S_3 are inertial

47. A bird of mass 0.1 kg rising vertically with an acceleration 0.2 ms^{-2}. The muscular force exerted by it is (g = 9.8 ms^{-2})

a) 1 Newton b) 9.8 kg.wt

c) 1/9.8 kg.wt d) 980 newtons.

1) a is correct 2) a and b are correct

3) a and c are correct 4) b and d are correct

<u>ASSERTION & REASON</u> :

These Questions consist of two statements each printed as Assertion and Reason.

While answering these questions you are required to choose any one of the following four responses.

1) Both (A) and (R) are true and (R) is the correct explanation of (A)

2) Both (A) and (R) are true and (R) is not the correct explanation of (A)

3) (A) is true but (R) is false

4) (A) is false but (R) is true

48. **(A) :** A player lowers his hands while catching a cricket ball.

(R) : Increase in the time of catch, decreases the impulsive force.

49. **(A) :** Sportsman runs some distance before taking a long jump

(R) : Because of inertia of motion body remains in state of motion and enables him to have greater velocity to jump.

50. **(A) :** A body can have acceleration even if its velocity is zero at a given instant of time

(R) : A body is momentarily at rest when it reverses its direction of motion

51. **(A) :** A rocket moves foward by pushing the surrounding air backwards.

(R) : Rocket derives the necessary thrust to move forward according to Newton's third law of motion

52. **(A) :** When stationary bomb explodes into two pieces their speeds are in the inverse ratio of their massess.

(R) : Explosion does not violate Law of conservation of linear momentum .

53. **(A) :** When bullet is fired from a gun, the gunner should exert force on gun in the direction of motion of bullet.

(R) : The gun recoils when bullet is fired.

54. **(A) :** A man standing on top of tower throws a ball downwards. He moves up when gravity is absent.

(R) : Momentum of system is conserved in absence of external force.

55. A cork is submerged in water by a spring attached to the bottom of a pail. When the pail is kept in elevator moving with acceleration downwards, the spring length (1995)

1) increases 2) decreases

3) remains unchanged 4) data insufficient

56. In a tug of war contest, two men pull on a horizontal rope from opposite sides. The winner will be the man who

1) Exerts greater force on the rope

2) Exerts greater force on the ground

3) Exerts force on the rope which is greater than tension in the rope

4) Makes smaller angles with vertical

57. The linear momentum 'p' of a body moving in one dimension varies with time according to the equation $p = a + bt^2$, where a and b are positive constants. The net force on the body is

1) Proportional to t^2 2) Proportional to t

3) constant 4) inversly proportional to t

58. When train stops suddenly, the passengers in the train feel jerk in the forward direction. This is because

1) the inertia of rest stops the train on takes the body forward.

2) the upper part of the body continues to be in the state of motion where as the lower part of the body which is in contact with seat comes to rest

3) the lack of seat pushes the body forward

4) due to some reason other than above

59. An object will continue accelerating till

1) the resultant force on it begins to decrease

2) the resultant force on it zero

3) the resultant force is at right angles to its motion

4) the resultant force is increasing continuously

ACADEMY TEXT BOOK LINES :

60. Which of the following forces are called fundamental forces :

a) gravitational force

b) frictional force c) electrostatic force

d) elastic force e) nuclear forces

f) force due to surface tension

1) b, d, f 2) a, c, e

3) a, b, d, f 4) a, b, c, d, e, f

61. Statement A : Action and reaction act on two different bodies

Statement B : Action, reaction never cancel each other

1) A & B are correct

2) A & B are wrong

3) A is correct and B is wrong

4) A is wrong and B is correct

62. Statement A : When a person is on the floor of a lift which is at rest, the resultant force on him is equal to his weight

Statement B : When the lift is moving with uniform velocity, then the apparent weight of the man is zero

1) A & B are correct 2) A & B are wrong

3) A is correct and B is wrong

4) A is wrong and B is correct

63. A man of mass m is on the floor of a lift. The lift moving up with acceleration 'a', then :

a) the net unbalanced force on him is 'ma'

b) the normal reaction exerted by the floor on the man is m(g + a)

c) the apparent weight is greater than his true weight

1) a, b, c are correct 2) a, b, c are wrong

3) a & c are correct 4) b & c are correct

64. A man of mass m is on the floor of a lift and lift is moving up with acceleration 'a' then :

a) normal reaction is more than his weight by ma

b) man appears to be heavier by (a/g) times his true weight

1) a & b are correct 2) a & b are wrong

3) a is correct & b is wrong

4) a is wrong & b is correct

65. A person of mass m is on the floor of a lift. The lift is moving down with an acceleation 'a'. Then :

a) the net force is acting in downword direction and is equal to mg

b) the force mg must be greater than reaction force

c) the man appears to be lighter than his true weight by a factor (a/g)

1) a, b, c are correct

2) b and c are correct

3) a and c are correct

4) only b is correct

66. Statement A : If the lift is falling freely then the man on its floor experiences no reaction from the floor.

Statement B : If the lift moves down with an acceleration a > g, then the normal reaction becomes negative and the man feels floating up in the lift.

1) A and B are correct

2) A and B are false

3) A is true and B is false

4) A is false and B is true

67. A man of mass m is on the floor of a lift then match the following :

LIST – I	LIST – II
a) lift is moving up with acceleration a	e) apparent weight is greater than true weight
b) lift is moving down with acceleration a	f) apparent weight is zero
c) lift is moving with uniform velocity	g) apparent weight is equal to true weight
d) lift is freely falling	h) apparent weight is less than true weight

1) a-e, b-h, c-f, d-g
2) a-h, b-e, c-g, d-f
3) a-e, b-h, c-g, d-f
4) a-e, b-f, c-g, d-h

68. Statement A : If the force varies with time in a complicated way then the average force is measured by the total change inmomentum of the body

Statement B : Change in momentum and impulsive force are numerically equal

1) A & B are true
2) A & B are false
3) A is true & B is false
4) A is false & B is true

69. Statement A : Shock absorbers reduce the magnitude of change in momentum.

Statement B : Shock absorbers increase the time of action of impulsive force

1) A & B are true
2) A & B are false
3) A is true & B is false
4) A is false & B is true

ANSWERS

1) 3	2) 2	3) 4	4) 3	5) 1
6) 2	7) 4	8) 1	9) 4	10) 3
11) 3	12) 2	13) 1	14) 2	15) 2
16) 3	17) 4	18) 2	19) 1	20) 4
21) 2	22) 2	23) 2	24) 4	25) 1
26) 3	27) 1	28) 1	29) 2	30) 2
31) 4	32) 2	33) 4	34) 2	35) 3
36) 3	37) 3	38) 3	39) 2	40) 2
41) 1	42) 2	43) 2	44) 1	45) 3
46) 2	47) 3	48) 1	49) 1	50) 1
51) 4	52) 1	53) 1	54) 1	55) 2
56) 2	57) 2	58) 2	59) 2	60) 2
61) 1	62) 2	63) 1	64) 1	65) 2
66) 1	67) 3	68) 3	69) 4	

EXERCISE - IIA

CHANGE IN MOMENTUM, FORCE, IMPULSE

1. Due to the application of a force on a body of mass 100 kg that is initially at rest, the body moves with an acceleration of $20ms^{-2}$ in the direction of the force. Find the magnitude of the force.

1) 1000N 2) 2000 N 3) 3000 N 4) 4000 N

2. A force is applied for a duration of 10sec on a body of mass 5kg that is at rest. As a result the body acquires a velocity of $2ms^{-1}$. Find the magnitude of the force applied.

1) 1 N 2) 2 N 3) 4 N 4) 5 N

3. A vehicle of mass 120kg is moving with a velocity of 90kmph. What force should be applied on the vehicle to stop it in 5s.

1) 240 N 2) 480 N 3) 600 N 4) 480 N

4. A vehicle of mass 20kg is moving with a velocity of $4ms^{-1}$. Find the magnitude of the force that is to be applied on the vehicle so that the vehicle have a velocity of $1ms^{-1}$ after travelling a distance of 20m.

1) 4 N 2) 6 N 3) 7.5 N 4) 9.5 N

5. A force is applied on a body of mass 0.9kg that is at rest. The force is applied for a duration of 5s and as a result the body covers a distance of 250 m. Find the magnitude of the force.

1) 9 N 2) 18 N 3) 27 N 4) 46 N

6. A ball of mass 600 gm strikes a wall with a velocity of $5ms^{-1}$ at an angle 30^0 with the wall and rebounds with the same speed at the same angle with the wall. The change in momentum of the ball is, (in kg ms^{-1})

1) 15 2) 10 3) 5 4) 3

7. When a force 1N acts on 1kg mass at rest for 1s, its final momentum is P. When 1N force acts on 1kg mass at rest through a distance 1m, its final momentum is $P^{|}$. The ratio of P to $P^{|}$ is

1) 1 : 1 2) $1:\sqrt{2}$ 3) 1 : 2 4) 2 : 1

8. A force of 25 N acts on a body at rest for 0.2s and a force of 70 N acts for the next 0.1s in opposite direction. If the final velocity of the body is $5ms^{-1}$, the mass of the body is

1) 1 kg 2) 2 kg 3) 0.8 kg 4) 0.4 kg

9. A player caught a cricket ball of mass 150g moving at a rate of 20 m/s. If the catching process is completed in 0.1s, the force of the blow exerted by the ball on the hand of the player is equal to : (AIEEE 2006)

1) 150N 2) 3N 3) 30N 4) 300N

10. A body of mass 3 kg is moving along a straight line with a velocity of 24 ms^{-1}. When it is at a point 'P' a force of 9 N acts on the body in a direction opposite to its motion. The time after which it will be at 'P' again is,

1) 8s 2) 16s 3) 12s 4) 24s

11. A balloon with its contents weighing 160 N is moving down with an acceleration of g/2 ms^{-2}. The mass to be removed from it so that the balloon moves up with an acceleration of g/3 ms^{-2} is

1) 5kg 2) 10 kg 3) 6 kg 4) 3 kg

12. A block of mass 5 kg is at rest on a smooth horizontal surface. Water coming out of a pipe horizontally at the rate of 2 kg s^{-1}, hits the block with a velocity of $6ms^{-1}$. The initial acceleration of the block is,

1) Zero 2) 1.2 ms^{-2} 3) 2.4 ms^{-2} 4) 0.6 ms^{-2}

13. A horizontal jet of water coming out of a pipe of area of cross – section 20 cm^2 hits a vertical wall with a velocity of 10 ms^{-1} and rebounds with the same speed. The force exerted by water on the wall is,

1) 0.2 N 2) 10 N 3) 400 N 4) 200 N

14. A body of mass 2kg is moving along positive X – direction with a velocity of 5 ms^{-1}. Now a force of $10\sqrt{2}$ N is applied at an angle 45^0 with X – axis. Its velocity after 3s is,

1) 20 ms^{-1} 2) 15 ms^{-1}
3) 25 ms^{-1} 4) 5 ms^{-1}

15. A disc of mass 5 kg is kept floating horizontally in mid air by firing 10 bullets per second vertically up. If the mass of each bullet is 50 gm and bullets rebound with same speed, the speed of each bullet is, ($g = 10ms^{-2}$)

1) 100 ms^{-1} 2) 200 ms^{-1}
3) 50 ms^{-1} 4) 10 ms^{-1}

16. A body of mass 5 kg moving on a horizontal surface with a velocity of 10 ms^{-1} comes to rest in 2s. The force required to make this body move with a velocity of 10 ms^{-1} on the same surface is

1) Zero 2) 10 N 3) 25 N 4) 50 N

17. A machine gun fires a bullet of mass 40 g with a velocity 1200 ms^{-1}. The man holding it, can exert a maximum force of 144N on the gun. How many bullets can he fire per second at the most ? (AIEEE-2004)

1) One 2) Four 3) Two 4) Three

18. Gravel is dropped on a conveyor belt at the rate of 2 kg/s. The extra force required to keep the belt moving at $3ms^{-1}$ is

1) 1 N 2) 3 N 3) 4 N 4) 6 N

19. A hammer of mass 1 kg strikes on the head of a nail with a velocity of 2 ms^{-1}. It drives the nail 0.01 m into a wooden block. Find the force applied by the hammer and the time of impact.

1) 200 N ; 10^{-2} sec. 2) 100 N ; 10^{-3} sec.
3) 300 N ; 10^{-2} sec. 4) 400 N ; 10^{-3} sec.

20. A force of 5 N acts on a body for 2 malliseconds. Calculate the impulse. If the mass of the body is 5 g, calculate the change of velocity.

1) 2 m/s 2) 4 m/s 3) 6 m/s 4) 10 m/s

21. The apparent weight of man inside a lift moving up with certain acceleration is 900N. When the lift is coming down with the same acceleration apparent weight is found to be 300N. The mass of the man is ($g = 10\ ms^{-2}$)

1) 45 kg 2) 60 kg 3) 75 kg 4) 80 kg

22. A 60kg man is inside a lift which is moving up with an acceleration of 2.45 ms^{-2}. The apparent percentage change in his weight is,

1) 20% 2) 25% 3) 50% 4) 75%

23. When an empty lift is moving down with an acceleration of $\frac{g}{4}ms^{-2}$, the tension in the cable is 9000N. When the lift is moving up with an acceleration of $\frac{g}{3}ms^{-2}$, the tension in the cable is,

1) 16,000N 2) 18,000N
3) 12,000N 4) 15,000N

24. A body of mass 40 kg stands on a weighing machine in an accelerated lift. The reading on the scale of the weighing machine is 300 N. Find the magnitude and direction of acceleration.

1) 2.3 m/s^2 upwards 2) 2.3 m/s^2 downwards
3) 4.6 m/s^2 upwards 4) 4.6 m/s^2 downwards

25. A fireman wants to slide down a rope. The breaking load for the rope is 3/4th of the weight of the man. With what minimum acceleration sholud the fireman slide down? Acceleration due to gravity is g.

1) zero 2) g/4 3) 3g/4 4) g/2

26. How can you lower a 100 kg body from the roof of a house using a cord with a breaking strenght of 80 kg weight without breaking the rope? (1 kg weight = g N)

1) 1.96 m/s^2 downwards 2) 9.8 m/s^2 downwards
3) 4.9 m/s^2 downwards 4) 19.6 m/s^2 downwards

EQUILIBRIUM

27. Two 10 kg bodies are attached to a spring balance as shown in figure. The reading of the balance will be

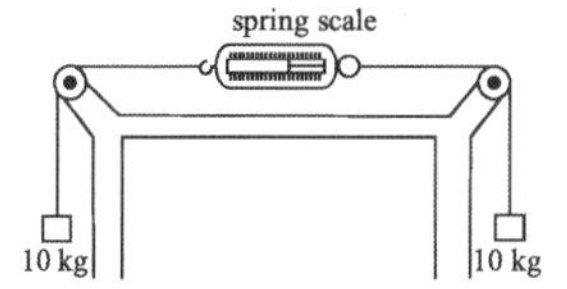

1) 20 kg-wt 2) 10 kg-wt
3) Zero 4) 5 kg-wt

28. A mass 'M' is suspended by a rope from a rigid support. It is pulled horizontally with a force F. If the rope makes an angle 'θ' with vertical in equilibrium, then the tention in the string is

1) $F\sin\theta$ 2) $\frac{F}{\sin\theta}$

3) $F\cos\theta$ 4) $\frac{F}{\cos\theta}$

29. In the given arrangement, for the system to remain under equilibrium, the 'θ' should be **(IIT 2001)**

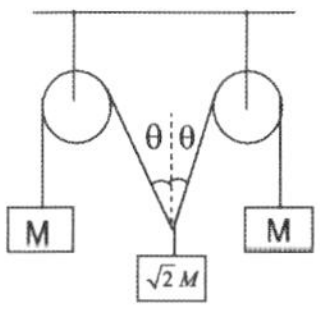

1) 0^0 2) 30^0 3) 45^0 4) 60^0

30. A uniform rope of length 5m is on a smooth horizontal surface. It is being pulled by a horizontal force of 20 N at one end. The ratio of tension at a distance of 2m from force end to tension at a distance of 2m from free end is,

1) 1 : 1 2) 2 : 3 3) 1 : 2 4) 3 : 2

31. Two block of masses 3 kg and 1 kg are kept in contact with rach other on a frictionless horizontal surface. If a force of 10 N is applied on the larger block what is the acceleration of the system? What is the contact force between the two blocks?

1) 2.5 m/s^2; 2.5N 2) 5 m/s^2; 5N
3) 5 m/s^2; 2.5N 4) 2.5 m/s^2; 5N

32. Two blocks of masses 10 kg and 20 kg are connected by a massless spring and are placed on a smooth horizontal surface. A force of 200 N is applied on 20 kg mass as shown in the diagram. At the instant, the acceleration of 10 kg mass is 12 ms^{-2}, the acceleration of 20kg mass is,

1) 4 ms^{-2} 2) 12 ms^{-2} 3) 20 ms^{-2} 4) 8 ms^{-2}

33. A block of mass 10 kg lying on a smooth horizontal surface is being pulled by means of a rope of mass 2 kg. If a force of 36N is applied at the end of the rope, the tension at the mid point of the rope is,

1) 33 N 2) 30 N 3) 24 N 4) 12 N

34. Ten coins each of mass 10 gm are placed one above the other. The reaction force exerted by 7th coin from the bottom on the 8th coin is (g = 10 m/s^2)

1) 0.3N 2) 0.2N 3) 0.4N 4) 0.7N

PSEUDO FORCE

35. A small mass is suspended by a string from the ceiling of a car. As the car accelerates at a rate 'a' the string makes an angle θ with the vertical. Then the tension in the string

1) $mg\cos\theta$ 2) $ma\sin\theta$

3) $m\sqrt{a^2+g^2}$ 4) $ma\sqrt{a^2+g^2}$

36. A pendulum of mass m hangs from a support fixed to a trolley. The direction of the string (i.e.., angle θ) when the trolley rolls up a plane of inclination α with acceleration 'a' is

1) Zero 2) $\tan^{-1}\alpha$ 3) $\tan^{-1}\frac{a+g\sin\alpha}{g\cos\alpha}$ 4) $\tan^{-1}\frac{a}{g}$

37. **All surfaces are smooth. The acceleration of mass m relative to wedge is**

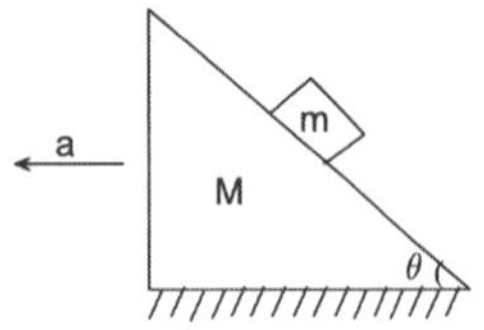

1) $g\sin\theta$ 2) $g\sin\theta + a\cos\theta$

3) $g\sin\theta - a\cos\theta$ 4) $a\cos\theta$

PULLEY

38. **A boy of mass 40 kg climbs up a rope with an acceleration of 2 ms^{-2}. What is the tension in the rope ?**

1) 472 N 2) 435 N 3) 232 N 4) 568 N

39. **Two bodies of masses 1 kg and 2 kg are connected by a very light string passed over a clamped light smooth pulley. If the system is released from rest, find the acceleration of the two masses and the tension in the string**

1) 3.27 m/s^2; 13N 2) 4.5 m/s^2; 26N

3) 1.27 m/s^2; 13N 4) 0.327 m/s^2; 13N

40. **A block of mass 3 kg which is on a smooth inclined plane making an angle of 30^0 to the horizontal is connected by a cord passing over a light frictionless pulley to a second block of mass 2 kg hanging vertically. What is the acceleration of each block and what is the tension of the cord?**

1) 0.98 m/s^2; 17.6N 2) 1.98 m/s^2; 19.6N

3) 0.49 m/s^2; 9.8N 4) 1.47 m/s^2; 4.9N

41. **Two blocks of masses 2 kg and 4 kg are connected by a light string passing over a light smooth pulley clamped to the edge of a horizontal table. The 2 kg block is on the smooth horizontal table and the other block is hanging vertically. (i) Find the acceleration of the stem if it is released from rest. (ii) Find the tension in the string.**

1) 6.54 m/s^2; 26N 2) 2.6 m/s^2; 6.54 N

3) 6.54 m/s^2; 13N 4) 13 m/s^2; 26N

42. **A constant force F = $m_2g/2$ is applied on the block of mass m_1 as shown. The string and the pulley are light and the surface of the table is smooth. Find the acceleration of m_1.**

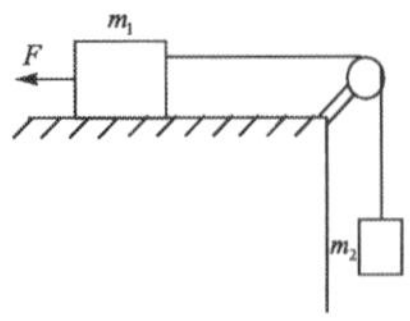

1) $\dfrac{m_2 g}{2m_1}$ 2) $\dfrac{m_2 g}{2(m_1+m_2)}$

3) $\dfrac{3m_2 g}{2(m_1+m_2)}$ 4) $\dfrac{3m_2 g}{2m_1}$

CONSERVATION OF LINEAR MOMENTUM

43. **A 50kg man is standing at one end of a 25m long boat. He starts running towards the other end. On reaching the other end his velocity is 2ms^{-1}. If the mass of the boat is 200 kg, final velocity of the boat is (in ms^{-1}) (2006 E)**

1) $\frac{2}{5}$ 2) $\frac{2}{3}$ 3) $\frac{8}{5}$ 4) $\frac{8}{3}$

44. **A wedge shaped block 'A' of mass M is at rest on a smooth horizontal surface. A small block 'B' of mass 'm' placed at the top edge of inclined plane of length 'L' as shown in the figure. By the time, the block 'B' reaches the bottom end, the wedge A moves a distance of**

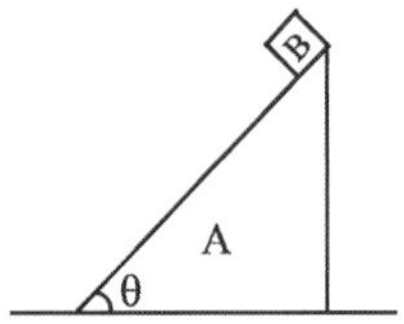

1) $\dfrac{mL}{M\cos\theta}$ 2) $\dfrac{mL\cos\theta}{m+M}$ 3) $\dfrac{mL}{m+M}$ 4) Zero

45. **A stationary shell explodes into two fragments, having masses in the ratio of 1:2. The heavier fragment attains a Kinetic energy of 100J. The Kinetic energy released in the explosion is**

1) 200 J 2) 150J 3) 300 J 4) 600 J

46. **A particle of mass 1 kg is thrown vertically upward with speed 100 m/s. After 5 sec it explodes into two parts. One part of mass 400 g comes back with speed 25 m/s , what is the speed of the other part just after explosion? (g=10 m/s^2)**

1) 600 m/s upward 2) 100 m/s upward

3) 100 m/s down ward 4) 300 m/s upward

47. **A body of mass m is projected at an angle θ to the horizontal, the projectile at the heighest point breaks into two fragments of equal masses. One of the fragments retraces its path to the point of projection. The velocity of the other fragment just after explosion is**

1) V Cos θ 2) 2 Cos θ

3) 3V Cos θ 4) (3v/2) Cos θ

48. A gun mounted on the top of a moving truck is aimed in the backward direction at an angle of 30^0 to the vertical. If the muzzle velocity of the gun is 4 ms^{-1}, the value of speed of the truck that will make the bullet come of out vertically is [EAMCET-98 M]

1) 1 ms^{-1} 2) $\frac{\sqrt{3}}{2} ms^{-1}$

3) 0.5ms^{-1} 4) 2ms^{-1}

49. A gun of mass M fires a bullet of mass m with a Kinetic energy E. The velocity of recoil of the gun is

1) $\frac{\sqrt{2ME}}{m}$ 2) $\frac{\sqrt{2mE}}{M}$

3) $\frac{\sqrt{2mE}}{M+m}$ 4) $\frac{\sqrt{2ME}}{(M+m)}$

50. A gun of mass M fires a bullet of mass m with a velocity v relative to the gun. The average force required to bring the gun to rest in 0.5 sec. is

1) $\frac{2Mmv}{M+m}$ 2) $\frac{Mmv}{2(M+m)}$

3) $\frac{3Mmv}{2(M+m)}$ 4) $\frac{Mmv}{(M+m)}$

51. A shell of mass 0.01 kg fired by a gun of mass 10 kg. If the muzzle speed of the shell is 50 ms^{-1}, what is the recoil speed of the gun?

1) 0.01 m/s 2) 0.02 m/s

3) 0.03 m/s 4) 0.05 m/s

ROCKET PROPULSION

52. A rocket consumes 20 kg fuel per second. The exhaust gases escape at a speed of 1000 ms^{-1} relative to the rocket. Calculate the upthrust received by the rocket. Also calculate the velocity acquired. When its mass is 1/100 of the initial mass.

1) 2×10^4 N ; 4.6 Km/s

2) 2×10^6 N ; 5.3 Km/s

3) 4×10^2 N ; 4.6 Km/s

4) 4×10^6 N ; 5.3 Km/s

53. A rocket of initial mass 6000kg ejects mass at constant rate of 200 kg/sec. with constant relative speed of 800 m/sec. The acceleration of the rocket after 5 sec is (neglect gravity)

1) 50 m/s^2 2) 16 m/s^2

3) 60 m/s^2 4) 32 m/s^2

54. A rocket of mass 20kg has 180 kg of fuel. The exhaust velocity of fuel is 1.6 km/sec. Calculate the ultimate velocity of the rocket gained, when the rate of consumption of the fuel is 2kg/sec. (neglect gravity)

1) 3.7 km/sec 2) 2 km/sec

3) 10 km/sec 4) 5 km/sec

ANSWERS

1) 2	2) 1	3) 3	4) 3	5) 2
6) 4	7) 2	8) 4	9) 3	10) 2
11) 2	12) 3	13) 3	14) 3	15) 3
16) 3	17) 4	18) 4	19) 1	20) 1
21) 2	22) 2	23) 1	24) 2	25) 2
26) 1	27) 2	28) 2	29) 3	30) 4
31) 1	32) 1	33) 1	34) 1	35) 3
36) 3	37) 2	38) 1	39) 1	40) 1
41) 3	42) 2	43) 1	44) 2	45) 3
46) 2	47) 3	48) 4	49) 2	50) 1
51) 4	52) 1	53) 4	54) 1	

EXERCISE-II(B)

CHANGE IN MOMENTUM, FORCE, IMPULSE

1. A force of 50 N acts on a body of mass 10 kg at rest. What is its acceleration? What is its velocity after 5 seconds if the same force acts?

1) 15 m/s 2) 20 m/s 3) 25 m/s 4) 30 m/s

2. A force of 10 N inclined to the horizontal at an angle of 60^0 acts on a body of mass 2 kg. If the body can move in horizontal direction only find its acceleration.

1) 2.5 m/sec^2 2) 5.5 m/sec^2

3) 4.5 m/sec^2 4) 3.5 m/sec^2

3. A constant retarding force of 20 N acts on a body of mass 5 kg moving initially with a speed of 10 ms^{-1}. How long does the body take to stop?

1) 1.5 s 2) 2.5 s 3) 3.5 s 4) 4.5 s

4. The driver of a three wheeler moving with a speed of 10 ms^{-1} sees a child standing in the middle of the road at a distance of 10 m. He applies brakes and brings the vehicle to rest just infornt of the child. If the mass of the three wheeler is 400 kg and the driver is 50 kg find the brake force.

1) 1500 N 2) 2000 N

3) 2250 N 4) 2500 N

5. **A bullet of mass 0.005 kg moving with a speed of 200 ms^{-1} enters a heavy wooden block and is stopped after a distance of 50 cm. What is the average force exerted by the block on the bullet ?**

1) –200 N 2) + 200 N 3) 400 N 4) –400 N

6. **A body of mass m falls from a height h_1 and rises to a height h_2. The magnitude of the change in momentum during the impact with the ground.**

1) mg (h_1+h_2) 2) $m\left(\sqrt{2gh_1}+\sqrt{2gh_2}\right)$

3) $m\left(\sqrt{2gh_1}-\sqrt{2gh_2}\right)$ 4) zero

7. **Two bodies of masses 4 kg and 16 kg at rest to acted upon by same force. The ratio of times required to attain the same speed is,**

1) 1 : 1 2) 4 : 1 3) 1 : 4 4) 1 : 2

8. **In a football match a ball is kicked by a player with a force of 25N for 0.2 sec and then by another player with a force of 70 N for 0.1 sec in the same direction. If the foot ball gains a velocity of 24 m/s after the two kicks, the mass of the foot ball is nearly**

1) 5kg 2) 10 kg 3) 2.5 kg 4) 0.5 kg

9. **A ball of mass 10 gm dropped from a height of 5 m hits the floor and rebounds to a height of 1.25 m. If the ball is in contact with the ground for 0.1s, the force exerted by the ground on the ball is (g = 10 m/s^2)**

1) 0.5 N 2) 1.5 N

3) 0.15 N 4) 2.5 N

10. **A body of mass 5 kg is at rest on a smooth horizontal surface. A horizontal force 'F' acts on it for 4s and then ceases to act. If the body travels 40 m in the next 5s, the value of F is,**

1) 5 N 2) 20 N 3) 15 N 4) 10 N

11. **A balloon of total mass 1000 Kg float motionless over the earth's surface. If 100 kg of sand ballast are thrown over board, the balloon starts to rise with an acceleration of**

1) 10 m/s^2 2) 9.8 m/s^2

3) 1.09 m/s^2 4) 4.9 m/s^2

12. **The horizontal speed of a jet of water is 100 cm/sec and 50 cm^3 of water hits the plate each second. Assume that the water moves parallel to the plate after striking it. The force exerted on the stationary plate if it is held perpendicular to the jet of water is :**

1) 5×10^{-2}N 2) 5×10^{2}N

3) 5×10^{-1}N 4) 5N

13. **A stream of water flowater flowing horizontally with a speed of 15 ms^{-1} gushes out of a tube of cross – sectional area 10^{-2} m^2 and hits a vertical wall near by. What is the force exerted on the wall by the impact of water, assuming that it does not rebound? (Density of water = 1000 kg m^{-3})**

1) 1250 N 2) 2250 N

3) 4500 N 4) 2550 N

14. **A body of mass 2 kg has an initial velocity of 3 ms^{-1} along OE and it is subjected to a force of 4 newton in OF direction perpendicular to OE. The distance of the body from O after 4 s will be**

1) 12 m 2) 20 m 3) 28 m 4) 48 m

15. **A disc of mass 0.5kg is kept floating horizontally in mid air by firing bullets of mass 5 g each, vertically at it at the rate of 10 per second. If the bullets drop dead. The speed of the bullet striking the disc is (g=10ms^{-2})**

1) 100ms^{-1} 2) 10ms^{-1} 3) 10 cm/s 4) 1 cm/s

16. **A constant horizontal force of 20N acts on a body on a smooth horizontal plane. The body starts from rest and is observed to move 20 m in two seconds. The mass of the body is**

1) 4 kg 2) 2 kg 3) 1 kg 4) 0.5 kg

17. **A gun of mass 10Kg fires four bullets per second. The mass of each bullet is 20g and the velocity of the bullet when it leaves the gun is 300 ms^{-1}. The force required to hold the gun while firing is [EAMCET-2000M]**

1) 6N 2) 8N 3) 24N 4) 240N

18. **A balloon has 5 gm of air. A small hole is pierced into it. The air escapes at a uniform rate with a velocity of 4 cm s^{-1}. If the balloon shrinks completely in 2.5 second, then the average force acting on the balloon is**

1) 2 dyne 2) 50 dyne 3) 8 dyne 4) 8 N

19. **A 1.5 kg hammer moving with a velocity of 10 m/s strikes a nail for 0.005 s. The average force exerted on the nail is**

1) 5000 N 2) 3000 N 3) 1500 N 4) 750 N

20. **A force of 10 N acts on a body for 2 micro second. Calculate the impulse. If the mass of the body is 1 g, calculate the change of velocity.**

1) 0.04 m/s 2) 0.02 m/s

3) 0.01 m/s 4) 0.2 m/s

21. The apparent weight of a mass in a lift moving up is 80 kg when its weight in the stationary lift is 60 kg. If the same lift now moves up with same retardation, the weight of the mass will be

1) 40 kg 2) 60 kg 3) 80kg 4) 100 kg

22. A 60 kg man stands on an elevator floor The elevator is going up with constant acceleration of 1.96 m/s^2. Percentage change in the apparent weight of the person is

1) 10 2) 15 3) 20 4) 25

23. A spring balance is attached to the ceiling of a lift. A man hangs his bag on the spring and the spring reads 49N, when the lift is stationary. If the lift moves downward with an acceleration of 5 m/s^2, the reading of the spring balance will be: (AIEEE-2003)

1) 24N 2) 74N 3) 15N 4) 49N

24. An elevator weighing 6000kg is pulled upward by a cable with an acceleration of 10 m/s^2. Taking g to be 10 ms^{-2}, find the tension in the cable

1) 120 KN 2) 60 KN 3) 30 KN 4) 90 KN

25. A 10 kg stone is suspended with a rope of breaking strength 30 kg-wt. The minimum time in which the stone can be raised through a height 10 m starting from rest is (Take g = 10 N/kg).

1) 0.5 s 2) 1.0 s 3) $\sqrt{2/3}$ s 4) 2 s

26. A body of mass 50 kg and specific gravity force is to be lifted using a rope of breaking strength of 60 kg wt. The acceleration with which it can be pulled up in least interval of time

1) zero 2) g/2 ms^2 3) g/4 ms^2 4) g/5 ms^2

EQUILIBRIUM

27. A block of mass 20 kg is suspended from two light spring balances as shown in the figure.

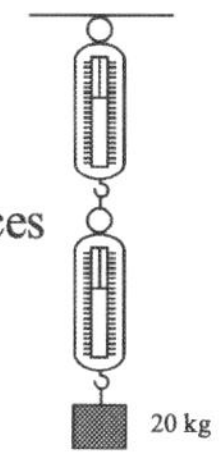

1) Both the scales will read 10 kg

2) Both the scales will read 20 kg

3) The sum of readings of two balances will be 20 kg

4) None of the above

28. A simple pendulum is carrying a bob of mass 'M'. It is pulled through an angle 45^0 from its mean position. The horizontal force required to keep it under equilibrium at this new position is

1) Mg 2) Mg/2 3) $Mg/\sqrt{2}$ 4) $\sqrt{2}Mg$

29. Two masses M_1 and M_2 connected by means of a string which is made to pass over light, smooth pulley are in equilibrium on a fixed smooth wedge as shown in figure. If $\theta = 60^0$ and $\alpha = 30^0$, the ratio of M_1 to M_2 is

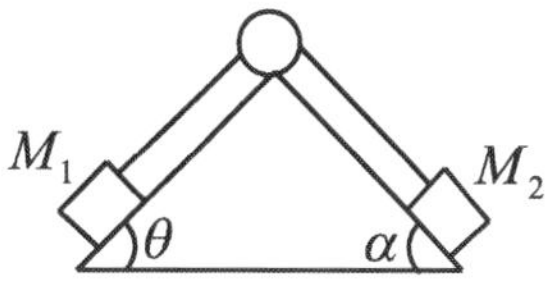

1) 1 : 2 2) $2:\sqrt{3}$

3) $1:\sqrt{3}$ 4) $\sqrt{3}:1$

TENSION, NORMAL REACTION

30. A uniform rope of length L, resting on a frictionless horizontal table is pulled at one end by a force F. What is the tension in the rope at a distance x form the end where the force is applied?

1) F(L – x) 2) Fx/L 3) (L – x)/F 4) F(1 – x/L)

31. Two blocks of masses 5 kg and 2 kg are kept in contact with each other on a frictionless horizontal surface. If a force of 14 N is applied on the larger block what is the acceleration of the system? What is the contact force between the two blocks?

1) 2 m/s^2 ; 4 N 2) 1 m/s^2 ; 2 N

3) 1 m/s^2 ; 4 N 4) 2 m/s^2 ; 2 N

32. A block of mass m is connected to another block of mass M by a spring (massless) of spring constant k. The blocks are kept on a smooth horizontal plane. Initially the blocks are at rest and the spring is unstretched. Then a constant force F starts acting on the block of mass M to pull it. Find the force on the block of mass m. (AIEEE-2007)

1) $\frac{mF}{M}$ 2) $\frac{(M+m)F}{m}$

3) $\frac{mF}{(m+M)}$ 4) $\frac{MF}{(m+M)}$

33. A chain of length 'L' and mass 'M' is hanging by fixing its upper end to a rigid support. The tension in the chain at a distance 'x' from the rigid support is

1) Zero 2) Mg

3) $\frac{Mg(L-x)}{L}$ 4) $\frac{Mg(L-x)}{x}$

34. Ten coins are arranged on above the other. The mass of each coin is 10 g. Reaction force of the sixth coin from the bottom on the seventh coin

1) 0.392N 2) 0.49N 3) 0.98N 4) 0.784N

PSEUDO FORCE

35. A small mass is suspended by a string from the ceiling of a car. As the car accelerates with an acceleration a horizontal road, the angle made by the string with the vertical is

1) $\tan^{-1}g/a$ 2) $\tan^{-1}a/g$ 3) $\sin^{-1}g/a$ 4) $\sin^{-1}a/g$

36. A pendulum is hanging from the ceiling of a trolley which moves down a smooth inclined plane. Then

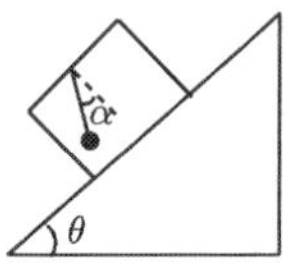

1) $\theta = \alpha$ 2) $\alpha = 0^0$

3) $\theta = 2\alpha$ 4) $\alpha = 2\theta$

37. Acceleration of block 'm' is $(\theta < 45^0)$

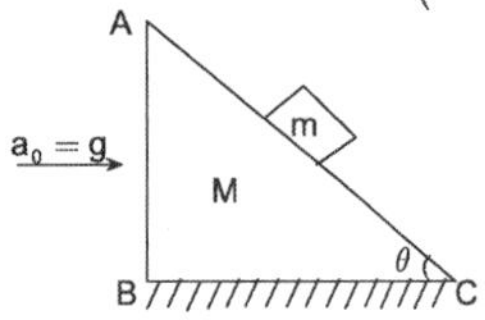

1) $g\sin\theta$ 2) $g\cos\theta$

3) $g(\cos\theta + \sin\theta)$ 4) $g(\cos\theta - \sin\theta)$

PULLEY

38. Two masses of 3 kg and 4 kg are connected at the two ends of a light inextensible string that passes over a frictionless pulley. Find the acceleration of the masses and the tension in the string, when the masses ae released.

1) 1.4 m/s^2 ; 33.6 N 2) 2.5 m/s^2 ; 16.3 N

3) 1.4 m/s^2 ; 18.5 N 4) 3.5 m/s^2 ; 32.5 N

39. Figure shows a weight of 30kg suspended at one end of cord and a weight of 70kg applied at other end of the cord passing over a pulley. Neglecting weight of rope and pulley find the tension in the cord and acceleration of the system (g=10ms^{-2})

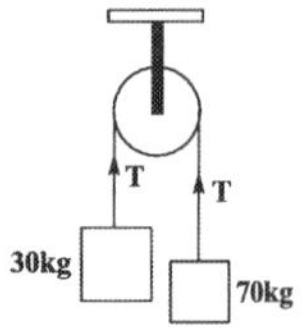

1) 2 m/s^2 ; 120 N 2) 4 m/s^2 ; 420 N

3) 6 m/s^2 ; 180 N 4) 4 m/s^2 ; 240 N

40. A block of mass m_1 = 4kg lying on a plane inclined at an angle of 30^0, is connected to another freely suspended block of mass m_2 = 6kg with the help of a string passing over a smooth pulley as shown in the figure. The acceleration of each block is (g =10 m/s^2)

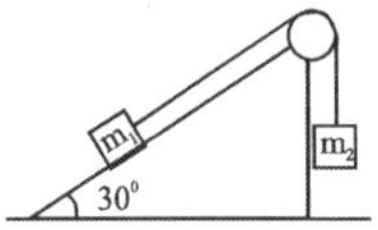

1) 8 m/s^2 2) 5 m/s^2

3) 4 m/s^2 4) 1 m/s^2

41. A block A of mass 6kg is applied on a frictionless table. A thread tied to it passes over a frictionless pulley and carries a body B of mass 4kg at the other end. Find the acceleration of the system and tension in the thread (given g=10ms^{-2})

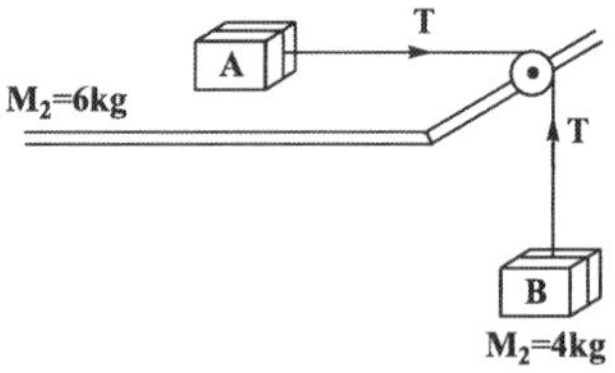

1) 2 m/s^2 2) 4 m/s^2

3) 6 m/s^2 4) 8 m/s^2

42. A block A of mass 7Kg is placed on a frictionless table. A thread tied to it passes over frictionless pulley and carries a body B of mass 3Kg at the other end. The acceleration of the system is (Given g=10ms^{-2}) (Kerala PET 2002)

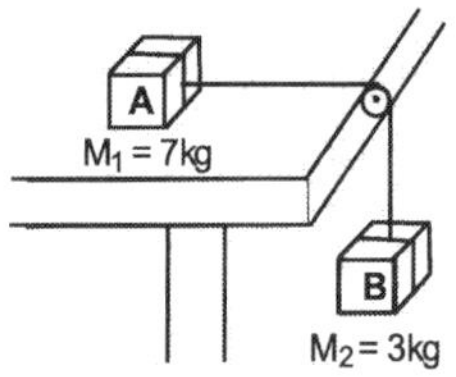

1) 100 ms^{-2} 2) 3 ms^{-2}

3) 10 ms^{-2} 4) 30 ms^{-2}

CONSERVATION OF LINEAR MOMENTUM

43. A man of mass 62 kg is standing on a stationary boat of mass 238 kg. The man is carrying a sphere of mass 0.5 kg in his hands. If the man throws the sphere horizontally with a velocity of 12 ms^{-1}, find the velocity with which the boat will move (in magnitude)

1) 0.02 ms^{-1} 2) 0.5 ms^{-1}

3) 0.04 ms^{-1} 4) 0.06 ms^{-1}

44. A block of mass 'M' is placed on the top of a wedge of mass '4M'. All the surfaces are frictionless. The system is released from rest. The distance moved by the wedge at the instant, the block reaches the bottom will be

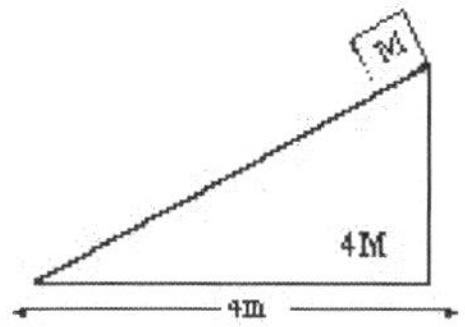

1) 0.2 m 2) 0.4 m 3) 0.8 m 4) Zero

45. A mass of 12 Kg at rest explodes into two pieces of masses 4 Kg and 8 Kg which move in opposite directions. If the velocity of 8 Kg peice is 6 ms^{-1}, then the kinetic energy of the other piece in joules is [96 M]

1) 64 2) 128 3) 144 4) 288

46. A stationary shell breaks into three fragments. The momentum of two of the fragment is P each and move at 60° to each other. The momentum of the third fragment is

1) P 2) 2P 3) $\frac{P}{\sqrt{3}}$ 4) $\sqrt{3}$ P

47. A cannon ball is fired with a velocity of 200ms^{-1} at an angle of 60° with the horizontal. At the heighest point of its flight it explodes into 3 equal fragments. One fragment is going vertically upwards with a velocity of 100m/s and second going vertically downwards with a velocity of 100ms^{-1}. The third fragment will be moving with a velocity of

1) 600 ms^{-1} in the horizontal direction
2) 300 ms^{-1} in the horizontal direction
3) 300 ms^{-1} in the direction making 60° with the horizontal
4) 200 ms^{-1} in the direction making 60° with the horizontal

48. A machine gun of mass 10kg, fires 10gm bullets at the rate of two every second. Each bullet comes out with a velocity of 200 m/s. The velocity of recoil of the gun at the end of the fourth second, after firing starts is

1) 1.6 m/s 2) 0.8 m/s 3) 3.2 m/s 4) 2.0 m/s

49. A nucleus of mass 218 amu is in free state decays to emit an α-particle. Kinetic energy of α-particle emitted is 6.7Mev. The recoil energy in (MeV) emitted of the daughter nucleus is (2005 M)

1) 1.0 2) 0.5 3)0.25 4) 0.125

50. A 2kg shot is fired from a cannon of mass 198 kg with a velocity 50 ms^{-1} w.r.t. the gun. Then the velocity of recoil of the gun is

1) 0.5ms^{-1} 2) 0.1ms^{-1}
3) 0.25ms^{-1} 4) 1 ms^{-1}

51. A shell of mass 0.02 kg is fird by a gun of mass 100 kg. If the muzzle speed of the shell is 60 ms^{-1}, what is the recoil speed of the gun?

1) 0.12 m/s 2) 0.16 m/s
3) 0.3 m/s 4) 0.4 m/s

ROCKET PROPULSION

52. A 3000 kg rocket is fired. If the exhaust speed is 500 ms^{-1}, how much gas must be ejected per second to supply the thrust needed, (a) to overcome the weight of the rocket, (b) to give the rocket an initial acceleration of 19.6 ms^{-2}?

1) 58.8 Kg/s ; 176.4 Kg/s
2) 132.8 Kg/s ; 82.5 Kg/s
3) 72.5 Kg/s ; 138.2 Kg/s
4) 93.2 Kg/s ; 185.6 Kg/s

53. A 500kg rocket has to be fired vertically. Exhaust velocity of the gases is 1.96 km/s. Minimum mass of the fuel to be released in kg per second is

1) 250 kg 2) 25 kg 3) 2.5 kg 4) 50 kg

54. A rocket of mass 6000kg is set for vertical firing. If the exhaust speed is 1000m/sec. The rate of mass of gas ejected to supply a thrust needed to give the rocket an initial upward acceleration of 20.2 m/s^2 is

1) 150 kg/sec 2) 160kg/sec
3) 28kg/sec 4) 180kg/sec

ANSWERS

1) 3	2) 1	3) 2	4) 3	5) 1
6) 2	7) 3	8) 4	9) 3	10) 4
11) 3	12) 1	13) 2	14) 2	15) 1
16) 2	17) 3	18) 3	19) 2	20) 2
21) 1	22) 3	23) 1	24) 1	25) 2
26) 3	27) 2	28) 1	29) 3	30) 4
31) 1	32) 3	33) 3	34) 1	35) 2
36) 2	37) 4	38) 1	39) 2	40) 3
41) 2	42) 2	43) 1	44) 3	45) 4
46) 4	47) 2	48) 1	49) 4	50) 1
51) 1	52) 1	53) 3	54) 4	

EXERCISE - III

CHANGE IN MOMENTUM, FORCE, IMPULSE

1. A body of mass 1 kg is moving with velocity 30 ms^{-1} due north. It is acted on by a force of 10 N due west for 4 seconds. Find the velocity of the body after the force ceases to act.

1) 50 m/s ; $\tan^{-1}\left(\frac{3}{4}\right)$ 2) 100 m/s ; $\tan^{-1}\left(\frac{4}{3}\right)$

3) 50 m/s ; $\tan^{-1}\left(\frac{3}{5}\right)$ 4) 100 m/s ; $\tan^{-1}\left(\frac{4}{5}\right)$

2. A force of 20 N acts on a body of mass 5kg at rest. What is the acceleration of the body ? What is its velocity after 5 seconds if the same force acts? After 5 seconds if the force ceases to act how will the body move?

1) $4m/s^2$; 5s ; 20 m/s 2) $2m/s^2$; 10s ; 15 m/s
3) $2m/s^2$; 8s ; 10 m/s 4) $3m/s^2$; 15s ; 20 m/s

3. A vehicle of mass 500 kg is moving with a velocity of $15ms^{-1}$. It is brought to rest by a retarding force. Find the distance moved by the vehicle before coming to rest, if the sliding friction between the tyres and the road is 3000N.

1) 18.75 m 2) 9.2 m 3) 10.5 m 4) 16.8 m

4. A ball of mass 400 gm is dropped from a height of 5m. A boy on the ground hits the ball vertically upwards with a bat with an average force of 100N so that it attains a vertical height of 20m. Find the time for which the ball remains in contact with the bat

1) 0.12 s 2) 0.24 s 3) 0.36 s 4) 0.48 s

5. A batsman deflects a ball by an angle of 60^0 without changing its initial speed of 20 ms^{-1}. what is the impulse imparted to the ball if its mass is 0.15 kg?

1) 9 Js 2) 6 Js 3) 3 Js 4) 1 Js

6. Three forces $20\sqrt{2}N$, $20\sqrt{2}N$ and 40N are acting along X, Y and Z – axis respectively on a $5\sqrt{2}kg$ mass at rest at origin. The magnitude of its displacement after 5s is,

1) 50m 2) 25m 3) 60m 4) 100m

7. A body of mass 5 kg is at rest. Three force F_1=10N due North F_2=10N along East and $F_3=10\sqrt{2}N$ along N–W act on it simulta-neously. The acceleration produced in the body is,

1) 4 ms^{-2} along North 2) 2 ms^{-2} along North
3) 4 ms^{-2} along East 4) 2 ms^{-2} along East

8. The momentum of a body in two perpendicular direction at any time 't' are given by $P_x=2t^2+6$ and $P_y=\frac{3t^2}{2}+3$. The force acting on the body at t = 2 sec is

1) 5 units 2) 2 units 3) 10 units 4) 15 units

9. A bullet is fired from a gun. The force on the bullet is given by $F=600-2\times10^5t$, where F is in newton and 't' is in second. The force on the bullet becomes zero as soon as it leaves the barrel. The average impulse imparted to the bullet is

1) 9Ns 2) Zero 3) 0.9Ns 4) 1.8Ns

10. Five persons A, B, C, D & E are pulling a cart of mass 100kg on a smooth surface and cart is moving with acceleration $3m/s^2$ in east direction. When person 'A' stops pulling, it moves with acceleration $1m/s^2$ in the west direction. When person 'B' stops pulling, it moves with acceleration $24m/s^2$ in the north direction. The magnitude of acceleration of the cart when only A & B pull the cart keeping their directions same as the old direction is

1) $26 m/s^2$ 2) $3\sqrt{71} m/s^2$
3) $25 m/s^2$ 4) $30 m/s^2$

11. Two bodies of masses 4 kg and 6 kg connected by means of a light string are lying on a smooth horizontal surface. A horizontal pulling force is applied on the lighter body. Two seconds later the string connecting the two masses is cut. After two more seconds if the velocity of the heavier mass is 2 ms^{-1}, the force initially applied is

1) 10N 2) 25N 3) 20N 4) 40N

12. A person of mass 60 kg stands on a weighing machine in a lift which is moving
(a) upwards with a uniform retardation of 2.8 ms^{-2}
(b) downwards with a uniform retardation of 2.2 ms^{-2}. Find the reading shown by the weighing machine in each case.

1) 420 N; 720 N 2) 320 N ; 520 N
3) 720 N ; 420 N 4) 640 N ; 720 N

13. A man is inside a lift. When the lift is moving up with certain acceleration, the apparent change in weight is 25%. When the lift moves down with double the acceleration, the apparent change in weight is

1) 25% 2) 50% 3) 75% 4) 40%

14. A body is kept on the floor of a topless lift at rest. The lift starts descending at an acceleration 'a' then

A) If a>g, the displacement of the body in a time 't' is $\frac{1}{2}gt^2$ w.r.t ground.

B) If a < g, the displacement of the body in a time 't' is $\frac{1}{2}at^2$ w.r.t ground.

1) Both A & B are true 2) A is true & B is false
3) A is false & B is true 4) Both A & B are false

15. The maximum tension a rope can with stand is 60kgwt. The ratio of maximum accelerations with which two boys of masses 20kg and 30kg can climb up the rope is,

1) 1 : 2 2) 2 : 1 3) 2 : 3 4) 3 : 2

16. A uniform rope of mass m hangs freely from a ceiling. A bird of mass M climbs up the rope with an acceleration a. The force exerted by the rope on the ceiling is

1) Ma + mg 2) M(a+g)+mg 3) M(a+g)
4) dependent on the position of bird on the rope

17. A lift of mass *m* is supported by a cable that can with stand a force of 3mg. Find the shortest distance in which the lift can be stopped when it is descending with a speed of g/4

1) g/8 m 2) g/16 m 3) g/32 m 4) g/64 m

18. A 15Kg mass is accelerated from rest with a force of 100N. As it moves faster, friction and air resistance create an oppositely directed retarding force given by F_R=A+BV. where A=25 N and B= $0.5\frac{N}{m/s}$ At what velocity does the acceleration equal to one half of the initial acceleration ?

1) 25ms^{-1} 2) 50ms^{-1} 3) 75ms^{-1} 4)100ms^{-1}

19. A body of mass 5kg starts from the origin with an initial velocity $\vec{u} = 30\hat{i}+40\hat{j}$ ms^{-1}. If a constant force $\vec{F} = -(\hat{i}+5\hat{j})$N acts on the body, the time in which the y–component of the velocity becomes zero is :

1) 5s 2) 20s 3) 40s 4) 80s

20. Two particles of masses m_1 and m_2 in projectile motion have velocities $\vec{v}_1$ and $\vec{v}_2$ respectively at time t = 0. They collide at time t_0. Their velocities become $\vec{v}'_1$ and $\vec{v}'_2$ at time $2t_0$ while still moving in air. The value of $\left|(m_1\vec{v}'_1 + m_2\vec{v}'_2) - (m_1\vec{v}_1 + m_2\vec{v}_2)\right|$ is (IIT 2001)

1) zero 2) $(m_1+m_2)gt_0$

3) $2(m_1+m_2)gt_0$ 4) $\frac{1}{2}(m_1+m_2)gt_0$

21. A particle of mass M is moving with acceleration a_0 as measured by an observer 1 standing in a frame of reference moving with uniform velocity. Another observer 2 is standing in a frame of referenece moving with acceleration a

1) Observer 1 measures the force acting on the body as Ma_0.
2) Observer 2 measures the force acting on the body as Ma.
3) Observer 2 measures the force acting on the body as Ma_0+Ma.
4) Observer 1 is standing in non-inertial frame of reference and observer 2 is standing in an inertial frame of reference

22. A dynamometer D is attached to two blocks of masses 6 kg and 4 kg. Forces of 20 N and 10N are applied on the blocks as shown in Fig. The dynamometer reads

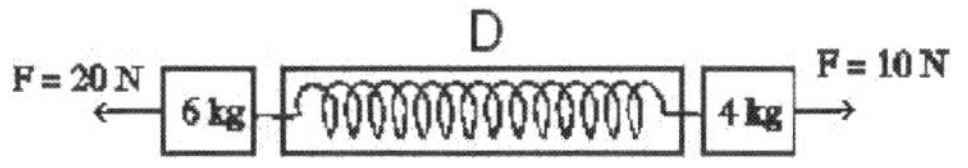

1) 10N 2) 20N 3) 6N 4) 14N

23. Two blocks of masses 7 kg and 3 kg are connected by a spring of stiffness 1000Nm^{-1} and placed on a smooth horizontal surface. They are acted by horizontal forces of 72 N and 32 N in opposite directions as shown in the figure. When accelerations of the blocks are equal and constant, the extension of the spring is

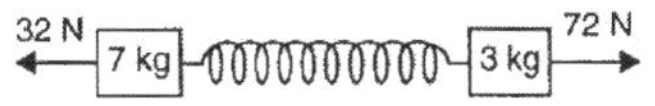

1) 4 cm 2) 5 cm 3) 6 cm 4) 8 cm

24. Two blocks of masses '3*m*' and '2*m*' are in contact on a smooth table. A force P is first applied horizontally on block of mass '3m' and then on mass '2m'. The contact forces between the two blocks in the two cases are in the ratio

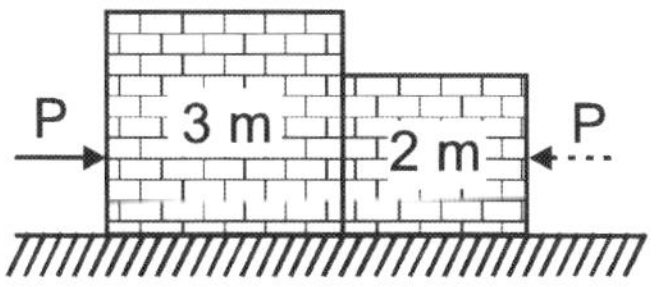

1) 1 : 2 2) 2 : 3 3) 3 : 2 4) 5 : 3

25.A horizontal force of 500 N pulls two masses 10 kg and 20 kg connected by a light string and placed on a frictionless horizontal table. (i) What is the tension in the string if the force is applied on the 20 kg mass? (ii) What is the tension in the string if the force is applied on the 10 kg mass?

1) 333.3 N; 266.7 N   2)166.7 N; 333.3 N
3) 233.3 N; 433 N   4) 217 N; 327 N

26. In the arrangment shown in the figure, the ratio of tensions $T_1 : T_2 : T_3$ is,

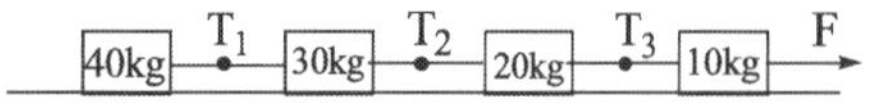

1) 2 : 3 : 4   2) 4 : 3 : 2   3) 9 : 7 : 4   4) 4 : 7 : 9

27. Three identical blocks of masses m=2 kg are drawn by a force F = 10.2N on a frictionless suface. Then the tension in string between blocks B and C is (AIEEE 02)

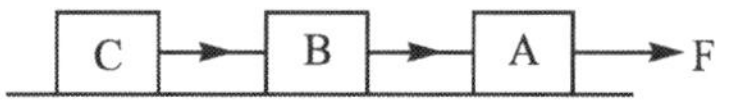

1) 9.2N   2) 9.8 N   3) 4N   4) 3.4 N

28. The monkey B shown in figure is holding on to the tail of the monkey A which is climbing up a rope. The masses of the monkeys A and B are 5 kg and 2 kg respectively. If A can tolerate a tension of 30 N in its tail, what force should it apply on the rope in order to carry the monkey B with it ? [Take g =10 ms^{-2}].

1) Between 70 N and 105 N
2) Between 50 N and 69 N
3) Between 30 N and 50 N
4) Between 106 N and 116 N

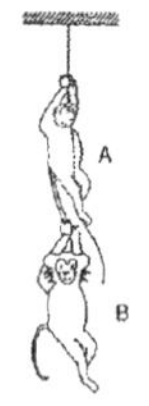

29. A massless string passes around a frictionless pulley whose axis is horizontal. A monkey holds on to one end of the string and a body with the same mass as the monkey is attached to the other end. The monkey starts climbing upwards with an acceleration a. Then the other body mvoes

1) downward with an acceleration a
2) downward with a uniform velocity
3) upward with an acceleration a
4) upward with an acceleration a/2

30. A homogeneous rod of length 'L' is acted upon by two forces F_1 and F_2 applied to its ends and directed opposite to each other. If $F_2 > F_1$, then with what force 'F' will the rod be stretched at the crossection at a distance l from the end where F_1 is applied

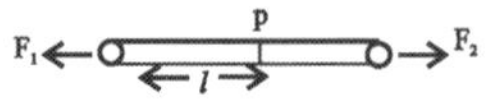

1) $\frac{(F_2 - F_1)l}{L}$ 2) $\frac{(F_2 - F_1)l}{L} + F_1$

3) $\frac{(F_2 + F_1)l}{2L}$ 4) $\frac{(F_2^2 - F_1^2)l}{L}$

31. A railway engine of mass 50 tons is pulling a wagon of mass 40 tons with a force of 4500N. The resistance force acting is 1N per ton. The tension in the coupling between the engine and the wagon is

1) 1600 N 2) 2000 N 3) 2000 N   4) 1500N

32. A chain consisting of 5 links each of mass 0.1 kg is lifted vertically up with a constant acceleration of 2.5m/s^2. The force of interaction between 1st and 2nd links as shown

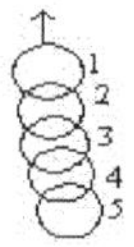

1) 6.15 N   2) 4.92 N 3) 9.84N   4) 2.46N

33. Ten one – rupee coins are put on top of each other on a table. Each coin has a mass of m kg. Find the magnotude and direction of (a) the force on the 6th coin (counted from the bottom) due to all the coins above it, (b) the force on the 6th coin by the 7th coin on it and (c) the reaction of the 5th coin on the 6th coin.

1) 4 mg; 4 mg; 5 mg   2) 5 mg; 5 mg ; 6 mg
3) 4 mg ; 5 mg ; 6 mg 4) 6 mg ; 6 mg ; 8 mg

34. A body is placed on a smooth inclined plane of inclination 1 in x. The horizontal acceleration to be given to the inclined plane so that the body on it remains at rest with respect to inclined plane is

1) $g\sqrt{x^2-1}$ 2) $\frac{g\sqrt{x^2-1}}{x}$

3) $\frac{gx}{\sqrt{x^2-1}}$ 4) $\frac{g}{\sqrt{x^2-1}}$

35. A block is placed on a smooth inclined plane at an angle 'θ' to the horizontal. With what acceleration must the plane be moved horizontally for the block to fall freely ?

1) $g\tan\theta$ 2) $g\cot\theta$ 3) $g\sin\theta$ 4) $g\cos\theta$

36. In the above case, what must be the acceleration of theplane horizontally for the block to slide up the plane with an acceleration of 'g' relative to the inclined plane ?

1) $g\cos\theta$ 2) $g\sin\theta$

3) $g\cot\theta$ 4) $g(\sec\theta+\tan\theta)$

37. In given figure all surfaces are smooth. The ratio of forces exerted by the wedge on mass 'M' when force 'F' is not applied and when 'F' is applied such that 'M' is at rest with respect to wedge is

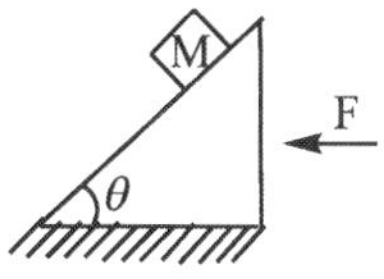

1) 1 2) 1 : 2

3) $\sec^2\theta$ 4) $\cos^2\theta$

38. Consider the shown arrangement. Assume all surfaces to be smooth. If N represents magnitudes of normal reaction between block and wedge, then acceleration of M along horizontal is equal to

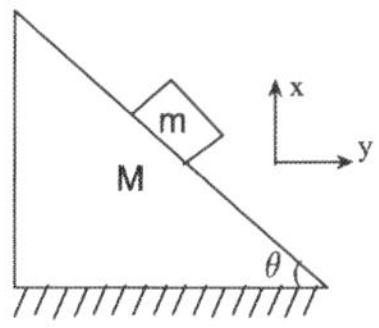

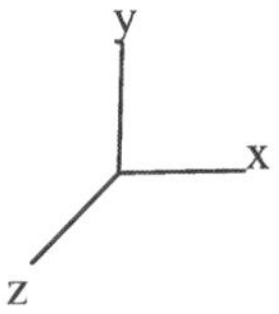

1) $\frac{N\sin\theta}{M}$ along + ve x – axis

2) $\frac{N\cos\theta}{M}$ along – ve x – axis

3) $\frac{N\sin\theta}{M}$ along – ve x – axis

4) $\frac{N\sin\theta}{m+M}$ along – ve x – axis

39. A smooth inclined plane of inclination 60^0 is fixed in a lift which is moving down with an acceleration 'g/3'. If the base of the inclined plane is 'L', the time taken by a body released at the top of the inclined plane to reach the bottom is

1) $\sqrt{\frac{16L}{3\sqrt{3}\,g}}$ 2) $\sqrt{\frac{4L}{\sqrt{3}\,g}}$

3) $\sqrt{\frac{4\sqrt{3}\,L}{g}}$ 4) $\frac{2L}{\sqrt{3}\,g}$

40. A pendulum is hanging from the ceiling of a cage. When the cage is moving up with certain acceleration and when it is moving down with the same acceleration, the tensions in the string are T_1 and T_2 respectively. When the cage moves horizontally with the same acceleration, the tension in the string is,

1) $\sqrt{2(T_1^2+T_2^2)}$ 2) $\sqrt{\frac{T_1^2+T_2^2}{2}}$

3) $\sqrt{2(T_1^2-T_2^2)}$ 4) $\sqrt{T_1^2+T_2^2}$

41. A car is moving in a circular horizontal track of radius 10m with a constant speed of 10 ms^{-1}. A plumb bob is suspended from the roof car by a light rigid rod of length 1m. The angle made by rod with the track is (IIT 1992)

1) zero 2) 30^0 3) 45^0 4) 60^0

42. A light stright fixed at one end to a wooden clamp on a ground passes overa fixed pulley and hangs at the other side. It makes an angle 30^0 with the ground. A boy weighing 60 kg climbs up the rope. The wooden clamp can tolerate upto vertical force 360N. Find the, maximum acceleration in the upward direction with which the boy can climb safely. The friction of pulley and wooden clamp may be ignored (g = 10 m/s²)

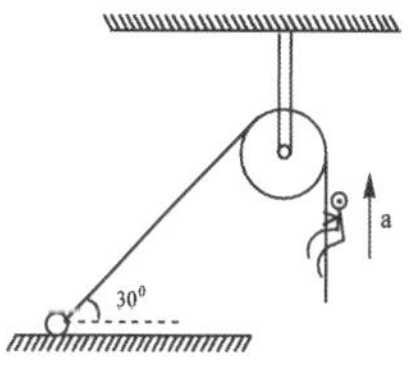

1) 1 m/s² 2) 2 m/s² 3) 3 m/s² 4) 4 m/s²

43. Two block of masses *m* and *2m* are kept on a smooth inclined plane and the system is pushed using force 3mg as shown. Find the contact force between those two blocks

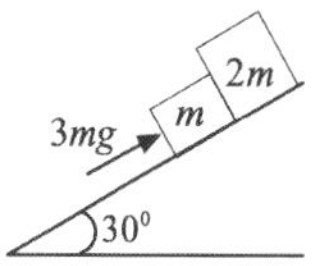

1) 1 mg 2) 2mg 3) 3 mg 4) 4 mg

PULLEY

44. Three equal weights of mass m each are hanging on a string passing over a fixed pulley as shown in Fig. What are the tensions in the string connecting weights A to B and B to C?

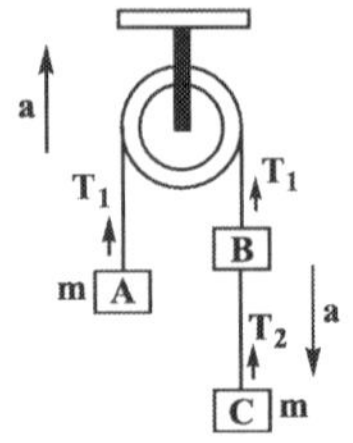

1) 7/4 mg ; 3/4 mg 2) 7/3 mg ; 4/3 mg
3) 4/3 mg ; 2/3 mg 4) 3/4 mg ; mg/4

45. Three blocks S, R and P connected by string are arranged as shown in the figure. The masses of S, R and P are 5 kg, 5 kg and 10 kg respectively. The tension in the string connecting R and P is (ignore friction)

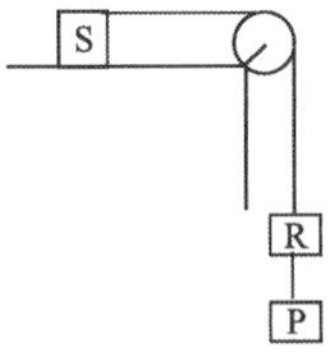

1) 10N 2) 25N 3) 20N 4) 30N

46. The force excrted by string on the pulley is ($g=10ms^{-2}$)

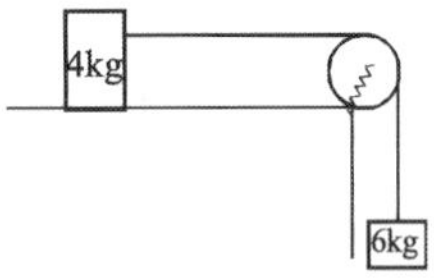

1) 24N 2) $24\sqrt{2}$N 3) $12\sqrt{2}$N 4) 12N

47. Two masses 5 kg and 3 kg are suspended from the ends of an unstretchable light string passing over a frictionless pulley. When the masses are released, the thrust on the pulley is ($g = 10ms^{-2}$)

1) 80 N 2) 37.5 N 3) 150 N 4) 75 N

48. In the figure the blocks A, B and C of mass m each, have acceleration a_1, a_2 and a_3 respectively. F_1 and F_2 are external forces of magnitudes 2 mg and mg respectively.

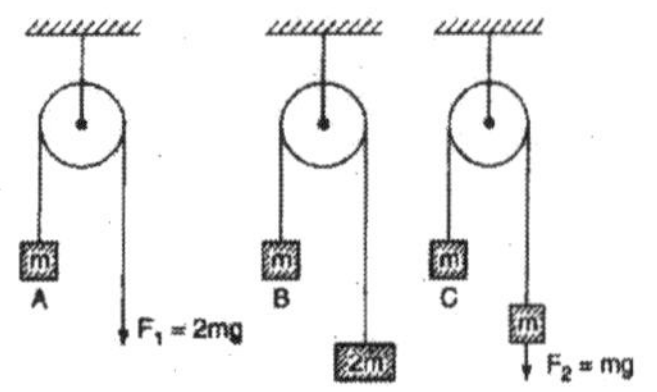

1) $a_1=a_2=a_3$ 2) $a_1> a_3> a_2$
3) $a_1= a_2$, $a_2> a_3$ 4) $a_1> a_2$, $a_2= a_3$

49. If m_1=10kg, m_2=4kg and m_3 = 2kg the tension in the string connecting m_1 and m_2 is,

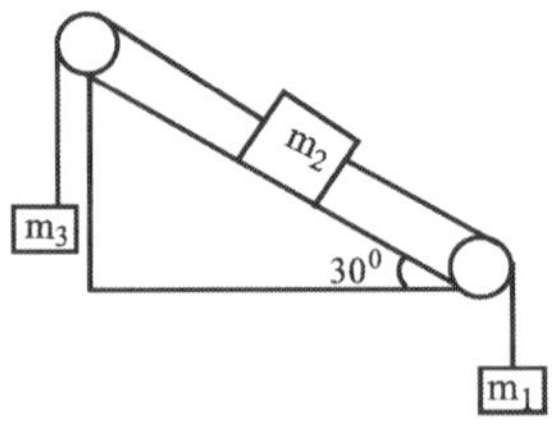

1) $\frac{15g}{4}$ 2) $\frac{5g}{8}$ 3) $\frac{10g}{3}$ 4) $\frac{20g}{3}$

50. Two masses M_1 and M_2 are attached to the ends of a light string which passes over a massless pulley attached to the top of a double inclined smooth plane of angles of inclination α and β. The tension in the string is :

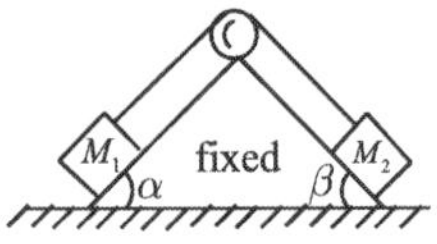

1) $\frac{M_2(\sin\beta)g}{M_1+M_2}$ 2) $\frac{M_1(\sin\alpha)g}{M_1+M_2}$

3) $\frac{M_1M_2(\sin\beta+\sin\alpha)g}{M_1+M_2}$ 4) Zero

51. The reading in the spring balance is

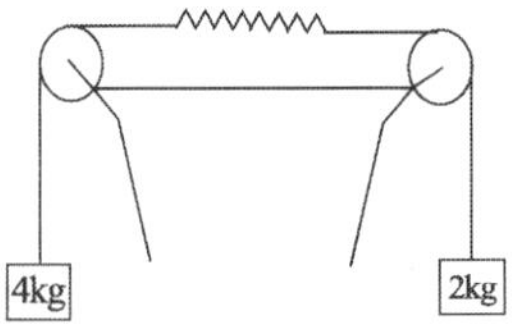

1) 2.6 kgwt 2) 2 kgwt 3) 6 kg wt 4) 4.3 kgwt

52. In the figure, a smooth pulley of negligible weight is suspended by a spring balance. Weights of 1 kg and 5 kg are attached to the opposite ends of a string passing over the pulley and move with acceleration because of gravity. During their motion, the spring balance reads a weight of

1) 6 kg 2) less than 6 kg 3) more than 6 kg
4) may be more or less than 6 kg

53. Two bodies of masses 15gm and 25gm are connected by means of a string of length 40cm and mass 10gm. The string is made to pass over a smooth massless pulley. Initially the two bodies are kept at the same level and released. By the time the heavier mass descends by 10cm, the acceleration of the system,

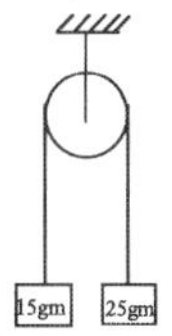

1) Remains constant 2) Decreases
3) Increases by a factor 1.5
4) Increases by a factor 3

54. In the system shown in figure, the pulley is smooth and massless, the string has a total mass 5 gm, and the two suspended blocks have masses 25 gm and 15 gm. The system is released from state $l = 0$ and is studied upto stage $l^1 = 0$. During the process, the acceleration of block A will (with g_E for acceleration due to gravity) be

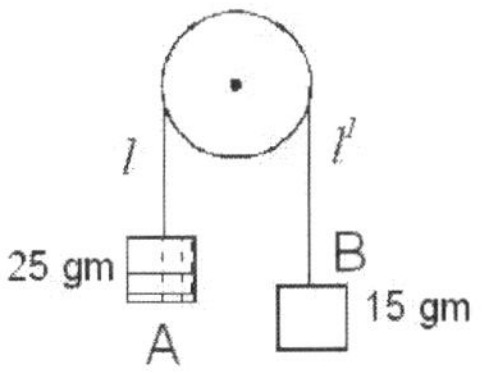

1) Constant at $g_E/9$ 2) constant at $g_E/4$
3) increasing by factor of 3
4) increasing by factor of 2

55. A man of mass 60 kg is standing on a weighing machine kept in a box of mass 30 kg as shown in the diagram. If the man manages to keep the box stationary the reading in the weighing machine is

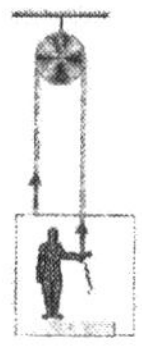

1) 150N 2) 600N 3) 300N 4) 400N

56. A man of mass 60 kg is standing on a weighing machine kept in a box of mass 30 kg as shown in the diagram. What force should be exert on the rope to get his correct weight on the machine ?

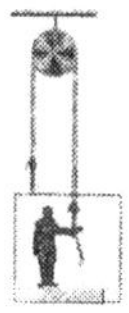

1) 1000N 2) 900N 3) 3600N 4) 1800N

57. In the arrangement shown in the figure, the acceleration of the mass is, (Ignore friction)

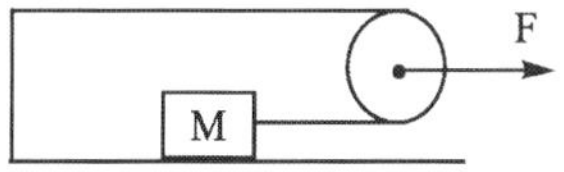

1) $\frac{F}{4M}$ 2) $\frac{F}{M}$ 3) Zero 4) $\frac{F}{2M}$

58. A string of negligible mass going over a clamped pulley of mass m supports a block of mass M as shown in the figure. The force on the pulley by the clamp is given by (IIT 2001)

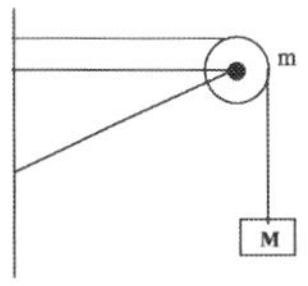

1) $\sqrt{2}$ Mg 2) $\sqrt{2}$ mg

3) $g\sqrt{(M+m)^2+m^2}$ 4) $g\sqrt{(M+m)^2+M^2}$

59. Two equal masses are connected by a light string which is made to pass over a smooth, light pulley. The percentage of mass to be transferred from one body to another so that the bodies move 50m in 5s is(g=10ms^{-2}) (nearly)

1) 20% 2) 40% 3) 25% 4) 50%

60. The vertical distance between the masses is $\frac{20}{3}$ m, at t = 0. The distance becomes zero at t = (g = 10 m/s^2)

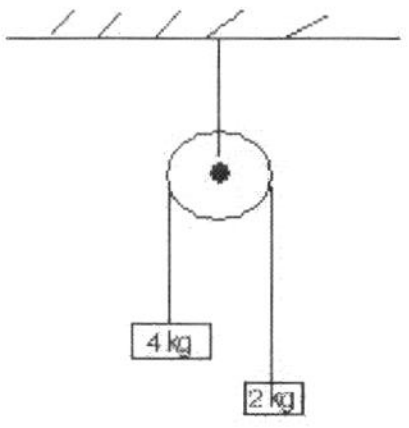

1) 2 2) $\sqrt{2}$ 3) 4 4) $\sqrt{6}$

61. A railway flat car, whose mass together with the artillery gun is M, moves at a speed V. The gun barrel makes an angle α with the horizontal. A shell of mass m leaves the barrel at a speed v, relative to the barrel. The speed of the flat car in order that it may stop after the firing is

1) $\frac{mv\cos\alpha}{M-m}$ 2) $\frac{mv}{M+m}$

3) (M + m) v cos α 4) $\frac{Mv\cos\alpha}{M+m}$

62. A machine gun is mounted on a flate rail car. The gun is firing bullets at the rate of 10 bullets per second each of mass 10 g. The bullets come out with velocity 500 ms^{-1} from the ear. Calculate the acceleration of the car at the instant when its mas is 200 kg. Also calculate the force at that instant.

1) 0.25 m/sec^2 ; 100 N   2) 0.5 m/sec^2 ; 50 N
3) 0.25 m/sec^2 ; 50 N   4) 0.5 m/sec^2 ; 100 N

63. A gun of mass M fires a bullet of mass m horizontally. The energy of firing is such that it is sufficient to project the bullet vertically to a height h. The velocity of recoil of the gun is

1) $\sqrt{\frac{2m^2gh}{M(M+m)}}$ 2) $\sqrt{\frac{2m^2gh}{m(M+m)}}$

3) $\sqrt{\frac{2Mgh}{(M+m)}}$ 4) $\sqrt{\frac{2mgh}{(M+m)}}$

64. A gun of mass M fires a bullet of mass m, with a Kinetic energy E. The total kinetic energy in the firing process is

1) $\left[\frac{m+M}{M}\right]E$ 2) $\frac{[m+M]E}{m}$

3) $\left(\frac{M}{M+m}\right)E$ 4) $\left(\frac{m}{M+m}\right)E$

65. A body of mass 8kg at rest explodes into three parts whose masses are in the ratio 1 : 3 : 4. The bigger parts travel with a velocity 10 m/s in perpendicular directions after the explosion. Velocity of the third part is

1) 25 m/s   2) 35 m/s   3) 50 m/s   4) 70 m/s

66. A bomb falling freely bursts after 10 sec. ($g=10ms^{-2}$) into two fragments of masses in the ratio of 2:1. The velocity of heavier fragment immediately after the explosion is 200 ms^{-1} vertically downwards. The velocity of the lighter fragment immediately after the explosion is

1) 50ms^{-1} upward   2) 75ms^{-1} downward
3) 100ms^{-1} upward   4) 400ms^{-1} upward

67. An object initially at rest explodes into three fragments A,B and C. The momentum of A is P and that of B is $\sqrt{3}$ P perpendicular to A. the momentumof C will be

1) $(1+\sqrt{3})$P in a direction making 120° with that of A

2) $(1+\sqrt{3})$P in a direction making 150° with that of B

3) 2P in a direction making 150° with that of A

4) 2P in a direction making 150° with that of B

68. A shell of mass m is fired from a cannon at angle θ to the horizontal with a velocity V. The shell at the heighest point breaks into two fragments having masses in the ratio 2:3. The lighter fragment has a velocity zero immediately after explosion and falls vertically downward. the velocity of the other fragment just after explosion is

1) 5V Cos θ   2) 3V Cos θ

3) $\frac{3}{5}$V Cos θ   4) $\frac{5}{3}$V Cos θ

69. In gravity free space, a man of mass 'M' standing at a height h above the floor throws a ball of mass m straight down with a speed u. When the ball reaches the floor, the distacne of the man above the floor will be

1) $h\left(1+\frac{m}{M}\right)$ 2) $h\left(2+\frac{m}{M}\right)$

3) 2h

4) A function of m, M, h and u

70. A 500kg boat has an initial speed of 10ms^{-1} as it passes under a bridge. At that instant a 50 kg man jumps straight down into the boat from the bridge. The speed of the boat after the man and boat attain a common speed is

1) $\frac{100}{11}ms^{-1}$ 2) $\frac{10}{11}ms^{-1}$

3) $\frac{50}{11}ms^{-1}$ 4) $\frac{5}{11}ms^{-1}$

71. A cart loaded with sand moves along a horizontal plane due to a constant force F coinciding in direction with the cart's velocity vector. In the process, sand spills through a hole in the bottom with a constant velocity μ kg/s. Find the acceleration and the velocity of the cart at the moment t, if at the initial moment t = 0, the cart with the loaded sand has a mass m_0 and its velocity was equal to zero. The friction is to be neglected.

1) $\frac{F}{m_0-\mu t}$ 2) $\frac{m_0-\mu t}{F}$

3) $\frac{m_0+\mu t}{F}$ 4) $\frac{F}{m_0+\mu t}$

72. **A box is put on a scale which is adjusted to read zero, when the box is empty. Astream of pebbles is then poured into the box from a height h above its bottom at a rate of n pebbles. Each pebble has a mass m. If the pebbles collide with the box such that they immediately come to rest after collision, find the scale reading at time t after the pebbles begin to fill the box.**

1) $mn\left\{\sqrt{(2gh)}+gt\right\}$ 2) $\left\{\sqrt{(2gh)}+gt\right\}$

3) $\dfrac{\left\{\sqrt{(2gh)}+gt\right\}}{mn}$ 4) $mn\{(2gh)+gt\}$

ANSWERS

1) 1	2) 1	3) 1	4) 1	5) 3
6) 4	7) 1	8) 3	9) 3	10) 3
11) 1	12) 1	13) 2	14) 1	15) 2
16) 2	17) 3	18) 2	19) 3	20) 3
21) 1	22) 4	23) 3	24) 2	25) 2
26) 4	27) 4	28) 1	29) 3	30) 2
31) 2	32) 2	33) 1	34) 4	35) 2
36) 4	37) 4	38) 3	39) 3	40) 2
41) 3	42) 2	43) 2	44) 3	45) 2
46) 2	47) 4	48) 2	49) 1	50) 3
51) 1	52) 2	53) 3	54) 3	55) 1
56) 4	57) 4	58) 4	59) 2	60) 2
61) 1	62) 3	63) 1	64) 1	65) 3
66) 3	67) 4	68) 4	69) 1	70) 1
71) 1	72) 1			

COMPREHENSION QUESTIONS

Comprehension - I : (for Questions 1–4)

A 5.0 kg block and a 4.0 kg block are connected by a 0.6 kg rod. The links between the blocks and the rod are denoted by A and B. A force F is applied to the upper block. (g = 9.8 m/s²)

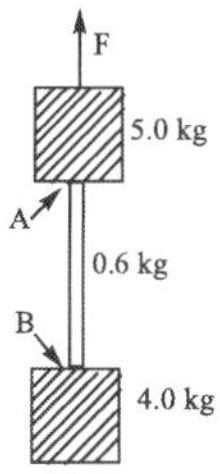

1. **In figure, the blocks and rod assembly moves downward at constant velocity. The applied force F is closest to :**

1) 88N 2) 90N 3) 92N 4) 94N

2. **In figure, the blocks and rod assembly moves upward at constant velocity. The force in link A is closest to :**

1) 39N 2) 41N 3) 43N 4) 45N

3. **In figure, the applied force F equals 150N. The foce in link B is closest to :**

1) 54N 2) 56N 3) 58N 4) 62N

4. **In figure, the force in link B is 40N. The acceleration of the blocks and rod assembly, including direction, is closest to :**

1) Zero 2) 1.2 m/s², upward

3) 2.4 m/s², upward 4) 1.2 m/s², downward

Comprehension - II : (for questions 5–7)

Two 100g blocks hang at the ends of a light flexible cord passing over a small frictionless pully. A 40g block rests on the block on right and removed after 2 sconds.

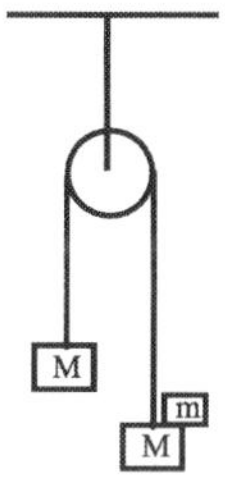

5. **The acceleration of masses after 40g is palced**

1) g 2) $\frac{g}{3}$ 3) $\frac{g}{6}$ 4) 0

6. **Tension in the cord before 40g is removed**

1) zero 2) 1.14 N 3) 5 N 4) 3.4 N

7. **Tension in the cord after 40g is removed**

1) zero 2) 0.98 N 3) 1.96 N 4) 3.4 N

Comprehension - III : (for questions 8–10)

Two blocks of masses m_1 = 2kg, m_2 = 5kg hang over a massless pully as shown in figure. A force f_0 = 100 N acting at the axis of the pulley accelerates the system upwards

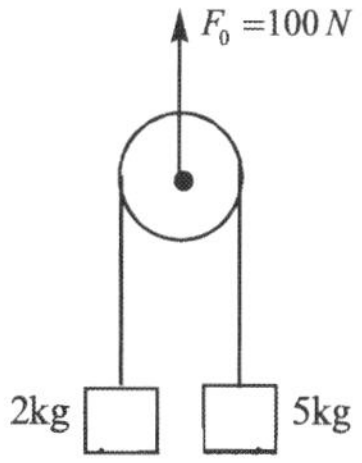

8. **The acceleration of 2kg mass is**

1) 4.2 m/s² upwards 2) 4.2 m/s² downwards

3) 15.2m/s² upwards 4) 15.2 m/s² downwards

9. The acceleration of 5kg mass is

1) 1 m/s² upwards 2) 1 m/s² downwards
3) 0.2 m/s² upwards 4) 0.2 m/s² downwards

10. The tension in the string is

1) 10 N 2) 28.2 N 3) 50 N 4) 80.4 N

Hint : $F_0 = T + T = 2T$

$T - m_1 g = m_1 a_1$ $m_2 g - T = m_2 a_2$

Comprehension - IV : (for questions 11–14)

Three blocks are connected by strings and are pulled by a force T_3=60 N. If m_1 = 10 kg, m_2 = 20 kg and m_3 = 30 kg

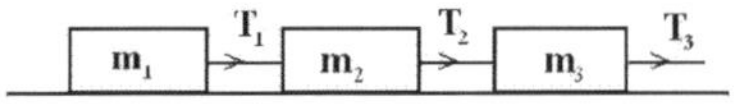

11. The acceleration of system is

1) 1 m/s² 2) 2 m/s²
3) 3 m/s² 4) 4 m/s²

12. The tension T_1 in string between masses m_1 & m_2 is

1) 5 N 2) 10 N 3) 15 N 4) 30 N

13. The tension T_2 in string between masses m_2 & m_3 is

1) 20 N 2) 25 N
3) 30 N 4) 35 N

14. The net force on m_2 is

1) 10 N 2) 12 N
3) 18 N 4) 20 N

Comprehension - V : (for questions 15–17)

A mass A of 0.5 kg is pulled on a smooth horizontal table with a string attached to it. The string goes over a frictionless pulley and is connected to another mass B of 0.2 kg. At t=0 the mass is at a distance 2m from the end and moving with a speed of 0.5 m/s towards left.

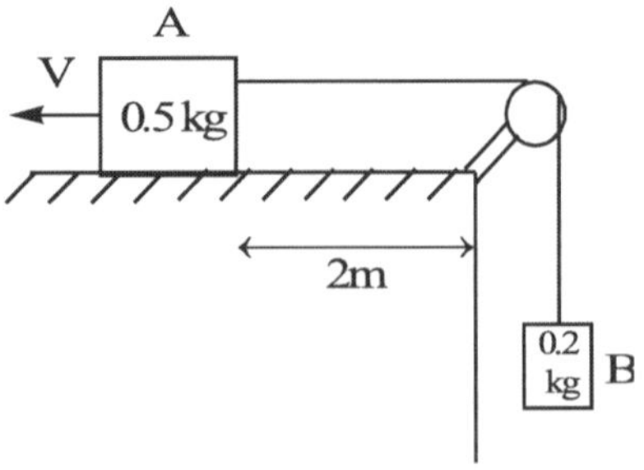

15. The acceleration of system is

1) 1 m/s² 2) 2.8 m/s²
3) 1.6 m/s² 4) 3.2 m/s²

16. The velocity of body after 1 sec. is

1) 1.4 m/s 2) 2.3 m/s
3) 5.6 m/s 4) 6.1 m/s

17. Tension in the string is

1) 10 N 2) 14 N
3) 18 N 4) 20 N

Comprehension - VI : (for questions 18–20)

Alift is going up. The total mass of the lift and the passengers is 1500 kg. The variation in the speed of the lift is given by the graph as shown in figure.

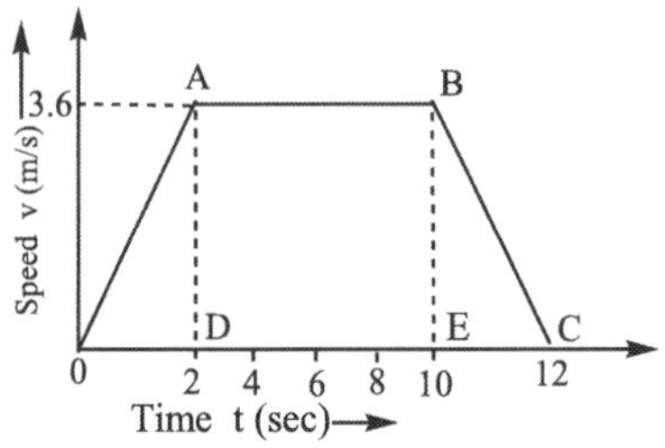

18. What will be the tension in the rope pulling the lift at time t equal to

i) 1 sec ii) 6 sec iii) 11 sec?

1) i) 1.8 m/s² ; ii) 0 ; (iii) –1.8 m/s²
2) i) 1.3 m/s² ; ii) 1 ; (iii) –2.1 m/s²
3) i) 2.1 m/s² ; ii) 0 ; (iii) 1.8 m/s²
4) i) 1.1 m/s² ; ii) 1 ; (iii) 0 m/s²

19. What is the height to which the lift lakes the passengers?

1) 41 m 2) 36 m 3) 18 m 4) 9 m

20. What will be the average velocity and the average acceleration during the course of the entire motion?

1) 0 2) 1 3) 2 4) 4

ANSWERS

1) 4	2) 4	3) 4	4) 1	5) 3
6) 2	7) 2	8) 3	9) 3	10) 3
11) 1	12) 2	13) 3	14) 4	15) 2
16) 2	17) 2	18) 1	19) 2	20) 1

Chapter - 6

GRAVITATION

GRAVITATION

SYNOPSIS

Newton's Law of Gravitation:

The gravitational force of attraction between two particles is directly proportional to the product of their masses and inversely proportional to the square of the distance between them.

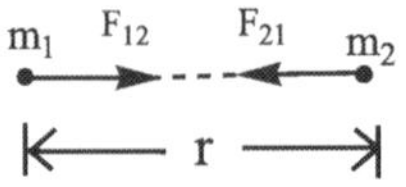

$$\vec{F}_{12} = \frac{Gm_1m_2}{r^2}\hat{r}_{12} = \frac{Gm_1m_2}{r^3}\vec{r}_{12}$$

$F = \frac{Gm_1m_2}{r^2}$, $G \rightarrow$ universal gravitational constant

$G = 6.67 \times 10^{-11}\,Nm^2\,kg^{-2}$

G is a Scalar quantity.

The value of 'G' is constant for any two bodies, in any medium and at any place in the universe.

* Modification of space by mass particle is called gravitational field.
* Law of gravitation is a consequence of "action at a distance concept".

Gravitational field strength:

It is the Gravitational force experienced by a unit mass kept at a point in a gravitational field.

$I = \frac{F}{m} Nkg^{-1}$ (It is a vector)

a) The Gravitational field strength at a point due to a point mass

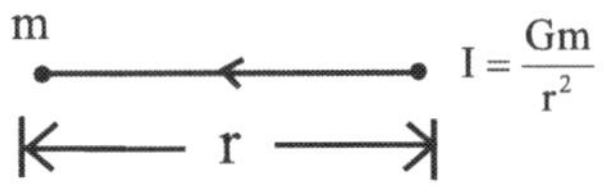

b) The Gravitational field strength at a point due to a uniform sphere of mass M

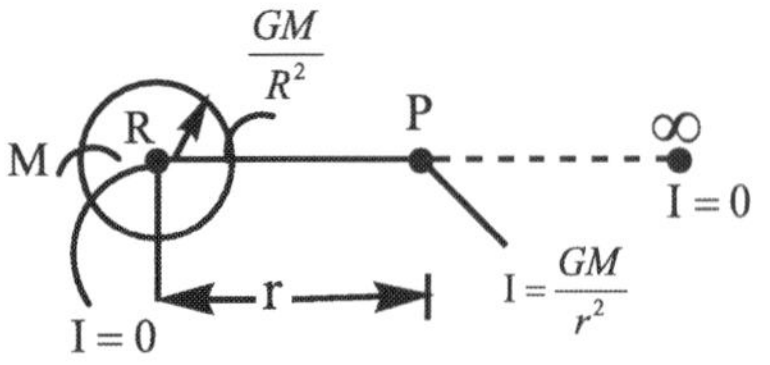

i) If $r \geq R$, $I = \frac{GM}{r^2}$ $\therefore I \propto \frac{1}{r^2}$

ii) If $r = R$; $I = \frac{GM}{R^2}$

iii) If $r \leq R\ (r \neq 0)$, $I = \frac{GM}{R^3}(r)$, $I \propto r$

c) The Gravitational field strength at a point is equal to the acceleration due to gravity at that point.

d) Location of null point due to two point masses

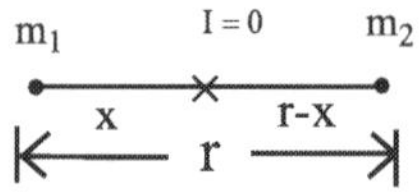

$$x = \frac{r}{\sqrt{\frac{m_2}{m_1}} + 1}$$

from the centre of weaker mass ($m_1 < m_2$)

Kepler's Laws of Planetary motion:

I Law (Law of orbits):

i) Every planet revolves around the sun in an elliptical orbit, with the sun at one of the foci.

ii) During the orbital motion of a planet around the sun in an elliptical orbit, at a particular instantt h e planet is very close to the sun (the point is called perogy or periheleon) and at another particular instant it is far away from the sun (the point is called epogy or epiheleon).

II Law (Law of areas):

i) The line joining the sun and the planet sweeps out equal areas in equal intervals of time or areal velocity of radius vector is constant.

Areal velocity, $V_A = \frac{dA}{dt} = \frac{d}{dt}\left(\frac{1}{2}r^2 d\theta\right)$

$$= \frac{1}{2}r^2\frac{d\theta}{dt} = \frac{1}{2}r^2 w = \frac{L}{2m}$$

ii) Where 'L' is the angular momentum of the planet of mass 'm' in the given orbit Keepler's II law is a consequence of the law of conservation of Angular momentum

iii) According to II law, a planet moves faster when it is near to sun and moves slower when it is far away from the sun.

iv) According to II law $V_{max}r_{min} = V_{min}r_{max}$

III Law (law of periods):

i) The square of the time period of revolution of a planet around the sun is proportional to the cube of the mean distance of the planet from the sun or semi-major axis,

$T^2 \propto R^3$ or $T^2/R^3 = \text{constant}$

If '2a' is length of major axis then $T^2 \propto a^3$

The proportionality constant is $K = \frac{4\pi^2}{GM}$

ii) According to III law, as the distance of the planet increases, then the duration of the year of the planet increases.

Relation Between g & G

$$g = \frac{GM}{R^2} = \frac{4}{3}\pi RG\rho$$

M – Mass of the planet or heavenly body
R – Radius of the planet or heavenly body
ρ – Uniform density of the planet or heavenly body

i) $g \propto \frac{M}{R^2}$

ii) $g \propto R\rho$

iii) $W \propto g \propto \frac{M}{R^2} \propto R\rho$

W – weight of an object

iv) $h \propto \frac{1}{g}$ (the height to which a person can jump on the surface of a planet.)

Variation of 'g' with altitude:

If R is the radius of the earth, M is the mass of the earth and h is the altitude, then

i) $g_h = \frac{GM}{r^2} = \frac{GM}{(R+h)^2} (\because r = R+h)$

ii) If h<<R, then $g_h = g\left(1 - \frac{2h}{R}\right)$

iii) Fractional decrease of 'g' for smaller altitude is $\frac{\Delta g}{g} = \frac{2h}{R}$

iv) Fractional loss of weight for smaller altitude is $\frac{\Delta W}{W} = \frac{2h}{R}$

v) As the altitude increases, 'g' decreases

Varification of 'g' with depth:

i) $g_d = g\left(1 - \frac{d}{R}\right)$ $\left(\because g = \frac{4}{3}\pi RG\rho\right)$

(ρ is the mean density of the earth)

ii) Fractional decreases of 'g' and weight of an object according to depth 'd' is

$$\frac{\Delta W}{W} = \frac{\Delta g}{g} = \frac{d}{R}$$

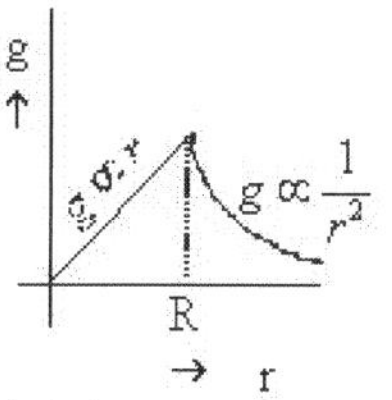

iii) As depth increases, 'g' decreases

iv) at the centre of the earth, $g_d = o \Rightarrow W = 0$

v) If $g_h = g_d$ and (h<<R, d<<R), then 2h=d.

Variation of 'g' with latitude (effect of rotation of earth)

i) If 'g' is the acceleration due to gravity when the earth does not rotate and g_φ is the acceleration due to gravity at an attitude 'φ' due to rotation of earth, then $g_\varphi = g - R\omega^2\cos^2\varphi$ (ω is angular velocity of rotation of earth)

ii) As φ increases, g also increases

iii) At equator $g_0 = g - R\omega^2$ (min)
the value of g at equator depends on angular velocity of earth

iv) At poles $g_{90} = g$ (max)

v) The value of 'g' at the poles is independent of the rotation of the earth.

vi) The value of 'g' at the equator become zero when the angular velocity of rotation increases to 17 times of the present value.

Variation of g according to shape of the earth.

As $g = \frac{GM}{R^2} \Rightarrow g \propto \frac{1}{R^2}$

i) g_{min} at equator, because R_{max}

ii) g_{max} at poles, because R_{min}

iii) The value of 'g' gradually increases from equator to poles.

Variation of g according to local conditions:

i) The value of g is slightly more at the location of mineral deposits.

ii) The value of g is slightly less on the top of mountains and also inside the mines.

iii) If g_1, g_2 and g_3 are the acceleration due to gravities on the surface of the earth, on the top of a mountain and inside a mine, then $g_2 > g_1 > g_3$.

Gravitational Potential:

i) It is the amount of workdone in bringing a unit mass from infinity to a point in the gravitational field.

ii) It is always- Ve and its maximum value is zero at infinity

iii) The Gravitational potential at a point due to a point mass

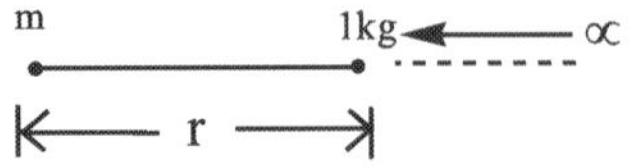

$V = \frac{-Gm}{r}$

iv) The gravitational PE of the system of two particles

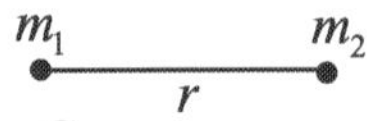

$U = \frac{-Gm_1m_2}{r}$

v) The gravitational potential energy of a body at a point in the gravitational field of earth

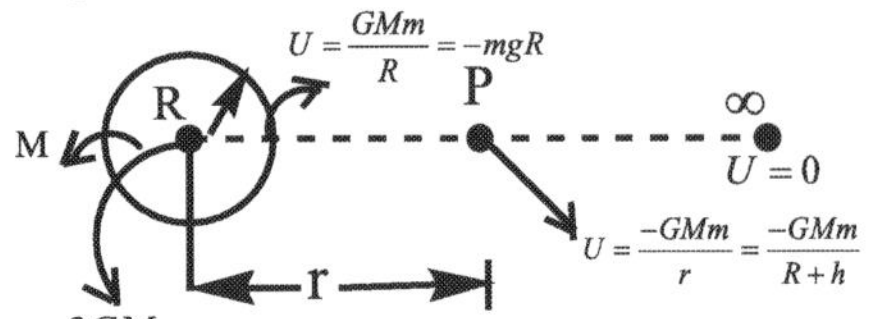

$U = \frac{-3GMm}{R}$

a) The potential energy on the surface of earth is

$$U = -\frac{GMm}{R} = -mgr$$

b) The potential energy at distance "R+h" from the center of earth is

$$U = \frac{-GMm}{R+h} = \frac{-GMm}{r} \quad (r = R+h)$$

c) The Potential energy at "∞" is U=0

d) Potential energy at the center of earth is

$$U = -\frac{3GMm}{2R}$$

vi) Workdone in lifting a body from the surface of the earth to a height 'h' against the gravitational forces is

$$W = \left(\frac{-GMm}{R+h} + \frac{GMm}{R}\right) = \frac{GMm\ h}{(R+h)R}$$

$$= \frac{mgh}{(1+h/R)}$$

If h<<<R, W= mgh

Orbit Velocity (V_0):

i) It is the minimum velocity required by a satellite to revolve in a particular orbit around a planet.

ii) Direction of V_0 is always tangential to the orbit

iii) If satelite revolving at a height h from the surface of the planet is $V_0 = \sqrt{\frac{GM}{R+h}}$

iv) If the orbit is very near the surface of the planet (for a surface satellite)

$$V_0 = \sqrt{\left(\frac{GM}{R}\right)} = \sqrt{gR}$$

v) $V_0 \propto \sqrt{\frac{M}{R}} \propto R\sqrt{\rho}$ $\rho \rightarrow$ mean density of the planet

vi) V_0 is independent of the mass of the orbiting body

vii) V_0 depends on mass of the planet and radius of the orbit

viii) For an earth's surface satellite V_0= 7.98kms^{-1}

ix) Orbital angular velacity satellite is

$$\omega_0 = \sqrt{\frac{GM}{(R+h)^3}}$$

x) Time periode of satellite is $T = 2\pi\sqrt{\frac{(R+h)^3}{GM}}$

Escape Velocity (V_e)

i) It is the minimum velocity of an object to escape from the gravitational field.

ii) From the surface of a planet

$$V_e = \sqrt{\frac{2GM}{R}} = \sqrt{2gR}, \qquad V_e \propto \sqrt{\frac{M}{R}} \propto R\sqrt{\rho}$$

iii) $(V_e)_{moon}$ = 2.38 kms^{-1}, $(Ve)_{earth}$ = 11.2kms^{-1} & $(V_e)_{sun}$=42kms^{-1}

iv) It depends on the mass and radius of the planet

v) It is independent of the mass of the body and the direction of projection

vi) There is no atmosphere on the surface of the moon, because r.m.s. velocity of the air moleculus is greater than the escape velocity on the moon.

vii) $V_e = \sqrt{2}\,V_0$

viii) If the speed of the orbital satellite is made $\sqrt{2}$ times or increased by 41.4% or its KE is doubled, then it will escape to infinity.

Geo-stationary or Communication or parking satellite:

i) Its time period of revolution around the earth is 24hr

ii) Its relative velocity w.r.t the earth is zero

iii) Its orbit is called parking orbit and its height from the surface of the earth is 36,000km or from the center of the earth is 42,400km.

iv) It revolves from west to east in the equitorial plane

Energies of an orbiting satellite

i) PE+KE=T.E (Binding energy of a satellite)

ii) $PE = \frac{-GMm}{r} = \frac{-GMm}{R+h}$

iii) $KE = \frac{GMm}{2r} = \frac{GMm}{2(R+h)}$

iv) $TE = \frac{-GMm}{2r} = \frac{-GMm}{2(R+h)}$

v) PE : KE : TE = – 2 : 1 : – 1

Angular momentum of a satellite

$L = mv_0 r = m\sqrt{\frac{GM}{r}}r = \sqrt{GMm^2 r}$

$= \sqrt{GMm^2(R+h)}$ ($\because$ r=R+h)

for a surface satellite (h<<R)

$L = \sqrt{GMm^2R} = \sqrt{gR^3m^2}$

* **For a Satellite with increase in height of the orbit from the surface of the planet, then its**

i) PE incresases

ii) KE decreases

iii) Orbital velocity decreases

iv) Total energy increases

v) Period of revolution increases

* **For a Satellite**

i) If $V < V_0$ then it falls to the surface of earth in parabolic path (TE<0)

ii) If $V=V_0$ then it revolves around the earth in circular orbit (TE<0)

iii) If $V_0<V<V_e$, then it revolves in an elliptical orbit (TE<0)

iv) If $V=V_e$, then it just escapes from the gravitational field by parabolic path. (TE=0)

v) If $V>V_e$, then it escapes to infinity and enters into interstellar space with some velocity in hyperbolic path.

$V_\infty = \sqrt{V_P^2 - V_e^2}$, $v_e \rightarrow$ escape velocity

(TE > 0) $V_\infty \rightarrow$ velocity at interstellar space or at infinity

EXERCISE - I

Newton's Law of Gravitation and Gravitational Intensity

1. **The value of G depends upon**

1) The mass of bodies

2) The medium between the bodies

3) The temperature of bodies

4) It is an independent constant

2. **Two identical spheres of gold are in contact with each other. The gravitational force of attraction between them is**

1) directly proportional to the square of their radius

2) directly proportional to the cube of their radius

3) directly proportional to the fourth power of their radius

4) inversely propostional to the square of their radius.

3. **If there were a small alteration in gravitational force, then which of the following forces do you think would alter in some respect**

1) Viscous forces

2) Archimedies principle

3) Electrostatic force

4) all the above

4. **Find the false statement (ATL)**

1) Gravitational force acts along the line joining the two interacting particles

2) Gravitational force is independent of medium

3) Gravitational force forms an action- reaction pair

4) Gravitational force does not obey the principle of superposition.

5. **Among the following find the wrong statement (ATL)**

1) Law of gravitation is framed using Newton's third law of motion

2) Law of gravitation cannot explain why gravity exists.

3) Law of gravitation does not explain the presense of force even when the particles are not in physical contact.

4) When the range is long, gravitational force becomes repulsive

6. **Law of gravitation is not applicable if (ATL)**

A) Velocity of moving objects are comparable to velocity of light

B) Gravitational field between objects whose masses are greater than the mass of sun.

1) A is true, B is false 2) A is false, B is true

3) Both A & B are true 4) Both A & B are false

7. **Statement A: Modification of space by a mass particle is called gravitational field (ATL)**
Statement B: Law of gravitation is a consequence of "Action at a distance concept".
1) A is true, B is false 2) A is false, B is true
3) Both A & B are true 4) Both A & B are false

8. **The force responsible for the motion of planets around sun is**
1) gravitational 2) magnetic
3) electrostatic 4) frictional

9. **The tidal Waves in the sea are primarily due to**
1) The gravitational effect of the moon on the earth
2) The atmospheric effect of the earth
3) The gravitational effect of the sun on the earth
4) 2 & 3

10. **The atmosphere is held to the earth by**
1) The rotation of the earth
2) The attraction of the sun
3) The gravity 4) All

11. **Which of the following is the evidence to show that there must be a force acting on earth and directed towards the sun?**
1) Deviation of falling bodies towards east
2) Revolution of the earth around the sun
3) Apparent motion of the sun round the earth
4) Phenomenon of day and night

12. **Among the following choose the wrong statement**
1) According to geo-centric theory, earth is the centre of universe
2) Copernicus proposed helio-centric theory of the universe
3) Ptolemy proposed geo-centric theory of the universe
4) According to helio-centric theory, earth is the centre of the universe

13. **F_g, F_e and F_n represent the gravitational, electro-magnetic and nuclear forces respectively, then arrange them in increasing order of their strengths**
1) F_n, F_e, F_g 2) F_g, F_e, F_n
3) F_e, F_g, F_n 4) F_g, F_n, F_e

14. **$F = \dfrac{Gm_1m_2}{r^2}$ is valid**
1) Between bodies with any shape
2) Between particles
3) Between any bodies with uniform density
4) Between any bodies with same shape

15. **During a journey from earth to the moon and back, the greatest energy required for the speceship rocket to overcome**
1) The earth's gravity at take-off
2) The moon's gravity at lunar landing
3) The moon's gravity at lunar take off
4) The force at the point where the pull of the earth and moon are equal and opposite

16. **The torque acting on a planet due to gravitational force of attraction is**
1) $\propto$ 2) 0
3) moderate 4) very high

17. **Intensity of gravitational field inside the hollow spherical shell is**
1) Variable 2) minimum
3) maximum 4) zero

18. **Planets move round the sun due to**
1) Mutual attraction and repulsion between sun and planet
2) frictional force 3) centrifugal force
4) gravitational attraction between the sun and planets

Kepler's Laws :

19. **As the distance of a planet from the sun increases, then the time period of revolution of the planet**
1) increases 2) decreases
3) constant 4) none

20. **According to Kepler, the period of revolution of a planet (T) and its mean distance from the Sun (R) are related by the equation**
1) T^2R^3=constant 2) T^2R^{-3}=constant
3) TR^3=constant 4) T^2R=constant

21. **The earth revolves round the sun in an elliptical orbit, its speed is**
1) going on decreasing continuousely
2) greatest when it is closest to the sun
3) greatest when it is farthest from the sun
4) constant at all the points on the orbit

22. **The planet mercury is revolving in an Elliptical orbit around the sun as shown in the figure. The kinetic energy of mercury will be greatest at**

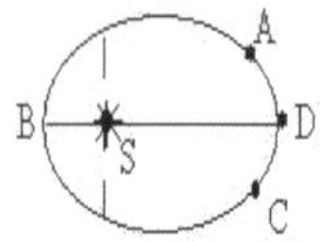

1) A 2) B 3) C 4) D

23. As shown in figure, a planet revolves round the sun in Elliptical orbit with sun at the focus. The shaded areas can be assumed to be equal. If t_1 and t_2 represent the time taken for the planet to move from A to B and C to D respectively, then

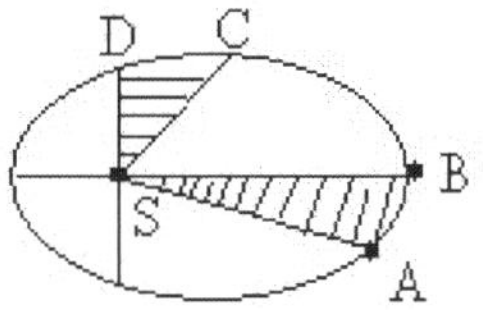

1) $t_1 < t_2$ 2) $t_1 > t_2$ 3) $t_1 = t_2$ 4) none

Relation between g & G, variation of 'g':

24. Consider the earth to be a homogeneous sphere. Scientist A goes deep down in a mine and scientist B goes high up in a balloon. The gravitational field measured by

1) A goes on decreasing and that by B goes on increasing

2)B goes on decreasing and that by A goes on increasing

3) each goes on decreasing

4) each remains unchanged

25. In order to find time, an astronant orbiting in an earth satellite should use

1) A watch having flat spiral spring

2) pendulum clock

3)Neither pendulum clock nor a spiral spring clock

4)either pendulum clock or a spring clock

26. A hydrogen balloon released on the moon

1) climbs up with an acceleration of $6x9.8ms^{-2}$

2) falls down with an acceleration of $9.8x6ms^{-2}$

3) falls with acceleration of $9.8\ ms^{-2}$

4) falls with an acceleration of $\frac{9.8}{6}\ ms^{-2}$

27. The unit of the quantity g/G in SI will be

1) kgm^{-2} 2) mkg^{-2} 3) m^2kg^{-1} 4) kg^2m^{-1}

28. The mass of a body at the centre of the earth is

1) infinite 2) zero

3) same as at other places 4) greater at poles

29. The weight of an object in the coal mine, sea level and at the top of the mountain are respectively W_1, W_2 and W_3, then

1) $W_1 < W_2 > W_3$ 2) $W_1 = W_2 = W_3$

3) $W_1 < W_2 < W_3$ 4) $W_1 > W_2 > W_3$

30. When a body is taken from the equator to the poles, its weight

1) remains same 2) increases

3) decreases

4) increases at N-pole and decreases at S-pole

31. If the earth shrinks to half of its radius and mass remains constant, then the weight of an object on earth will become

1) doubled 2) halved 3) fourtimes 4) same

32. Assuming the earth to be a uniform sphere of mass M and radium R. which of the following graphs represents the variation of acceleration due to gravity with distance 'r' from the centre of the earth.

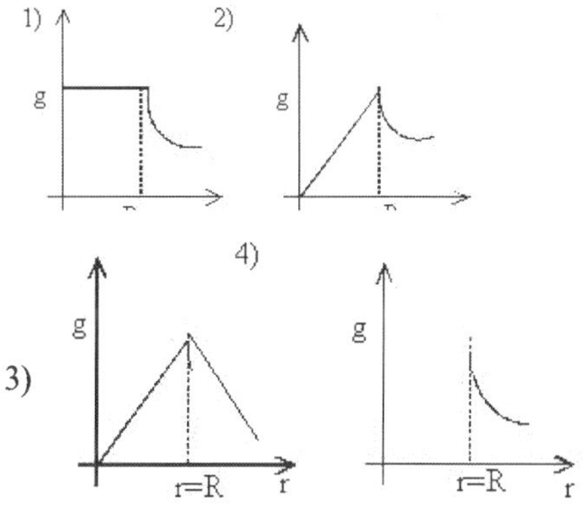

Gravitational Potential & Gravitational P.E:

33. Three particles each of mass 'm' are placed at the corners of an equilateral triangle of side 'd'then

A) potential energy of the system is $\frac{-3Gm^2}{d}$

B) the workdone on this system if the side of the triangle is changed from d to 2d is $\frac{3Gm^2}{d}$

1) both A and B are correct

2) Both A and B are wrong

3) A is true, B is wrong 4) A is wrong, B is true.

Satellites, Orbital and escape velocity:

34. A space-ship entering the earths atmosphere is likely to catch fire, this is due to

1) surface tension of air

2) viscosity of air

3) greater proportional of O_2 in the atmosphere at higher height

4) high temperature of upper atmosphere

35. Two satellites A&B move round the earth in the same orbit. The mass of B is twice that of A,then

1) speed of A & B are equal
2) PE of (earth+A) = PE of (earth +B)
3) KE of A& B are equal
4) Total energy of (earth +A) = Total energy of (earth +B)

36. Two artificial satellites A and B of same mass are revolving round the earth at heights h_1 and h_2 in circular orbits. If $h_2 < h_1$, the satellite possessing greater KE is

1) A 2) B 3) Both A & B
4) cannot be determined

37. A satellite appears to be at rest when seen from the equator of the earth. The height of the satellite from the surface of the earth is

1) 36000km 2) 6400km 3) 40000km 4) 1000km

38. If an astronaut comes out of the artificial satellite, then

1) He flies off tangentially 2) He falls to the earth
3) He performs SHM
4) He continues to move along the satellite in the same orbit.

39. A satellite launching station should be

1) near the equator region 2) near the polar region
3) on the polar axis 4) on any place

40. A satellite is orbiting at a certain height in a circular orbit. If the mass of the planet is reduced to half the initial value, then the satellite would

1) fall on the planet
2) go to the orbit of smaller radius
3) go to the orbit of larger radius
4) escapes from the planet

41. If a satellite is orbiting in space having air and no energy is being supplied, then the path of that satellite would be

1) circular 2) spiral of increases radius
3) spiral of decreasing radius 4) elliptical

42. There is no atmosphere on moon because

1) it is closer to earth
2) it revolves round the earth
3) it gets light from the earth
4) RMS speed of any gas moleculer is greater than the escape velocity on the surface of the moon.

43. When a satellite is orbitting round a planet in circular orbit, workdone by the gravitational force acting on the satellite is

1) zero on completing one revolution only
2) zero always 3) infinite 4) negative

44. A satellite in vacuum

1) is kept in orbit by solar energy
2) previous energy from gravitational field
3) is kept in an orbit by remote control
4) does not require any energy for revolving

45. Weightlessness experienced by spaceman in a space-ship orbiting the earth is the result of

1) inertia 2) acceleration
3) zero gravity 4) surface tension

46. A satellite is revolving around the earth in a circular orbit with a uniform speed. If the gravitational force suddenly disappears, then the satellite will (AIEEE 2002)

1) continue to move in the same orbit with that speed
2) move tangentially to the orbit with that speed
3) move away from the earth normally to the orbit
4) fall down on the earth

47. To have an earth satellite synchronous with the rotation of the earth, it must be launched at a proper height

1) from west to east in equatorial plane
2) from south to north in equatorial plane
3) from east to west in equatorial plane
4) from north to south in equatorial plane

48. An earth satellite is moving round the earth in a circular orbit. For such a satellite which of the following statement is wrong

1) it is a freely falling body
2) it is moving with a constant speed
3) it suffers no acceleration
4) its angular momentum remains constant

49. An artificial satellite orbiting the earth doesnot falldown, because the earth's attraction

1) is balanced by the attraction of the moon
2) produces the necessary acceleration for its motion in a curved path
3) vanishes at such distances
4) is balanced by viscous drag produced by the atmosphere.

50. An astronaut orbiting the earth in a circular orbit 120km above the surface of the earth gently drops a spoon out of the space-ship. It will

1) fall vertically down to the earth
2) move towards the moon
3) will more in an irregular way and then fall down to earth
4) move along the space ship

51. A man inside an artificial satellite feels weightlessness, because the force of attraction due to earth is

1) zero at that place
2) is balanced by the force of attraction due to moon
3) is balanced by centrifugal force
4) non-effective due to particular design of satellite

52. If an orbiting satellite comes to stand still suddenly

1) The satellite will move along the tangent
2) The satellite will move radially towards the centre of the orbit
3) The satellite will go outerspace and will be lost
4) The satellite will continue to move in the same orbit

53. For satellite to appear stationary at a position

1) it must rotate with a period equal to that of the earth
2) its orbit shoudl be in the equitorial plane
3) it should revolve in the direction in which the earth is rotating
4) all the above

54. Two heavenly bodies S_1 and S_2, not far off from each other are seen to revolve in orbit

1) around their common centre of mass
2) which are orbitrary
3) with S_1 fixed and S_2 moving rounds S_1
4) with S_2 fixed and S_1 moving round S_2

55. The period of geostationary artificial satellite of the earth is

1) 24hr 2) 48hr 3) 12hr 4) zero

56. A missile in launched with a velocity less than the escape velocity. The sum of its kinetic and potential

1) Positive 2) negative 3) zero
4)may be positive or negative depending on its initial velocity

57. When a projectile attains the escape velocity, then on surface of planet its

1) KE is greater than PE numarically
2) PE is greater than KE numerically
3) Both PE and KE are numerically equal
4) none

58. An earth satellite is moved from one stable circular orbit to a farther stable circular orbit. Which one of the following quantities increases

1) gravitational force 2) gravitational PE
3) linear orbital speed
4) centripetal acceleration

59. When a satellite is going round the earth in a circular orbit of radius 'r' and with a velocity V. If it loses some of the energy, then
(2004 M & 2001 M)

1) r and v both will increase
2) r and v both will decrease
3) r will decrease and v will increase
4) r will increase and v will decrease

60. Orbital velocity of a satellite revolving in a circular path close to the planet

1) is directly propotional to the density of the planet
2) is directly propotional to the square root of density of the planet
3) is directly propotional to the square of density of planet
4) is inversely proportional to the square root of density of planet

61. The time period of an earth satellite in circular orbit is independent of (AIEEE 2004)

1) The mass of the satellite
2) Neither the mass of the satellite nor the radius of its orbit
3) both the mass of the satellite and radius of the orbit
4) radius of the orbit

Propagation of Gravitational fields, black hole inertial mass and gravitational mass

62. The ratio of the inertial mass to gravitational mass is equal to

1) 0.5 2) 1
3) 2 4) no fixed number

63. According to the principle of equivalence, it is not possible to distinguish between

1) inertial and non-inertial frames of reference
2) the gravitational and inertial masses
3) both 1 & 2 4) none

64. Gravitational field propagates with a speed

1) lessthan the speed of light
2) equal to the speed of light
3) equal to the speed of sound
4) less than speed of sound

65. The mass of a white dwarf is

1) Lessthan 10Ms but more than $1.4M_S$
2) lessthan 1.4Ms 3) morethan 10Ms
4) lessthan 3Ms but more than $1.4M_S$

66. The density of matter in a whitedwarf stage is

1) very high 2) very low
3) medium 4) we can't say

67. The expression for the schwarz child's radius is

1) $\frac{2G}{MC^2}$ 2) $\frac{2MC^2}{G}$ 3) $\frac{2C^2}{GM}$ 4) $\frac{2GM}{C^2}$

68. Degenerate electron pressure temporarily halts core collapse in (ATL)

1) Red giant stage 2) white dwarf stage
3) Super nova stage 4) Neutron star stage

69. 'White dwarf' stage (ATL)

1) contains all the mass of original star but size becomes $\frac{1th}{10^6}$ of original star
2) contains more than the mass of original star but its layer expand
3) contains lass mass than the original star but its layers expand
4) contains less mass than the original star but size becomes $\frac{1th}{10^6}$ of original star

70. Electrons are forced to occupy higher energy levels in the stage of (ATL)

1) Red giant 2) white dwarf
3) neutron star 4) black hole

71. A 'star' may reach black hole stage, if its mass is more than (ATL)

1) 1.4times the mass of sun
2) 3 times the mass of sun
3) 5 times the mass of sun
4) 10times the mass of sun

72. An explosive shock wave in supernova is due to (ATL)

1) Repulsive force between nuclii overcomes gravity
2) Gravity overcomes the repulsive force between nuclii
3) Magnetic force overcomes gravitational force
4) gravitational force overcomes magnetic force

73. According to Einstein (ATL)

1) only the laws of mechanics will be the same in every inertial frame
2) all the laws of physics will be the same in every inertial frame
3) only the laws of mechanics will be the same in every non-inertial frame
4) all the laws of physics will be same in every non-inertial frame

74. The weight of a body (W) is measured using a spring balance, then the ratio $\frac{W}{g}$ gives

1) Inertial mass 2) gravitational mass
3) non-inertial mass 4) relativistic mass

75. According to size, identify the correct decreasing order in

a) original star b) red giant c) white dwarf
1) a,b,c, 2) b,c,a 3) c,a,b 4) b,a,c

76. Give the correct order of formation of black holes

a) Red gaint stage b) Neutron star
c) Whit dwarf stage
1) a,b,c 2) a,c,b 3) c,a,b 4) b,c,a

MATCHING TYPE QUESTIONS

77. Match the items in List-I with those in

List-I	List-II
a) Universal gravitational constant	**e) Central force**
b) Gravitational force	**f) Scalar**
c) 'g' at equator	**g) does not change If earth stops rotating**
d) 'g' at poles	**h) increases by $R\omega^2$ if earth stops rotating**

1) a-f, b-e, c-h, d-g 2) a-f, b-g, c-h, d-e
3) a-h, b-e, c-h, d-g 4) a-g, b-e, c-h, d-f

78. Match list I with list II

List-I	List-II
a) White dwarf	**e) 1.4times solar mass**
b) Nuetron star	**f) mass of star must be greater than 10times the solar mass**
c) Black holes	**g) Mass of the star must be less than10times solar mass**
d) Chandra Sekhar limit	**h) Mass of remaining star is greater than 3 solar masses**

1) a-h, b-e, c-h, d-g 2) a-f, b-g, c-h, d-e
3) a-g, b-f, c-h, d-e 4) a-g, b-e, c-h, d-f

More than one option type Questions

79. In which of the following cases the planet will have same value of 'g' as that of the value on the earth

a) with mass double to that of earth and radius $\sqrt{2}$ times that of earth

b) with mass fourtimes that of earth and radius $\sqrt{2}$ times that of earth

c) with mass one fourth that of the earth and radius half that of earth

d) with mass twice that of earth and radius 1/3 that of earth

1) a and b only 2) a and c only
3) b and c only 4) b and d only

80. Which of the following is true for a satellite in a circular orbit

a) it is a freely falling body

b) its speed is constant

c) it suffers no acceleration

d) it does not require energy for motion in the orbit

1) a, b and c 2) a, b and d
3) a and b 4) a,b,c and d

81. With usual notation, the time period of rotation of a satellite orbiting close to the surface of earth

a) $T = 2\pi\sqrt{\frac{R}{g}}$ b) 84.6minutes

c) $T = 2\pi\sqrt{\frac{R^3}{GM}}$ d) $T = 2\pi\sqrt{\frac{GM}{R}}$

1) a is only correct 2) a & b are correct
3) a,b & c are correct 4) a,b & d are correct

82. When a satellite is lifted from a lower orbit to a heigher orbit

a) Gravitational potential energy increases
b) KE increses

c) Gravitational PE decreases
d) KE increases

1) a is only correct 2) a & d are correct
3) a and c are correct 4) a & b are correct

83. When a planet moves around the sun

a) Its angular momentum remains constant

b) its moves faster when it is near to the sun

c) its total energy increases when it goes near to the sun

d) its potential energy decreases when it goes near to the sun

1) only a&b are true 2) only b&c are true
3) only a,b & d are true 4) all are true

84. The magnitudes of the gravitational fields at distances r_1 and r_2 from the centre of a uniform sphere of radius R and mass M are F_1 and F_2 respectively then

a) $\frac{F_1}{F_2} = \frac{r_1}{r_2}$ if $r_1<R$ and $r_2<R$

b) $\frac{F_1}{F_2} = \left(\frac{r_2}{r_1}\right)^2$ if $r_1>R$ and $r_2>R$

c) $\frac{F_1}{F_2} = \frac{r_1}{r_2}$ if $r_1>R$ and $r_2>R$

d) $\frac{F_1}{F_2} = \left(\frac{r_1}{r_2}\right)^2$ if $r_1<R$ and $r_2<R$

1) only a and b are true 2) only b anc c are true
3) only a,b and d are true 4) all are true

85. Suppose universal gravitational constant starts to decrease, then

a) length of the year will increase

b) earth will follow a spiral path of decreasing order

c) kinetic energy of earth will decrease

1) only a and b are true 2) only a and c are true
3) only b and c are true 4) all are true

86. A satellite is revolving round the earth in an elliptical orbit, then

a) gravitational force exerted by the earth is equal to centripetal force at same point only

b) power associated with gravitational force is zero at every point

c) workdone by gravitational force is zero in some small parts of the orbit

d) at some points, the magnitude of gravitational force is greater than that of centripetal force

1) only a and b are true 2) only b&c are true
3) only a,b & d are true 4) only a,c,d are true

87. In case of an orbiting satellite, if the radius of orbit is decreased, then

a) its KE decreases b) its PE decreases

c) its mechanical energy decreases

1) only a & b are true 2) only b & c are true
3) only a & c are true 4) all a, b & c are true

88. **Let V,T,L,K and r denote the speed, time period, angular momentum, kinetic energy and radius of satellite in circular orbit, then**

a) $V \propto r^{-1}$ **b)** $L \propto r^{1/2}$

c) $T \propto r^{3/2}$ **d)** $K \propto r^{-2}$

1) only a & b are true 2) only b & c are true

3) only a, b & d are true 4) only a, b & c are true

89. **Assume that earth is a spherical planet of uniform density ρ, radius R_e, mass M_e and acceleration due to gravity g. Then the gravitational constant G can be written as**

a) $\frac{3g}{4\pi\rho R_e}$ b) $\frac{gR_e^2}{M_e}$ c) $\frac{3g}{4\pi\rho R_e^2}$ d) $\frac{12g}{4\pi\rho R_e}$

1) only a & b are true 2) only b & c are true

3) only a, b & d are true

4) only a & d are true

90. **The acceleration due to gravity on moon is only one sixth that on earth. Ratio of moon radius (R_m) to earth's radius (R_e) should be:**

a) $\frac{6}{1}$ **if both are assumed to have same density**

b) $\frac{1}{6}$ **if both are assumed to have same density**

c) $\frac{5}{18}$ *if* $\frac{\rho_e}{\rho_m}$ **is given as 5/3**

d) $9/5$ **if** $\frac{\rho_m}{\rho_e}$ **is given as 3/5**

1) only a & b are true 2) only b & c are true

3) only a, b & d are true 4) all are true

91. **If a body is projected with speed less than escape velocity, then**

a) the body can reach a certain height and may fall down following straight line path

b) the body can reach a certain height and may fall down following a parabolic path

c) the body may orbit the earth in a circular path

d) the body may orbit the earth in an elliptical path.

1) only a & b are true 2) only b & c are true

3) only a,b & d are true 4) all a,b,c & d are true

92. **Two spherical planets have the same mass but densities in the ratio 1:8. For these planets, the**

a) acceleration due to gravity will be in the ratio 4:1

b) acceleration due to gravity will be in the ratio 1:4

c) escape velocities from their surfaces will be in the ratio $\sqrt{2}$:1

d) escape velocities from their surfaces will be in the ratio 1:$\sqrt{2}$

1) only b&d are true 2) only b&c are true

3) only a,c&d are true 4) only a&b are true

93. **Below are four equations expressing relationships between physical qualities. 'g' gravitational field strengths, r radius, ρ a density, G the universal constant of gravitation, m a mass and T a time period. Which of the equations are true.**

a) $gT^2=4\pi r^2$ b) $\frac{4}{3}\pi\rho Gr^2 = g$

c) $gr^2 = Gm$ d) $g = \frac{4}{3}\pi\rho Gr$

1) only a&b are true 2) only b & c are true

3) only c&d are true 4) only a,b & c are true

94. **An orbiting satellit.e will escape if**

a) its speed increases by 41.4%

b) its speed in the orbit is made $\sqrt{1.414}$ times that of its initial value

c) its kinetic energy increases by 100%

d) its kinetic energy is made $\sqrt{2}$ times that of its initial value

1) only a & b are correct 2) only a & c are correct

3) only b & c are correct 4) only c is correct

Assersion and Reason type Qestions

The question constist of two statements each printed as Assertion and Reason. While answering these questions you are required to choose any one of the following four responses.

1) Both (A) and (R) are true and (R) is the correct explanation of (A)

2) Both (A) and (R) are true and (R) is not the correct explanation of (A)

3) (A) is true but (R) is false

4) (A) is false but (R) is true

95. **A: Escape velocity from the surface of a black hole is greater than speed of light**

R: Light cannot escape from black hole

96. **A: White dwarf is relatively a dim star**

R: Hydrogen available is completely exhausted

97. A: The gravitational mass and inertial mass are equal

R: Physical laws are equivalent in an appropriately accelerated frame of reference and in an inertial frame

98. A: Two satellite of masses 100kg and 4000kg are orbiting in the same orbit with orbital velocities in the ratio 1:2

R: Orbital velocity is independent of the mass of satellite

99. A: A body is first taken along a smooth inclined plane through certain height. Inthe next case it is lifted vertically upwards to same height, workdone by gravity in both cases is same

R: Gravitational force is conservative force

100. A: For a satellite to escape from its orbit, its velocity should be increased by 41.4%

R: Orbital velocity and escape velocity are related as $V_e = \sqrt{2}\, v_0$.

101. A: Inertial and gravitational masses differ when the speed of the particle (u) is comparable to the speed of light (c)

R: Gravitational mass is measured when the body is at rest and inertial mass is measured when the body is in motion. Relativistic variation of mass is significant when 'u'is comparable to C.

102. A: The length of year on the planet pluto is much longer than of the earth

R: Pluto is much farther from the sun than the earth

103. A: When a body is dropped in a tunnel dug along the diameter of the earth, it executes simple hormonic motion.

R: At any position, magnitude of Gravitational force experienced by the body is directly proportional to the displacement of the body as measured from centre of the earth

104. A: When a body is projected from the surface of the earth with a velocity greater than the escape velocity, its total mechanical energy is negative.

R: The body is not under the influence of the earth

105. A: The bodies in the satellites revolving round the earth are in the state of weightlessness

R: The net acceleration due to gravity due to the earth is zero in the satellite

106. A: There is no atmosphere on the moon surface

R: RMS speed of the gas molecules is greater than the escape velocity on moon

107. A: Spaceship while entering atmosphere catches fire

R: The temperature of upper atmosphere is very high

108. A: Two point masses freely positioned at rest move under mutual gravitational force of attraction. Velocity of C.M. of the system is zero at any instant.

R: Gravitational force of attraction is an internal force of the system

109. A: An object would be black hole of all of its mass in inside a sphere with a radius equal to schwarz child radius.

R: At the schwarzchild radius, escape velocity is equal to the speed of light

110. A: A particle of mass 'm' dropped into a hole made along the diameter of the earth from one end to the other and possesses simple harmonic motion. (2006 E)

R: Gravitational force between any two particles is inversely proportional to the square of the distance between them

ANSWERS

1) 4	2) 3	3) 2	4) 4	5) 4
6) 3	7) 3	8) 1	9) 1	10) 3
11) 2	12) 4	13) 2	14) 1	15) 1
16) 2	17) 4	18) 4	19) 1	20) 2
21) 2	22) 2	23) 3	24) 3	25) 1
26) 4	27) 1	28) 3	29) 1	30) 2
31) 3	32) 2	33) 3	34) 2	35) 1
36) 2	37) 1	38) 4	39) 1	40) 4
41) 3	42) 4	43) 2	44) 4	45) 2
46) 2	47) 1	48) 3	49) 2	50) 4
51) 3	52) 2	53) 4	54) 1	55) 1
56) 2	57) 3	58) 2	59) 3	60) 2
61) 1	62) 2	63) 3	64) 2	65) 2
66) 1	67) 4	68) 2	69) 4	70) 2
71) 4	72) 1	73) 2	74) 2	75) 4
76) 2	77) 1	78) 3	79) 2	80) 4
81) 3	82) 4	83) 3	84) 1	85) 2
86) 4	87) 2	88) 2	89) 1	90) 2
91) 4	92) 1	93) 3	94) 2	95) 1
96) 1	97) 1	98) 4	99) 1	100) 2
101) 1	102) 1	103) 1	104) 4	105) 1
106) 1	107) 3	108) 1	109) 1	110) 2

EXERCISE - II(A)

Newton's Universal law of Gravitation & Gravitational Intensity :

1. **You are given total mass M. How do you divide it into two parts so that the gravitational force between them at a given distance is maximum**

   1) $\frac{M}{4},\frac{3M}{4}$ 2) $\frac{M}{3},\frac{2M}{3}$

   3) $\frac{M}{5},\frac{4M}{5}$ 4) $\frac{M}{2},\frac{M}{2}$

2. **Two spherical balls each of mass 1 kg are placed 1 cm apart. The gravitational force of attraction between them - newtons**

   1) 6.67×10^{-9} 2) 6.67×10^{-7}

   3) 6.67×10^{-8} 4) 6.67×10^{-5}

3. **The gravitational force between two bodies is F. If the mass of each body is doubled and the distance between them is halved, then the force between them will be**

   1) $\frac{4F}{9}$ 2) $\frac{9F}{4}$ 3) 16F 4) $F/16$

4. **When two bodies are saparated by certain distance ,the gravitational force between then is F. If distance is increased by 100%,then gravitational force becomes --**

   1) $\frac{F}{4}$ 2) $4F$ 3) $\frac{F}{2}$ 4) 2F

5. **An infinite number of particles each of mass 1kg are placed on the positive x-axis at 1m, 2m, 4m, 8m.... from the origin. The magnitude of the resultant gravitational force on 1kg mass kept at the origin is**

   1) 0 2) G 3) $\frac{3G}{4}$ 4) $\frac{4G}{3}$

6. **Three equal masses m are placed at the three vertices of an equilateral triangle of side a.The gravitational force exerted by this system on another particle of mass m placed at the mid-point of a side**

   1) $\frac{Gm^2}{3a^2}$ 2) $\frac{4Gm^2}{3a^2}$ 3) $\frac{Gm^2}{a^2}$ 4) $\frac{Gm^2}{3a^2}$

7. **Two metal spheres of same material and radius 'r' are in contact with each other. The gravitational force of attraction between the spheres is given by (k in a constant).**

   1) $F=kr^4$ 2) $F = k/r^2$ 3) $F= k/4r^2$ 4) $F = kr^2$

8. **Three uniform spheres each having mass 'm' and radius 'R' kept in such a way that each two touches the other. The magnitude of the gravitational force on any sphere, due to the other two is**

   1) $\sqrt{3}\frac{Gm^2}{4R^2}$ 2) $\frac{Gm^2}{4R^2}$ 3) $\frac{\sqrt{3}Gm}{2R^2}$ 4) $\sqrt{3}\frac{Gm}{R^2}$

9. **Two bodies of masses m_1 and m_2 are separated by certain distance. If $\vec{F}_{12}$ is the force on m_1 due to m_2 and $\vec{F}_{21}$ is the force on m_2 due to m_1, then**

   1) $F_{12}=F_{21}$ 2) $\vec{F}_{12}=\vec{F}_{21}$

   3) $\vec{F}_{12}=-\vec{F}_{21}$ 4) both 1& 3

10. **Two spheres of masses m and M are situated in air and the gravitational force between them is F. The space around the masses is now filled with a liquid of specific gravity 3. The gravitational force will now be**

    1) F/9 2) 3F 3) F 4) F/3

11. **Three spherical balls of masses 1 kg, 2kg and 3 kg are placed at the corners of an equilateral triangle of side 1 m.. The magnitude of the gravitational force exerted by 2 kg and 3 kg masses on 1 kg mass.**

    1) 19 G 2) $\sqrt{17}G$ 3) $\sqrt{19}G$ 4) $\frac{G}{\sqrt{17}}$

12. **Two particles each of mass 'm' move in a circle of radius 'r' under the action of their mutual gravitational attraction. Then speed of each particle is**

    1) $\sqrt{\frac{Gm}{r}}$ 2) $\sqrt{\frac{Gm}{4r}}$ 3) $\sqrt{\frac{Gm}{2r}}$ 4) $\sqrt{\frac{2Gm}{r}}$

13. **Two spherical balls of mass 1 kg and 4 kg are separated by a distance of 12 cm. Thedistance from 1 kg at which the gravitational force on any mass becomes zero.**

    1) 4 cm 2) 12 cm 3) 6 cm 4) 8 cm

14. **Mass of the earth is 81 times that of the moon. If the distance between their centers is d, then the point on the line joining their centers at which the gravitational field due to them is zero is**

    1) d/10 from the center of moon
    2) d/10 from the center of earth
    3) 9d/10 from the center of moon
    4) d/2 from each other

Kepler's laws :

15. A planet revolves round the sun in an elliptical orbit of minor and major axes x and y respectively. Then the time period of revolution is proportional to

1) $(x+y)^{3/2}$ 2) $(y-x)^{3/2}$ 3) $x^{3/2}$ 4) $y^{3/2}$

16. A and B are two satellites revolving round the earth in circular orbits have time periods 8hr and 1hr respectively. The ratio of their radius of orbits

1) $8^{3/2}:1$ 2) 8:1 3) 4:1 4) 2:1

17. The period of revolution of a surface satellite around a planet of radius R is T. The period of another satellite around the same planet in an orbit of radius 3R is

1) T 2) 3T 3) 9T 4) $3\sqrt{3}$ T

18. If a planet of mass 'm' is revolving around the sun in a circular orbit of radius 'r' with time period T, then the mass of the sun is

1) $\frac{4\pi^2 r^3}{GT}$ 2) $\frac{4\pi^2 r^3}{GT^2}$ 3) $\frac{4\pi^2 r}{GT}$ 4) $\frac{4\pi^2 r^3}{G^2T^2}$

19. In a planetary motion the areal velocity of a position vector of a planet depends on angular velocity 'w' and distance of the planet from the sun (r). If so the correct relation for areal velocity

1) $\frac{dA}{dt} \propto \omega r$ 2) $\frac{dA}{dt} \propto \omega^2 r$

3) $\frac{dA}{dt} \propto \omega r^2$ 4) $\frac{dA}{dt} \propto \sqrt{\omega r}$

20. The distances of two planets from the sun are 10^{12} m and 10^{10}m respectively. The ratio of the time periods of these planets is

1) 1000:1 2) 100:1 3) 10:1 4) 1:1

21. Taking that earth revolves round the sun in circular orbit of radius 15 x 10^{10}m with a time period of 1year. The time taken by another planet which is at a distance of 540 x 10^{10}m, to revolve round the sun in circular orbit once will be (99E)

1) 216years 2) 144years

3) 72years 4) 36years

22. A satellite is launched into circular orbit of radius R around the earth while a second satellite is launched into an orbit of Radius 1.02R. The percentage change in the time periods of the two satellite is (2003 E)

1) 0.7 2) 1.0 3) 1.5 4) 3

23. Two satellites of masses m_1 and m_2 ($m_1 > m_2$) are revolving the earth in circular orbits of radius r_1 and r_2 ($r_1 > r_2$) respectively. Which of the following statement is true regarding their velocities v_1 and v_2.

1) $v_1 = v_2$ 2) $v_1<v_2$ 3) $v_1>v_2$ 4) $\frac{v_1}{r_1} = \frac{v_2}{r_2}$

24. Let 'A' be the area swept by the line joining the earth and the sun during Feb 2008. The area swept by the same line during the first week of that month is

1) A/4 2) 7A/29 3) A 4) 7A/30

Relation between g & G :

25. A spherical planet far out in space has a mass M_0 and diameter D_0. A particle of mass 'm' falling freely near the surface of this planet will experience an acceleration due to gravity which is equal to

1) $\frac{GM_0}{D_0^2}$ 2) $4m\frac{GM_0}{D_0^2}$

3) $4\frac{GM_0}{D_0^2}$ 4) $\frac{GmM_0}{D_0^2}$

26. g_e and g_p denote the acceleration due to gravity on the surface of earth and another planet whose mass and radius are twice that of the earth, then

1) $g_p=g_e$ 2) $g_p={g_e}/{2}$

3) $g_p=2g_e$ 4) $g_p=g_e$

27. Mass of the moon is 1/81 of the earth but gravitational pull is 1/6 of the earth. It is due to the fact that

1) Radius of the moon is 81/6 of the earth

2) Radius of the Earth is $\sqrt{\frac{81}{6}}$ of the moon

3) Moon is the satellite of the earth 4) none

28. A body weighs 700gm on the surface of the earth. How much will it weigh on the surface of a planet whose mass is 1/7 and radius half of the earth

1) 200gm wt 2) 400 gm wt

3) 500gm wt 4) 300gm wt

29. Weight of a body on the surfaces of two planets is the same. If their densities are d_1 and d_2, then the ratio of their radius

1) $\frac{d_1}{d_2}$ 2) $\frac{d_2}{d_1}$ 3) $\frac{d_1^2}{d_2^2}$ 4) $\frac{d_2^2}{d_1^2}$

30. **The radius of the earth is about 6400km and that of mars is about 3200km. The mass of the earth is about 10 times that of mars. If an object weighs 200 N on the surface of earth its weight on the surface of mars would be**

1) 40N 2) 80N 3) 8N 4) 20N

31. **Average density of earth is (AIEEE 2006)**

1) is directly proportional to g

2) is inversely proportional to g

3) is complex function of g

4) does not depend on g

Variation of 'g' with altitude, depth and speed of rotation of earth.

32. **The value of acceleration due to gravity on the surface of earth is x. At alttitude of 'h' from the surface of earth, its value is y. If 'R' is the radius of earth, then the value of h is**

1) $\left(\sqrt{\frac{x}{y}}-1\right)R$ 2) $\left(\sqrt{\frac{y}{x}}-1\right)R$

3) $\sqrt{\frac{y}{x}}R$ 4) $\sqrt{\frac{x}{y}}R$

33. **If the change in the value of g at a height h above the surface of the earth is same as at depth 'x' below the surface of the earth, then (h<<R)**

1) x = h 2) x = 2h

3) $x = \frac{h}{2}$ 4) $x = h^2$

34. **The alttitude at which the weight of a body is only 64% of its weight on the surface of the earth is (Radius of the earth is 6400km)**

1) 1600m 2) 16m 3) 160km 4) 1600km

35. **At a certain height above the earth's surface, the gravitational acceleration is 4% of its value at the surface of the earth. Find the height. (R is the surface of the earth)**

1) 2R 2) 4R 3) R 4) R/2

36. **The acceleration due to gravity at a depth of 1 600 km inside the earth.**

1) 0 2) 9.8 m/s^2

3) 7 m/s^2 4) 7.35 m/s^2

37. **If R is radius of the earth, the height above the surface of the earth where the weight of a body is 36% less than its weight on the surface of the earth is**

1) 4R/5 2) R/5 3) R/6 4) R/4

38. **The angular velocity of the earth with which it has to rotate so that the acceleration due to gravity on 60^0 latitude becomes zero.**

1) $3\sqrt{\frac{g_p}{R}}$ 2) $4\sqrt{\frac{g_p}{R}}$ 3) $\sqrt{\frac{g_p}{R}}$ 4) $2\sqrt{\frac{g_p}{R}}$

39. **The angular velocity of the earth about its polar axis so that the weight of the body at the equator will be zero is**

1) 1.25 x 10^{-3} rad s^{-1} 2) 2 rad s^{-1}

3) 1 rad s^{-1} 4) 0

Gravitational potential and Gravitational Potential Energy :

40. **The amount of work done in lifting a body of mass 'm' from the surface of the earth to a height equal to twice the radius of the earth is**

1) $\frac{2GMm}{3R}$ 2) $\frac{3GMm}{2R}$ 3) $\frac{5GMm}{3R}$ 4) $\frac{3GMm}{5R}$

41. **Find the PE of three objects of masses 1 kg, 2 kg and 3 kg placed at the three vertices of anequilateral triangle of side 20 cm.**

1) 50J 2) – 55J 3) 5J 4) 60J

42. **The magnitude of gravitational potential energy of a body at a distance 'r' from the center of the earth is V. Its weight at a distance '2r' from the center of the earth is**

1) $\frac{V}{r}$ 2) $\frac{V}{4r}$ 3) $\frac{V}{2r}$ 4) $\frac{4V}{r}$

43. **A particle of mass 10g is kept on the surface of a uniform sphere of mass 100kg and radius 10cm. Find the work to be done against the gravitational force between them to take the particle far away from the sphere is (G=6.67x 10^{-11} Nm^{2}kg^{-2}) (AIEEE 2005)**

1) 6.67 x 10^{-10}J 2) 6.67 x 10^{-9}J

3) 3.33 x 10^{-10}J 4) 13.34 x 10^{-10}J

Satellites, Orbital and escape velocities :

44. **A satellite is revolving very close to a planet of density D. The time period of revolution of that planet is**

1) $\sqrt{\frac{3\pi}{DG}}$ 2) $\left(\frac{3\pi}{DG}\right)^{3/2}$ 3) $\sqrt{\frac{3\pi}{2DG}}$ 4) $\sqrt{\frac{3\pi G}{D}}$

45. **A satellite orbits the earth at a height of R/6, its orbital speed ?**

1) $\sqrt{\frac{2GM}{R}}$ 2) $\sqrt{\frac{GM}{R}}$

3) $\sqrt{\frac{GM}{7R}}$ 4) $\sqrt{\frac{6GM}{7R}}$

46. A satellite is in a circular orbit around a planet. Its period of revolution is T, radius of the orbit is R, orbital velocity V and acceleration 'a', then

1) V = at and a = $\frac{V^2}{R}$ 2) V = $\frac{2\pi R}{T}$ and V = a T

3) V = $\frac{2\pi R}{T}$ and a = $\frac{V^2}{R}$ 4) V = $\frac{1}{2}aT^2$

47. A satellite is revolving near the earth's surface. Its orbital velocity depends on its mass'm' as

1) m^0 2) m 3) m^2 4) m^3

48. The escape velocity of a body from the surface of the earth is nearly

1) 7.98kms^{-1} 2) 8.92kms^{-1}
3) 10.48kms^{-1} 4) 11.2kms^{-1}

49. A satellite is revolving in a circular orbit of radius R. Ratio of KE and its PE is (only magnitude)

1) 1:4 2) 2:1 3) 3:2 4) 1:2

50. The PE of a satellite is –U. Its KE is

1) U 2) 2U 3) -U 4) U/2

51. If the mass of the earth is M, its radius is R and gravitational constant G, then the speed required to be given to a projectile so that it does not return to earth is

1) $\sqrt{\frac{GM}{R}}$ 2) $\sqrt{\frac{2GM}{R}}$ 3) $\sqrt{\frac{GR}{M}}$ 4) $\sqrt{GRM}$

52. A satellite is revolving in a circular orbit round the earth at a height of h. If the acceleration due to gravity at that height is g. The orbital speed of the satell ite.

1) $\sqrt{gh}$ 2) $\sqrt{2gh}$ 3) $\sqrt{g(R+h)}$ 4) $\sqrt{gR}$

53. The relation between the escape velocity v_e and orbital velocity v_0 is given by

1) $V_e = \sqrt{2}V_0$ 2) $V_0 = \sqrt{2}V_e$

3) $\sqrt{V_e} = 2V_0$ 4) $V_e = V_0$

54. The ratio of the time periods of two satellites at heights h_1 , h_2 from the surface of the earth.

1) $\left(\frac{h_1}{h_2}\right)^{3/2}$ 2) $\left(\frac{R+h_1}{R+h_2}\right)^{3/2}$

3) $\left(\frac{R+h_2}{R+h_1}\right)^{3/2}$ 4 $\left(\frac{h_2}{h_1}\right)^{3/2}$

55. The KE required to project a body of mass 'm' from the earth's surface to infinity is

1) mgR/4 2) $\frac{mgR}{2}$

3) mgR 4) 2mgR

56. A spaceship is launched into a circular orbit of radius R close to the surface of the earth. The additional velocity to be imparted to the spaceship in the orbit to overcome the earth's gravitational pull is (g $\rightarrow$ acceleration due to gravity)

1) 1.414Rg 2) $1.414\sqrt{gR}$

3) 0.414Rg 4) $0.414\sqrt{gR}$

57. The escape velocity for a body projected vertically upwards from the surface of earth is 11kms^{-1}. If the same body is projected at an angle of 45^0 with the vertical, the escape velocity will be (AIEEE 2003)

1) $11\sqrt{2}$ kms^{-1} 2) 22kms^{-1}

3) 11kms^{-1} 4) $11/\sqrt{2}$ kms^{-1}

58. The mass of a planet is half that of the earth and the Radius of planet is one fourth that of the earth. If we plan to send an artificial satellite from the planet, the escape velocity will be nearly (V_e of earth is 11kms^{-1})

1) 11kms^{-1} 2) 5.5kms^{-1}
3) 15.5kms^{-1} 4) 7.7kms^{-1}

59. The escape velocities on the surface of two planets of masses m_1 and m_2 and having same radius are v_1 and v_2 respectively. Then

1) $\frac{v_1}{v_2} = \sqrt{\frac{m_1}{m_2}}$ 2) $\frac{v_2}{v_1} = \frac{m_1}{m_2}$

3) $\frac{v_1}{v_2} = \sqrt{\frac{m_2}{m_1}}$ 4) $\frac{v_1}{v_2} = \left(\frac{m_1}{m_2}\right)^2$

60. If the escape velocity on earth is 11.2kms^{-1}, its value for a planet having double the radius and 8 times the mass of the earth is

1) 10.2kms^{-1} 2) 22.4kms^{-1}
3) 10kms^{-1} 4) 0

61. The radius in kilometers to which the present radius of the earth (R=6400 km) is to be compressed so that the escape velocity is increases to ten times is (2003M)

1) 6.4 2) 64 3) 640 4) 4800

Black hole :

62. The expression for the Schwarz child radius around a mass 'M' is (C → is the velocity of light)

1) $\frac{2G}{MC^2}$ 2) $\frac{2MC^2}{G}$ 3) $\frac{2C^2}{GM}$ 4) $\frac{2GM}{C^2}$

63. For the formation of black hole, the initial mass of original star should be graterthan n times the solar mass, the value of n is (2006M)

1) 2 2) 6 3) 8 4) 10

ANSWERS

1) 4	2) 2	3) 3	4) 1	5) 4
6) 2	7) 1	8) 1	9) 4	10) 3
11) 3	12) 2	13) 1	14) 1	15) 4
16) 3	17) 4	18) 2	19) 3	20) 1
21) 1	22) 4	23) 2	24) 2	25) 3
26) 2	27) 2	28) 2	29) 2	30) 2
31) 4	32) 1	33) 2	34) 4	35) 2
36) 4	37) 4	38) 4	39) 1	40) 1
41) 2	42) 2	43) 1	44) 1	45) 4
46) 3	47) 1	48) 4	49) 4	50) 4
51) 2	52) 3	53) 1	54) 2	55) 3
56) 4	57) 3	58) 3	59) 1	60) 2
61) 2	62) 4	63) 4		

EXERCISE-II(B)

Newton's Universal Law of Gravitation and Gravitational Intensity :

1. Mass 1 kg is divided into two parts 'x' m and (1-x) m for a given seperation, the value of'x' for which the gravitational attraction between the two pieces becomes maximum is (2001 E)

1) 1/2 2) 3/5 3) 1 4) 2

2. The mass of a ball is four times the mass of another ball. When these balls are separated by a distance of 10 cm, the gravitational force between them is 6.67 x 10^{-7}N. Then the mass of the each ball in Kg is

1) 0.5 , 2 2) 50 , 20 3) 2 , 5 4) 5 , 20

3. If the distance between the earth and moon were doubled, then the gravitational force between them will be

1) Halved 2) Doubled
3) Quadrupled 4) Reduced to 1/4th

4. The force between two objects decreases by 36% when the distance between them is increasedby 4 m. The original distance between them in meter is

1) 4 2) 16 3)10 4) 12

5. Two particles of masses 4kg and 8kg are kept at x=-2m and x=4m respectively. Then the net gravitational force acting on a third particle of mass 1kg kept at the origin is

1) G 2) 2G 3) G/2 4) G/4

6. Three equal masses m are placed at the three vertices of an equilateral triangle of side a.The gravitational force exerted by this system on another particle of mass m placed at the centroid of triangle

1) $\frac{Gm^2}{3a^2}$ 2) $\frac{Gm^2}{a^2}$ 3) $\frac{Gm^2}{a^2}$ 4) 0

7. When two identical spheres are kept in contact, the gravitational force between them is F. If two spheres of same material but with twice the radius are radius are kept in contact, then the gravitational force between them would be

1) F/4 2) 4F 3) F/16 4) 16F

8. Three uniform spheres each of mass 'm' and diameter D are kept in such a way that each touches the other two, then the magnitude of the gravitational force on any one sphere due to other two is

1) $\frac{3Gm^2}{D^2}$ 2) $\frac{2\sqrt{3}Gm^2}{D^2}$

3) $\frac{\sqrt{3}Gm^2}{4D^2}$ 4) $\sqrt{3}\frac{Gm^2}{D^2}$

9. Gravitational force between masses m_1 and m_2 is F. If a third mass m_3 is placed near m_1 and m_2, then the force between m_1 and m_2 would be

1) Greaterthan F 2) Less than F
3) Equal to F 4) Zero

10. The gravitational force between two particles of masses m_1 and m_2 seperated by certain distance in air medium is F. If they are taken to vaccum and seperated by the same distance, then the gravitational force between them will be

1) greater than F 2) Less than F
3) F 4) Zero

11. Three objects each of mass 'm' are placed at the corners of an equilateral triangle of side 'l'. What is the gravitational force acting on each mass due to other masses is

1) $\frac{Gm^2}{\sqrt{3}l^2}$ 2) $\frac{\sqrt{3}Gm^2}{2l^2}$

3) $\frac{\sqrt{3}Gm^2}{l^2}$ 4) $\frac{Gm^2}{l^2}$

12. **Two particles each of mass m seperated by a distance d and move in a uniform circle under the action of their mutual force of attraction. The speed of each particle is**

1) $\sqrt{\frac{Gm}{d}}$ 2) $\sqrt{\frac{Gm}{4d}}$ 3) $\sqrt{\frac{Gm}{2d}}$ 4) $\sqrt{\frac{2Gm}{d}}$

13. **Masses 1kg and 4kg are 6cm apart. The Point along the line joining the masses where the gravitational field due to them is zero is**

1) 2cm from 1kg mass
2) 2cm from 4kg mass
3) 4cm from 1kg mass
4) 3cm from each mass

14. **The mass of a planet is $\frac{1}{25}$ th of the mass of the earth. The distance between the planet and the earth is 6×10^8m. At what distance from the earth, the gravity becomes zero**

1) 1×10^8m 2) 0.32×10^8m
3) 5×10^8m 4) 0.032×10^8m

Kepler's Laws :

15. **The period T and mean distance R of a planet from the sun are related by**

1) $T^2 \propto R^2$ 2) $T \propto R^3$
3) $T^3 \propto R^2$ 4) $T^{2/3} \propto R$

16. **The ratio of the distance of two planets from the sun is 1:2. The ratio of their periods of revolution is**

1) 1:4 2) 1:$\sqrt{2}$ 3) 1:2 4) 1:2$\sqrt{2}$

17. **The period of a satellite in circular orbit of radius R is T. The period of an other satellite in a circular orbit of radius 4R is**

1) 4T 2) T/4 3) 8T 4) T/8

18. **If a planet of radius R and density D is revolving around the sun in a circular orbit of radius r with time period T, then the mass of the sun is**

1) $\frac{4\pi^2 r^3}{GT}$ 2) $\frac{4\pi^2 r^3}{GT^2}$

3) $\frac{4\pi^2 r}{GT}$ 4) $\frac{4\pi^2 r^3}{G^2T^2}$

19. **If 'A' is the A real velocity of a planet of mass M, its angular momentum is**

1) M/A 2) 2MA 3) A^2M 4) AM^2

20. **The distance of Neptune and Saturn from the sun are respectively 10^{13} and 10^{12} metres and their periodic times are respectively T_N and T_S. If their orbits are assumed to be circular, then the value of T_N/T_S.**

1) 100 2) $10\sqrt{10}$ 3) $\frac{1}{10\sqrt{10}}$ 4) 10

21. **The time period of revolution of a planet A around the sun is 8 times that of another planet B. The distance of planet A from the sun is how many times greater than that of the planet B from the sun (AIEEE 2002)**

1) 2 2) 3 3) 4 4) 5

22. **Which of the following graph represents the motion of a satellite revolving about a planet (T is period of revolution and R is radius of orbit)**

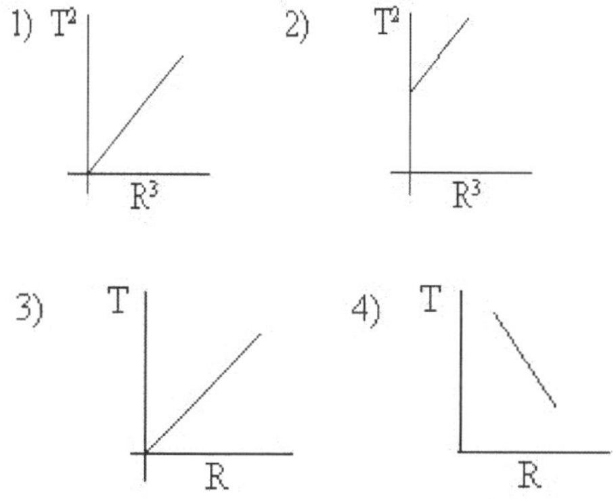

23. **The earth moves round the sun in an elliptical orbit. O is the focus of ellipse and A &B are two points in the path such that $\frac{OA}{OB} = K$ The ratio of the speed of the earth at B and at A is nearly**

1) $\sqrt{K}$ 2) K 3) K^2 4) $K^{3/2}$

24. **Let A be the Area Swept by the line joining the earth and the sun during Feb-2007. The Area Swept by the same line during the first week of that month is**

1) A/4 2) 7A/29 3) A 4) 7A/30

Relation between g & G :

25. **If R is the radius of the earth and 'g' the acceleration due to gravity, the mass of the earth is**

1) $\frac{gR^2}{G}$ 2) $\frac{gG}{R^2}$ 3) $\frac{GR^2}{g}$ 4) GgR^2

26. The value of acceleration due to gravity 'g_p' on the surface of a planet with radius double that of earth and same mean density as that of the earth is

($g_e \rightarrow$ acceleration due to gravity on the surface of earth)

1) $g_p=2g_e$ 2) $g_p=g_e/2$ 3) $r_p=g_e/4$ 4) $g_p=4g_e$

27. The radius and acceleration due to gravity of moon are $\frac{1}{4}$ *and* $\frac{1}{5}$ that of earth. The ratio of the mass of earth to mass of moon is

1) 1:80 2) 80:1 3) 1:20 4) 20:1

28. A man weighs 75kg on the surface of the earth. His weight on a geostationary satellite is

1) infinity 2) 150kg 3) zero 4) $\frac{75}{2}$kg

29. Two planets of radii r_1 and r_2 are made from the same material. The ratio of the acceleration due to gravities at the surface of the two planets is

1) $\frac{r_1}{r_2}$ 2) $\frac{r_2}{r_1}$ 3) $\left(\frac{r_1}{r_2}\right)^2$ 4) $\left(\frac{r_2}{r_1}\right)^2$

30. R and r are the radii of the earth and moon respectively. ρ_e and ρ_m are the densities of earth and moon respectively. The ratio of the accelerations due to gravity on the surfaces of earth to moon is

1) $\frac{R}{r}\frac{\rho_e}{\rho_m}$ 2) $\frac{r}{R}\frac{\rho_e}{\rho_m}$ 3) $\frac{r}{R}\frac{\rho_m}{\rho_e}$ 4) $\frac{r}{R}\frac{\rho_m}{\rho_e}$

31. If the Radius of earth were shrink by 10% its mass remaining same, the acceleration due to gravity on earth's surface would

1) decrease 2) remain unchanged
3) increase 4) none

Variation of 'g' with altitude, depth and Speed of Rotation of earth :

32. If the radius of the earth is 6400 km, the height above the surface of the earth, where the value of acceleration due to gravity will be 1% of its value from the surface of the earth is

1) 6400km 2) 64km 3) 711km 4) 57600km

33. The change in the value of 'g' at a height 'h' above the surface of the earth is the same as at a depth 'd' below the surface of the earth. When both d & h are very much smaller than the radius of the earth, then which one of the following is correct (AIEEE 2005)

1) d=h 2) d=2h 3) d=$\frac{3h}{2}$ 4) d=$\frac{h}{2}$

34. If R is the radius of earth, the height at which the weight of a body becomes $\frac{1}{4}$ th its weight on the surface of the earth is

1) 2R 2) R 3) R/2 4) R/4

35. Find the depth at which g becomes 4% lessthan value at surface

1) 0.2 R 2) 0.04 R 3) 0.3 R 4) 0.03 R

36. If the radius of the earth is 6400km, the depth at which the acceleration due to gravity will be 5% less than the value on the surface of the earth is

1) 320km 2) 2.0km 3) 480km 4) 500km

37. A man weighs 60kg on the surface of the earth. The altitude where his weight will be 30kg is (R is the radius of the earth)

1) h= $2\sqrt{R}$ 2) h=2R
3) h=$\frac{R}{2}$ 4) h=$(\sqrt{2}-1)$R

38. The acceleration due to gravity at the poles is $10ms^{-2}$ and equitorial radius is 6400km for the earth. Then the angular velocity of rotation of the earth about its axis so that the weight of a body at the equator reduces to 75% is

1) $\frac{1}{1600}$ $rads^{-1}$ 2) $\frac{1}{800}$ $rads^{-1}$
3) $\frac{1}{400}$ $rads^{-1}$ 4) $\frac{1}{200}$ $rads^{-1}$

Gravitational potential and Gravitational potential energy :

39. If 'g' is acceleration due to gravity on the earth's surface, then the gain in PE of an object of mass 'm' raised from the surface of the earth to a height equal to the radius R of the earth is

1) $\frac{mgR}{2}$ 2) 2mgR 3) mgR 4) $\frac{mgR}{4}$

40 The gravitational potential at centre of earth is

1) $-\frac{2GM}{R}$ 2) $-\frac{3GM}{2R}$ 3) $\frac{3GM}{2R}$ 4) $-\frac{GM}{2R}$

41. If the gravitational force has varied as $r^{-5/2}$ instead r^{-2}, then the P.E. of a particle at a distance 'r' from the earth would be propostional to

1) r^{-1} 2) r^{-2} 3) $r^{-3/2}$ 4) $r^{-5/2}$

42. Two particles each of mass 1kg are seperated by a distance of 1m in a gravitational field. The workdone to increase the seperation between them to 2m is

1) G 2) G/2 3) G/4 4) 2G

Satellites, orbital and escape velocities :

43. The orbiting time 'T' of a surface satellite is given by the relation

1) $2\pi\sqrt{\frac{R}{g}}$ 2) $2\pi\sqrt{\frac{g}{R}}$ 3) $2\pi\sqrt{2gR}$ 4) $2\pi\sqrt{\frac{2R}{g}}$

44. An earth satellite is revolving in a circular orbit of radius 7000km. If its period of revolution is 5500 sec, its orbital velocity is

1) $1kms^{-1}$ 2) $8kms^{-1}$ 3) $2kms^{-1}$ 4) $2.2kms^{-1}$

45. Thes orbital speed of satellite revolving very close to earth

1) $\sqrt{\frac{2GM}{R}}$ 2) $\sqrt{\frac{GM}{R}}$ 3) $\sqrt{\frac{GM}{7R}}$ 4) $\sqrt{\frac{6GM}{7R}}$

46. The escape velocity of a body depends upon its mass 'm' as

1) m^0 2) m^1 3) m^2 4) m^3

47. A satellite is revolving near the earth's surface. Its orbital velocity is nearly (98E)

1) $5.8kms^{-1}$ 2) $18.4kms^{-1}$
3) $11.4kms^{-1}$ 4) $8kms^{-1}$

48. An artificial satellite moving in a circular orbit around the earth has a total energy E_0 (PE+KE) its potential energy is

1) E_0 2) $3/2E_0$ 3) $2E_0$ 4) E_0

49. The K.E. of a satellite is 2MJ. Its total energy is

1) -0.5MJ 2) -2MJ 3) -1MJ 4) -4MJ

50. The escape velocity of a projectile from the earth's surface is given by the formula ($R \rightarrow$ radius of the earth and 'g' is acceleration due to gravity)

1) $\sqrt{2gR}$ 2) $\sqrt{gR}$ 3) $\sqrt{\frac{g}{R}}$ 4) $\sqrt{\frac{gR}{2}}$

51. The Ratio of the escape velocity to the orbital velocity of a surface satellite of the earth is

1) $\sqrt{2}$ 2) $\frac{1}{\sqrt{2}}$ 3) $\frac{1}{2}$ 4) 2

52. Two satellites are at heights h_1, h_2 The ratio of their orbital angular speeds is

1) $\left(\frac{h_1}{h_2}\right)^{3/2}$ 2) $\left(\frac{R+h_1}{R+h_2}\right)^{3/2}$

3) $\left(\frac{R+h_2}{R+h_1}\right)^{3/2}$ 4) $\left(\frac{h_2}{h_1}\right)^{3/2}$

53. The KE of a satellite in its orbit around the earth is E. The KE of the satellite so as to enable to escape from gravitational pull of the earth is

1) 4E 2) 2E 3) $\sqrt{2}$E 4) E

54. For a satellite to attain escape velocity, the percentage increase in its velocity should be

1) 50% 2) 100% 3) 41.4% 4) 1.414%

55. The escape velocity of a body of mass 1kg when projected from the surface of the earth vertically is $11.2kms^{-1}$. The escape velocity of a body of 1gm when projected at an angle 30^0 with the horizontal is

1) $\frac{11.2}{\sqrt{2}}$ kms-1 2) $11.2kms^{-1}$

3) $11.2kms^{-1}$ 4) $\frac{11.2}{2}$ m/s

56. The mass of earth is 9 times of that of Mars. The radius of the earth is twice that of Mars. The escape velocity on the earth is $12kms^{-1}$, then the escape velocity on the Mars is

1) $5.65kms^{-1}$ 2) $6kms^{-1}$ 3) $7kms^{-1}$ 4) $8kms^{-1}$

57. The escape velocities on the surface of two planets of masses m_1 and m_2 and having the radius r_1 and r_2 are v_1 and v_2 respectively, then

1) $\frac{v_1}{v_2} = \frac{m_1 r_1}{m_2 r_2}$ 2) $\frac{v_2}{v_1} = \frac{m_1}{m_2}\frac{r_1}{r_2}$

3) $\frac{v_2}{v_1} = \sqrt{\frac{m_2}{m_1}\frac{r_1}{r_2}}$ 4) $\frac{v_2}{v_1} = \sqrt{\frac{m_1 r_1}{m_2 r_2}}$

58. If the escape velocity on the earth's surface is $11.2kms^{-1}$, its value on a planet having double the mass and double the radius that of the earth is

1) $5.6kms^{-1}$ 2) $11.2kms^{-1}$
3) $22.4kms^{-1}$ 4) $11.2\sqrt{2}\ kms^{-1}$

59. The escape velocity on a planet is v. If the radius of the planet contracts to $\frac{1}{4}^{th}$ the present value without any change in its mass, then the escape velocity becomes

1) v 2) v/2 3) 2v 4) $\sqrt{2}$ v

Black hole :

60. A star having mass three times that of sun turns into black hole. The radius of that star is approximately (Mass of sun = 2×10^{30}kg)

1) 10km 2) 7km 3) 8km 4) 9km

61. **Degenerate electron pressure will not be sufficient to prevent core collapse of white dwarf if its mass becomes 'n' times of our solar mass, n value is equal to**

1) 0.5 2) 0.8 3) 1 4) 1.4

ANSWERS

1) 1	2) 4	3) 4	4) 2	5) 3
6) 4	7) 4	8) 4	9) 3	10) 3
11) 3	12) 3	13) 1	14) 3	15) 4
16) 4	17) 3	18) 2	19) 2	20) 2
21) 3	22) 1	23) 2	24) 1	25) 1
26) 1	27) 2	28) 3	29) 1	30) 1
31) 3	32) 4	33) 2	34) 2	35) 2
36) 1	37) 4	38) 1	39) 1	40) 2
41) 3	42) 2	43) 1	44) 2	45) 2
46) 1	47) 4	48) 3	49) 2	50) 1
51) 1	52) 3	53) 2	54) 3	55) 2
56) 1	57) 3	58) 2	59) 3	60) 4
61) 4				

EXERCISE III(A)

Universal of Gravitation and Gravtiational Intensity:

1. **A mass M is broken into two parts m and (M-m). What should be the ratio between m and M. So that the force of gravitation between them is maximum**

1) 2:1 2) 1:2 3) 1:1 4) 1:3

2. **Two metal spheres each of radius 'r' are kept in contact with each other. If d is the density of the material of the sphere, then the gravitational force between those spheres is propositional to**

1) d^2r^6 2) d^2r^4 3) $\frac{d^2}{r^4}$ 4) $\frac{r^4}{d^2}$

3. **In a double star system, two stars of masses m_1 and m_2 seperated by a distance d rotate about their centre of mass. Then their common angular velocity would be**

1) $\sqrt{\frac{Gm_1}{d^3}}$ 2) $\sqrt{\frac{Gm_2}{d^3}}$

3) $\sqrt{\frac{G(m_1+m_2)}{d^3}}$ 4) $\sqrt{\frac{d^3}{G(m_1+m_2)}}$

4. **Three bodies each of mass 'm' are placed at the three corners of a square of side 'a'. The gravitational force on unit mass kept at the fourth corner is**

1) $\frac{Gm}{3a^2}$ 2) $\frac{Gm}{\sqrt{3}a^2}$ 3) $\frac{3Gm}{a^2}$ 4) $\frac{Gm}{a^2}\left(\sqrt{2}+\frac{1}{2}\right)$

5. **Three particles each of mass 'm' are located at the three vertices of an equilateral tirangle of side 'a'. If the system is revolving in a uniform circle due to their mutual force of attraction, then the angular velocity of each particle is**

1) $\sqrt{\frac{Gm}{3a^3}}$ 2) $\sqrt{\frac{3Gm}{a^3}}$

3) $\sqrt{\frac{Gm^2}{3a^3}}$ 4) $\sqrt{\frac{Gm^2}{a^3}}$

6. **The gravitational field at some point in space is $\vec{g}=3\hat{i}+4\hat{j}$ N/kg. The force exerted on a 2kg mass placed at that point is**

1) 5N, 53^0 with x-axis 2) 5N, 37^0 with x-axis

3) 10N, 53^0 with x-axis 4) 10N, 37^0 with x-axis

Kepler's Laws :

7. **A satellite takes 1/8 years to move round the earth in its permissible orbit of radius R. The period when it revolves round the earth in an orbit of radius '2R' is**

1) $\frac{1}{2\sqrt{2}}$ years 2) $\frac{2\sqrt{2}}{1}$ years

3) $\frac{4}{1}$ years 4) $\frac{1}{4}$ years

8. **Imagine a light planet is revolving round a very massive star in a circular orbit of radius R with a time period of revolution T. If the gravitational force of attraction between the star and planetis propostional to R^{-n}, then T^2 is propostional to (AIEEE-2004)**

1) R^{n+1} 2) R^{n+2} 3) R^{n-1} 4) R^{n-2}

9. **A geostationary satellite is orbiting the earth at a height of 6R above the surface of the earth. The time period of another satellite at a height of 2.5R from the surface of the earth is**

1) 6hr 2) 8hr 3) 12hr 4) $6\sqrt{2}$ hr

10. **A geo-stationary satellite orbits around the earth in a circular orbit which is at a height of 36000km from the surface of earth. Then the period of spy satellite orbiting a few hundred km above the earth's surface (R_{earth}= 6400km) will approximately be**

1) $\frac{1}{2}$hr 2) $1\frac{1}{2}$hr 3) 2hr 4) 4hr

11. The minimum and maximum distance of a satellite from the centre of the earth 2R and 4R respectively. Where R is the radius of the earth and M is the mass of earth. Its maximum speed is

1) $\sqrt{\frac{GM}{6R}}$ 2) $\sqrt{\frac{2GM}{3R}}$

3) $\sqrt{\frac{GM}{3R}}$ 4) $\sqrt{\frac{2GM}{5R}}$

Relation between g and G :

12. The mass and diameter of a planet are two times those of earth. If a seconds pendulum is taken to it, the time period of the pendulum in seconds is (2002 M)

1) $\frac{1}{\sqrt{2}}$ 2) $\frac{1}{2}$ 3) $2\sqrt{2}$ 4) $\sqrt{2}$

13. Assume that the acceleration due to gravity on the surface of the moon is 0.2times the acceleration due to gravity on the surface of the earth. If R_e is the maximum range of a porjectile on the earth's surface then what is the maximum range on the surface of the moon for the same velocity of projection [Kelala PET-2001]

1) $0.2R_e$ 2) $2R_e$ 3) $0.5R_e$ 4) $5R_e$

Variation of 'g' with altitude, depth and speed of rotation of earth :

14. The period of a simple pendulum on the surface of earth is T. At an attitude of half of the radius of earth from the surface, its period will be

1) $\sqrt{\frac{3}{2}}T$ 2) $\frac{3T}{2}$ 3) $\frac{2T}{3}$ 4) $\sqrt{\frac{2}{3}}T$

15. The value of 'g' at a particle point on the surface of the earth is $9.8ms^{-2}$. Suppose the earth suddenly shrinks uniformly to half its present size without losing any mass, the value of g at the same point (assuming that the distance of the point from the centre of the earth does not shrink) will now be

1) $4.9ms^{-2}$ 2) $3.1ms^{-2}$ 3) $9.8ms^{-2}$ 4) $19.6ms^{-2}$

16. The difference in the value of 'g'at pole and at a lattitude in $\frac{3}{4}Rw^2$ then latitude angle is

1) 60^0 2) 30^0 3) 45^0 4) 95^0

17. If 'g' is acceleration due to gravity on the earth and R is radius of the earth and angular speed of rotation of the earth about its axis is made to be $\sqrt{\frac{2g}{5R}}$, then weight of a body at the equator decreases by

1) 60% 2) 50% 3) 40% 4) 75%

18. A body on the surface of the earth at latitude 60^0 will be under weight lessness state if the angular velocity of rotation of the earth becomes

1) $\sqrt{\frac{GM}{R^3}}$ 2) $\sqrt{\frac{2GM}{R^3}}$ 3) $\sqrt{\frac{GM}{2R^3}}$ 4) $\sqrt{\frac{4GM}{R^3}}$

19. Weight of a body of mass 'm' decreases by 1% when it is raised to a height 'h' above the earth's surface. If the body is taken to a depth 'h' in a mine, the change in its weight is

1) decreases by 2% 2) decreases by 0.5%

3) increases by 1% 4) increases by 0.5%

Gravitational Potential and Gravitational P.E.:

20. Four particle each of mass 1kg are at the four corners of a square of side 1m. The workdone to move one of the particles to infinity

1) $2\sqrt{2}G$ 2) $\left(\frac{2\sqrt{2}+1}{\sqrt{2}}\right)G$

3) $(2\sqrt{2}+1)G$ 4) 3G

21. The velocity with which a body should be projected from the surface of the earth such that it reaches a maximum height equal to n times the radius R of the earth is

1) $\sqrt{\frac{n}{n+1}\frac{GM}{R}}$ 2) $\sqrt{\frac{nGM}{R}}$

3) $\sqrt{\frac{2n}{n+1}\frac{GM}{R}}$ 4) $\sqrt{\frac{GM}{nR}}$

22. A body starts from rest from a point distance R_0 from the centre of the earth. The velocity acquired by the body when it reaches the surface of the earth will be

($R \rightarrow$ Radius of the earth) (R_0>R)

1) $2GM\left(\frac{1}{R}-\frac{1}{R_0}\right)$ 2) $\sqrt{2GM\left(\frac{1}{R}-\frac{1}{R_0}\right)}$

3) $GM\left(\frac{1}{R}-\frac{1}{R_0}\right)$ 4) $2GM\sqrt{\left(\frac{1}{R}-\frac{1}{R_0}\right)}$

23. The minimum speed with which a body must be thrown to reach a height $\frac{R}{4}$ above the surface of the earth is
(R$\rightarrow$Radius of the earth)

1) $\sqrt{\frac{gR}{2}}$ 2) $\sqrt{gR}$ 3) $\sqrt{\frac{2gR}{5}}$ 4) $\sqrt{\frac{gR}{5}}$

24. A satellite of mass 'm' revolves around the earth of radius R at a height x from its surface. If 'g' is the acceleration due to gravity on the surface of the earth, than the orbital speed of the satellite is (AIEEE 2004)

1) gx 2) $\left(\frac{gR^2}{R+x}\right)^{1/2}$ 3) $\frac{gR^2}{R+x}$ 4) $\frac{gR}{R-x}$

Satellites, orbital and escape velocities :

25. Two satellites of masses 100kg and 200kg are revolving the earth at attitudes 6400 km and 44800km. The ratio of orbital velocity of the satellites is

1) 1:2 2) 2:1 3) 1:4 4) 4:1

26. The energy required to shift a satellite from orbital radius 'r' to radius 2r is E. What energy will be required to shift the satellite from orbital radius 2r to 3r?

1) E 2) E/2 3) 3E/4 4) E/3

27. A satellite is moving in a circular orbit around the earth with a speed equalt to half the escape speed from the earth. If 'R' is the radius of the earth, then the height of the satellite above the surface of the earth is

1) R/2 2) 2R/3 3) R 4) 2R

28. Ratio of the radius of a planet A to that of planet B is 'r'. The ratio of acceleration due to gravity on the surface of the two planets is x. The ratio of the escape velocities from the two planets is

1) $\sqrt{rx}$ 2) $\sqrt{\frac{r}{x}}$ 3) $\sqrt{r}$ 4) $\sqrt{\frac{x}{r}}$

29. Two particles of same mass fall down on to the surface of the earth one from infinity and the other from an attitude 3R. The ratio of velocities on reaching the earth is

1) 2:3 2) 3:2 3) 2:$\sqrt{3}$ 4) $\sqrt{3}$:2

30. The escape velocity of a body on the earth's surface is V_e. A body is thrown vertically up with a speed of (kV_e) (k<1). The maximum height reached by the body above the earth is

1) $R\left(\frac{k^2}{1-k^2}\right)$ 2) $R^2\frac{k^2}{(1-k^2)}$

3) $R\left(\frac{1-k^2}{k}\right)$ 4) $\frac{R}{k^2}$

31. A body projected vertically upwards from the surface of the earth with a velocity equal to $\frac{1}{3}$ rd of the escape velocity. The height reached by it if the radius of the earth is 6400km is

1)3200km 2) 7200km 3) 6400km 4) 800km

32. A particle is projected vertically up with a velocity $\sqrt{gR}$. Where R is the radius of the earth, g is the acceleration due to gravity, then the maximum height reached is

1) R/2 2) R 3) 2R 4) 5R/4

33. The escape velocity of a body on the earth's surface is v_e. A body is thrown up with a speed of kv_e, where k>1, Assuming that the sun and planets do not influence the motion of the body, then the velocity of the body at the infinite distance in

1) $v_e/\sqrt{k^2-1}$ 2) $v_e\sqrt{k^2-1}$

3) $v_e(k^2-1)$ 4) $\frac{v_e}{k-1}$

34. A satellite moves around the earth in a circular orbit with speed V. If 'm' is the mass of the satellite, its total energy is

1) $-\frac{1}{2}mv^2$ 2) $\frac{1}{2}mv^2$ 3) $\frac{3}{2}mv^2$ 4) $-mv^2$

35. A body is projected with a velocity 2/3rd of escape velocity from the surface of the earth in order to orbit the earth. Its orbital radius will be (R_e is radius of the earth)

1) $\left(\frac{9}{8}\right)R_e$ 2) $2R_e$ 3) $\left(\frac{3}{2}\right)R_e$

4) It cannot be put into any orbit

ANSWERS

1) 2	2) 2	3) 3	4) 4	5) 2
6) 3	7) 1	8) 1	9) 4	10) 2
11) 2	12) 3	13) 4	14) 2	15) 3
16) 2	17) 3	18) 4	19) 2	20) 2
21) 3	22) 2	23) 3	24) 2	25) 2
26) 4	27) 3	28) 1	29) 3	30) 1
31) 4	32) 2	33) 2	34) 1	35) 4

EXERCISE - III(B)

Universal Law of Gravitation and Gravitational Intensity:

1. A mass M is broken into two parts m and (M-m) what should be the ratio between m and (M-m) so that the force of attraction between them is maximum

1) 2:1 2) 1:2 3) 1:1 4) 1:3

2. Two metal spheres each of radius 'r' and made of same material are seperated by a certain fixed distance or (d>2r) in a medium. The gravitational force between those spheres is proportional to

1) $\frac{1}{r^2}$ 2) $\frac{1}{r^4}$ 3) r^4 4) r^6

3. In double star system, two stars of masses m and 2m are separted by a distance d rotate about their centre of mass. Then their common angular velocity would be

1) $\sqrt{\frac{Gm}{d^3}}$ 2) $\sqrt{\frac{3Gm}{d^3}}$

3) $\sqrt{\frac{2Gm}{d^3}}$ 4) $\sqrt{\frac{Gm}{3d^3}}$

4. Three bodies each of mass 'm' are placed at the three corners of a square of side 'a'. The intensity of the gravitational field at the fourth corner is

1) $\frac{Gm}{3a^2}$ 2) $\frac{Gm}{\sqrt{3}a^2}$

3) $\frac{Gm}{a^2}$ 4) $\frac{Gm}{a^2}\left(\sqrt{2}+\frac{1}{2}\right)$

5. Four particles each of mass 'm' are located at the four corners of a square of side 'a'. If the system is revolving in a uniform circle due to their is mutual force if attraction, then the angular velocity of each particle is

1) $\sqrt{\frac{Gm\left(2\sqrt{2}+1\right)}{a^3}}$ 2) $\sqrt{\frac{Gm}{a^3}\left(\sqrt{2}+1\right)}$

3) $\sqrt{\frac{Gm}{a^3}\left(\sqrt{2}+\frac{1}{2}\right)}$ 4) $\sqrt{\frac{Gm}{a^3}\left(2+\frac{1}{\sqrt{2}}\right)}$

6. The gravitational field at some point in space is $\vec{g}=8\hat{i}+6\hat{j}$ N/kg. The force exerted on a 1kg mass placed at that point is

1) 10N, 53^0 with x-axis 2) 10N, 37^0 with x-axis

3) 20N, 53^0 with x-axis 4) 20N, 37^0 with x-axis

Kepler's Law s :

7. In an atom two electrons move round the nucleus in circular orbits of radii R and 4R. The ratio of their time periods is

1) 8:1 2) 1:8 3) 4:1 4) 1:4

8. Imagine a satellite revolving round a very massive planet in circular orbit of radius R with period T. If the gravitational force of attraction between the satellite and planet is inversely propotional to $R^{5/2}$, then T^2 is propostional to

1) R^3 2) $R^{7/2}$ 3) $R^{3/2}$ 4) $R^{5/2}$

9. The mean radius of the orbit of a satellite is 4 times as that of parking orbit of the earth. Then its period of revolution around the earth is

1) 96hr 2) 108hr 3) 144hr 4) 192hr

10. The time period of a satellite of earth is 5hr. If the seperation between earth and the satellite is increased to 4 times the previous value, then the new period will become (AIEEE 2003)

1) 10hr 2) 80hr 3) 40hr 4) 20hr

11. The minimum and maximum distances of a satellite from the centre of earth are 2R and 4R respectively. Where R is the radius of the earth and M is the mass of the earth, its minimum speed is

1) $\sqrt{\frac{GM}{6R}}$ 2) $\sqrt{\frac{2GM}{3R}}$ 3) $\sqrt{\frac{GM}{3R}}$ 4) $\sqrt{\frac{2GM}{5R}}$

Relation between g & G :

12. A body released from certain height reaches the surface of the earth after 't' seconds. If the same body is released from the same height on the moon, it reaches the surface after $\left(g_{moon} = {g_e}/{6}\right)$

1) $\frac{t}{6}S$ 2) $\frac{t}{\sqrt{6}}S$ 3) $\sqrt{6}$t S 4) 6t S

13. A person can jump safely from a height of 2m on the earth. On a planet where acceleration due to gravity is 2.45ms^{-2}, the maximum height from which he can jump safely

1) 8m 2) 4m 3) 16m 4) 64m

14. The period of oscillation of a simple pendulum on the surface of earth is T. At an altitude of $\left(\frac{1}{100}\right)^{th}$ radius of earth, its time period will be (h<<R)

1) $\frac{5\sqrt{2}}{7}T$ 2) $\frac{5}{7}T$

3) $\frac{2\sqrt{5}}{7}T$ 4) $\frac{5T}{7\sqrt{2}}$

15. The value of 'g' at a particular point on the surface of the earth is 9.8ms^{-2}. Suppose the earth suddenly shrinks uniformly to half its present size without losing mass, then the value of 'g' on the surface of new earth will be

1) 9.8ms^{-2} 2) 19.6ms^{-2}

3) 4.9ms^{-2} 4)39.2ms^{-2}

16 The diference in the value of ' g ' at poles and at a place lattitude 45^0 is

1) $R\omega^2$ 2) $\frac{R\omega^2}{2}$ 3) $\frac{R\omega^2}{4}$ 4) $\frac{R\omega^2}{3}$

17. The acceleration due to gravity at the latitude 45^0 on the earth becomes zero if the angular velocity of rotation of earth is equal to

1) $\sqrt{\frac{2}{gR}}$ 2) $\sqrt{2gR}$

3) $\sqrt{\frac{2g}{R}}$ 4) $\sqrt{\frac{5R}{2}}$

18. If 'R' is the radius of the earth and 'w' is angular velocity with which it rotates about its axis, the variation in the value of 'g' at 45^0 latitude of the earth stops its rotation will be

1) increases by $\frac{Rw^2}{2}$ 2) decreases by $\frac{Rw^2}{2}$

3) increases by $\frac{Rw^2}{4}$ 4)decreases by $\frac{Rw^2}{4}$

19. The radius of the earth is R. The depth of a point vertically below the earth's surface at which the acceleration due to gravity becomes 1% of its value at the surface is

1) 8R 2) 9R 3) $\frac{99R}{100}$ 4) 10R

Gravitational Potential and Gravitational PE :

20. If three particles each of mass m are placed at the three corners of an equilational triangle of side l, the workdone to increase the side of that triangle 2l is

1) $\frac{3Gm^2}{l}$ 2) $\frac{3Gm^2}{2l}$ 3) $\frac{Gm^2}{2l}$ 4) $\frac{Gm^2}{l}$

21. A body of mass 'm' is raised from the surface of earth to a point which is at a height nR from the surface of the earth. The change in PE is

1) nmgR 2) $\frac{2mgR}{n}$ 3) $\frac{n}{n+1}mgR$ 4) $\frac{mgR}{n}$

22. A body starts from rest from a point distance R_0 from the surface of the earth. The velocity acquired by the body when it reaches the surface of the earth will be

(R→ Radius of the earth) (R_0>R)

1) $\sqrt{\frac{GMR_0}{R+R_0}}$ 2) $\sqrt{\frac{GMR}{R+R_0}}$

3) $\sqrt{\frac{2GMR_0}{R(R+R_0)}}$ 4) $\sqrt{\frac{2GMR}{R+R_0}}$

23. A particle is projected from the surface of the earth with an initial speed of 4kms^{-1}. The maximum height attains by the particles

(R=6400km/g=9.8ms^{-2})

1) 935km 2) 1080km 3) 640km 4) 1560km

Satellites, orbital and escape velocities :

24. An earth satellite of mass 'm' revolves in a circular orbit at a height 'h' from the surface of the earth. If 'R' is the radius of the earth, then the velocity of the satellite on that orbit is given by

1) $\sqrt{\frac{gR^2}{R-h}}$ 2) $\sqrt{gR}$

3) $\sqrt{\frac{gR}{R+h}}$ 4) $\sqrt{\frac{gR^2}{R+h}}$

25. Two earth satellites of same mass are revolving round the earth in circular orbits of radii 6400km and 19, 200km from the centre of the earth. Ratio of their orbital velocities will be

1) $1:\sqrt{3}$ 2) $\sqrt{3}:1$ 3) 1:3 4) 3:1

26. Energy required to move a body of mass 'm' from orbit of radii 2R to 3R is (AIEEE 2002)

1) $\frac{GMm}{2R^?}$ 2) $\frac{GMm}{3R^?}$

3) $\frac{GMm}{8R}$ 4) $\frac{GMm}{6R}$

27. Two satellites of masses 50kg and 100kg revolve around the earth in circular orbits of radii 9R and 16R respectively, where R is the radius of the earth. The speeds of the two satellites will be in the ratio (99M)

1) 3/4 2) 4/3 3) $\frac{9}{16}$ 4) $\frac{16}{9}$

28. The ratio of the radii of planet A and B is K_1 and the ratio of acceleration due to gravity on them is K_2. The ratio of escape velocities from them will be

1) K_1K_2 2) $\sqrt{K_1K_2}$ 3) $\sqrt{\frac{K_1}{K_2}}$ 4) $\sqrt{\frac{K_2}{K_1}}$

29. A particle falls towards the earth from infinity. The velocity with which it reaches earth's surface (g$\rightarrow$ acceleration due to gravity on earth's surface and R is the radius of the earth)

1) 2gR 2) $\sqrt{2gR}$

3) $\sqrt{gR}$ 4) R/g

30. A body is projected up with a velocity equal to $\frac{3}{4}^{th}$ of the escape velocity from the surface of the earth. The height if reaches is

1) $\frac{10R}{9}$ 2) $\frac{9R}{7}$ 3) $\frac{9R}{8}$ 4) $\frac{10R}{3}$

31. A body is projected from earth's surface with a velocity equal to half the escape velocity of earth. The maximum height it will reach is (R$\rightarrow$ Radius of earth)

1) 3R 2) R/2 3) R/3 4) R

32. The velocity acquired by a body in falling to the surface of earth from a point at a height equal to the radius of the earth is approximately

1) $4kms^{-1}$ 2) $8kms^{-1}$

3) $4\sqrt{2}$ kms^{-1} 4) $8\sqrt{2}$ kms^{-1}

33. The escape velocity of a body on the earth's surface is v_e. A body is thrown up with a speed of $\sqrt{5}$ V_e. Assuming that the sun and planets do not influence the motion of the body, then the velocity of the body at the infinite distance is

1) 0 2) V_e 3) $\sqrt{2}V_e$ 4) $2V_e$

34. A satellite moves around the earth in a circular orbit with speed V. If 'm' is the mass of the satellite, its potential energy is

1) mv^2 2) mv^2 3) mv^2 4) $-mv^2$

35. An object is projected with a velocity $\sqrt{\frac{8gR}{3}}$ from earth. The velocity of the object, at the maximum height reached by it will be

1) $\sqrt{\frac{2gR}{3}}$ 2) $\sqrt{\frac{8R}{3}}$

3) $\sqrt{2gR}$ 4) zero

ANSWERS

1) 3	2) 4	3) 2	4) 4	5) 4
6) 2	7) 2	8) 2	9) 4	10) 3
11) 1	12) 3	13) 1	14) 1	15) 4
16) 2	17) 3	18) 1	19) 3	20) 2
21) 3	22) 3	23) 1	24) 4	25) 2
26) 4	27) 2	28) 2	29) 2	30) 2
31) 3	32) 2	33) 4	34) 4	35) 1

COMPREHENSION QUESTIONS

Comprehension - I (1-5) :

The basic forces in nature can be classified into four categories depending upon their nature and relative strength. They are Gravitational force, Electromagnetic force, Strong Nuclear force and weak Nuclear force.

Gravitational force is responsible for attraction between particles of various bodies having some mass. The Electromagnetic force is responsible for holding atoms together that make up the molecules. This strong Nuclear force is responsible for holding protons and neutrons in the nucleus. The weak nuclear force is responsibel for a type of radioactive decay known as beta decay.

1. **Mention the type of strong basic attrative force involved in between two protons in the nucleus is**

1) Gravitational 2) Electromagnetic
3) Weak Nuclear force 4) Strong nuclear force

2. **Among the basic forces, the weaker force is**

1) Gravitational 2) Electromagnetic
3) Weak Nuclear force 4) Strong nuclear force

3. **The type of basic force involved in electron revolving around the nucleus is**

1) Gravitational 2) Electromagnetic
3) Weak Nuclear force 4) Strong nuclear force

4. **The type of basic force involved in the process of beta decay is**

1) Gravitational 2) Electromagnetic
3) Weak Nuclear force
4) Strong nuclear force

5. **Among basic forces, the strongest force is**

1) Gravitational 2) Electromagnetic
3) Weak Nuclear force 4) Strong nuclear force

Comprehension - II (6-9) :

A frame of reference discribes the positon of an object in space.

The reference frame in which Newton's Laws hold good is called inertial referece frame. Ex: The reference frame at rest

The reference frame in which Newton's Laws do not hold good is called Non inertial referece frame. Ex: The linearly accelerating frame.

The property of a body that measures its resistance to acceleration is called inertial mass. The property of a body responsible for gravitational force it exerts on another body is called its gravitational mass.

The Principle of equivalence states that no experiment performed inside an accelerated, closed system can distinguish between the effects of gravitational field and the effect of accelerated motion.

6. **The reference frame that represents the motion of bus starting from rest and moves with acceleration is**

1) Inertial frame 2) Non inertial frame
3) Rotating frame 4) We can not explain

7. **According to stationary observer on the earth, the following that represents inertial frame of reference is**

1) Merry-go round 2) Gaint wheel
3) A vehicle moving with uniform velocity
4) A car starts from rest

8. **Statement -A : Principle of equivalence established that the inertial mass and gravitational mass are completely equivalent.**

Statement - B : Principle of equivalence is applicable to electromagnetic phenomena

1) only A is true 2) only B is true
3) both A and B are false 4) both A and B are true

9. **According to principle of equivalence, it is not possible to distinguish between**

1) Inertial and non inertial frames of reference
2) The gravitational and inertial mass
3) both 1 and 2 4) mass and weight

Comprehension - III (10-13) :

If the period of revolution of an artifical satellite is equal to be period of rotation of earth then such satellite is called Geo-stationary satellite. It appears stationary for an observer on the earth. These satellite have a variety of uses right from studying the atomosphere to the transmission of TV programmes.

10. **The height of geo-stationary satellite above the surface of earth is nearly**

1) 36000 km 2) 3600 km
3) 360 km 4) 36 km

11. The time period of revolution and the sense of revolution of a geo-stationary satellite respectively are

1) 365 days from west to east

2) 30 days from west to east

3) 24 hrs. from west to east

4) 24 hrs. from east to west

12. The orbital speed of geo-stationary satellite is about

1) 30 m/sec 2) 30 km/sec

3) 3 km/sec 4) 300 m/sec

13. Geo-stationary satellites are used for

1) Study the upper layers of atomosphere

2) Transmit the TV programmes to distant places

3) To know about the planets, satellites and comets etc.

4) All the above

Comprehension - IV (14-17) :

The minimum velocity required for an object to escape from gravitational influence of a planet is known as escape velocity V_e. The expression for escape velocity is obtained by equating the work required to move the body from surface of a planet to infinity with the initial kinetic energy given to the body. The escape velocity from a planet of mass M and radius R is given by $V_e = \sqrt{\frac{2GM}{R}} = \sqrt{2gr}$ where g is acceleration due to gravity on the surface of the planet and G is universal gravitational constant.

14. Choose the only incorret statement from the following, the escape velocity from a planet is

1) Independent of mass of the body which is projected

2) Independent of direction in which body is projected

3) Depends on mass and radius of planet

4) Depends on mass and direction of projection of body

15. The mass of Jupiter is above 319 times that of the earth and its radius is about 11 times that of the earth. The ratio of escape velocity on Jupiter to that of the earth is

1) $\sqrt{29}$ 2) 29 3) $\frac{1}{\sqrt{29}}$ 4) 2.9

16. If R is the radius of the earth and g is acceleration due to gravity on its surface, the escape velocity of a body projected from a satellite orbiting the earth at a height h=R from the surface of the earth will be

1) $\sqrt{gr}$ 2) $\sqrt{2gr}$ 3) $\sqrt{3gr}$ 4) $2\sqrt{gr}$

17. A body is dropped from a height at equal to half the radius of the earth. If V_e is the escape velocity on the earth and air resistance is neglected, it will strike the surface of the earth with a speed

1) $\frac{V_e}{\sqrt{2}}$ 2) $\frac{V_e}{2}$ 3) $\frac{V_e}{\sqrt{3}}$ 4) $\frac{V_e}{3}$

ANSWERS

1) 4	2) 1	3) 2	4) 3	5) 4
6) 2	7) 3	8) 4	9) 3	10) 1
11) 3	12) 3	13) 4	14) 4	15) 1
16) 1	17) 3			

Chapter - 7 - (PART - A)

WORK POWER ENERGY

COLLISIONS

SYNOPSIS

1. Collision is the interaction between two (or) more particles where exchange of momentum takes place.

2. In case of collisions as the impulsive force acting during collision is internal, the total momentum of system always remains conserved.

3. In a collision, "total energy" is also conserved.

4. If the velocities of the colliding particles are along the same line before and after the collision then the collision is said to be one dimensional collision.

5. In a collision, if the motion of colliding particles before and after the collision are not along the initial line of motion, then the collision is said to be oblique collision.

6. In elastic collision both the total momentum and K.E are conserved before and after the collision.

7. In inelastic collision only momentum is conserved.

8. If velocity of separation along the line of impact just after collision becomes zero then the collision is perfectly inelastic.

9. The line passing through the common normal to the surfaces in contact during impact is called line of impact. The force during collision acts along this line on both the bodies.

10. If u_1 and u_2 are the velocities of m_1 and m_2 before collision and v_1 and v_2 are their velocities after collision, then for elastic collision.

$$\frac{1}{2}m_1u_1^2+\frac{1}{2}m_2u_2^2=\frac{1}{2}m_1v_1^2+\frac{1}{2}m_2v_2^2$$

and $m_1\vec{u}_1+m_2\vec{u}_2=m_1\vec{v}_1+m_2\vec{v}_2$

11. In one dimensional elastic collision, the coefficient of restitution $e = 1$

12. In one dimensional elastic collision, relative velocity of separation after collision is equal to the relative velocity of approach before collision.

i.e $(v_2 - v_1) = (u_1 - u_2)$

13. In One dimensional elastic collison, formulas for final velocities.

$$V_1=\left(\frac{m_1-m_2}{m_1+m_2}\right)u_1+\left(\frac{2m_2}{m_1+m_2}\right)u_2$$

$$V_2=\left(\frac{2m_1}{m_1+m_2}\right)u_1+\left(\frac{m_2-m_1}{m_1+m_2}\right)u_2$$

14. In one dimensional elastic collision, if $m_1 = m_2 = m$ then $v_1 = u_2$ and $v_2 = u_1$ i.e., the particles mutually exchange their velocities.

15. In one dimensional elastic collision if m_2 is at rest before collision ($u_2 = 0$) then

$$V_1=\left(\frac{m_1-m_2}{m_1+m_2}\right)u_1 \text{ and}$$

$$V_2=\left(\frac{2m_1}{m_1+m_2}\right)u_1$$

16. If m_2 is at rest and kinetic energy of m_1 before collision with m_2 is E, then kinetic energies of m_1 and m_2 after collision are

$$K_1=\left(\frac{m_1-m_2}{m_1+m_2}\right)^2E,\ K_2=\frac{4m_1m_2E}{(m_1+m_2)^2}$$

17. In the above case, fraction of KE retained by m_1 is

$$\left(\frac{m_1-m_2}{m_1+m_2}\right)^2$$

Fraction of KE transferred by m_1 to m_2 (or) fraction of KE gained by m_2 is $\frac{4m_1m_2}{(m_1+m_2)^2}$

18. In the above case if the target particle is massive and at rest ($m_2 >> m_1$ & $u_2 = 0$) then m_1 rebounds with almost the same speed it had before collision, and the target particle continues to remain at rest.

$V_1 \approx -u_1$ and $V_2 \approx 0$

19. If target particle is very light ($m_1 >> m_2$) and initially at rest ($u_2 = 0$), then

$V_1 = u_1$ and $V_2 \approx 2u_1$

20. Transfer of K.E in a head - on elastic collision is maximum (100%) when $m_1 = m_2$ and m_2 is initially at rest.

21. **Coefficient of restitution**

$$e=\frac{\text{relative velocity of separation after collision along the line of impact}}{\text{relative velocity of approach before collision along the line of impact}}$$

$$e = \frac{v_2 - v_1}{u_1 - u_2}$$

u_1, u_2 are the **components** of the velocities of m_1 and m_2 before collision **along the line of impact** and v_1,v_2 are the **components** of the velocities of m_1 and m_2 after collision **along the line of impact.**

22. The value of coefficient of restitution (e) is independent of masses and velocities of the colliding bodies. It depends on their materials.
23. e has no units and no dimensions.
24. For a perfectly elastic collision, e = 1

    For a perfectly inelastic collision, e = 0

    For other collisions, e lies between 0 and 1
25. If a body falls from a height 'h' and strikes the level ground with velocity 'V' in time 't' seconds and rebounds with velocity V_1 upto a height h_1 in time t_1 seconds.

    The coefficient of restitution is given by

    $$e = \frac{v_1}{v} \quad \text{(or)} \quad e = \sqrt{\frac{H_1}{H}} \quad \text{(or)} \quad e = \frac{t_1}{t}$$

    For a perfectly elastic collision, $H_1 = H$

    For a perfectly inelastic collision, $H_1 = 0$

    For other collisions, $H_1 < H$

    For any collision, H_1 cannot be greater than h
26. A small metal sphere falls freely from a height 'H' upon a fixed horizontal plane. If e is the coefficient of restitution, then
    a) the height to which it rebounds after n collisions is $H_n = e^{2n} H$
    b) the velocity with which it rebounds from the ground after n[th] collision is $v_n = e^n v$. Where v is the velocity of the sphere just before first collision.
    c) the total distance travelled by it before it stops rebounding is

    $$d = H\left(\frac{1+e^2}{1-e^2}\right)$$

    d) the total time taken by it to come to rest is

    $$T = \sqrt{\frac{2H}{g}}\left(\frac{1+e}{1-e}\right)$$

    Note : In the above discussion the net displacement of the body = h

    In the above discussion average speed =

    $$\frac{\text{Total distance}}{\text{Total time}} = \frac{H\left[\frac{1+e^2}{1-e^2}\right]}{\sqrt{\frac{2H}{g}}\left[\frac{1+e}{1-e}\right]}$$

    In the above discussion average velocity =

    $$\frac{\text{Net displacement}}{\text{Total time}} = \frac{H}{\sqrt{\frac{2H}{g}}\left[\frac{1+e}{1-e}\right]}$$
27. In one dimensional semi elastic (or) inelastic collisions, Linear momentum is conserved but kinetic energy of the system is not conserved.
28. In above collisions, there is loss of kinetic energy from the system in the form of heat, sound, light etc.,
29. In one dimensional semi elastic collision relative velocity of seperation = e x realtive velocity of approach

    i.e., $v_2 - v_1 = e(u_1 - u_2)$
30. Formulas for final velocities in case of one dimensional semi elastic collision are

    $$v_1 = \left(\frac{m_1 - em_2}{m_1 + m_2}\right)u_1 + \left(\frac{1+e}{m_1 + m_2}\right)m_2 u_2$$

    $$v_2 = \left(\frac{m_2 - em_1}{m_1 + m_2}\right)u_2 + \left(\frac{1+e}{m_1 + m_2}\right)m_1 u_1$$

31) Loss in Kinetic energy of the system in one dimensional semi elastic collision is

    $$\Delta E_K = \frac{1}{2}\left[\frac{m_1 m_2}{m_1 + m_2}\right]\left[1 - e^2\right]\left[u_1 - u_2\right]^2$$

32. In one dimensional Perfectly inelastic collision the two particles stick together after the collision and move with common velocity.
33. For a perfectly inelastic collision, e = 0
34. The formula for common velocity of compound body after perfectly inelastic collision is

    $$\vec{v} = \frac{m_1 \vec{u_1} + m_2 \vec{u_2}}{m_1 + m_2}$$

35. The loss of kinetic energy in a perfectly inelastic collision (e = 0) is given by

$$\Delta E_K = \frac{1}{2}\left[\frac{m_1 m_2}{m_1 + m_2}\right](u_1 - u_2)^2$$

36. In a perfectly inelastic collision, the ratio of loss of energy of the system and its initial energy, if u_2 is zero, is given by

$$\frac{\Delta E_k}{E_i} = \frac{m_2}{m_1 + m_2}$$

37. In a perfectly inelastic collision, the ratio of final energy to initial energy of the system, if u_2 is zero, is given by

$$\frac{E_f}{E_i} = \frac{m_1}{m_1 + m_2} (\because m_1 < (m_1 + m_2) \Rightarrow E_f < E_i)$$

38. **Ballistic Pendulum**
A bullet of mass m is fired towards a freely suspended woorden block of mass M. Let the bullet gets embeded into the block and consequently the whole system (block + bullet) rises to a height h. According to law of conservation of momentum

$$P_{bullet} + P_{block} = P_{system}$$

$$mv_{bullet} + 0 = (M+m)\sqrt{2gh}$$

$$\therefore v_{bullet} = \frac{M+m}{m}\sqrt{2gh}$$

ii) If the bullet emerges out of the block with velocity 'v' then

$$mu = mv + MV$$

where $V = \sqrt{2gh}$

39. In an oblique elastic collision, if $m_1 = m_2$ and m_2 is initially at rest, then after the collision the two masses will move in directions inclined at 90° to each other.

EXERCISE - I

1. **Two spheres A and B collide. If the momentum of A is doubled in the collision then,**
1) The initial momenta of A and B are equal
2) The initial momentum of A is greater than that of B
3) The initial momentum of B is double that of A
4) The total momentum of A and B is constant

2. **Internal force can change**
1) Kinetic energy
2) total momentum of system
3) total energy of the system
4) All of the above

3. **In head on elastic collision of two bodies of equal mass**
1) the faster body speeds up further and the slower body slows down
2) the faster body slows down and the slower body speeds up
3) both of the above 4) none of the above

4. **Two bodies A and B of masses m_1 and m_2 moving with velocities V_1 and V_2 respectively collide. If $m_2 > m_1$ then the force exerted by B on A during the collision is**
1) greater than the force exerted by A on B
2) less than the force exerted by A on B
3) same as the force exerted by A on B
4) same as the force exerted by A on B but opposite in direction

5. **When two bodies of same mass undergo head on elastic collision,**
1) their velocities are interchanged
2) their speeds are interchanged
3) their momenta are interchanged
4) all the above are true

6. **A lead ball strikes a wall and falls down. A tennis ball having the same mass and velocity strikes the same wall and bounces back. Which of the following is correct**
1) the tennis ball suffers a greater change in momentum
2) the lead ball suffers greater change in momentum
3) both the balls suffer same change in momentum
4) the momentum of lead ball is greater than that of tennis ball

7. **A heavier sphere moving eastward with a certain velocity 'v' collides with a lighter sphere at rest. If it is perfect elastic head on collision, then after collision**
1) heavier sphere moves west ward with same speed
2) heavier sphere comes to rest
3) lighter sphere moves eastward with velocity 2v approximately
4) lighter sphere moves eastward with velocity approximately

8. **Two particles of equal masses have an elastic collision, the target particle being initially at rest. If it were not a head - on collision, the directions of their motion after the collision are**
1) in the same direction 2) in opposite directions
3) in mutually perpendicular directions
4) none of the above

9. A bullet hits a block at rest on a frictionless horizontal surface and gets embedded into it. The physical quantity associated with the block that remains unchanged is

1) linear momentum 2) kinetic energy
3) temperature
4) gravitational potential energy

10. Choose the correct statement

1) In an elastic collision, momentum is conserved but K.E. is not conserved
2) In an elastic collision, K.E. is conserved but momentum is not conserved
3) In an inelastic collision, both momentum and K.E. are conserved
4) In an inelastic collision, momentum is conserved but K.E. is not conserved

11. Choose the false statement

1) In a perfect elastic collision, the relative velocity of approach is equal to the relative velocity of separation
2) In an inelastic collision, the relative velocity of approach is less than the relative velocity of separation
3) In an inelastic collision, the relative velocity of separation is less than the relative velocity of approach
4) none of the above

12. When a ball hits a floor and rebounds after an inelastic collision,

1) the momentum of the ball just before collision is equal to its momentum just after collision
2) the total mechanical energy of the ball just before collision is equal to its total mechanical energy after collision
3) the total momentum of the ball and earth is conserved
4) all the above are ture

13. Coefficient of restitution depends upon

1) the relative velocities of approach and separation
2) the masses of the colliding bodies
3) the materials of the colliding bodies
4) all the above

14. If the coefficient of restitution is 0.5, the collision is

1) perfectly elastic collision
2) perfectly inelastic collision
3) semielastic collision 4) none of the above

15. Two particles of different masses collide head on. Then for the system

1) Loss of K.E. is zero, if it were perfect elastic collision.
2) If it were perfect inelastic collision, the loss of K.E. of the bodies moving in opposite directions is more than that of the bodies moving in the same direction
3) loss of momentum of the system is zero for both elastic and inelastic collisions
4) all the above are true

16. A particle moving with certain velocity collides elastically with another particle at rest. If it were head on collision, the K.E. transferred by the colliding particle is 100% when its mass is equal to

1) the mass of the stationary particle
2) twice the mass of the stationary particle
3) half the mass of the stationary particle
4) 1/4 times the mass of the stationary particle

17. Two bodies of equal masses moving mutually at right angles to each other, collide elastically. After the collision, the particles will move in

1) mutually perpendicular directions
2) any arbitiary directions
3) a direction at 30° from x-axis
4) a direction at 30° from y-axis

18. During the collision a relatively large force acts on each colliding particle for a relatively — time

1) Short 2) Moderately short
3) Large 4) Depends on a particular case

19. During elastic collision

1) Law of conservation of momentum is valid
2) Law of conservation of kinetic energy is valid
3) Both 1 and 2 4) Neither 1 and 2

20. During inelastic collision

1) Law of conservation of momentum is valid
2) Law of conservation of kinetic energy is valid
3) Both 1 and 2 4) Neither 1 and 2

21. Choose the wrong statement :

1) Collision between grossy bodies are always inelastic
2) The loss of K.E in elastic collision is converted into other forms of energy like heat (or) sound energy
3) During inelastic collision, law of conserv ation of momentum is valid
4) Collision between vehicles is inelastic collsion

22. The collision in which the bodies stick together after collision is

1) Perfectly elastic collision
2) Completely inelastic collision
3) Inelastic collision 4) Both 2 and 3

23. Which of the following is an example of completely inelastic collision

1) The collision between nuclei and fundamental particles
2) The collision between ivory (or) glass balls
3) The collision between grossy bodies
4) Collision between the bullet and target when the bullet remains embeded in the target

24. If the velocities of the objects involved in a collision before and after the collision are limited to any one of the co-ordinate axes, such a collision is known as

1) One dimensional collision
2) Two dimensional collision
3) Three dimensional collision 4) Oblique collision

25. The collision in which the relative velocity of approach before collision is equal to relative velocity of separation after collision is

1) Completely inelastic collision
2) Elastic collision 3) Inelastic collision
4) Both elastic and inelastic collision

26. During head on elastic collision, two bodies of masses m_1 and m_2 completely exchange their velocities when

1) $m_1 << m_2$ 2) $m_1 >> m_2$ 3) $m_1 = m_2$ 4) $m_1 \leq m_2$

27. A body collides head on and elastically with another identical body at rest then after collision

1) First body comes to rest, then second body begins to move with the initial velocity of first body
2) Both the bodies moves with same intial velocity of first body
3) First body reversed in direction and move with same speed but second body continues to be at rest
4) Both the bodies comes to rest after collision

28. A lighter body collides head - on and elastically with heavier body at rest then after collision

1) Lighter body comes to rest, then heavier body begins to move with initial velocity of lighter body
2) The velocity of lighter sphere is reversed and the final velocity of heavier sphere approximately remains zero
3) Both lighter and heavier bodies exchange their velocities
4) Both lighter and heavier bodies come to rest

29. In an inelastic collision between two bodies, the physical quantity that is conserved is[99M]

1) Kinetic energy 2) Momentum
3) Potential energy
4) Kinetic energy and momentum

Questions on more than one choice

30. A) When a body slides on rough surface, its momentum is not conserved.

B) When a ball falls from a height, momentum of earth ball system is conserved

C) In case of explosion of flying bomb, momentum is conserved.

1) A,B are true 2) A,B,C are true
3) A,C are false 4) A is false

31. A) When a ball falls on the earth, its momentum is not conserved.

B) When bullet is fired from a gun, gun moves in the direction of the bullet

C) When a stationary bomb explodes, linear momentum is conserved.

1) A and C are true 2) A & B are true
3) A,B,C are true 4) A,B,C are false

32. A) For an inelastic collision e= 1

B) For elastic collision e = 0

C) For partially elastic collision e= -1

1) A and C are true 2) A & B are true
3) A,B,C are true 4) A,B,C are false

33. A) Collision doesn't require physical contact

B) Collision between sub atomic particles is elastic

C) Collision between macroscopic bodies is generally inelastic.

1) A and C are true 2) A & B are true
3) A,B,C are true 4) A,B,C are false

34. A) For any collision, momentum transfer takes place.

R) In any collision, energy transfer takes place.

C) Impulsive forces are involved in collisions.

1) A and C are true 2) A & B are true
3) A,B,C are true 4) A,B,C are false

35. A) Momentum is conserved in oblique collisions

B) Rocket propulsion is based on conservation of momentum

C) When coefficient of restitution is one, then K.E is conserved.

1) A and C are true 2) A & B are true

3) A,B,C are true 4) A,B,C are false

36. A) When external forces are involved, momentum of the system is not conserved.

B) When stationary bomb explodes, energy released is converted to K.E of fragments

C) When rubber ball hits the wall, its kinetic energy is not conserved.

1) A and C are true 2) A & B are true

3) A,B,C are true 4) A,B,C are false

37. A) When a ball hits a wall (stationary) it returns back with same velocity, if collision is elastic.

B) A ball hits a bus moving towards it. If the collision is inelastic, they move together.

C) Inelastic collision arises due to loss in K.E during collision

1) A,B are true 2) A,C are false

3) A,B,C are true 4) A,B,C are false

38. A) When two spheres collide, momentum transfer takes place along normal to tangent drawn at contact surface

B) When a ball hits a wall at an angle 'θ' to horizontal its momentum along the wall remains constant

C) When a freely dropped ball bounces 'n' times before coming to rest, the total time elapsed is $t=\sqrt{\frac{2h}{g}}\left(\frac{1+e}{1-e}\right)$

1) A,B are true 2) A,C are false

3) A,B,C are true 4) A,B,C are false

39. During collision

a) If the velocities of colliding objects before and after the collision are limited to a straight line but may be opposite in their direction such collisions are known as head on collisions.

b) If the velocities before and after the collision are not along the same straight line, such collisions are known as oblique collisions.

1) a is ture b is false 2) a is false b is ture

3) Both a and b are true 4) Both a and b are false

40. For a system to follow the law of conservation of linear momentum during a collision, the condition is

a) total external force acting on the system is zero

b) total external force acting on the system is finite and time of collision is negligible

c) total internal force acting on the system is zero (2006 E)

1) (a) only 2) (b) only

3) (c) only 4) (a) or (b)

41. Consider the following statements A and B. Identify the correct choice in the given answers [EAMCET 2003 M]

A) In a one-dimensional perfectly elastic collision between two moving bodies of equal masses, the bodies exchange their velocities after the collision.

B) If a lighter body at rest suffers head - on and perfectly elastic collision with a very heavy body moving with a certain velocity, after the collision both travel with same velocity

1) A and B are correct

2) Both A and B are wrong

3) A is correct and B is wrong

4) A is wrong and B is correct.

42. Consider following statements A and B and identify the correct answer

A : Coefficient of restitution varies between '0' and '1'.

B : In inelastic collision, the law of conservation of energy is satisfied. (2005 M)

1) A and B are true 2) A and B are false

3) A is true but B is false

4) A is false but B is true

43. Consider the following statements 'A' and 'B' and identify the correct answer.

A: In an elastic collision, if a body suffers a head on collision with another of same mass at rest, the first body comes to rest while other starts moving with the velocity of first one.

B: Two bodies of equal masses suffering a head on elastic collision exchange their velocities. (2005 E)

1) Both A and B are true 2) Both A and B are false

3) A is true but B is false 4) A is false but B is true

Assertion & Reason

These Questions consist of two statements each printed as Assertion and Reason. While answering these questions you are required to choose any one of the following four responses.

1) Both (A) and (R) are true and (R) is the correct explanation of (A)
2) Both (A) and (R) are true and (R) is not the correct explanation of (A)
3) (A) is true but (R) is false
4) (A) is false but (R) is true

44. **A:** Collision between two particles is not necessarily associated with physical contact between them.
R: Only in physical contact, momentum transfer takes place.

45. **A:** In an elastic collision, K.E also is conserved along with momentum.
R: There is no deformation after elastic collision.

46. **A:** In an inelastic collision, K.E is not conserved.
R: For an inelastic collision, body permanently deforms

47. **A:** For a perfectly elastic collision, coeffcient of restitution is one
R: Relative velocity of seperation = relative velocity of approach, for an elastic collision

48. **A:** Collision between two fundamental particles is elastic
R: Fundamental particles lose their shape in collision.

49. **A:** When two bodies of equal masses collide, their velocities are always interchanged.
R: momentum is conserved in any collision

50. **A :** Law of conservation of momentum is the consequence of Newtons third law.
R: Action, Reaction pairs are internal forces in collisions.

MATCHING TYPE

51. Match the following

List - I	**List - II**
a) elastic collision	**e) e= 0**
b) Inelastic collision	**f) 0 < e < 1**
c) Explosion	**g) e = 1**
d) Plastic collision	**h)Final K.E. > Initial K.E.**

1) a-e, b-f,c-g,d-h 2) a-g,b-e,c-f,d-h
3) a-h,b-g,c-f,d-e 4) a-g, b-f, c-h, d-e

52. Match the following

List - I	**List - II**
a) Impulse	**e) elastic collision**
b) coefficient of restitution	**f) change in momentum**
c) conservation of momentum	**g) relative velocity of seperation/ relative velocity of approach**
d) conservation of kinetic energy	**h) internal forces**

1) a-h,b-g,c-f,d-e 2) a-f,b-g,c-h,d-e
3) a-e, b-f,c-g,d-h 4) a-e,b-g,c-f,d-h

53. Match the following

List - I	**List - II**
a) Impulse	**e) Area under f-t graph**
b) $\left[\frac{m_1 - m_2}{m_1 + m_2}\right]^2$	**f) elastic collision between similar bodies**
c) $\frac{4m_1m_2}{(m_1 + m_2)^2}$	**g) fraction of K.E retained in an elastic collision**
d) Interchange of velocities	**h) fraction of K.E transfer in an elastic collision**

1) a-h,b-g,c-f,d-e 2) a-f,b-g,c-h,d-e
3) a-e, b-f, c-h, d-g 4) a-e,b-g,c-h,d-f

54. Match the following.

List - I	**List - II**
a) Elastic collision	**e) Collision between grossy balls**
b) Inelastic collision	**f) Collision between ivory balls**
c) Completely inelastic	**g) Collision between bullet and its target**

1) a - f, b - g, c - e 2) a - e, b - f, c - g
3) a - e, b - g, c - f 4) a - f, b - e, c - g

55. A sphere of mass M_1, moving with a velocity v_0 collides head on with a stationary sphere of mass M_2. The collision is elastic. V_1 and V_2 are respectively their velocities immediately after collision.

Match the following

List - I	List - II
a) $M_1 = M_2$	e) $V_1 = -V_0, V_2 = 0$
b) $M_1 << M_2$	f) $V_1 = 0,\ V_2 = V_0$
c) $M_1 >> M_2$	g) $V_0 < V_2 < 2V_0$
d) $2M_1 > M_1 + M_2$	h) $V_1 = V_0, V_2 = 2V_0$

1) a-g, b-e, c-h, d-f 2) a-f, b-h, c-e, d-g
3) a-f, b-e, c-h, d-g 4) a-e, b-f, c-h, d-g

ANSWERS

1) 4	2) 1	3) 2	4) 4	5) 4
6) 1	7) 3	8) 3	9) 1	10) 4
11) 2	12) 3	13) 3	14) 3	15) 4
16) 1	17) 1	18) 1	19) 3	20) 1
21) 2	22) 2	23) 4	24) 1	25) 2
26) 3	27) 1	28) 2	29) 2	30) 1
31) 1	32) 4	33) 3	34) 1	35) 3
36) 3	37) 3	38) 3	39) 3	40) 1
41) 3	42) 1	43) 1	44) 3	45) 1
46) 1	47) 1	48) 3	49) 4	50) 1
51) 4	52) 2	53) 4	54) 4	55) 3

EXERCISE - IIA

Elastic collisions

1. A body of mass 6kg travelling with a velocity 10 m/s collides head - on and elastically with a body of mass 4kg travelling at a speed 5 m/s in opposite direction. The velocity of the second body after the collision is

1) 0 m/s. 2) 6 m/s 3) 8 m/s 4) 13 ms^{-1}

2. If a ball of mass 0.4 kg moving with a velocity of 3 ms^{-1} collides elastically with another ball of mass 0.6 kg which is at rest, find their velocities after collision.

1) – 0.6ms^{-1}, 2.4ms^{-1}
2) 2.4 ms^{-1}, –0.6 ms^{-1}
3) 0.6 ms^{-1}, –2.4 ms^{-1}
4) – 2.4 ms^{-1}, 0.6 ms^{-1}

3. A particle experiences a perfectly elastic collision with a stationary particle. Then the ratio of their masses, if after this head-on collision the particles fly apart in opposite directions with equal velocities.

1) 1 : 3 2) 3 : 1 3) 3 : 2 4) 2 : 3

4. A body of mass 1 kg makes an elastic collision with another body at rest and continues to move in the original direction after collision with a velocity equal to 1/5th of its original velocity. Find the mass of the second body.

1) $\frac{2}{3}$kg 2) $\frac{3}{2}$kg 3) $\frac{5}{2}$kg 4) $\frac{2}{5}$kg

5. A sphere of mass 0.3 kg moving with a velocity of 4 m/s collides with another sphere of mass 0.5 kg which is at rest. Assuming the collision to be elastic, their velocities after the impact are

1) 4 m/s and 0 ms^{-1} 2) -1 m/s and 3 m/s
3) 2 m/s and 2 m/s 4) 4 m/s and 8 ms^{-1}

6. A steel sphere of mass 100 gm moving with a velocity of 4 m/s collides with a dust particle elastically moving in the same direction with a velocity of 1 m/s. The velocity of the dust particle after the collision is

1) 8 m/s 2) 7 m/s 3) 6 m/s 4) 9 m/s

7. A neutron moving with a certain kinetic energy collides head on with an atom of mass number A. The fractional kinetic energy retained by it is

1) $\frac{A-1}{A+1}$ 2) $\left(\frac{A+1}{A-1}\right)^2$

3) $\frac{A+1}{A-1}$ 4) $\left(\frac{A-1}{A+1}\right)^2$

8. A perfectly elastic ball P_1 of mass m moving with velocity v collides elastically with three exactly similar balls P_2, P_3, P_4 lying on a smooth table as shown. Velocities of the four balls after the collision are

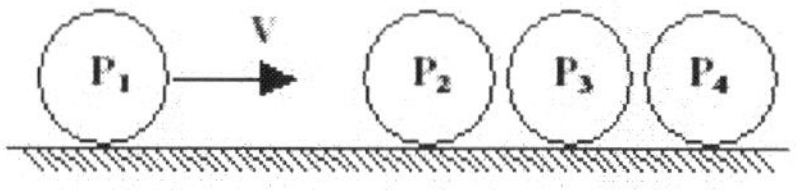

1) v, v, v, v, 2) 0, 2v, 3v, 4v
3) 0, 0, 0, v 4) 0, 0, 0, 0

9. Two spheres A and B moving in opposite directions have velocities of 10ms^{-1} and 20ms^{-1}. The two spheres collide with each other elastically. If A continues to move in the same direction at 4 ms^{-1}, the velocity of sphere B just after the collision is

1) 34 m/s in the same direction
2) 34 m/s in the opposite direction
3) 26 m/s in the same direction
4) 26 m/s in the opposite direction

10. A massive ball moving with speed v collides head on with a fine ball having mass very much smaller than the mass of the first ball. The collision is elastic. Then immediately after the impact, the second ball will move with a speed approximately equal to

1) v 2) 2v 3) v/3 4) Infinte

Coefficient of restitution

11. A ball of mass 0.2 kg is dropped down vertically from a height of 1m above the ground. If it rebounds to a height of 0.64m, find the coefficient of restitution between the surface of the ball and ground.

1) 0.32 2) 0.64 3) 0.16 4) 0.8

12. A ball falls from a height of 10m on to a horizontal plane. If the coefficient of restitution is 0.6, the height to which it rebounds after 3 collisions is approximately

1) 2.16 m 2) 0.47 m 3) 0.3 m 4) 0.6 m

13. A ball striking a horizontal plane with velocity v rebounds with velocity 256v/625 after 4 th collision. The coefficient of restitution is

1) 0.4 2) 0.6 3) 0.2 4) 0.8

14. A ball is dropped on to a floor. If the coefficient of restitution is 0.5, the percentage loss of energy of the ball on rebounding from the floor is

1) 12.5% 2)25% 3) 50% 4) 75%

15. In two separate collisions, the coefficients of restitutions 'e_1' and 'e_2' are in the ratio 3 : 1. In the first collision, the relative velocity of approach is twice the relative velocity of seperation. Then the ratio between the relative velocity of approach and relative velocity of seperation in the second collision is (2006 M)

1) 1 : 6 2) 2 : 3 3) 3 : 2 4) 6 : 1

16. A body is dropped from a height of 80m on to a horizontal floor and impings repeatedly. If it stops in 6 seconds, the total distance travelled by it before stops is (nearly) (g =10ms^{-2})

1) 86.6m 2) 95m 3) 100m 4) 110m

Inelastic collisions

17. A 1kg ball moving at 12 ms^{-1} collides head on with a 2kg ball moving in the opposite direction at 24 m/s. The velocity of each ball after the impact, if the coefficient of restitution is 2/3, is

1) -28m/s; -4 m/s 2) -20 m/s; -4 m/s

3) 20 m/s; 24 m/s 4) 28m/s, –4m/s

18. A ball impings directly upon another ball at rest and is itself brought to rest by the impact. If half of initial kinetic energy is destroyed in the collision, the coefficient of restitution is

1) 0.3 2) 0.4 3) 0.5 4) 0.6

19. A block of mass m moving with speed v collides with another block of mass 2m at rest. The lighter block comes to rest after the collision. The coefficient of restitution is

1) $\frac{1}{2}$ 2) $\frac{1}{3}$ 3) $\frac{1}{4}$ 4) $\frac{1}{5}$

20. A ball A moving with a certain velocity collides, with another ball B of the same mass at rest. If the coefficient of restitution is e, the ratio of the velocities of A and B just after the collision is

1) $\frac{1+e}{1-e}$ 2) $\frac{1+e}{2}$ 3) $\frac{1-e}{2}$ 4) $\frac{1-e}{1+e}$

Perfectly inelastic collisions

21. A neutron of mass 1.67 x 10^{-27} kg moving with a velocity 10^6 ms^{-1} collides with a deuteron of mass 3.34 x 10^{-27} kg at rest. After collision, if both move as a single particle find its velocity.

1) 3.19 x 10^5 m/s 2) 1.25 x 10^5 m/s

3) 1.39 x 10^5 m/s 4) 2.15 x 10^5 m/s

22. A 2 kg mass moving with a velocity of 10 ms^{-1} collides with another 6 kg mass moving in opposite direction with a velocity of 20 ms^{-1}. During the collision, they stick together. Find their common velocity and the momentum.

1) 12.5 m/s,100 kgms^{-1}

2) 12.5 m/s,50 kgms^{-1}

3) 7.5 m/s,100 kgms^{-1}

4) 7.5 m/s,50 kgms^{-1}

23. A 60kg box sledge is travelling horizontally on ice at a speed of 10 ms^{-1}. A 90 kg packet is dropped on it vertically. The subsequent velocity of the sledge is

1) 1ms^{-1} 2) 2 ms^{-1} 3) 3 ms^{-1} 4) 4 ms^{-1}

24. A 6kg mass collides with a body at rest. After the collision, they travel together with a velocity one third the velocity of 6 kg mass. The mass of the second body is

1) 6 kg 2) 3 kg 3) 12 kg 4) 18 kg

25. A particle of mass m moving with velocity 2v collides with another particle of mass 3m moving with velocity v in the same direction. If it is perfect inelastic collision, the loss of K.E. of the system is

1) $\frac{3}{8}$ mv^2 2) $\frac{27}{8}$mv^2 3) $\frac{3}{4}$ mv^2 4) zero

26. A body of mass 100 gm moving with an initial velocity of 4ms^{-1} collides with another body of mass 150 gm moving in opposite direction with velocity 6 ms^{-1}. If the collision is perfectly inelastic, then after the collision,

1) the two bodies inter change their velocities
2) the two bodies combine and move with a common velocity of 2ms^{-1} in the initial direction of lighter body
3) the two bodies combine and move with a common velocity of 5.2 ms^{-1} in the initial direction of heavier body
4) the two bodies combine and move with a common velocity of 2ms^{-1} in the initial direction of heavier body

27. Two spheres each of mass 'm', each moving with velocity u moving in directions at 30° S of E and 30° N of E respectively collide at point O and stick to each other. The velocity of the combination later becomes

1) $\frac{\sqrt{3}u}{2}$ due east 2) $\sqrt{3}u$ due east

3) $\frac{2u}{\sqrt{3}}$ due north 4) $\frac{\sqrt{3}u}{2}$ due west

28. A bullet of mass 2 g travelling with a velocity of 500 ms^{-1} is fired into a block of wood of mass 1 kg suspended from a string of 1m length. With the bullet penetrates the block of wood and comes out with a velocity of 100 ms^{-1}. Find the vertical height through which the block of wood will rise (assuming the value of g to be 10 ms^{-2})

1) 0.032 m 2) 0.32m 3) 3.2 m 4) 0.0032m

29. A ball of mass 0.4 kg moving with a uniform speed of 2 ms^{-1} strikes normally a wall and rebounds. Assuming the collision to be elastic and the time of contact of the ball with the wall as 0.4s, find the force exerted on the ball.

1) 1 N 2) 2 N 3) 3 N 4) 4 N

30. A ^{238}U nucleus emits an $\alpha-$ particle and is converted into ^{234}Th nucleus. If velocity of the $\alpha-$ particle is 1.4 x 10^7 ms^{-1} calculate the velocity of recoil of remaining ^{234}Th nucleus.

1) 2.393 x 10^5 m/s 2) 3.293 x 10^5 m/s
3) 1.393 x 10^5 m/s 4) 4.293 x 10^5 m/s

ANSWERS

1) 4	2) 1	3) 1	4) 1	5) 2
6) 2	7) 4	8) 3	9) 1	10) 2
11) 4	12) 2	13) 4	14) 4	15) 4
16) 1	17) 1	18) 3	19) 1	20) 4
21) 1	22) 1	23) 4	24) 3	25) 1
26) 4	27) 1	28) 1	29) 4	30) 1

EXERCISE - IIB

Elastic collisions

1. A particle of mass m moving with velocity u collides elastically with a particle of mass 3 m at rest. After collision the heavier particle moves with velocity

1) 3 u 2) $\frac{u}{3}$ 3) $\frac{u}{2}$ 4) 2u

2. In the above problem the velocity of the colliding particle after collision is

1) $\frac{u}{2}$ 2) $\frac{-u}{2}$ 3) 2u 4) $\frac{-3u}{2}$

3. Two bodies of masses m_1 and m_2 are moving with velocities 1 ms^{-1} and 3 ms^{-1} respectively in opposite directions. If the bodies undergo one dimensional elastic collision, the body of mass m_1 comes to rest. Find the ratio of m_1 and m_2.

1) 7 : 1 2) 1 : 7 3) 1 : 3 4) 3 : 1

4. A sphere of mas 2 kg moving with a velocity collides elastically with another sphere which is at rest. If after the collision the first sphere moves in the same direction with its velocity reduced to one third of its initial velocity, find the mass of the second sphere.

1) 1kg 2) 0.5 kg 3) 0.25kg 4) 0.75kg

5. Two perfectly elastic spheres of masses 4kg and 6kg moving in opposite directions with velocities of 4 m/s and 6 m/s collide with each other. Their velocities after head on collision are

1) 0.4 m/s and -3.6 m/s 2) 0.4 m/s and 0.8 m/s
3) 0.5 m/s nd 1 m/s 4) -8 m/s and 2 m/s

6. A rubber ball dropped from a height of 20m ($g=10m/s^2$) falls on the roof of an elevator going up at 2 m/s. If the collision of the ball with the elevator is elastic, the ball rebounds with a velocity of

1) 20m/s 2) 22m/s 3) 24m/s 4) 26m/s

7. A particle of mass m moving with certain velocity collides elastically head on with a particle of mass 4m at rest. The percentage of K.E. transferred is

1) 75% 2) 25% 3) 64% 4) 32%

8. A body P moving with a velocity of $20Kms^{-1}$ collides with another body of Q of the same mass at rest. After collison the velocity of Q if P comes to rest after collision is [EAMCET 96 M]

1) 10 Kms^{-1} 2) 20 Kms^{-1}

3) 30 Kms^{-1} 4) 5 Kms^{-1}

9. A ball of mass 'm' approaches a wall of mass 'M'(>>m) with speed 4 m/s along the normal to the wall. The speed of wall is 1 m/s towards the ball. The speed of the ball after an elastic collision with the wall is

1) 5 m/s away from the wall

2) 9 m/s away from the wall

3) 6 m/s away from the wall

4)3 m/s away from the wall

10. A heavy steel ball of mass greater than 1kg moving with a speed of 2 m/s collides head on with a stationary ping-pong ball of mass less than 0.1gm. The collision is elastic. after the collision the ping pong ball moves approximately with speed of

1) 2 m/s 2) 4 m/s 3) 2×10^4 m/s 4) 2×10^3 m/s

Coefficient of restitution

11. A ball strikes the floor vertically and rebounds with a co–efficient of restitution 0.6. If the mass of the ball is 0.01 kg and its velocity is 5 ms^{-1} just before it hits the ground what is the velocity and momentum of the ball as it leaves the floor?

1) 3m/s, 0.03 $kgms^{-1}$ 2) 6 m/s, 0.06 $kgms^{-1}$

3) 2 m/s, 0.15 $kgms^{-1}$ 4) 8 m/s, 0.18 $kgms^{-1}$

12. A ball hits the ground and loses 20% of its momentum. Coefficient of restitution is

1) 0.2 2) 0.4 3) 0.6 4) 0.8

13. A ball falls from a height of 10 m on to a horizontal plane. If the coefficient of restitution is 0.4, the velocity with which it rebounds from the plane after second collision is

1) 2.24 ms^{-1} 2) 5.6 ms^{-1}

3) 2.8 ms^{-1} 4) 0.9 ms^{-1}

14. A tennis ball bounces down a flight of stairs striking each step in turn and rebounding to the height of the step above. The coefficient of restitution is :

1) 1/2 2) $1/\sqrt{2}$ 3) 1/4 4) 1

15. Sphere B approaches relative to sphere A with a velocity 10 m/s due east. If they collide elastically, speed of B relative to A is

1) 10 m/s due east 2) 5 m/s due east

3) 10 m/s due west 4) 20 m/s due east

16. A particle falls from a height 'h' upon a fixed horizontal plane and rebounds. If 'e' is the coefficient of restitution, the total distance travelled before rebounding has stopped

1) $h\left(\frac{1+e^2}{1-e^2}\right)$ 2) $h\left(\frac{1-e^2}{1+e^2}\right)$ [2001 E]

3) $\frac{h}{2}\left(\frac{1-e^2}{1+e^2}\right)$ 4) $\frac{h}{2}\left(\frac{1+e^2}{1-e^2}\right)$

Inelastic collisions

17. A 2kg ball moving at $24ms^{-1}$ undergoes inelastic head - on collision with a 4kg ball moving in the opposite direction at $48ms^{-1}$. If the coefficient of restitution is 2/3, their velocities in ms^{-1} after impact are (2004E)

1) –56,–8 2) -28,–4 3) –14,–2 4) –7, –1

18. A 1kg ball moving at $12ms^{-1}$ collides headon with a 2kg ball moving in the opposite direction at 24 m/s. If the coefficient of restitution is 2/3, then the energy lost in the collision is

1) 60J 2) 120 J 3) 240 J 4) 480 J

19. A particle A approach relative to a particle B with a velocity of 20m/s. If they collide head on and the coefficient of restitution is 1/4, the velocity with which A moves away from B after impact is

1) 10m/s 2) 15 m/s 3) 5 m/s 4) 2.5m/s

20. A ball A moving with certain velocity collides with another body B of the same mass at rest. If the coefficient of restitution is 2/3. The ratio of velocity of A and B after collision is

1) 5 :1 2) 1 : 5 3) 2 : 5 4) 3 : 5

Perfectly inelastic collisions

21. A sphere of mass 50×10^{-3} kg moving with a velocity of 2 ms^{-1} hits another sphere which is at rest. Assuming the collision to be head–on collision and if they stick together after collision and move in the same direction with a velocity of 0.5 ms^{-1}, find the mass of the second sphere.
1) 150×10^{-3}kg 2) 50×10^{-3} kg
3) 100×10^{-3} kg 4) 200×10^{-3} kg

22. Two balls each of mass 0.06 kg moving in opposite directions each with a velocity of 8 ms^{-1} collide and move back with equal velocities. Then the change in momentum of each ball due to collision.
1) 0.96 $kgms^{-1}$ 2) 0.48 $kgms^{-1}$
3) 1.96 $kgms^{-1}$ 4) 0.24 $kgms^{-1}$

23. A shot moving with a velocity 140 ms^{-1} collides a wooden block and comes to rest in it. If mass of the block is 13 times the mass of the shot, velocity of the block is
1) 14ms^{-1} 2) 10 ms^{-1} 3) 4 ms^{-1} 4) 2.5ms^{-1}

24. Two particles each of mass m, collide head on when their speeds are 2u and u. If they stick to each other on impact, when moving in opposite directions, their combined speed is
1) u 2) 2u 3) u/2 4) 3u/2

25. A body of mass 10 gm moving with velocity 5ms^{-1} collides with another body of mass 15 gm moving with 3ms^{-1} in the same direction. After collision if they stick and move together, the loss of kinetic energy of the system is
1) 12 J 2) 12×10^{-3} J
3) 24×10^{-3} J 4) 192×10^{-3} J

26. 10 bullets each of mass 10 gm are fired in succession into a block of mass 450 gm at rest. If velocity of each bullet is 110 ms^{-1} and all the bullets are embeded in the block, the velocity of the block is
1) 5 ms^{-1} 2) 10 ms^{-1} 3) 15 ms^{-1} 4) 20 ms^{-1}

27. A truck of mass 1200 kg is moving with a speed of 7 ms^{-1} when it collides with a second truck of mass 1600kg which is stationary. If the two trucks are automatically coupled together on impact, the speed with which combination moves is
1) 2 m/s 2) 3 m/s 3) 1.5 m/s 4) 6 m/s

28. A bomb at rest explodes into two pieces of masses 1 kg and 2 kg. If the velocity of the lighter piece is 200 ms^{-1}, find the velocity of the heavier piece.
1) 100 ms^{-1} 2) 200 ms^{-1} 3) 300 ms^{-1} 4) 400 ms^{-1}

29. A bullet of mass 5 g moving with a velocity of 300 ms^{-1} strikes a plank of wood of mass 1.995 kg suspended by a thread and gets embedded in it. Find the velocity of the plank with bullet immediately after collision and vertical height through which the combined mass rises.
1) 0.75m/s, 0.028 m 2) 0.5 m/s, 0.014m
3) 1 m/s, 0.056 m 4) 0.25 m/s. 0.082m

30. A sphere of mass 3 kg moving with a velocity of 2 ms^{-1} collides head–on with another sphere of mass 4 kg moving in opposite direction with a velocity of 1 ms^{-1}. After collision if they stick together and move with a velocity of v then the total initial momentum and velocity v are
1) 2$kgms^{-1}$, $\frac{2}{7}$m/s 2) 1 $kgms^{-1}$, $\frac{1}{7}$m/s
3) 3 $kgms^{-1}$, $\frac{3}{7}$m/s 4) 4 $kgms^{-1}$, $\frac{4}{7}$m/s

ANSWERS

1) 3	2) 2	3) 1	4) 4	5) 4
6) 3	7) 3	8) 2	9) 3	10) 2
11) 1	12) 4	13) 1	14) 2	15) 3
16) 1	17) 1	18) 3	19) 3	20) 2
21) 1	22) 1	23) 2	24) 3	25) 2
26) 4	27) 2	28) 1	29) 1	30) 1

EXERCISE - IIIA

1. Two steel spheres approach each other headon with the same speed and collide elastically. After the collision one of the sphere's of radius r comes to rest. the radius of the other sphere is
1) $\frac{r}{(3)^{1/3}}$ 2) $\frac{r}{3}$ 3) $\frac{r}{9}$ 4) $(3)^{1/2}$ r

2. A body of mass 50 gm collides elastically with another body of mass 30 gm at rest. Then the percentage loss of the velocity of the colliding body during collision is
1) 25% 2) 75% 3) 50 % 4) 67 %

3. A particle of mass 5 kg strikes a body of mass 2.5 kg at rest elastically. The 2.5 kg mass after collision moves with a kinetic energy of 5J. The Kinetic energy of 5 kg mass before collision is
1) 0.62 J 2) 5.625 J 3) 7.68J 4) 4.20 J

4. **A fast moving neutron collides elastically and head on with a nucleus of N_7^{14}. The percentage of energy lost by the neutron during elastic collision is nearly**

1) 20% 2) 25% 3) 50% 4) 80%

5. **Three balls A,B and C are placed on a smooth horizontal surface. Given that $m_A=m_C=4m_B$. Ball B collides with ball A with an initial velocity V as shown in figure. Find the total number of collisions between the balls. All collisions are elastic.**

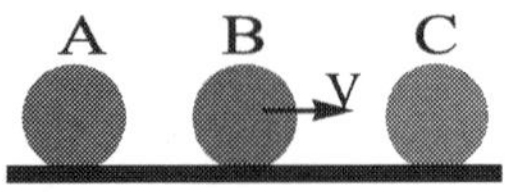

1) 1 2) 2 3) 3 4) infinity

6. **A particle moving with kinetic energy of 3 J makes a head – on collision with a stationary particle which has twice the mass. During the impact,**

a) the minimum kinetic energy of the system is IJ

b) the maximum elastic potential energy of the system is 2 J

c) momentum and total energy are conserved at every instant.

d) the ratio of kinetic energy to potential energy of the system first decreases and then increases.

1) only a, b are true 2) onlya, b, c are true

3) a, b, c, d are false 4) a,b,c,d are true.

7. **Five spheres of masses 1, 2, 3, 4, 5 kg are moving along a straight line in the same direction with velocities 5, 4, 3, 2, 1 m/s respectively. The first sphere collides with the second and both stick together. The compound mass collides with the third sphere and so on. the velocity of the compound mass when all stick together is**

1) 3/7 m/s 2) 7/3 m/s 3) 1 m/s 4) 4/3m/s

8. **A simple pendulum of length 0.2m has bob of mass 5gm, it is pulled aside through an angle 60^0 from the vertical. A spherical body of mass 2.5 gm is placed at the lowest position of the bob. When the bob is released it strikes the spherical body and comes to rest. What is the velocity of the spherical body ? ($g = 9.8\ ms^{-2}$) (in m/s)**

1) 1.4 2) 2.8 3) 3.5 4) 4.9

9. **A particle traveling horizontally with speed u collides and sticks with a particle of equal mass hanging at rest at the end of a light inextensible string of length $2l$. If the string rotates through an angle of 60° before its velocity becomes zero, then u is**

1) $\sqrt{4gl}$ 2) $\sqrt{8gl}$ 3) $\sqrt{2gl}$ 4) $\sqrt{gl}$

10. **A ball of mass 100gm is projected vertically upward from the ground with a velocity of $50ms^{-1}$. At the same time another identical ball is dropped from a height of 100m to fall freely along the same path as that followed by the first ball. After some time the two balls collide, stick together and finally fall to the ground. The time taken by the combined mass to fall to the ground is approximately ($g = 10ms^{-2}$)**

1) 4.5 s 2) 6.5 s 3) 9 s 4) 13 s

11. **A body 'x' with a momentum 'p' collides with another identical stationary body 'y' one dimensionally. During the collision 'y' gives an impulse 'J' to the body 'x'. Then the coefficient of restitution is (2004 M)**

1) $\frac{2J}{p}-1$ 2) $\frac{J}{p}+1$

3) $\frac{J}{p}-1$ 4) $\frac{J}{2p}-1$

12. **A moving sphere P collides another sphere Q at rest. If the collision takes place along the line joining their centres of mass such that their total kinetic energy is conserved and the fraction of K.E transferred by the colliding particle is 8/9 , the ratio of mass of P and the mass of Q is**

1) $\sqrt{8}:3$ 2) 9 : 8 3) 2: 3 4) 1: 2

13. **Two bodies of masses m_1 and m_2 moving on the same direction with velociities u_1 and u_2 collide. The velocities after collision are V_1 and V_2. If each sphere loses the same amount of kinetic energy, then**

1) $u_1 + u_2 + V_1 - V_2 = 0$ 2) $u_1 - u_2 + V_1 + V_2 = 0$

3) $u_1 + u_2 + V_1 + V_2 = 0$ 4) $u_1 + u_2 - V_1 - V_2 = 0$

14. **A ball is dropped onto a fixed horizontal surface from a height h. The coefficient of restitution is e. The average speed of the ball is**

1) $\left[\frac{1+e}{1-e}\right]^2\sqrt{\frac{gh}{2}}$ 2) $\left[\frac{(1+e)^2}{1-e^2}\right]\sqrt{\frac{gh}{2}}$

3) $\left[\frac{1+e^2}{(1+e)^2}\right]\sqrt{\frac{gh}{2}}$ 4) $\left[\frac{(1-e)^2}{1-e^2}\right]\sqrt{\frac{gh}{2}}$

15. **A body of mass 'm' falls from a height 'h' and rebounds. If 'e' is the coefficient of restitution between the ground and the body, the change in linear momentum of the body is**

1) $(2me)\sqrt{2gh}$ 2) $(me)\sqrt{2gh}$
3) $m(1+e)\sqrt{2gh}$ 4) $m(1-e)\sqrt{2gh}$

16. **If an ivory ball of mass 2m moving with a velocity 'v' strikes head on to a close row of three identical ivory balls each of mass m as shown in the figure. Then after the colisions which may be assumed as elastic, which of the following would occur?**

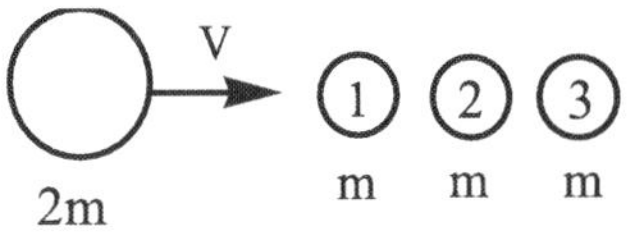

1) The balls numbered 2 and 3 will move with velocity v, the others staying at rest.
2) The ball numbered 3 will move with velocity 2v, all others staying at rest.
3) The ball numbered 3 moves with velocity 4v/3 and the striking ball moves with velocity v/3
4) The balls numbered 1, 2 and 3 move with velocity v/3 and the striking ball stops dead.

17. **'n' smooth spheres have masses 'm', 'm/e', 'm/e^2', 'm/e^3', etc.... are placed in a row with all their centres on a straight line. Initially all the spheeres are at rest. The first sphere of mass 'm' is projected with a velocity 'u' towards the second. The velocity of the last sphere after the $(n-1)^{th}$ sphere has made a collision with it is (take 'e' as the coefficient of restitution for all the collision)**

1) $e^n u$ 2) $e^{n-1}u$ 3) $e^{n+1}u$ 4) $e^{n/2}u$

18. **Two identical balls are dropped from the same height onto a hard surface, the second ball being released exactly when the first ball collides with the surface. If the first ball has made two more collisions by the time the second one collides. Then the coefficient of restitution between the ball and the surface satisfies.**

1) $e > 0.5$ 2) $e = 0.5$
3) $e = \frac{\sqrt{3}-1}{2}$ 4) $e < \frac{\sqrt{3}-1}{2}$

19. **An elastic ball is dropped from a height h and it rebounds many times from the floor. If the coefficient of restitution is e, the time interval between the second and the third impact is**

1) $\frac{ev}{g}$ 2) $e^2\sqrt{\frac{8h}{g}}$ 3) $e^2\sqrt{\frac{h}{g}}$ 4) $e^2\sqrt{\frac{2h}{g}}$

ANSWERS

1) 1	2) 2	3) 2	4) 2	5) 2
6) 4	7) 2	8) 2	9) 2	10) 1
11) 1	12) 4	13) 3	14) 3	15) 3
16) 3	17) 2	18) 4	19) 2	

EXERCISE - IIIB

1. **A solid sphere is moving and it makes an elastic collision with another stationary sphere of half of its own radius. After collision it comes to rest. The ratio of the densities of the materials of second sphere and first sphere is**

1) 2 2) 4 3) 8 4) 16

2. **A body of mass m moving with velocity u collides head on elastically with another body of mass 2m at rest. The ratio of kinetic energy of the colliding body before and after collision is**

1) 4:1 2) 9:1 3) 2:3 4) 5:4

3. **A moving particle of mass 'm' makes head-on elastic collision with a particle of mass '2m' which is initialy at rest. The fraction of K.E. lost by colliding particle is**

1) 1/3 2) 1/9 3) 8/9 4) 2/3

4. **A ball of mass 'm', moving with uniform speed, collides elastically with another stationary ball. The incident ball will lose maximum kinetic energy when the mass of the stationary ball is**

(ROORKEE-96)

1) m 2) 2m 3) 4m 4) infinity

5. **Two perfectly elastic balls of same mass m are movkng with velocities u_1 and u_2. They collide elastically n times. The kinetic energy of the system finally is**

1) $\frac{1}{2}\frac{m}{n}u_1^2$ 2) $\frac{1}{2}\frac{m}{n}\left(u_1^2+u_2^2\right)$
3) $\frac{1}{2}m\left(u_1^2+u_2^2\right)$ 4) $\frac{1}{2}mn\left(u_1^2+u_2^2\right)$

6. **A ball A, moving with kinetic energy E, makes a head on collision with a stationary ball with mass *n* times that of A. The maximum potential energy stored in the system during the collision is**

1) $nE/(n+1)$ 2) $(n+1)E/n$
3) $(n-1)E/n$ 4) E/n

7. **After perfectly inelastic collision between two identical particles moving with same speed in different directions, the speed of the particles become half the initial speed. The angle between the velocities of the two before collision is**

1) 60^0 2) 45^0 3) 120^0 4) 30^0

8. **A pendulum consists of a wooden bob of mass m and length l. A bullet of mass m_1 is fired towards the pendulum with a speed V_1. The bullet energes out of the bob with a speed $V_1/3$ and the bob just completes motion along a vertical circle, then V_1 is**

1) $\left(\frac{m}{m_1}\right)\sqrt{5gl}$ 2) $\frac{3}{2}\left(\frac{m}{m_1}\right)\sqrt{5gl}$

3) $\frac{2}{3}\left(\frac{m_1}{m}\right)\sqrt{5gl}$ 4) $\frac{m_1}{m}\sqrt{gl}$

9. **A bullet of mass 0.04kg travelling horizontally at 100 m/s hits a stationary block of wood of mass 8kg, passes through it and emerges horizontally with a speed of 40 m/s. If the block is free to move on a smooth horizontal plane, the speed with which it is moving after the bullet had passed through it is**

1) 0.2 m/s 2) 0.3 m/s 3) 0.6 m/s 4) 0.4 m/s

10. **From the top of a tower of height 100m a 10gm block is dropped freely and a 6gm bullet is fired vertically upwards from the foot of the tower with velocity $100ms^{-1}$ simultaneously. They collide and stick together. The common velocity after collision is ($g = 10ms^{-2}$)**

1) 27.5 ms^{-1} 2) 150 ms^{-1}

3) 40 ms^{-1} 4) 100 ms^{-1}

11. **A ball falls vertically onto a floor, with momentum p, and then bounces repeatedly. The coefficient of restitution is e. The total momentum imparted by the ball to the floor is**

1) $p(1+e)$ 2) $\frac{p}{1-e}$

3) $p\left(1+\frac{1}{e}\right)$ 4) $p\left(\frac{1+e}{1-e}\right)$

12. **Two bodies move towards each other and collide inelastically. The velocity of the first body before impact is 2m/s and of the second is 4m/sec. The common velocity after collision is 1m/s in the direction of the first body. How many times did the K.E. of the first body exceed that of the second body before collision.**

1) 4.25 2) 3.25 3) 2.25 4) 1.25

13. **Two bodies of equal masses moving with momentum of first double that of the second collide completely inelastically. If the total final kinetic energy is 9 joule, the initial kinetic energy of the body of greater momentum is**

1) 5 joule 2) 6 joule 3) 8 joule 4) 10 joule

14. **A ball is droped on to a fixed horizontal surface from a height 'h'. The coefficient of restitution is 'e'. The average velocity of the ball from the instant it is dropped till it goes to maximum height after first impact with ground is**

1) $\left(\frac{1+e}{1-e}\right)\sqrt{2gh}$ 2) $\left(\frac{1+e^2}{1+e}\right)\sqrt{\frac{gh}{2}}$

3) $\left(\frac{1-e}{1+e}\right)\sqrt{2gh}$ 4) $(1-e)\sqrt{\frac{gh}{2}}$

15. **A bag of mass M hangs by a long thread and a bullet (mass m) comes horizontally with velocity v and gets caught in the bag. Then for the combined system (bag + bullet)**

1) Momenum is m Mv/(M + m)

2) KE is (1/2) Mv^2

3) Momentum is mv (M + m)/M

4) KE is $m^2v^2/2$ (M + m)

16. **Three balls are shown in figure. Ball A has mass m and initial velocity v to the right. Ball B with mass m and ball C with mass 4m are both initially at rest. Ball A collides elastically with ball B, which in turn collides with ball C. After the collision, ball B is reduced to rest. The final velocities of balles A and C aqre**

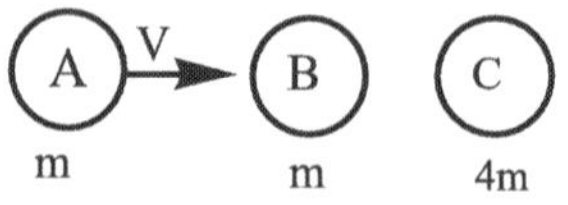

1) Ball A: 0.6v to the left and ball C:0.4 to the right

2) Ball A: 2.6 v to the left and ball C: 0.4v to the right

3) Ball A: at rest and ball C: 0.5v to the right

4) Ball A: at rest and ball C: 0.25v to the righte

17. Two spheres P and Q moving in opposite direction with velocities v_1 and v_2 collide with each other. If 'e' is the coefficient of restitution, then their distance of separation at time 't' after collision is

1) $(v_1 - v_2)te$ 2) $(v_2 - v_1)te$

3) $(v_1 + v_2)te$ 4) $(v_1 / v_2)te$

18. Two identical spheres are placed at two diametrically opposite points of a smooth circular groove in a horizontal plane. One sphere starts with a speed 'u' and collide the other after time 't'. If the coefficient of restitution of colliding bodies is 'e'. The minimum time after which they again collides in

1) $\frac{t}{4e}$ 2) $\frac{t}{2e}$ 3) $\frac{t}{e}$ 4) $\frac{2t}{e}$

19. A body falls from height 20m. If coefficient of restitution is 1/2 . The time interval between starting point and second bounce is [$g = 10\ ms^{-2}$]

1) 1s 2) 2s 3) 3s 4) 4s

ANSWERS

1) 3	2) 2	3) 3	4) 1	5) 3
6) 1	7) 3	8) 2	9) 2	10) 1
11) 4	12) 4	13) 3	14) 4	15) 4
16) 4	17) 3	18) 4	19) 4	

COMPREHENSION QUESTIONS

Comprehension - I (1-3) :

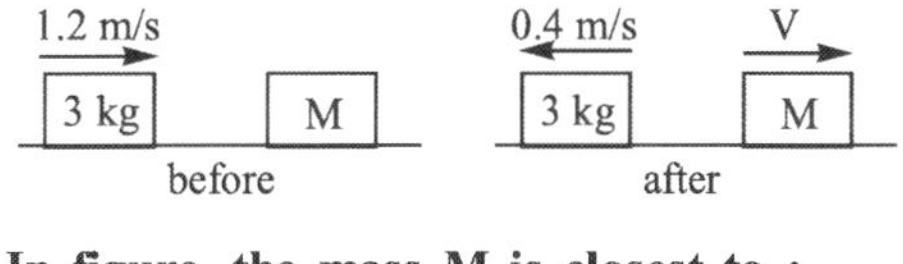

1. In figure, the mass M is closest to :

1) 1.5kg 2) 3.0kg 3) 4.5kg 4) 6.0kg

2. In figure, the speed of the block of mass M after the collision is closest to :

1) 0.8 m/s 2) 1.0 m/s 3) 1.2 m/s 4) 1.4 m/s

3. In figure, the blocks are in contact for 0.20s. The average force on the 3.0 kg block, while the two blocks are in contact, is closest to :

1) 8N 2) 12N 3) 16N 4) 24N

Comprehension - II (4-6) :

A 10 g bullet is shot vertically into an 8kg block .The block lifts upward 3mm.The bullet penetrates the block in a time interval of 0.001s. Assume the force on the bullet is constant during penetration .

4. In figure, the initial kinetic energy of the bullet is closest to :

1) 110J 2) 130J 3) 150J 4) 190J

5. In figure , the impulse on the block due to capture of the bullet is closest to :

1) 1.5N.s 2) 2.0N.s 3) 2.5N.s 4) 3.5N.s

6. In figure , the penetration of the bullet into the block, in cm, is closest to:

1) 7 2) 8 3) 10 4) 11

ANSWERS

1) 4 2) 1 3) 4 4) 3 5) 2 6) 3

EXERCISE - IVA

Problems in additonal syllabus for aieee

1. A 50 gm. ball collides with another ball of mass 150 gm. The two balls move at an angle 30° with their initial direction. Ratio of their velocities after collisions is

1) 3 :1 2) 1 : 3 3) 2 : 3 4) 1 : 1

2. A ball A is falling vertically downwards with velocity v_1. It strikes elastically with a wedge moving horizontally with velocity v_2 as shown in figure. What must be the ratio v_1 / v_2 so that the ball bounces back in vertically upward direction relative to the wedge.

1) $\sqrt{3}$

2) $\frac{1}{\sqrt{3}}$

3) 2

4) $\frac{1}{2}$

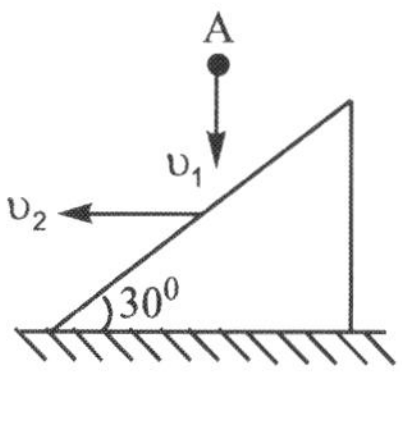

3. A ball falls with a velocity 'u' in a direction making an angle 'θ' with the normal to the surface. The ball goes back with a velocity 'v' in a direction making an angle 'ϕ' with the normal. The coefficient of restitution is

1) $\frac{\tan\phi}{\tan\theta}$ 2) $\frac{\tan\theta}{\tan\phi}$ 3) $\frac{\sin\phi}{\sin\theta}$ 4) $\frac{\sin\theta}{\sin\phi}$

4. Two small identical spheres each of mass 'm' are projected slant wise from the points A and B on the ground with equal velocities 'u' and 'u' making same angles 'θ' and 'θ' as shown in the fugure. They collide with each other at the highest point 'C' of the common path. If the collision is completely inelastic, the time after the collision the particles come back to the ground is

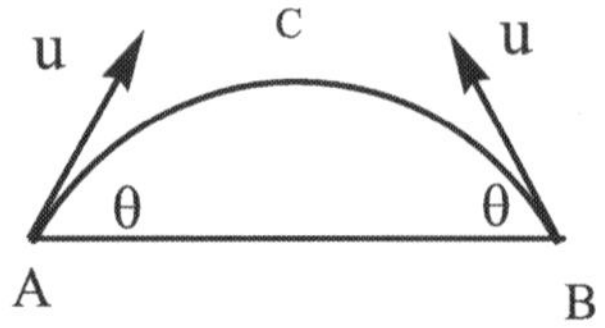

1) $2u\sin\theta/g$
2) $2u\cos\theta/g$
3) $u\cos\theta/g$
4) $u\sin\theta/g$

5. A ball is projected with a velocity 'u' at an angle 'θ' with the horizontal from a point on the ground. The coefficient of restitution for the ball and ground is 'e' the total range of the ball is

1) $\frac{u^2\sin^2\theta}{g}$ 2) $\frac{u^2\sin^2\theta(1+e)}{g}$

3) $\frac{u^2\sin 2\theta}{g(1+e)}$ 4) $\frac{u^2\sin 2\theta}{g(1-e)}$

6. A ball A collides elastically with an another identical ball B with velocity 10 m/s at an angle of 30^0 from the line joining their centres C_1 and C_2 select the false statement

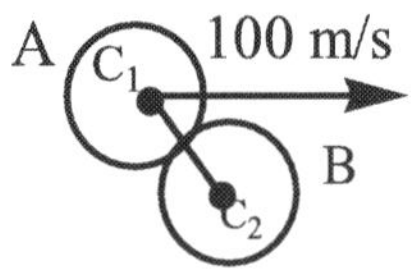

1) Velocity of ball A after collision is 5 m/s
2) Velocity of ball B after collision is $5\sqrt{3}$ m/s
3) Both the balls move at right angles after collision
4) Velocity of ball A after collision is $5\sqrt{3}$ m/s

7. A smooth sphere is moving on horizontal surface with a velocity vector $\left(2\vec{i}+2\vec{j}\right)$ m/s immediately before it hit a vertical wall. The wall is parallel to vector $\vec{j}$ and coefficient of restitution between the sphere and the wall is e = 1/2. The velocity of the sphere after it hits the wall is

1) $\vec{i}-\vec{j}$ 2) $-\vec{i}+2\vec{j}$

3) $-\vec{i}-\vec{j}$ 4) $2\vec{i}-\vec{j}$

ANSWERS

1) 1	2) 2	3) 2	4) 4	5) 4
6) 4	7) 2			

EXERCISE - IVB

1. Particles A and B of masses 20 gm and 40 gm are simultaneously projected at angles 45° and 135° from two points p and q with a velocity 49 m/s. They collide in their path, after collision A retraces its path. Velocity of B is

1) 49 m/s 2) 24.5 m/s
3) zero 4) 98 m/s

2. A particle strikes a horizontal frictionless floor with a speed u, at an angle θ to the vertical, and rebounds with a speed υ, at an angle θ to the vertical. The coefficient of restitution between the particle and the floor is e. The magnitude of υ is

1) eu 2) (1 – e)u

3) $u\sqrt{\sin^2\theta+e^2\cos^2\theta}$

4) $u\sqrt{e^2\sin^2\theta+\cos^2\theta}$

3. A ball is thrown at an angle of incidence 'θ' on a horizontal plane such that the incident direction and the reflected direction are at right anlges to each other if the coefficient of restitution is 'e' then 'θ' is equal to

1) $\text{Tan}^{-1}(e)$ 2) $\text{Tan}^{-1}(2e)$

3) $\text{Tan}^{-1}\left(\sqrt{2}\,e\right)$ 4) $\text{Tan}^{-1}\left(\sqrt{e}\right)$

4. A ball is projected with initial velocity u at an angle θ to the horizontal. Then horizontal distance covered by ball as it collides third time to the ground would be, if coefficient of restitution is e

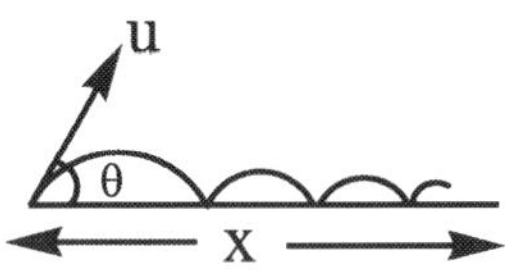

1) $\frac{u^2 \sin 2\theta}{g}$ 2) $e\frac{u^2 \sin 2\theta}{g}$

3) $(1-e)\frac{u^2 \sin 2\theta}{g}$ 4) $(1+e)\frac{u^2 \sin 2\theta}{g}$

5. A projectile is fixed on a horizontal ground. Coefficient of restitution between the projectile and the ground is e. If a, b and c be the ratio of time of flight $\left[\frac{T_1}{T_2}\right]$ maximum height $\left[\frac{H_1}{H_2}\right]$ and horizontal range $\left[\frac{R_1}{R_2}\right]$ in first two collisions with the ground. Then

1) $a=\frac{1}{e}$ 2) $b=\frac{1}{e^2}$

3) $c=\frac{1}{e}$ 4) All the above

6. After perfectly inelastic collision between two identical particles moving with same speed in different directions, the speed of the particles become half the initial speed. The angle between the velocities of the two particles before collision is

1) 60^0 2) 45^0 3) 120^0 4) 30^0

7. A ball is projected from ground with a velocity V at an angle 'θ' to the vertical. On its path it makes an elastic collision with a vertical wall and returns to ground. The total time of flight of the ball is (g = acceleration due to gravity)

1) $\frac{2V\sin\theta}{g}$ 2) $\frac{2V\cos\theta}{g}$

3) $\frac{V\sin 2\theta}{g}$ 4) $\frac{U\sin\theta + U\cos\theta}{g}$

ANSWERS

1) 3	2) 3	3) 4	4) 4	5) 4
6) 3	7) 2			

Chapter - 7 - (PART - B)

WORK POWER ENERGY

WORK POWER ENERGY

SYNOPSIS

1) Work is said to be done by a force if the pointof application of force undergoes displacement either in the direction of force or in the direction of component of force.

2) Amount of work done is equal to the dot product of force and displacement. If $\overline{F}$ is the force acting on a body and $\overline{S}$ is displacement, then the work done by that force is $W = \overline{F}.\overline{S} = FSCos\,\theta$

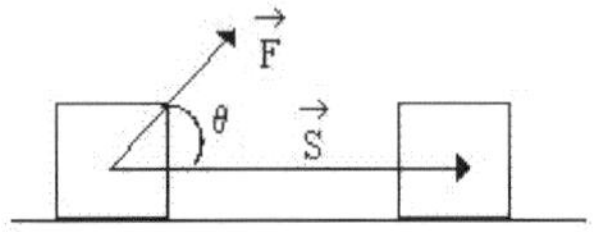

3) Work is a scalar

4) Work done by a force can be positive (or) negative (or) zero

(a) If the displacement is in the direction of force or in the direction of component of force, work done is positive

Ex : When a body is lifted vertically up from ground, work done by the lifting force on the body is positive. W = F x h.

(b) If the displacement is in a direction opposite to that of force or component of force, work done is negative.

Ex : When a body is lifted vertically up from ground, work done by the gravitational force acting on the body is negative. W = – mgh.

c) If the displacement and force are perpendicular to each other, work done is zero

Ex : When a body is in circular motion work done by the centripetal force acting on the body is zero. W = Fcos90 x s = 0.

5) If the work done in moving a body between two points is independent of path along which it is moved, then the force acting on the body is known as conservative force.

e.g: gravitational force, electrostatic force. The work done by a conservative force in moving a body along a closed path is zero.

6) If the work done in moving a body by the applicaton of force depends on the path along which it is moved, such force is known as non conservative force.

e.g : frictional force, viscous force.

If a body is moved along a closed path by applying a non conservative force, work done by such force is not equal to zero.

7) When a body of mass m is lifted to a height h above the ground without any acceleration,

a) work done by the applied force = mgh

b) work done by the gravitational force=-mgh

c) work done by the resultant force = O

8) The work done in lifting a body of mass m having density d_1 inside a liquid of density d_2 through a height h is (a=0)

$$W = mgh\left(1 - \frac{d_2}{d_1}\right)$$

9) The work done by a gas against pressure during its expansion is dW = PdV

$$W = \int dW = \int PdV$$

(P is pressure and dV is change in volume)

10) A point sized sphere of mass m is suspended vertically using a string of length l . If the bob is pulled to a side till the string makes an angle θ with the vertical, work done against gravity is $W = mgl(1 - Cos\,\theta)$

11) A uniform rod of mass m and length l is suspended vertically. If it is pulled to a side till it makes an angle θ with the vertical, work done against gravity is $W = \frac{mgl}{2}(1 - Cos\theta)$

12) A uniform chain of mass m and length l is suspended vertically. If the lower point of the chain is lifted to the point of suspension, work done against gravity is $W = {mgl}/{4}$

13) A uniform chain of mass m and length 'l' is kept on a smooth horizontal table such that ${1}/{n}th$ of its length is hanging down vertically from the edge of the table. The work done against gravity to pull the hanging part on to the table is $W = \frac{mgl}{2n^2}$

14) Work done in stretching or compressing a spring is $W = \frac{1}{2}Kx^2$

15) A bucket full of water of total mass M is pulled up using a uniform rope of mass m and length l, work done is $W = Mgl + mg\,{}^{l}\!/_{2} = \left(M + {}^{m}\!/_{2}\right)gl$

(Bucket is treated as point mass)

16) When a body of mass m is moved on a rough horizontal surface through a distance 's', work done against friction is $W = \mu mgs$

17) If a body of mass m slides down through a distance S along a smooth inclined plane, work done by gravity is W = (mg Sinθ)S

18) A block of mass m is suspended vertically using a rope of negligible mass. If the rope is used to lift the block vertically up with uniform acceleration a, work done by the tension in the rope is W = m(g+a) h

In the above case if the block is lowered with acceleration a W = –m (g – a) h

19) The ability or capacity to do work is called energy

20) Energy is cause for doing work and work is effect.

21) Energy is a scalar. Units of energy are same as that of work.

22) Mechanical energy is of two types

a) potential energy b) kinetic energy

The energy possessed by a body by virtue of its state or position is known as potential energy.

e.g : Water on hill top and stretched or compressed spring possess potential energy.

The energy possessed by a body by virtue of its motion is called kinetic energy. Any body in motion possess kinetic energy.

23) Gravitational potential energy of a body which is at a height 'h' above the ground is mgh (h is small so that g remains same almost)

Here surface of the earth is taken as reference.

24) Kinetic energy of a body of mass 'm' moving with a velocity v is $\frac{1}{2}mv^2$.

25) A body can have both potential and kinetic energies e.g a flying bird, freely falling body

26) A body at rest cannot have kinetic energy but it can have potential energy A body in motion can have potential energy.

27) When a spring is compressed or stretched, elastic potential energy stored in it is $\frac{1}{2}Kx^2$

28) (i) The negative of the work done by an internal force on a system is equal to the change in its potential energy.

(ii) The work done by an external force on a system is equal to the change in its mechanical energy.

(iii) Work done by an unbalanced or resultant force on a body is equal to change in its kinetic energy. This is known as work-energy theorem.

$W = \frac{1}{2}mv^2 - \frac{1}{2}mu^2$

29) According to the law of conservation of energy, the total energy of a closed system remains constant. Energy can neither be created nor be destroyed but it can be transformed from one form to another form without any loss or gain.

30) In the case of a freely falling body or a body projected vertically up, the sum of potential and kinetic energies at any point in its path is constant.

31) Kinetic energy of a freely falling body
a) is directly proportional to height of fall
b) is directly proportional to square of time of fall.

32) If p is linear momentum of a body moving with kinetic energy E, we can write $E = \frac{p^2}{2m}$

$p = \sqrt{2mE} \Rightarrow E \propto p^2 \ and \ p \propto \sqrt{E}$

33) (i) If two bodies of different masses have same kinetic energy, then the heavier body will have greater momentum. $p \propto \sqrt{m}$

(ii) If two bodies of different masses have same momentum, then the lighter body will have greater kinetic energy. $KE \propto \frac{1}{m}$

eg: When a bullet is fired from a gun, bullet and gun have same linear momentum but kinetic energy of bullet is more than that of gun.

34) A body can have energy without momentum. But it can not have momentum without energy.

35) A bullet of mass m moving with velocity v stops in wooden block after penetrating through a distance x. If F is resistance offered by the block to the bullet, (Assuming F is constant inside the block)

$\frac{1}{2}mv^2 = Fx \Rightarrow F = \frac{mv^2}{2x} \ or \ v^2 \propto x$

36) A sphere of mass m is dropped from a height 'h' above the ground. On reaching the ground, it pierces through a distnace S and then stops finally.

Resistance offered by the ground is R. Then $R = mg\left(1+\frac{h}{s}\right)$

Here time of penetration is given by $t = \frac{2s}{\sqrt{2gh}}$

37) A body is projected vertically up from the ground. When it is at a height 'h' above the ground, its potential and kinetic energies are in the ratio x:y. If H is the maximum height reached by that body, then $\frac{x}{y} = \frac{h}{H-h}$ (or) $\frac{h}{H} = \frac{x}{x+y}$

38) A bullet of mass m is fired horizontally with a velocity V which strikes a wooden block of mass M suspended vertically. After striking the block, the bullet stops in the block by piercing. If the combined system rises to a height 'h' above the initial position, $V = \left(1+\frac{M}{m}\right)\sqrt{2gh} = \frac{(m+M)}{m}\sqrt{2gh}$

39) The rate of doing work by a force is called power

$\Rightarrow \text{power} = \frac{\text{work}}{\text{time}} = \frac{w}{t}$

1 H.P = 746 W, 1KW = 10^3W, 1MW = 10^6W, 1GW = 10^9W.

40) If $\vec{F}$ is the force acting on a body and $\vec{V}$ is its velocity at an instant, instantaneous power $P = \vec{F}.\vec{V}$

41) If a machine gun fires n bullets per second such that mass of each bullet is m and coming out with velocity v, then the power of that machine gun is

$P = n\left(\frac{1}{2}mv^2\right) = \frac{1}{2}mnv^2$

42) A conveyor belt moves horizontally with a constant speed 'v'. Gravel is falling on it at a rate $\frac{dm}{dt}$. Then extra power required to drive the belt is $\left(\frac{dm}{dt}\right)v^2$

43) If a pump/ crane is used to lift water/coal of mass m to a height h in time 't', then the power of that pump or crane is $p = \frac{mgh}{t}$

If efficiency of that crane or motor is $\eta\%$, then $\frac{\eta}{100}p = \frac{mgh}{t}$

44) A car of mass m is moving on horizontal road with constant acceleration 'a'. If R is resistance offered to its motion, then the power of the engine when its velocity is V will be P=(R+ma)V

45) A body of mass m is initially at rest. By the application of a constant force, its velocity changes to v_0 in time t_0. (i) The kinetic energy of that body at any time 't' is $\frac{1}{2}mv_0^2\left(\frac{t}{t_0}\right)^2$

(ii) Instantaneous power at any time 't' is $m\left(\frac{v_0}{t_0}\right)^2 t$

(iii) Average power during the time 't' is $\frac{m}{2}\left(\frac{v_0}{t_0}\right)^2 t$

46) If a car moves on a rough horizontal road with constant speed v., then the Power of its engine is $P = \mu mgv$ (μmg is frictional force)

47) When a lorry and a car moving with same momentum are brought to rest by applying the same retarding force, then (i) both comes to rest at the same time. (ii) car travels more distance before coming to rest.

48) When a lorry and a car are moving with same KE are brought to rest by applying same retarding force, then (i) car comes to rest earlier (ii) both travel equal distances before coming to rest.

EXERCISE - I

1. Work is said to be done by a force on a body if

1) the body has displacement in the direction of force
2) the angle between force and displacement vectors is other than 90^0
3) both 1 and 2
4) none of the above

2. The area under a 'force - displacement' curve gives :

1) impulse 2) Power 3) Work 4) time

3. The product of pressure and volume of a gas has the units of

1) Momentum 2) Force
3) Power 4) Work

4. A man pushes a wall and fails to displace it. Then

1) work done by him is positive
2) work done by him is negative
3) no work is done by him
4) any of the above

5. **The work done by gravity in lifting a body to a certain height depends upon**

1) time of its rise 2) path followed by the body

3) both 1 & 2 4) none

6. **When a body is falling freely from some height work done by gravity is**

1) negative 2) positive 3) zero

4) may be positive or negative

7. **A man is waiting at a bus stop by holding a box on his head. Work done by him is**

1) zero 2) positive

3) negative 4) infinity

8. **A block of mass 'm' is pulled by a force 'F' on a level surface through a distance 'S', then**

1) The work done on the body aganist gravitational force is zero

2) The work done on the body against gravitational force is mgS

3) The work done on the body against its weight is F.S

4) The work done on the body against friction is mgS

9. **In the case of conservative force**

1) Work done is indepenent of the path

2) Work done in a closed loop is zero

3) Work done against conservative force is stored is the form of potential energy

4) All the above

10. **Two identical bodies of same mass are raised to same heights by two persons X and Y in 10s and 20s respectively. The work done by**

1) X is greater 2) Y is greater

3) both X and Y are the same

4) none

11. **Two bodies of different masses have same linear momentum. The one having more K.E. is**

1) Lighter body 2) heavier body

3) both 4) none

12. **Two bodies of different masses have same K.E. The one having more momentum is**

1) Heavier body 2) lighter body

3) both 4) none

13. **The K.E. of a body is represented by 'Amv' where A is a constant. The dimensions of 'A' are**

1) L^1T^{-1} 2) L^1T^2 3) $M^1L^1T^{-2}$ 4) $M^1L^1T^{-1}$

14. **Two bodies of masses m_1 and m_2 have equal momentum. Their K.E. are in the ratio**

1) $\sqrt{m_2} : \sqrt{m_1}$ 2) $m_1 : m_2$

3) $m_2 : m_1$ 4) $m_1^2 : m_2^2$

15. **The total workdone on a particle is equal to change in it's kinetic energy**

1) Only if the forces acting on it are conservative

2) Only if gravitational force alone acts on it

3) Only if elastic force alone acts on it

4) For any kind of force acting on it

16. **If a body is released from a certain height, during its fall.**

1) its kinetic energy increases and potential energy decreases.

2) its potential energy increases and kinetic energy decreases

3) both potential and kinetic energy of that body increase

4) both potential and kinetic energy of the body decrease

17. **A particle moves along a circle with uniform acceleration. Workdone by the resultant force acting on that particle**

1) is always zero 2) may be zero

3) is not equal to zero 4) is infinity

18. **Two springs have their force constants k_1 and k_2 and they are stretched to the same extension. If $k_2 > k_1$ work done is**

1) same in both the springs

2) more in spring K_1

3) more in spring K_2 4) none

19. **Two springs have their force constants k_1 and k_2 (K_2>K_1). When they are stretched by the same force, work done is**

1) same in both the springs

2) more in spring K_1

3) more in spring K_2 4) none

20. **A lorry and a car moving with same KE are brought to rest by applying the same retarding force. Then**

1) Lorry will come to rest in a shorter distance

2) Car will come to rest in a shorter distance

3) Both come to ret in the same distance

4) None

21. A lorry and a car moving with same momentum are brought to rest by applying the same retarding force. Then

1) Lorry will come to rest in a shorter distance
2) Car will come to rest in a shorter distance
3) Both come to ret in the same distance
4) None

22. A shell is fired into air at an angle θ with the horizontal from the ground. On reaching the maximum height,

1) its kinetic energy is not equal to zero
2) its kinetic energy is equal to zero
3) its potential energy is equal to zero
4) both its potential and kinetic energies are zero

23. If the force acting on a body is inverely proportional to its speed, then the kinetic energy of the body is

1) Constant
2) Directly proportional to time
3) Inversely proportional to time
4) Directly proportional to square of time

24. The kinetic energy of freely falling body

1) is directly proportional to the height of its fall
2) is directly proportional to the square of the time of its fall
3) both 1 and 2
4) None

25. A body can have

1) zero momentum and finite kinetic energy.
2) zero kinetic energy and finite momentum
3) zero acceleration and increasing kinetic energy
4) finite acceleration and zero kinetic energy

26. When a wound spring is dissolved in an acid, the temperature of the acid

1) Increases
2) Decreases
3) Remains same
4) none

27. Same force acts on two bodies of masses m_1 and m_2 (m_2>m_1). After travelling the same distance

1) Kinetic energy of m_2 is greater than kinetic energy of m_1
2) Kinetic energy of m_1 is greater than kinetic energy of m_2
3) Ratio of kinetic energies of those two is equal to ratio of their momentum
4) Kinetic energy of m_1 is equal to kinetic energy of m_2

28. A constant force acts on a body which is intially at rest. If a graph is drawn by plotting its kinetic energy along Y-axis and displacement along X-axis, shape of the graph is

1) Straight line passing through origin
2) Parabola
3) Straight line parallel to X-axis
4) Straight line parallel to Y-axis

29. A body at rest can have

1) Kinetic energy
2) momentum
3) both momentum and kinetic energy
4) potential energy

30. A body is moved along a straight line by a machine delivering constant power. The distance moved by the body in time 't' is proportional to

1) $t^{1/2}$ 2) $t^{3/4}$ 3) $t^{3/2}$ 4) t^2

31. **Statement A : An engine A can perform a given work in 1 hr and engine B can prform the same work in ½hr. then B has greater power than A .**

Statement B : Power is the dot product of force and velocity

1) A & B is correct
2) A & B are wrong
3) A is correct and B is wrong
4) A is wrong & B is correct

32. **Statement A : Energy is the cause for doing work and work is the effect**

Statement B : Energy is a scalar but power is a vector quantity.

1) A & B is correct and B wrong
2) A & B are wrong
3) A is correct and B is wrong
4) A is wrong & B is correct

33. Two solid spheres of same material having radii in the ratio 1:2 are moving with same kinetic energy on a horizontal path. They are brought to rest by applying same retarding force. Then

i) their initial momenta are in the ratio $1:2\sqrt{2}$

ii) distance travelled are in the ratio 1:1 before coming to rest

iii) The times taken by them coming to rest is $1:2\sqrt{2}$

1) i) and (ii) are correct
2) ii) and (iii) are correct
3) All the three are correct
4) (i) and (iii) are correct

34. A bullet strikes a wooden block resting on a frictionless horizontal surface and stops in it. Later the combined system moves. In this case

i) only KE is conserved

ii) only momentum is conserved

iii) Both KE and momentum are conserved

1) i) is correct 2) iii) is correct

3) i) and (iii) are in correct

4) (ii) is correct

35. A) When a body is thrown into air, at its maximum height its KE is zero always.

B) If a body is moved along a rough circular path in the horizontal plane,work done by the force applied is zero

1) A is true, B is false 2) A is false, B is true

3) Both A and B are true

4) Both A and B are false

36. A) Work done by frictional force is always negative

B) A body at rest can have mechanical energy

C) Mechanical energy of freely falling body decrease gradually

1) Only A is true 2) Only B is true

3) Only C is true 4) All the three one true

37. Match the pairs in two lists given below

List - I	List - II
a) Gravitational force	e) decreases
b) Fractional force	f) conservative force
c) KE of a dropped body	g) nonconservative force
d) PE of a dropped body	h) increases

1) a-f , b-h , c-g, d-e 2) a-f, b-g,c-h,d-e

3) a-f,b-g,c-e,d-h 4) a-h, b-g,c-f,d-e

38.

List - I	List - II
a) 1HP	e) $10^7 ergs^{-1}$
b) 1KWH	f) 1.6×10^{-19} J
c) 1eV	g) 36×10^5 J
d) 1W	h) 746W

1) a-g, b-h,c-f,d-e 2) a-e,b-f,c-g,d-h

3) a-h,b-g,c-f,d-e 4) a-e,b-f,c-h,d-g

39.

List - I	List - II
a) Work	e) ML^2T^{-2}
b) Linear momentum	f) MLT^{-1}
c) Power	g) ML^2T^{-3}
d) Efficiency of motor	h) $M^0L^0T^0$

1) a-e,b-f,c-g,d-h 2) a-f,b-e,c-g,d-h

3) a-g,b-f,c-e,d-h 4) a-h,b-g,c-f,d-e

40. A body is allowed to fall from a height h above the ground. Then match the following

List - I	List - II
a) PE=KE	e) at height h/2
b) PE=2KE point	f) constant at any
c) KE = 2PE	g) at height 2h/3
d) PE +KE	h) at height h/3

1) a-e,b-g,c-h,d-f 2) a-g,b-e,c-f,d-h

3) a-f,b-g,c-e,d-h 4) a-e,b-h,c-g,d-f

ASSERTION & REASON TYPE

These Questions consist of two statements each printed as Assertion and Reason. While answering these questions you are required to choose any one of the following four responses.

1) Both (A) and (R) are true and (R) is the correct explanation of (A)

2) Both (A) & (R) are true but (R) is not correct explanation of (A)

3) (A) is true and (R) is false

4) (A) is false but (R) is true

41. A) : When a person is walking horizontally with a suitcase on his head, no work is done by him against gravitational force

R) : Gravitational force on suitcase acts vertically downwords and motion is in horizontal direction, hence dot product becomes zero

42. A) : No work is done by the centripetal force acting on a body moving along the circumference of a circle

R) : At any instant, the motion of the body is along the tangent to the circle where as the centripetal force is along the radius vector towards the centre of the circle.

43. A) Work done by gravitational force in moving a body is path independent

R) Gravitational force is non conservative force.

44. A) A body can possess mechanical energy without momentum.

R) A bent bow possess potential energy by virtue of deformation.

45. A) Work done by a force is a vector

R) Work done is measured as the dot product of force and displacement vectors

46. A) No work is done by gravitational force if a body moves along a horizontal surface.

R) In horizontal motion, angle between gravity and displacement is 90°.

47. A) Whenever a force acts on a body, its KE always increases

R) KE of a particle remains constant in uniform circular motion

48. A) Gravitational force is conservative

R) Frictional force is non conservative

49. Three spheres of different materials are in motion their radii, densities and speeds are as given

Sl.No	Radius	Density	Speed
1	r	d	υ
2	r/2	d	V/2
3	2r	d/2	υ

If K_1,K_2,K_3 are the kinetic energies of those three spheres respectively, those values can be arranged in descending order as given below

1) K_3, K_2,K_1 2) K_3,K_1,K_2

3) K_1,K_2,K_3 4) K_2,K_1,K_3

50. A body is in motion and it is stopped by applying retarding force. Initial speed magnitude of force applied are as given below

	Speed	Force
1	V	F
2	2V	F/2
3	V/2	2F

If S_1,S_2,S_3 are the distances travelled before coming to rest, those can be arranged in ascending order as given below

1) S_1,S_2,S_3 2) S_3,S_2,S_1

3) S_2,S_1,S_3 4) S_3,S_1,S_2

ANSWERS

1) 3	2) 3	3) 4	4) 3	5) 4
6) 2	7) 1	8) 1	9) 4	10) 3
11) 1	12) 1	13) 1	14) 3	15) 4
16) 1	17) 3	18) 3	19) 2	20) 3
21) 1	22) 1	23) 2	24) 3	25) 4
26) 1	27) 4	28) 1	29) 4	30) 3
31) 1	32) 3	33) 3	34) 4	35) 4
36) 2	37) 2	38) 3	39) 1	40) 1
41) 1	42) 1	43) 3	44) 2	45) 4
46) 1	47) 4	48) 2	49) 2	50) 4

EXERCISE - II(A)

1. The length of a smooth inclined plane of inclination 30^0 is 5m. The work done in moving a 10kg mass from the bottom of the inclined plane to top is

1) 245J 2) 490J 3) 980J 4) 500J

2. A body of mass 10kg is raised from certain depth. By the time it is raised by 10m, if its velocity is $2ms^{-1}$, work done during this time is,

1) 980J 2) 20J 3) 960J 4) 1000J

3. A bucket filled with water weighing 20 kg is raised from a well of depth 20m. If the linear density of the rope is 0.2kg per meter, the amount of work done is, ($g = 10ms^{-2}$)

1) 4000J 2) 4040J 3) 4400J 4) 4200J

4. A block of mass 'm' is lowered with the help of a rope of negligible mass through a distance 'd' with an acceleration of g/3. Work done by the rope on the block is

1) $\frac{2Mgd}{3}$ 2) $-\frac{2Mgd}{3}$

3) $\frac{Mgd}{3}$ 4) $-\frac{Mgd}{3}$

5. A force $\overline{F} = (5\hat{i} - 3\hat{j} + 2\hat{k})N$ moves a particle from $\overline{r_1} = (2\hat{i} + 7\hat{j} + 4\hat{k})m$ to $\overline{r_2} = (5\hat{i} + 2\hat{j} + 8\hat{k})m$. The work done by the force is,

1) 18J 2) 28J 3)38J 4) 48J

6. A uniform chain of length 2 m is kept on a table such that a length of 60 cm hangs freely from the edge of the table. The total mass of the chain is 4 kg. What is the work done in pulling the entire chain on to the table ? ($g = 10\ m/s^2$) (AIEEE 2004)

1) 7.2J 2) 3.6J 3) 120J 4) 1200J

7. n identical cubes each of mass 'm' and side '*l*' are on the horizontal surface. Then the minimum amount of work done to arrange them one on the other is_

1) nmg*l* 2) $\frac{mgl\,n^2}{2}$

3) $\frac{mgl\,n(n-1)}{2}$ 4) $\frac{mgl\,n(n+1)}{2}$

8. A person of mass 50kg carrying a load of 20kg walks up a staircase. If the width and height of each step are 0.25m and 0.2m respectively, the work done by the man in walking up 20 steps is, ($g = 10ms^{-2}$)

1) 1400J 2) 800J 3) 2000J 4) 2800J

9. A rectanglular block of dimensions 6m x 4m x 2m and of density 1.5 gm/c.c is lying on horizontal ground with the face of largest area in contact with the ground. The work done in arranging it with its smallest area in contact with the ground is, ($g=10ms^{-2}$)

1) 2880 kJ 2) 1440 kJ 3) 3800 kJ 4) 720 kJ

10. A body of mass 5kg moving along a straight line is accelerated from $4ms^{-1}$ to $8ms^{-1}$ with the application of a force of 10N in the same direction. Then,

1) Work done by the force is 120J
2) Displacement of the body is 12m
3) Time for the change in velocity is 2s
4) All the above

11. A uniform rod of length 2m and mass 5kg is lying on a horizontal surface. The work done in raising one end of the rod with the other end in contact with the surface until the rod makes an angle 30^0 with the horizontal is, ($g = 10ms^{-2}$)

1) 25J 2) 50J 3) $25\sqrt{3}\,J$ 4) $50\sqrt{3}\,J$

12. A ladder 'AB' of weight 300N and length 5m is lying on a horizontal surface. Its centre of gravity is at a distance of '2m' from end A. A weight of 80N is attached at end B. The work done in raising the ladder to the vertical position with end 'A' in contact with the ground is,

1) 500J 2) 1000J 3) 1150J 4) 1900J

13. Force acting on a particle is $\left(2\hat{i}+3\hat{j}\right)N$. Work done by this force is zero, when a particle is moved along the line 3y+kx = 5. Here the value of k is

1) 2 2) 4 3) 6 4) 8

14. A particle of mass 0.1 kg is subjected to a force which varies with distance as shown in figure. It starts its journey from rest at x = 0, then its velocity at x = 12 m is :

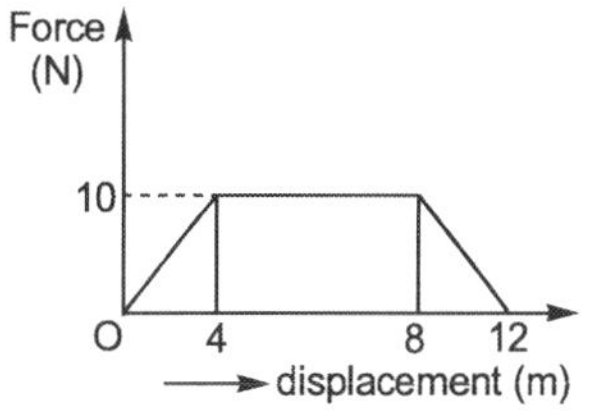

1) 0 2) $20\sqrt{2}\,m/s$

3) $20\sqrt{3}\,m/s$ 4) 40 m/s

15. A force of $(4x^2 + 3x)$ N acts on a particle which displaces it from x = 2m to x = 3m. The work done by the force is

1) 32.8 J 2) 3.28J 3) 0.328J 4) zero

16. Velocity – time graph of a particle of mass 2 kg moving in a straight line is as shown in figure. Work done by all the forces on the particle is

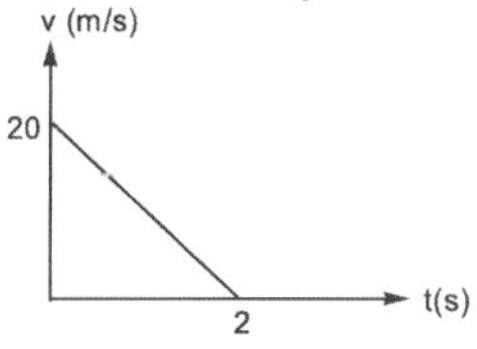

1) 400 J 2) – 400 J 3) – 200 J 4) 200 J

17. A force of 5N is applied on a 20kg mass at rest. The work done in the third second is,

1) $\frac{25}{8}J$ 2) $\frac{25}{4}J$ 3) 12 J 4) 25 J

18. A uniform metre scale of mass 2kg is suspended from one end. If it is displaced through an angle 60^0 from the vertical, the increase in its potential energy is,

1) 4.9J 2) 9.8 J

3) $9.8\sqrt{3}J$ 4) $4.9(2-\sqrt{3})J$

19. Two spheres whose radii are in the ratio 1 : 2 are moving with their velocities in the ratio 3 : 4. If their densities are in the ratio 3 : 2, the ratio of their kinetic energies is,

1) 27 : 64 2) 1 : 1 3) 27 : 256 4) 9 ; 64

20. A river of salty water is flowling with a velocity 2 m/sec. If the density of the water is 1.2 gm/c.c., then the kinetic energy of each of cubic metre of water is (2008-E)

1) 2.4 J 2) 24 J

3) 4.8 KJ 4) 2.4 KJ

21. When a spring is compressed by 3cm, the potential energy stored in it is U. When it is compressed further by 3cm, the increase in potential energy is,

1) 4 U 2) U 3) 2 U 4) 3 U

22. If the kinetic energy of a body increases by 125%, the percentage increase in its momentum is,

1) 50% 2) 62.5% 3) 250% 4) 200%

23. The kinetic energy of a body is 'K'. If one-fourth of its mass is removed and velocity is doubled, its new kinetic energy is,

1) K 2) 3 K 3) 4 K 4) $\frac{9K}{4}$

24. A train is moving with a uniform velocity of $(2\hat{i}-\hat{j}+4\hat{k})$ ms^{-1}. If the force required to overcome friction is $(\hat{i}-3\hat{j}+2\hat{k})N$, the power of the engine is

1) 7 W 2) 10 W 3) 13 W 4) 15 W

25. Sand is falling on a conveyor belt at the rate of $5kgs^{-1}$. The extra power required to move the belt with a veloctiy of $6ms^{-1}$ is,

1) 30 W 2) 180 W 3) 150 W 4) 1.2 W

26. A body of mass 6kg is under a force which causes displacement in it which is given by $s=\frac{t^2}{4}m$, where 't' is time. The work done by the force in 2s is

1) 12J 2) 9 J 3) 6 J 4) 3 J

27. A bus of mass 6 metric tons is pulled at a speed of 18 KMPH on a smooth incline of inclination 1 in 20. Power of the engine is, $(g = 10m^{-2})$

1) 15 kW 2) 1500W 3) 30 kW 4) 150 kW

28. A pump of 200 W power is lifting 2kg water per second from an average depth of 10m. The velocity with which water comes out of the pump is, $(g = 9.8\ ms^{-2})$

1) 2 ms^{-1} 2) 2.5 ms^{-1} 3) 4 ms^{-1} 4) 3 ms^{-1}

29. A man carries a load of 50 kg to a height of 40 m in 25s. If the power of the man is 1568W, the mass of the man is,

1) 20 kg 2) 50 kg 3) 100 kg 4) 75 kg

30. A particle of mass 100g is thrown vertically upwards with a speed of 5 m/s. The work done by the force of gravity during the time the particle goes up is $(g = 10ms^{-2})$ (AIEEE 2006)

1) – 0.5J 2) –1.25J 3) 1.25J 4) 0.5J

31. A bullet of mass 10gm hits a fixed target and penetrates through 8cm in it before coming to rest. If the average resistance offered by the target is 100 N, the velocity with which the bullet hits the target is,

1) 20 m/s 2) 25 m/s 3) 40 m/s 4) 50 m/s

32. A body of mass 5kg has linear momentum of 20kgm/s. If a constant force of 5N acts on it in the direction of its motion for 10s. The change in kinetic energy of the body is

1) 320 J 2) 600 J 3) 200 J 4) 450 J

33. The ratio of energy required to accelerate a car from rest to 20 ms^{-1} to the energy needed to accelerate from 20ms^{-1} to 40ms^{-1} is

1) 1 : 1 2) 1 : 3 3) 1 : 2 4) 1 : 4

34. A light string of length 20cm is carrying a bob of mass 100 gm. It is pulled through an angle 60^0 and released. Its maximum velocity is,

1) 1 ms^{-1} 2) 2 ms^{-1} 3) 1.4 ms^{-1} 4) 3 ms^{-1}

35. A particle is projected at 60^0 to the horizontal with a kinetic energy K. The kinetic energy at the highest point is : (AIEEE 2007)

1) K 2) Zero 3) K/4 4) K/2

36. A ball released from a height of 20m, hits the ground and rebounds to a height of 16m. The percentage loss of energy during collision is,

1) 20 % 2) 40 % 3) 25 % 4) 75 %

37. A car moving with a velocity of 72 KMPH stops engine just before ascending an inclined road. If 25% of energy is wasted in overcoming friction, the car rises to a height of $(g = 10ms^{-2})$

1) 6 m 2) 7.5 m 3) 12 m 4) 15 m

38. A body moves a distance of 20m along a straight line under the action of a force of 10N. If the work done is 100J, the angle between force and displacement vectors is,

1) 0^0 2) 30^0 3) 45^0 4) 60^0

39. The work done in lifting a body of mass 20kg and specific gravity 3.2 to a height of 8m in water is, $(g = 10\ ms^{-2})$

1) 110 J 2) 1600 J 3) 500 J 4) 1100 J

40. The work done in carrying a body from A to B is 20 J and from B to C is 15J. The work done in moving the body from A to C assuming the force involved in all the cases is conservative force is

1) 35 J 2) 25 J 3) 5 J 4) $\sqrt{175}J$

41. A simple pendilum is swinging in a vertical plane. The ratio of its potential energies when it is making angles 30^0 and 60^0 with the vertical is,

1) $1:(2-\sqrt{3})$ 2) $(2-\sqrt{3}):1$ 3) 1 : 2 4) 2 : 1

42. A 2kg block is dropped from a height of 0.4m on a spring of force constant 1960 Nm^{-1}. The maximum compression of the spring is

1) 10 cm 2) 20 cm 3) 30 cm 4) 40 cm

43. A boy is swinging on a swing such that his lowest and highest positions are at heights of 2m and 4.5m respectively. His velocity at the lowest position is,

1) 2.5 ms^{-1} 2) 7 ms^{-1} 3) 14 ms^{-1} 4) 20 ms^{-1}

44. A body of mass 5kg at rest is acted upon by two mutually perpendicular forces 6N and 8N simultanecusly. Its kinetic energy after 10s is,

1) 100 J 2) 200 J 3) 500 J 4) 1000 J

45. Two masses 10 gm and 40 gm are moving with kinetic energies in the ratio 9 : 25. The ratio of their linear momenta is,

1) 5 : 6 2) 3 : 10 3) 6 : 5 4) 10 : 3

46. If the kinetic energy of a body is fourtimes its momentum, its velocity is (1999 M)

1) 2 ms^{-1} 2) 4 ms^{-1} 3) 8 ms^{-1} 4) 16 ms^{-1}

47. A machine gun fires 360 bullets per minute. Each bullet moves with a velocity of 600 ms^{-1}. If the power of the gun is 5.4 kw, the mass of each bullet is,

1) 5 kg 2) 0.5 kg 3) 0.05 kg 4) 5 gm

48. A 1.0 HP motor pumps out water from a well of depth 20m and fills a water tank of volume 2238 litres at a height of 10m from the ground. The running time of the motor to fill the empty tank is (g = 10 ms^{-2})(2000E)

1) 5 min 2) 10 min 3) 15 min 4) 20 min

49. A force applied by the engine of a train of mass 2.05×10^6 kg changes its velocity from 5ms^{-1} to 25ms^{-1} in 5 minutes. The power of the engine

1) 1.025 MW 2) 2.05 MW 3) 5 MW 4) 6 MW

50. The average resisting force that must act on a 5kg mass to reduce its speed from 65cms^{-1} to 15 cms^{-1} in 0.2s is

1) 12.5 N 2) 25 N 3) 50 N 4) 100 N

51. A ball is projected vertically down with an initial velocity from a height of 20m on to a horizontal floor. During the impact it loses 50% of its energy and rebounds to the same height. The velocity of projection is (g = 10ms^{-2}) (2000 E)

1) 20 ms^{-1} 2) 15 ms^{-1} 3) 10 ms^{-1} 4) 5 ms^{-1}

52. A stone is projected vertically up to reach a maximum height 'h'. The ratio of its kinetic to potential energies at a height $\frac{4h}{5}$ will be (1999E)

1) 5 : 4 2) 4 : 5 3) 1 : 4 4) 4 : 1

53. Two bodies of masses 4kg and 16kg are at rest. The ratio of times for which the same force must act on them to produce the same kinetic energy in both of them is

1) 1 : 4 2) 2 : 1 3) 1 : 2 4) 4 : 1

54. A block of mass 'm' is connected to one end of a spring of 'spring constant' k. The other end of the spring is fixed to a rigid support. The mass is released slowly so that the total energy of the system is then constituted by only the potential energy, then 'd' is the maximum extension of the spring. Instead, if the mass is released suddenly from the same initial position, the maximum extension of the spring now is : (g–acceleration due to gravity) (2009-M)

1) $\frac{mg}{k}$ 2) $\frac{mg}{3k}$ 3) 2d 4) 4d

55. A man of mass 60kg lifts a 15kg mass to the top of a building of height 10m in 5minutes. His efficiency is

1) 10% 2) 20% 3) 30% 4) 40%

56. A motor of power P_0 is used to deliver water at a certain rate through a given horizontal pipe. To increase the rate of flow of water through the same pipe n times, the power of the motor is increased to P_1. The ratio of P_1 to P_0 is :

1) n : 1 2) n^2 : 1 3) n^3 : 1 4) n^4 : 1

57. A motor is used to lift water from a well of depth 10m and to fill a water tank of volume 30m^3 in 10 minutes. The tank is at a height of 20m above ground. If 20% of energy is wasted, the power of the motor is

1) 18.75kW 2) 20 kW
3) 22.5 kW 4) 37.5 kW

58. A car without passengers moving with certain velocity on a level ground can be stopped in a distance of 10m. If the passengers add 25% of its weight, its stopping distance for the same braking force and velocity is (ignore friction)

1) 15m 2) 10m 3) 7.5m 4) 12.5m

59. A bullet loses 25% of its initial kinetic energy after penetrating through a distance of 2cm in a target. The further distance it travels before coming to rest is

1) 1 cm 2) 2 cm 3) 8/3 cm 4) 6 cm

60. A particle of mass 'm' is moving in a horizontal circle of radius 'r' under a centripetal force equal to $-\frac{k}{r^2}$, where k is a constant. The total energy of the particle

1) $\frac{k}{r}$ 2) $-\frac{k}{2r}$ 3) $-\frac{k}{r}$ 4) $\frac{k}{2r}$

61. A person of mass M = 90 kg standing on a smooth horizontal plane of ice throws a body of mass m = 10 kg horizontally on the same surface. If the distance between the person and body after 10 seconds is 10 metres, the K.E of the person (in Joules) is (2008-M)

1) 0.45 2) 4.5 3) 0.90 4) zero

62. An athlete in the olympic games covers a distance of 100 m in 10 s. His kinetic energy can be estimated to be in the range (AIEEE-2008)

1) 200 J - 500 J 2) 2×10^5 J - 3×10^5 J
3) 20,000 J - 50,000 J 4) 2,000 J - 5,000 J

63. A particle is projected up from a point at an angle θ with the horizontal direction. At any time 't', if p is the linear momentum, y is the vertical displacement, 'x' is horizontal displacement, the graph among the following which does not represent the variation of kinetic energy 'k' of the particle is :

(2009 E & M)

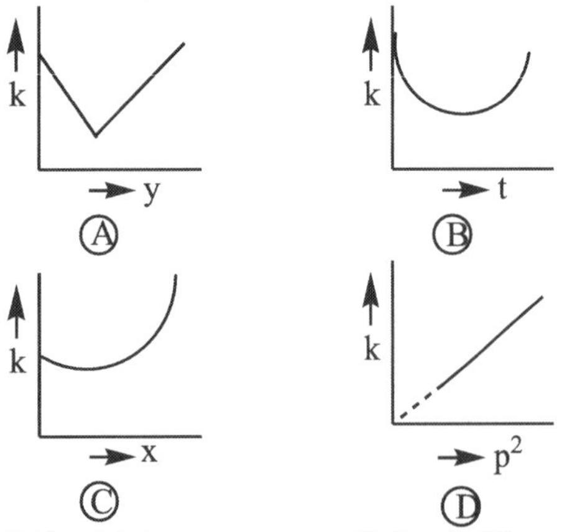

1) Graph(A) 2) Graph (B)
3) Graph (C) 4) Graph (D)

ANSWERS

1) 1	2) 4	3) 3	4) 2	5) 3
6) 2	7) 3	8) 4	9) 2	10) 4
11) 1	12) 2	13) 1	14) 4	15) 1
16) 2	17) 1	18) 1	19) 3	20) 4
21) 4	22) 1	23) 2	24) 3	25) 2
26) 4	27) 1	28) 1	29) 2	30) 2
31) 3	32) 4	33) 2	34) 3	35) 3
36) 1	37) 4	38) 4	39) 4	40) 1
41) 2	42) 1	43) 2	44) 4	45) 2
46) 3	47) 4	48) 3	49) 2	50) 1
51) 1	52) 3	53) 3	54) 3	55) 2
56) 3	57) 1	58) 4	59) 4	60) 2
61) 1	62) 4	63) 1		

EXERCISE - II(B)

1. The work done in moving a body of mass 5kg from the bottom of a smooth inclined plane to the top is 50J. If the angle of inclination of the inclined plane is 30^0, its length is (g = 10 ms^{-2})

1) 1/2m 2) 1m 3) 2m 4) 4m

2. A body of mass 15kg is raised from certain depth. If the work done in raising it by 10m is 1620J, its velocity at this position is (g = 10ms^{-2})

1) 2ms^{-1} 2) 4 ms^{-1} 3) 1 ms^{-1} 4) 8 ms^{-1}

3. A bucket filled with water having a mass of 15kg is raised from a well of depth 12m with the help of a rope. If the work done to do so is 2160J, the linear density of the rope is (g=10ms^{-2})

1) 0.2kg/m 2) 0.1 kg/m
3) 1 kg /m 4) 0.5 kg/m

4. A block of mass 6kg is lowered with the help of a rope of negligible mass through a distance of 20m with an acceleration of 1.96 ms^{-2}. Work done by the rope on the block is

1) 940.8J 2) 470.4J 3) 235.2J 4) –940.8J

5. A force $\overline{F} = \left(4\hat{i} - 3\hat{j} + 5\hat{k}\right)$N moves a body from $\left(3\hat{i} + 6\hat{j} + 3\hat{k}\right)$m to $\left(a\hat{i} - 8\hat{j} + 5\hat{k}\right)$m. If the work done by the force is 8J, the value of 'a' is

1) 6 2) 4 3) 8 4) – 8

6. A uniform rope of length 'L' and linear density 'μ' is on a smooth horizontal table with a length 'ℓ' lying on the table. The work done in pulling the hanging part on to the table is

1) $\frac{\mu g(L-\ell)^2}{2}$ 2) $\frac{\mu g(L-\ell)^2}{2\ell^2}$

3) $\frac{\mu g(L-\ell)^2}{2L^2}$ 4) $\frac{\mu gL}{2(L-\ell)}$

7. One–fourth chain is hanging down from a table. Work done to bring the hanging part of the chain on to the table is (mass of chain=M and length = L) (AIEEE - 2002)

1) $\frac{MgL}{32}$ 2) $\frac{MgL}{16}$ 3) $\frac{MgL}{8}$ 4) $\frac{MgL}{4}$

8. A person of mass 50kg carrying a box walks up a staircase. The height and width of each step is 20cm and 15cm. If the work done in walking up 50 steps is 7840J, the mass of the box is,

1) 20kg 2) 30kg 3) 40kg 4) 50kg

9. A rectangular block of dimensions 10m x 6m x 3m is lying on a horizontal surface with its smallest area in contact with the surface. If the work done in arranging it with its largest area in contact with the surface is 13.86MJ, the density of the body is ($g=10ms^{-2}$)

1) 2200 gm/c.c 2) 2.2 kg/m^3
3) 2.2 gm/c.c 4) 1800 kg/m^3

10. With the application of a force of 20N, the momentum of a body of mass 2kg changes from 16 kg ms^{-1} to 24 kg ms^{-1}. Then,

1) Time of application of force is 4s
2) Displacement of the body during this change is, 8m
3) Work done by the force is, 80J
4) Change in kinetic energy is 320J

11. A uniform rod of mass 2kg and length 1m is lying on a horizontal surface. If the work done in raising one end of the rod through an angle 45^0 is 'W', the work done in raising it further through 45^0 is,

1) W 2) $\sqrt{2}$W 3) W/$\sqrt{2}$ 4) ($\sqrt{2}$−1)W

12. A ladder 'AB' 2.5m long and of weight 150N with its centre of mass at a distance 1m from end 'A' is on the ground. A 40N weight is attached to the end B. The work to be done to arrange the ladder in vertical position with end 'A' contact with the ground is [1999M]

1) 190J 2) 250J 3) 285J 4) 475J

13. A body of mass m is accelerated uniformly from rest to a speed v in a time T. The instantaneous power delivered to the body as a function of time, is given by : (AIEEE 2005)

1) $\frac{mV^2}{T^2}t$ 2) $\frac{mV^2}{T^2}t^2$ 3) $\frac{1}{2}\frac{mv^2}{T^2}t$ 4) $\frac{1}{2}\frac{mv^2}{T^2}t^2$

14. The force – displacement graph is shown in the figure. The work done is

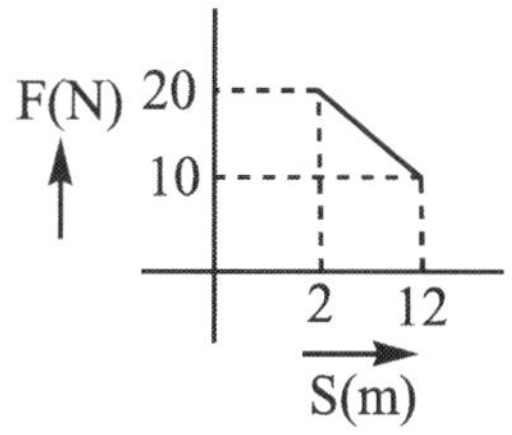

1) 252 N 2) 27.5 N
3) 150 J 4) 110 J

15. A locomotive of mass m starts moving so that its velocity varies as $v=K\sqrt{S}$, where K is a constant and S is the distance traversed. The total work done by all the forces acting on the locomotive during the first t second after the start of motion is

1) $\frac{1}{2}mK^4t^2$ 2) $\frac{1}{4}mK^4t^2$
3) $\frac{1}{8}mK^4t^2$ 4) $\frac{1}{16}mK^4t^2$

16. Velocity – time graph of a particle moving in a straight line is as shown in figure. Mass of the particle is 2 kg. Work done by all the forces acting on the particle in time interval between t = 0 to t = 10 is

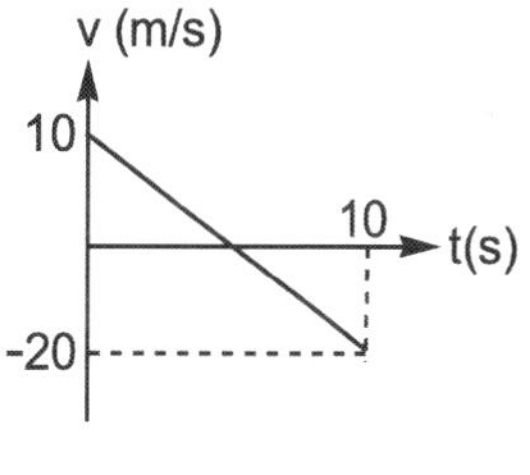

1) 300 J 2) – 300 J 3) 400 J 4) – 400 J

17. A constant force is applied on a 10kg mass at rest. The ratio of works done in 1st, 2nd and 3rd seconds is

1) 1 : 2 : 3 2) 1 : 3 : 5
3) 1 : 4 : 9 4) 1 : 1 : 1

18. A uniform metal rod is suspended from one end. When it is displaced through 60^0 from the vertical, the change in potential energy is 'U', the change in potential energy when it is displaced further through 60^0 is

1) U 2) U/2 3) 2U 4) 3U

19. Two spheres of same material are moving with kinetic energies in the ratio 108 : 576. If the ratio of their velocities is 2 : 3, the ratio of their radii is

1) 1 : 1 2) 2 : 3 3) 3 : 4 4) 4 : 3

20. A liquid of density 0.8 gm/c.c is flowing through a pipe. If the kinetic energy per cubic meter of the liquid is 14.4 KJ/m^3, the velocity of flow is

1) 4 ms^{-1} 2) 2ms^{-1} 3) 8 ms^{-1} 4) 6ms^{-1}

21. When a spring is compressed by a distance 'x', the potential energy stored is U_1. When it is further compressed by a distance 2x, the increase in potential energy is U_2. The ratio of U_1:U_2 is

1) 1 : 7 2) 1 : 4 3) 1 : 8 4) 1 : 3

22. If the momentum of a body decreases by 30% its kinetic energy decreases by

1) 60% 2) 51% 3) 69% 4) 90%

23. If the mass of a moving body is decreased by one - third of its initial mass and velocity is tripled, the percentage change in its kinetic energy is

1) 500% 2) 600% 3) 300% 4) 200%

24. A train of mass 800 metric tons is moving with a velocity of 72 KMPH. If the resistive force acting on the train is 10N per metric ton, the power of the engine is

1) 16 kw 2) 160 kw 3) 8 kw 4) 240 kw

25. Gravel is falling on a conveyor belt at the rate of 12 kg s^{-1}. The extra power required to move the belt with a velocity of 5ms^{-1} is

1) 15 w 2) 300 w 3) 30 w 4) 150 w

26. A car of mass 2 metric tons is moving up a smooth inclined plane with a velocity of 72KMPH. If the power of the engine is 8 kw, the angle of inclination of the inclined plane is

1) $\mathrm{Tan}^{-1}\left(\frac{1}{49}\right)$ 2) $\cos^{-1}\left(\frac{48}{49}\right)$

3) $\cos^{-1}\left(\frac{1}{49}\right)$ 4) $\sin^{-1}\left(\frac{1}{49}\right)$

27. A body of mass 10 kg moves according to the relation $x = t^2 + 2t^3$. The work done by the force in the first 2s is.

1) 7840 J 2) 1960 J

3) 3920 J 4) 4840 J

28. A motor pump lifts water from an average depth of 15 m at the rate of 5 kg per second. If the water is coming out of the pump with a velocity of 4 ms^{-1}, the power of the motor is ($g = 10ms^{-2}$)

1) 200 w 2) 790 w 3) 395 w 4) 1580 w

29. A man of mass 80 kg carrying a load of 20 kg walks up a stair case in 20s. If the number of steps is 40 and width and height of each step are 20 cm and 15 cm respectively. The efficiency of the man is

1) 20% 2) 25% 3) 40% 4) 50%

30. A body of mass 'm' dropped from a height 'h' reaches the ground with a velocity $\sqrt{\frac{gh}{2}}$. The work done by air resistance is

1) $\frac{mgh}{2}$ 2) $\frac{mgh}{4}$ 3) $\frac{2mgh}{4}$ 4) $\frac{3mgh}{4}$

31. A bullet of mass 10 gm moving with a velocity 'u' penetrates 4 cm in to a target before coming to rest. If the average resistance offered by the target is 200 N, the value of 'u' is

1) 20 ms^{-1} 2) 40 ms^{-1} 3) 60ms^{-1} 4) 80 ms^{-1}

32. A body of mass 20 kg is moving along a straight line with a linear momentum of 240 kg ms^{-1}. If a constant force of 20 N acts for 5s opposite to direction of motion, the work done by the force during this time is

1) 950 J 2) 1440 J 3) 1190 J 4) 1200 J

33. The energy required to accelerate a body from 30ms^{-1} to 60 ms^{-1} is three times the energy required to accelerate it from 10 ms^{-1} to 'v'. The value of 'v' is

1) 30 ms^{-1} 2) $20\sqrt{2}$ ms^{-1}

3) $30\sqrt{3}$ ms^{-1} 4) $10\sqrt{10}$ ms^{-1}

34. A simple pendulum of length 20 cm and of mass 0.1 kg is pulled to one side through an angle 'θ' and released. If its maximum velocity is 1.4 ms^{-1}, the value of 'θ' is

1) 30^0 2) 45^0 3) 60^0 4) 90^0

35. A body is projected at an angle 30^0 with a kinetic energy of 80J. The ratio of its potential to kinetic energies at half the maximum height is

1) 1 : 1 2) 1 : 3 3) 1 : 5 4) 1 : 7

36. A ball is released from a height of 10 m. If it loses 20% of its energy on hitting the ground, the height to which it bounces is

1) 7.5 2) 8 m 3) 9 m 4) 6 m

37. A car moving with a velocity of 72 KMPH stops its engine just at the foot of a hill. If the car ascends a height of 12m, the energy lost in overcoming friction is ($g = 10m/s^2$)

1) 40 % 2) 25 % 3) 60% 4) 50%

38. A force of 100 N acts on a body of mass 5 kg at rest for 4s. If the body moves through 10m in a direction making an angle 60^0 with the direction of force, the work done by the force is

1) 80 KJ 2) 500 J 3) 10 KJ 4) 160 KJ

39. A body of mass 12 kg is at the bottom of a lake of depth 5 m. If the work done in raising the body to the water surface is 350 J, the specifice gravity of the body is

1) 1.2 2) 4.8 3) 2.1 4) 2.4

40. The work done against friction in moving a body from A to B is 200 J. If the body is again brought back to A along the same path, the total work done for the trip against friction is

1) 0 J 2) 400 J 3) 200 J 4) 800 J

41. A simple pendulum is swinging in vertical plane. The ratio of its potential energies when it is making angles 45^0 and 90^0 with the vertical is

1) 1 : 1 2) $1 : (\sqrt{2}+1)$

3) $\sqrt{2} : 1$ 4) $(\sqrt{2}-1) : \sqrt{2}$

42. A block of mass 1 kg is dropped on a spring from a height of 20 cm. If the spring compresses by 5 cm, the force constant of the spring is ($g = 10\ ms^{-2}$)

1) 2000 Nm^{-1} 2) 1000 Nm^{-1} 3) 200 Nm^{-1} 4) 500 Nm^{-1}

43. A simple pendulum is swinging in a vertical plane. Its highest point above ground is 50cm.If the maximum velocilty of the pendulum is 1.4 ms^{-1}, its lowest position above ground is

1) 20 cm 2) 30 cm 3) 40 cm 4) 25 cm

44. Two forces 10 N each inclined to each other at angle 120^0 act on a 6 kg mass at rest.Its kinetic energy after 12 s is

1) 400 J 2) 1200 J 3) 600 J 4) 900 J

45. If two bodies of masses 3 gm and 5 gm are moving with linear momenta in the ratio 2 : 3, the ratio of their kinetic energies is

1) 4 : 15 2) 3 : 5 3) 5 : 6 4) 20 : 27

46. If the momentum of a body moving with constant velocity is four times its kinetic energy, its velocity is

1) 2 ms^{-1} 2) 1 ms^{-1} 3) 0.5 ms^{-1} 4) 0.25 ms^{-1}

47. A machine gun having a power 27 kw fires 'n' bullets per second each of mass 10 gm. If the velocity of each bullet is 300 ms^{-1}, the value of 'n' is

1) 60 2) 120 3) 600 4) 300

48. A motor of power 2.45 kw draws water from a well of depth 20 m. If it fills a tank of capacity 3000 lit in 7 minutes. The height of the tank above ground is

1) 5 m 2) 10 m 3) 15 m 4) 20 m

49. A train of mass 4000 metric tons is moving with a velocity of 18 KMPH. With the application of a force, its velocites changes to 'ϑ' in 2 minutes. If the power of the engine is 10 MW, the value of 'ϑ' is (in KMPH)

1) 72 2) 90 3) 36 4) 54

50. The force which changes the velocity of a 200 gm mass from 36 KMPH to 180 KMPH in 0.02s is

1) 40 N 2) 200 N 3) 400 N 4) 20 N

51. A ball is projected vertically down with a velocity of $10\sqrt{3}$ m/s from a height of 20 m. On hitting the ground if it loses 40% of its energy, the height to which it bounces is ($g=10ms^{-2}$)

1) 20 m 2) 30 m 3) 19 m 4) 21 m

52. When a body is projected vertically up from ground, the ratio of its potential to kinetic energies at a height 'h' above the ground is 3 : 4. The height above ground at which the ratio of potential to kinetic energies becomes 4 : 3 is

1) $\frac{4}{3}h$ 2) $\frac{4}{7}h$ 3) $\frac{3}{7}h$ 4) $\frac{7}{4}h$

53. Two bodies of masses 2 kg and 8 kg are at rest. If same force acts on them, the ratio of distances travelled by them before they attain same kinetic energy is

1) 1:1 2) 1 : 2 3) 4 : 1 4) 2 : 1

54. When a mass 'm' is attached to a vertical spring and slowly released, the extension in the spring is 'x'. When a mass '2m' is attached to the same spring and released suddenly, the extension in the spring is

1) x 2) 2x 3) 3x 4) 4x

55. Two persons A and B each of mass 60 kg are carrying loads 15 kg each. If they reach the top of a building of height 10 m in 20s and 40s respectively, the ratio of their efficiencies is

1) 1 : 1 2) 1 : 2 3) 2 : 1 4) 4 : 1

56. **A motor pump is delivering water at certain rate. In order to increase the rate of delivery by 100%, the power of the motor is to be increased by**

1) 300 % 2) 200 % 3) 400 % 4) 700 %

57. **A motor of efficiency 80 % draws water from a well of depth 20 m and fills a tank of dimensions 4m x 3m x 2m in 5 minutes. If the tank is at a height of 10 m above ground, the power of the motor is**

1) 0.3 kw 2) 24 kw 3) 120 kw 4) 30 kw

58. **A car without passengers moving with a constant velocity can be stopped in a distance of 20m. If the passengers add 40% of the weight, its stopping distance for the same braking force and velocity is (ignore friction)**

1) 14m 2) 28m 3) 16m 4) 24m

59. **A bullet emerges out from a wooden plank with 75% of its initial kinetic energy. The number of additional planks required to stop the bullet is**

1) 2 2) 3 3) 4 4) 5

60. **A particle of mass 'm' is moving in a horizontal circle of radius 'r' under a centripetal force equal to $-\frac{k}{r^2}$, where k is a constant. The potential energy of the particle**

1) $\frac{k}{r}$ 2) $-\frac{k}{2r}$ 3) $-\frac{k}{r}$ 4) $\frac{k}{2r}$

61. **A person of mass 40kg standing on a smooth horizontal plane of ice throws a body of mass 10kg with a velocity $8ms^{-1}$. Then the total kinetic energy of person and body.**

1) 100J 2) 200 J 3) 300 J 4) 400J

62. **The mass of an athlete is 40kg. He can cover a distance of 150m in maximum of 15 sec and minimum of 10 sec then the range of his kinetic energy is**

1) 2000J – 4500J 2) 4500 J – 9000J

3) 2000 J – 9000J 4) 4000 J – 6000 J

63. **A particle is projected up from a point at an angle 'θ' with the horizontal direction. Which of the following graph represents the variation kinetic energy (K) with time (t).**

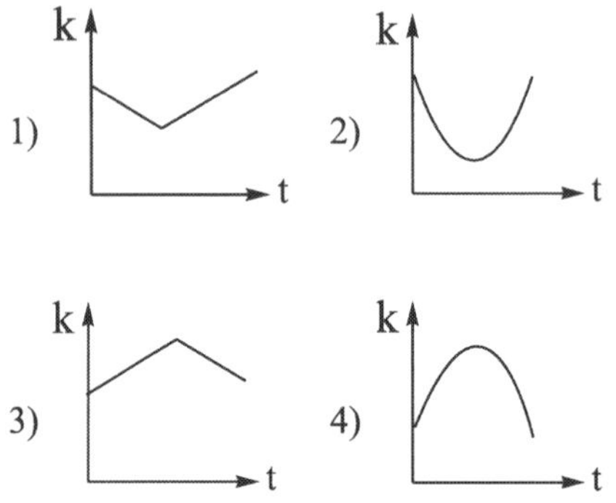

ANSWERS

1) 3	2) 2	3) 4	4) 4	5) 4
6) 1	7) 1	8) 2	9) 3	10) 3
11) 4	12) 2	13) 1	14) 3	15) 3
16) 1	17) 2	18) 3	19) 3	20) 4
21) 3	22) 2	23) 1	24) 2	25) 2
26) 4	27) 3	28) 2	29) 1	30) 4
31) 2	32) 1	33) 4	34) 3	35) 4
36) 2	37) 1	38) 2	39) 4	40) 2
41) 4	42) 1	43) 3	44) 2	45) 4
46) 3	47) 1	48) 3	49) 2	50) 3
51) 4	52) 1	53) 1	54) 4	55) 4
56) 4	57) 4	58) 2	59) 2	60) 3
61) 4	62) 1	63) 2		

EXERCISE - III(A)

1. **The radius of a well is 7m. Water in it is at a depth of 20m and depth of water column is 10m. Work done in pumping out water completely from the well is, ($g = 10ms^{-2}$)**

1) 38.5 MJ 2) 38.5 KJ

3) 46.2 MJ 4) 385 MJ

2. **A particle of mass 'm' is projected with a velocity 'u' at an angle 'α' with the horizontal. Work done by gravity during its descent from its highest point to, the position where its velocity vector makes an angle $\frac{\alpha}{2}$ with the horizontal is,**

1) $\frac{1}{2}mu^2\tan^2\alpha$ 2) $\frac{1}{2}mu^2\tan^2\frac{\alpha}{2}$

3) $\frac{1}{2}mu^2\cos^2\alpha\,Tan^2\frac{\alpha}{2}$

4) $\frac{1}{2}mu^2\cos^2\frac{\alpha}{2}\sin^2\alpha$

3. **Twenty identical cubical blocks each of mass 100gm and of side 10cm are lying on a horizontal surface. Work done in piling them one above the other is, (g = $10ms^{-2}$)**

1) 20 J 2) 2 J 3) 19 J 4) 10 J

4. **A uniform chain of mass 'm' and length 'L' is kept on a horizontal table with half of its length hanging from the edge of the table. Work done in pulling the chain on to the table so that only $\frac{1}{5}$th of its length now hangs from the edge is,**

1) $\frac{mgl}{8}$ 2) $\frac{mgl}{50}$ 3) $\frac{mgl}{18}$ 4) $\frac{21mgl}{200}$

5. **A uniform rope of length ℓ and mass 'm' hangs over a horizontal table with two third part on the table. The coefficient of friction between the table and the chain is μ. The work done by the friction during the period the chain slips completely off the table is**

1) $\frac{2}{9}\mu mgl$ 2) $\frac{2}{3}\mu mgl$ 3) $\frac{1}{3}\mu mgl$ 4) $\frac{1}{9}\mu mgl$

6. **A small block of mass 'm' is kept on a rough inclined surface of inclination θ fixed in an elevator. The elevator goes up with a uniform velocity V and the block does not slide on the wedge. The work done by the force of friction on the block in a time 't' will be**

1) zero 2) mgvt $\cos^2\theta$

3) mgvt $\sin^2\theta$ 4) $\frac{1}{2}$ mgvt $\sin 2\theta$

7. **The displacement of a body of mass 2 kg varies with time 't' as S = t^2 + 2t, where S is in metres and 't'is in seconds. The work done by all the foces acting on the body during the time interval t = 2 s to t = 4 s is**

1) 36 J 2) 64 J 3) 100 J 4) 120 J

8. **An engine develops 10KW of power. How much time will it take to lift a mass of 200kg to a height of 40m. (g = $10ms^{-2}$) (AIEEE-2004)**

1) 4s 2) 5s 3) 8s 4) 10s

9. **A rectangular plank of mass 'm_1' and height 'a' is on a horizontal surface. On the top of it another rectangular plank of mass 'm_2' and height 'b' is placed. The potential energy of the system is,**

1) $(m_1+m_2)\frac{(a+b)}{2}g$ 2) $\left[\frac{m_1+m_2}{2}.a+m_2\frac{b}{2}\right]g$

3) $\left[\left(\frac{m_1}{2}+m_2\right)a+m_2\frac{b}{2}\right]g$ 4) $\left[\left(\frac{m_1}{2}+m_2\right)a+m_1\frac{b}{2}\right]g$

10. **A box of mass 50kg at rest is pulled up on an inclined plane 12m long and 2m high by a constant force of 100N. When it reaches the top of the inclined plane if its velocity is $2ms^{-1}$, the work done against friction in Joules is (g = $10ms^{-2}$)**

1) 50 2) 100 3) 150 4) 200 **(2002 M)**

11. **Two idential cylindrical vessels each of area of cross – section A are on a level ground. Each contains a liquid of density 'ρ'. The heights of liquid columns are h_1 and h_2. If the two vessels are connected by means of a narrow pipe at the bottom, the work done by gravity in equalizing the liquid levels is,**

1) $\frac{A\rho g(h_1-h_2)^2}{2}$ 2) $\frac{A\rho g}{2}(h_1^2-h_2^2)$

3) $\frac{A\rho g}{4}(h_1^2-h_2^2)$ 4) $\frac{A\rho g}{4}(h_1-h_2)^2$

12. **The kinetic energy of a man is half that of a boy whose mass is half that of the man. When the man speeds up by $5ms^{-1}$, his kinetic energy is 100% more than that of the boy. The initial velocity of the man is,**

1) $(\sqrt{2}+1)$ m/s 2) 5 m/s

3) $2(\sqrt{2}-1)$ m/s 4) 2 m/s

13. **An open knife edge of mass M is dropped from a height 'h' on a wooden floor. If the blade penetrates a distance 's' into the wood, the average resistance offered by the wood to the blade is :**

1) Mg 2) $Mg\left(1+\frac{h}{s}\right)$

3) $Mg\left(1-\frac{h}{s}\right)$ 4) $Mg\left(1+\frac{h}{s}\right)^2$

14. **A shell of mass 'm' moving horizontally explodes in to two equal pieces at the instant its momentum is '3p'. One of the fragments attains a linear momentum of '4p' in upward direction. The kinetic energy gained by the system immediatley after explasion is,**

1) $\frac{25p^2}{m}$ 2) $\frac{16p^2}{m}$ 3) $\frac{41p^2}{m}$ 4) $\frac{73p^2}{2m}$

15. **A car of mass 1000kg moves on a horizontal surface. If the road friction amounts to 10% of weight of the car and air resistance amounts to 2% of the weight, the power required to run the car at a uniform speed of 36 KMPH is**

1) 112 kW 2) 56 kW 3) 12 kW 4) 6 kW

16. A dam is situated at a height of 300m above sea level. It supplies water to a power house which is at a height of 60m above sea level. 1000kg of water passes through the turbine per second. If the efficiency of the system is 80%, the electrical power out put is,(g = 10 ms^{-2})

1) 4 MW 2) 2.5 MW 3) 3 MW 4) 6 MW

17. A particle moves in a straight line with retardation proportional to its displacement. Its loss of kinetic energy for any displacement x is proportional to : (AIEEE 2004)

1) x^2 2) e^x 3) x 4) $\log_e x$

18. 800 kW power is supplied to a motor of 70% efficiency. This motor is in turn connected to a crane, whose efficiency is also 70%. If this crane lifts a 20 ton rock, its velocity is,

1) 2 ms^{-1} 2) 0.2 ms^{-1} 3) 20 ms^{-1} 4) 10 ms^{-1}

19. A car weighing 500kg climbs up a hill of slope 1 in 49 with a velocity of 36 KMPH. If the frictional force is 50N, the power delivered by the engine is,

1) 500W 2) 5 kW 3) 1.5 kW 4) 150 W

20. A truck moves up a hill of incline 1 in 50 with a speed of 20 KMPH. If the resistance to the motion of the truck is 4% of its weight, the speed of the truck while moving down hill utilising the same power is,

1) 20 KMPH 2) 40 KMPH
3) 60 KMPH 4) 80 KMPH

21. A block of mass 2 kg slides along a frictionless table with a speed of 10 m/sec. Directly in front of it and moving in the same direction is a block of mass 5 kg moving at 3 m/sec. A massless spring of spring constant k = 1120 N/m is attached to the back side of 5 kg mass as shown in figure. When the blocks collides, the maximum compression in the spring (if the spring does not bend) will be :

1) 0.25 m
2) 0.4 m
3) 0.33 m
4) 1.12 m

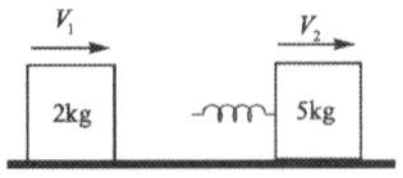

22. A spring of force constant 'k' is stretched by a small length 'x'. The work done in stretching it further by a small length 'y' is,

1) $\frac{1}{2}k(x^2+y^2)$ 2) $\frac{1}{2}k(x+y)^2$

3) $\frac{1}{2}k(y^2-x^2)$ 4) $\frac{1}{2}ky(2x+y)$

23. A body of mass 1kg is dropped from a height of 5m on to the ground.If the body penetrates 2cm in to the ground, the average resistance offered by the ground on to the body is (g = 10ms^{-2})

1) 2510 N 2) 35 N 3) 50.2 N 4) 2500 N

24. A mason worker throws the bricks to a height of 10m, at which the bricks reach with a velocity of 6ms^{-1}. The percentage of energy he is wasting is nearly (g = 10ms^{-2})

1) 15% 2) 18% 3) 220% 4) 36%

25. A body is projected vertically up with certain velocity. At a point 'P' in its path, the ratio of its potential to kinetic energies is 9 : 16. The ratio of velocity of projection to velocity at 'P' is,

1) 3 : 4 2) 5 : 4 3) 9 : 25 4) 25 : 16

26. When a body is projected vertically up, at a point 'P' in its path, the ratio of potential to kinetic energies is 3 : 4. If the same body were to be projected with two times the initial velocity, the ratio of potential to kinetic energies at the same point is

1) 3 : 25 2) 3 : 28 3) 1 : 20 4) 1 : 25

27. A unifrom chain of length 'L' is placed on a smooth table of height 'h' (h > L) with a length 'ℓ' hanging from the edge of the table. The chain begins to slide down the table. When the end of the chain is about to leave the edge of the table its velocity is

1) $\sqrt{\frac{g(L+\ell)}{L}}$ 2) $\sqrt{\frac{g(L-\ell)}{L}}$

3) $\sqrt{\frac{g(L^2-\ell^2)}{L}}$ 4) $\sqrt{2g(L-\ell)}$

28. A body is projected vertically up with a velocity of 60 m/s. The percentage of its initial kinetic energy converted into potential energy after 3s is (g = 10 ms^{-2})

1) 25 % 2) 75 % 3) 50 % 4) 87.5 %

29. A bullet of mass 10 gm is fired horizontally with a velocity of 1000 ms^{-1} from a height of 50m above the ground. If the bullet reaches the ground with a velocity of 500 ms^{-1}, the work done against air resistance in Joules is (g = 10ms^{-2}) (2006 E)

1) 5005 2) 3755 3) 3750 4) 17.5

30. A motor is pumping water with a velocity of 20ms^{-1} from a pump of area of cross section 10cm^2. Its average power is,

1) 8 kW 2) 12 kW 3) 16 kW 4) 4 kW

31. Human heart pumps 70 cc of blood at each beat against a pressure of 125 mm of Hg. If the pulse frequency is 72 per minute, the power of the heart is nearly

1) 1.2 W 2) 1.4 W 3) 1.6 W 4) 1.8 W

32. A body of mass 2 kg starts from rest and moves with uniform accleration. If it acquires a velocity of 20 ms^{-1} in 4s, the power consumed in two seconds is (2002 E)

1) 50W 2) 100W 3) 250W 4) 200W

33. A body of mass 100 gm is suspended from a massless spring of natural length 1 m. If the spring stretches through a vertical distance of 20 cm, the potential energy stored in the spring is ($g = 10\ ms^{-2}$)

1) 1.2J 2) 0.6J 3) 0.1J 4) 0.2J

34. A body of mass 2 kg is projected vertically up with a velocity of 100 ms^{-1}. If it rises to a height of 400 m, the energy utilized to overcome friction is ($g = 10\ ms^{-2}$)

1) 10 KJ 2) 8 KJ 3) 1 KJ 4) 2 KJ

35. A force of 10N acts on a body of mass 1 kg at rest for 10 s and the body acquires a kinetic energy K_1. A force of 10 N acts on a body of mass 1 kg at rest through a distance of 10 m and the body acquires a kinetic enegy K_2. The ratio of K_1 to K_2 is

1) 1 : 1 2) 2 : 1 3) 20 : 1 4) 50 : 1

36. The force required to tow a car at constant velocity is directly proportional to velocity. If it requires 160 W to tow a car with a velocity of 4 ms^{-1}, the power it required to move the car with a velocity of 9 ms^{-1} is

1) 900W 2) 180W 3) 810W 4) 640W

37. A body projected vertically up with a velocity of 20 ms^{-1} returns to the ground with a velocity of 18 ms^{-1}. The maximum height attained by the body is ($g = 10\ ms^{-2}$)

1) 12.1 m 2) 22.5 m

3) 25 m 4) 18.1 m

38. A block of mass 'm' moving with a speed 'υ' compresses a spring through a distance 'x' before its speed reduces by 25%. The spring constant of the spring is

1) $\frac{m\upsilon^2}{4x^2}$ 2) $\frac{3m\upsilon^2}{4x^2}$

3) $\frac{7m\upsilon^2}{16x^2}$ 4) $\frac{9m\upsilon^2}{16x^2}$

39. A small block of mass 'M' slides down from top edge 'A' of a smooth curved surface as shown in the figure. The surface becomes horizontal at edge 'B'. The maximum possible horizontal range for the body is

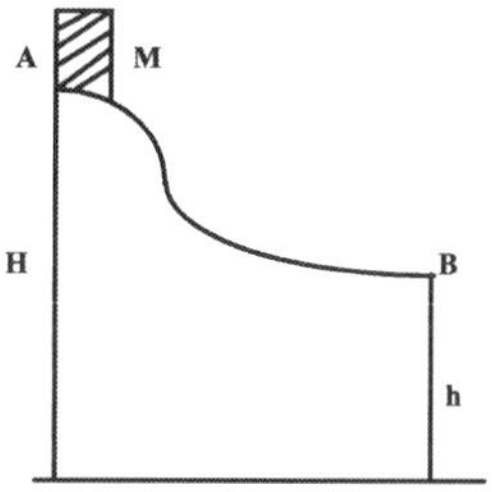

1) H 2) $\frac{H}{2}$

3) $\frac{H}{3}$ 4) $\frac{H}{4}$

40. In case of a freely falling body, the ratio of kinetic energy at the end of the third second to increase in kinetic energy in the next three seconds is

1) 1 : 1 2) 1 : 2 3) 1 : 3 4) 1 : 9

41. A car of mass 1250 kg experiences a resistance of 750 N when it moves at 30 ms^{-1}. If the engine can develop 30 KW at this speed, the maximum acceleration that the engine can produce is

1) 0.8 ms^{-2} 2) 0.2 ms^{-2}

3) 0.4 ms^{-2} 4) 0.5 ms^{-2}

42. A lorry and a car of mass ratio 4:1 are moving with KE in the ratio 3 : 2 on a horizontal road. Now brakes are applied and braking forces produced are in the ratio 1 : 2. Then the ratio of stopping timings of lorry and car is

1) 1 : 1 2) 1 : 3 3) $2\sqrt{6}:1$ 4) $6\sqrt{2}:1$

ANSWERS

1) 4	2) 3	3) 3	4) 4	5) 1
6) 3	7) 2	8) 3	9) 3	10) 2
11) 4	12) 2	13) 2	14) 4	15) 3
16) 3	17) 1	18) 1	19) 3	20) 3
21) 1	22) 4	23) 1	24) 1	25) 2
26) 1	27) 3	28) 2	29) 2	30) 4
31) 2	32) 2	33) 3	34) 4	35) 4
36) 3	37) 4	38) 3	39) 2	40) 3
41) 2	42) 3			

EXERCISE–III(B)

1. A well is in the form of a cylinder of radius 3.5m. Water surface in it is at depth of 15m and depth of water column is 10m. Work done in draining out half of water from it is, ($g = 10ms^{-2}$)

1) 33.7 MJ 2) 42.5 MJ

3) 67.375 MJ 4) 18.75 MJ

2. **A particle of mass 'm' is projected with a velocity 'u' at an angle 'θ' with the horizontal. Work done by gravity during its descent from its highest point to a point which is at half the maximum height is**

1) 0 2) $\frac{mu^2 \sin^2\theta}{4}$

3) $\frac{1}{2}mu^2 \sin^2\theta$ 4) $\frac{1}{2}mu^2 \cos^2\theta$

3. **'n' identical cubes each of mass 'm' and side 'a' are lying on a horizontal surface. Work done in piling them one above the other is W_1. Work done in arranging them in two identical columns is W_2. The ratio of $W_1 : W_2$ is**

1) 1 : 2 2) (n – 1) : (n – 2)

3) 2 (n – 1) : (n – 2) 4) (2n – 1) : (2n – 2)

4. **A uniform chain of mass 'm' and length 'l' is lying on a smooth horizontal table with one – third of its length hanging from the edge of the table. Work done in pulling the chain on to the table so that 5/6th of the chain is on the table is**

1) $\frac{mg\ell}{12}$ 2) $\frac{mg\ell}{48}$ 3) $\frac{mg\ell}{72}$ 4) $\frac{mg\ell}{24}$

5. **Power applied to a particle varies with time as P=(3t^2–2t+1) watt, where t is in second. The change in its kinetic energy between time t=2 sec. and t=4 sec. is**

1) 32 J 2) 46J 3) 61 J 4) 102 J

6. **A force $\vec{F} = -K(y\hat{i} + x\hat{j})$ acts on a particle moving in the xy–plane. Starting from the origin, the particle is taken along the positive x–axis to the point (a, 0) and then parallel to the y–axis to the point (a, a). The total work done by the force $\vec{F}$ on the particle is**

1) –2Ka^2 2) 2Ka^2 3) –Ka^2 4) Ka^2

7. **The work done by a force $\vec{F} = \left(-6x^3\hat{i}\right)N$ in displacing a particle from x = 4 m to x = – 2 m is**

1) – 240 J 2) 360 J 3) 420 J 4) 100 J

8. **If the speed of a vehicle increases by 2ms^{-1}, its kinetic energy is doubled, then original speed of the vehicle is (AIEEE - 2002)**

1) $(\sqrt{2}+1)$ms^{-1} 2) $2(\sqrt{2}-1)$ms^{-1}

3) $2(\sqrt{2}+1)$ms^{-1} 4) $2(\sqrt{2}+1)$ ms^{-1}

9. **On the top of a cube of mass 'M_1' and side 'a' another cube of mass 'M_2' and side 'b' is lying. It the upper block is removed, the decrease in potential energy of the system is**

1) M_2 (a + b)g 2) $M_2\left(a+\frac{b}{2}\right)g$

3) $M_2\left(\frac{a}{2}+b\right)g$ 4) $M_2\left(\frac{a+b}{2}\right)g$

10. **A block of mass 100kg is pulled up on an inclined plane of length 10m and 3m high by a constant force of 400N. If the work done against friction is 200J, the velocity of the block at the top of the inclined plane is (g = 10ms^{-2})**

1) 2ms^{-1} 2) 4ms^{-1} 3) 8ms^{-1} 4) 1ms^{-1}

11. **There are two identical containers each of area of cross section 'A' one has liquid of density 'd' upto a height 'h' and the other one is empty. If they are joined by means of narrow pipe at the bottom, the work done by gravity in equalizing the water levels is**

1) $\frac{Adgh^2}{4}$ 2) $\frac{Adgh^2}{2}$

3) $\frac{Adgh^2}{16}$ 4) $\frac{Adgh}{2}$

12. **When a man running on a level ground increases his speed by 4ms^{-1}, if his kinetic energy increase by 125% his initial velocity is**

1) 4ms^{-1} 2) 8ms^{-1} 3) 2ms^{-1} 4) 6ms^{-1}

13. **A man of mass 'M' standing on a perfect smooth surface throws a body with a velocity 'υ' and kinetic energy 'K'. By the time the body travels a distance 'x' the separation between the man and the body is**

1) $\frac{2Kx}{M\upsilon^2}$ 2) $\frac{2Kx}{M}$

3) $x\left[1+\frac{2K}{M\upsilon^2}\right]$ 4) $x\left[1+\frac{K}{\upsilon^2}\right]$

14. **A shell of mass 'm' moving horizontally with a momentum '3P' explodes in to two equal pieces. If one of the pieces flies off vertically up with a momentum '4P', the kinetic energy of the other fragment immediatley after explosion is**

1) $\frac{5P^2}{2m}$ 2) $\frac{P^2}{m}$ 3) $\frac{25P^2}{2m}$ 4) $\frac{25P^2}{m}$

15. A body of weight 'W' is moving on a level ground with a velocity 'υ'. If frictional force and air resistance are 7% and 1% of weight of the vehicle respectively, the power of the engine is

1) $\frac{2W\upsilon}{25}$ 2) $\frac{5W\upsilon}{7}$ 3) $\frac{8W\upsilon}{7}$ 4) $\frac{7W\upsilon}{8}$

16. In a hydel power project, 2000kg of water passes through the turbine every second. Water is falling from a height of 240m above ground. If the power output is 6.4 Mw, the efficiency of the power house is (g= $10ms^{-2}$)

1) 60% 2) 75% 3) 80% 4) 90%

17. A body is moving under the expense of a constant power .If 'X' is the displacement in time 't', then 'x' is proportional to

1) $t^{2/3}$ 2) $t^{3/2}$ 3) t^2 4) $t^{1/2}$

18. 400KW power is supplied to a motor of efficiency 60%. This motor is connected to a crane. Transmission loss of crane is 30%. If its is used to lift a 2.1 ton box, its velocity is (g = $10ms^{-2}$)

1) $4ms^{-1}$ 2) $2ms^{-1}$ 3) $6ms^{-1}$ 4) $8ms^{-1}$

19. A car weighing 1000kg climbs up a hill of slope 1 in 98 with a velocity of 72KMPH. If the power delivered by the engine is 3.5KW, the frictional force acting on the car is

1) 35N 2) 50N 3) 75N 4) 17.5N

20. A car moves up a hill of incline 1 in 50 with a velocity of 15KMPH. The resistance due to friction is $\frac{1}{25}$th of weight of the car. The speed of the car while moving down hill using same power is

1) 15KMPH 2) 30 KMPH
3) 45 KMPH 4) 60 KMPH

21. A block of mass 'M' moving on the firctionless horizontal surface collides with the spring of spring constant 'k'. The other end of the spring is rigidly fixed to a vertical wall. If the spring is compressed by L, the maximum momentum of the block after collision is : (AIEEE 2005)

1) $\sqrt{Mk}\,L$ 2) $\frac{kL^2}{2M}$ 3) Zero 4) $\frac{ML^2}{k}$

22. A spring of force constant $1000Nm^{-1}$ is stretched by 4cm. The work done in stretching it further by 4cm is

1) 12J 2) 0.8J 3) 4.8J 4) 2.4J

23. A body of mass 2kg is dropped from a height of 4m. If the average resistance offered by the ground is 4020N, the distance through which the body penetrates in to the ground before coming to rest is (g = $10ms^{-2}$)

1) 0.02cm 2) 0.2cm 3) 4cm 4) 2cm

24. A ball of mass 0.2 kg is thrown vertically upwards by applying a force by hand. If the hand moves 0.2 m while applying the force and the ball goes upto 2m height further, the magnitude of the force is, (g = 10 m/s^2) (AIEEE 2006)

1) 4N 2) 16N 3) 20N 4) 22N

25. A body is projected vertically up with certain velocity. The ratio of velocity of projection to velocity at a point in its path is 6:5. The ratio of potential to kinetic energies at that point is

1) 9 : 25 2) 1 : 4 3) 11 : 36 4) 11 : 25

26. For a vertically projected body at a point 'P' in its path the ratio kinetic to potential energies is 2 : 1. If the body is projected up with triple the previous velocity, the ratio of potential to kinetic energies at the same point 'P' is

1) 1 : 5 2) 1 : 26 3) 26 : 1 4) 25 : 26

27. A uniform chain of length 3m is placed on a smooth horizonal table of height 5m, with a length 1m hanging from the edge of the table. The chain begins to slide down from the table. When the end of the chain is about to leave the edge of the table, its velocity is (g = $10ms^{-2}$)

1) $2\sqrt{\frac{5}{3}}ms^{-1}$ 2) $\frac{4}{3}\sqrt{5}\,ms^{-1}$ 3) $\frac{3}{\sqrt{5}}ms^{-1}$ 4) $4\sqrt{\frac{5}{3}}ms^{-1}$

28. A freely falling body takes 4s to reach the ground. One second after release, the percentage of its initial potential energy, that is still retained is

1) 6.25% 2) 25% 3) 37.5% 4) 93.75%

29. A body of mass 0.1kg released from a height of 50m reaches the ground with a velocity of $10ms^{-1}$. The work done against air resistance is (g = 10 m/s^2)

1) 5J 2) 50J 3) 0 4) 45J

30. A motor pump of power 1H.P is discharging water with a velocity of $10ms^{-1}$. In order to discharge water with a velocity of $20ms^{-1}$, the power of the motor should be

1) 2 H.P 2) 4 H.P 3) 8 H.P 4) 16 H.P

31. An air pump is pumping air in to a large vessel against a pressure of 1 atmosphere. In each stroke 1000C.C of air enters in to the vessel. If the number of strokes made by the pump per minute is 30, the power of the pump is [1 atm = 10^5Pa)

1) 5KW 2) 50KW
3) 100KW 4) 50W

32. A machine rated as 150W, changes the velocity of a 10kg mass from $4ms^{-1}$ to $10ms^{-1}$ in 4s. The efficiency of the machine nearly is

1) 70% 2) 30% 3) 50% 4) 40%

33. A block of mass 2kg is allowed to fall from a height of 40cm on to a vertical spring of force constant 1960N/m. The maximum compression produced is nearly.

1) 1cm 2) 10cm 3) 0.1cm 4) 100cm

34. A body of mass 4kg is projected vertically up with a velocity of $20ms^{-1}$. If the energy spent in over coming air resistance is 200J, the maximum height attained by it is (g = $10ms^{-2}$)

1) 10m 2) 20m 3) 5m 4) 15m

35. A force of 1N acts on a 1kg mass at rest for 1s. In another case 1N force acts on 1kg mass at rest and moves it through 1m. The ratio of kinetic energies in the two cases is

1) 1 : 1 2) 1 : 2 3) 1 : 4 4) 1 : $\sqrt{2}$

36. The force required to move a body with a constant velocity is directly proportional to velocity. The ratio of powers required to move a body with velocity $20ms^{-1}$ and $30ms^{-1}$ is

1) 2 : 3 2) 3 : 2 3) 4 : 9 4) 9 : 4

37. A body projected vertically up with a velocity of $10ms^{-1}$ returns to the point of projection with a velocity of $8ms^{-1}$. The maximum height attained by it is (g = $10ms^{-2}$)

1) 4 .1m 2) 3.2m 3) 2.6m 4) 5.3m

38. A body of mass 1kg moving with $40ms^{-1}$ hits a spring of force constant 2800N/m as shown in the figure. The compression in the spring by the time the speed of the body reduces by one - fourth of initial speed is (ignore friction)

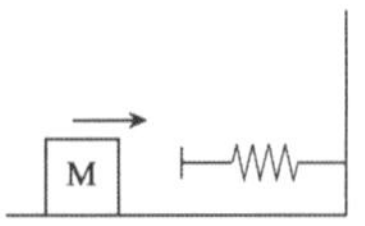

1) 10cm 2) 25cm
3) 40cm 4) 50cm

39. The minimum velocity required for a body at the bottom of a vertical circle to complete the circle of radius 'R' is $\sqrt{5gR}$. A body is released from the top edge of an inclined plane as shown in the figure. The minimum value of 'h' for which the body can complete the vertical circle is

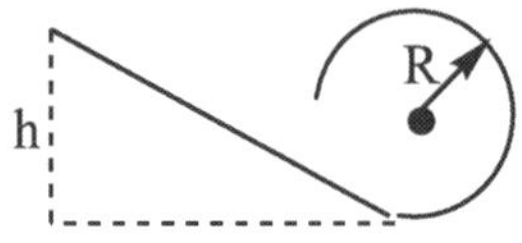

1) $\frac{2R}{5}$ 2) $\frac{R}{2}$ 3) $\frac{R}{5}$ 4) $\frac{5R}{2}$

40. A vertically projected body attains the maximum height in 6s. The ratio of kinetic energy at the end of 3rd second to decrease in kinetic energy in the next three seconds is

1) 1 : 1 2) 1 : 3 3) 3 : 1 4) 9 : 1

41. The power developed by the engine of a car of mass 1000kg at a speed of $20ms^{-1}$ is 10kw. If the maximum acceleration that can be produced is the car is $0.1ms^{-2}$, the force of resistance acting on the car at the speed of $20ms^{-1}$ is

1) 200N 2) 400N 3) 250N 4) 500N

42. Two blocks of masses 'm' and '3m' joined by means of a light spring are on a horizontal surface. They are pushed towards each other and released. If the frictional force acting on them is same, the ratio of distances travelled by them before coming to rest is

1) 1 : 3 2) 3 : 1 3) 1 : 9 4) 9 : 1

ANSWERS

1) 1	2) 2	3) 3	4) 4	5) 2
6) 3	7) 2	8) 3	9) 2	10) 2
11) 1	12) 2	13) 3	14) 4	15) 1
16) 2	17) 2	18) 4	19) 3	20) 3
21) 1	22) 4	23) 4	24) 4	25) 4
26) 2	27) 4	28) 4	29) 4	30) 3
31) 4	32) 1	33) 2	34) 4	35) 2
36) 3	37) 1	38) 4	39) 4	40) 1
41) 2	42) 2			

COMPREHENSION QUESTIONS

Comprehension - I (1 to 3) :

Mechanical energy exists in two forms : kinetic energy and potential energy. Kinetic energy is the energy possessed by a body by virtue of motion. Potential energy is the energy possessed by a body by virtue of its position or configuration. These two forms of energy are interconvertible. If no other form of energy is involved in a process, the sum of kinetic energy and potential energy always remains constant.

1. Two particles of masses m_1 and m_2 have equal linear momenta. The ratio of their kinetic energies is

   1) 1 2) $\sqrt{\frac{m_2}{m_1}}$ 3) $\frac{m_2}{m_1}$ 4) $\left(\frac{m_2}{m_1}\right)^2$

2. Two particles of masses m_1 and m_2 have equal kinetic energies. The ratio of their linear momenta is

   1) 1 2) $\sqrt{\frac{m_1}{m_2}}$

   3) $\frac{m_2}{m_1}$ 4) $\left(\frac{m_2}{m_1}\right)^2$

3. A particle of mass m has half the kinetic energy of another particle of mass m/2. If the speed of the heavier particle is increased by 2 ms^{-1}, its new kinetic energy equals the original kinetic energy of the lighter particle. The ratio of the original speeds of the lighter and heavier particles is

   1) 1 : 1 2) 1 : 2 3)2 : 1 4) 1 : 4

Comprehension - II (4 to 6) :

A light rod of length L having a body of mass M attached to its end hangs vertically. It is turned through 90^0 so that it is horizontal and then released.

4. The centripetal acceleration when the rod makes an angle θ with the vertical is

   1) g cos θ 2) 2g cos θ

   3) g sin θ 4) 2g sin θ

5. The tension in the rod when it makes an angle θ with the vertical is

   1) Mg cos θ 2) 2 Mg cos θ

   3) 3 Mg cos θ 4) zero

6. The value of θ when the tension in the rod equals the weight of the body is given by

   1) $\theta = \cos^{-1}\left(\frac{1}{2}\right)$ 2) $\theta = \cos^{-1}\left(\frac{1}{3}\right)$

   3) $\theta = \sin^{-1}\left(\frac{1}{2}\right)$ 4) $\theta = \sin^{-1}\left(\frac{1}{3}\right)$

Comprehension - III (7 to 9) :

The term kinetic energy is associated with bodies by virtue of motion. The term momentum is related to the ability of a body to communicate motion. The two are inter-related. A body with momentum has necessarily kinetic energy associated with it. Similarly a body with kinetic energy has moentum associated with it.

7. Two different bodies have equal momenta. Then

   1) The lighter body has greater kinetic energy

   2) the heavier body has greater kinetic energy

   3) both of them have equal kinetic energy

   4) information insufficient

8. Two different bodies have equal kinetic energy. Then

   1) The lighter body has greater momentum

   2) the heavier body has greater momentum

   3) both of them have equal momenta

   4) information insufficient

9. If the momentum of a body increases by 30%, the percentage increase in kinetic energy is

   1) 30% 2) 60% 3) 15% 4) 69%

Comprehension - IV (10 to 12) :

Two different bodies of masses m_1 and m_2 are initially at rest. The ratio of times for which the same force must act on both such that they have:

10. The same final kinetic energy is

    1) $\sqrt{\frac{m_1}{m_2}}$ 2) $\sqrt{\frac{m_2}{m_1}}$ 3) $\frac{m_1}{m_2}$ 4) $\frac{m_2}{m_1}$

11. The same final momenta is

    1) $\sqrt{\frac{m_1}{m_2}}$ 2) 1 : 1 3) $\frac{m_2}{m_1}$ 4) $\sqrt{\frac{m_2}{m_1}}$

12. The same final dispalcement is

    1) $\sqrt{\frac{m_1}{m_2}}$ 2) $\sqrt{\frac{m_2}{m_1}}$ 3) $\frac{m_1}{m_2}$ 4) $\frac{m_2}{m_1}$

Comprehension - V (13 to 15) :

The work done by a constant force acting on a body is given by W = F.r

where F is the force vector and r is displacement vector.

The displacement vector $r=r_2-r_1$ where r_1 is the initial position vector and r_2 is the final position vector.

If the force is variable, the work done in moving a body from a position r_1 to a position r_2 is given by

$$W=\int_{r_1}^{r_2} F.dr$$

where dr is an infinitesimally small displacement.

13. A particle is moved from a position $r_1 = (3\hat{i}+2\hat{j}-4\hat{k})$ metre to a position $r_2 = (5\hat{i}+6\hat{j}+9\hat{k})$ metre under the action of a force $F=(\hat{i}+3\hat{j}+\hat{k})$ newton. The work done is

1) zero 2) 13 J 3) 27 J 4) 35 J

14. A body of mass m is projected from a tower of height h with a velcoity u at an angle θ above the horizontal. The work done by the gravitational force during the time it takes to hit the ground is

1) $mgh(1+\cos\theta)$ 2) $mgh(1+\sin\theta)$

3) mgh 4) zero

15. A body of mass m is projected from the ground with a velocity u at an angle θ above the horizontal. The work done by the gravitational force in time $t=\frac{u\sin\theta}{g}$ from the ground is

1) $2mu^2\sin^2\theta$ 2) $mu^2\sin^2\theta$

3) $-\frac{mu^2\sin^2\theta}{2}$ 4) zero

Comprehension - VI (16 to 18) :

A Small roller coaster starts at point A with a speed u on a curved track as shown in Fig. The friction between theroller coaster and the track is negligible and it always remains in contact with the track.

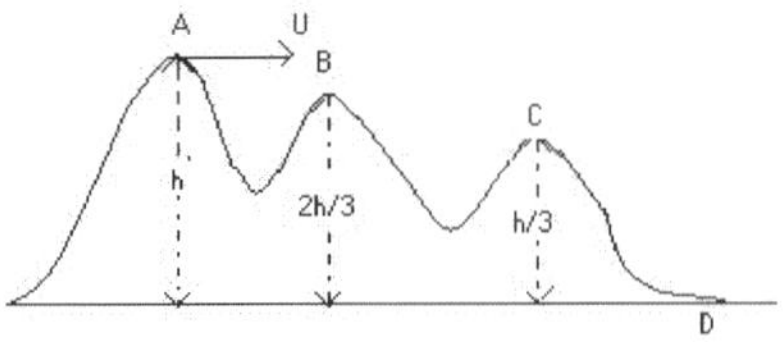

16. The speed of the roller coaster at point B on the track will be

1) $(u^2+gh)^{1/2}$ 2) $\left(u^2+\frac{2gh}{3}\right)^{1/2}$

3) $(u^2+2gh)^{1/2}$ 4) $\left(u^2+\frac{3gh}{2}\right)^{1/2}$

17. The speed of the roller coaster at point C on the track will be

1) $\left(u^2+\frac{gh}{3}\right)^{1/2}$ 2) $\left(u^2+\frac{2gh}{3}\right)^{1/2}$

3) $\left(u^2+\frac{4gh}{3}\right)^{1/2}$ 4) $(u^2+2gh)^{1/2}$

18. The speed of the roller coaster at point D on the track will be

1) $(u^2+gh)^{1/2}$ 2) $(u^2+2gh)^{1/2}$

3) $(u^2+3gh)^{1/2}$ 4) $(u^2+4gh)^{1/2}$

ANSWERS

1) 3	2) 2	3) 3	4) 2	5) 3
6) 2	7) 1	8) 2	9) 4	10) 1
11) 2	12) 1	13) 3	14) 3	15) 3
16) 2	17) 3	18) 2		

Chapter - 8

THERMAL PROPERTIES OF MATTER

EXPANSION OF GASES

SYNOPSIS

1. Gases do not have fixed size and fixed shape. But they have pressure.
2. While doing experiments, a gas must be taken in a closed container.
3. The volume of the space in which gas molecules can move around is taken as the volume of the gas. Hence the volume of the closed container itself must be taken as the volume of the gas irrespective of its mass
4. The real volume occupied by the molecules of the gas is several times lesser than measured volume of the gas.
5. Volume of a gas can be measured by a gas burette or Eudiometer.
6. Pressure of a gas means pressure applied by the gas molecules on the walls of the container. When the gas is taken in a cylindrical vessel provided with a frictionless piston, the pressure of the gas will be equal to the external pressure acting on the piston.
7. Pressure of a gas is measured by manometer, Bourden gauge (high pressures) and Mcleod gauge (low pressures)
8. The three parameters that determine the physical state of a gas are P,V&T.
9. Volume Coefficient of a gas at constant pressure (α):-

   At constant pressure, the ratio of change in the volume of a given mass of gas per degree rise in temperature to its volume at 0^0c is called volume coefficient of the gas.

$$\alpha = \frac{V_2 - V_1}{V_0(t_2 - t_1)} = \frac{V_t - V_0}{V_0 t} = \frac{V_2 - V_1}{V_1 t_2 - V_2 t_1}$$

10. Pressure Coefficient of a gas at constant volume (β) :-

$$\beta = \frac{P_2 - P_1}{P_0(t_2 - t_1)} \text{ (or) } \beta = \frac{P_t - P_0}{P_0 t} \text{ (or) } \beta = \frac{P_2 - P_1}{P_1 t_2 - P_2 t_1}$$

11. For any gas $\alpha = \beta = \frac{1}{273} / ^0 c = 0.0036/^0C$

13. Significance of Absolute zero :- At that temperature, the pressure and volume of a gas, both would be zero.

14. Boyle's law:- At constant temperature, the pressure of a given mass of gas is inversely proportional to the volume of the gas. (or) At constant temperature, the pressure of a given mass of gas is directly proportional to its density.

   ie, $P_1V_1 = P_2V_2 = \ldots$ (or) $\frac{p_1}{d_1} = \frac{p_2}{d_2} = \ldots$

15. At constant temperature the graph between pressure and volume of a gas is called "iso-therm" and the shape is rectangular hyperbola.

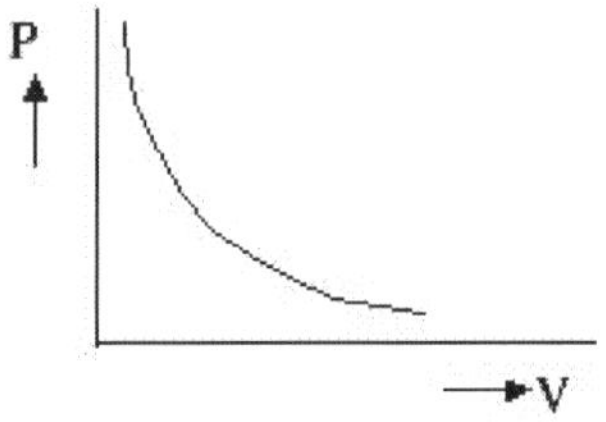

16. PV-V graph is a straight line parallel to volume axis.
17. P-1/V graph is a straight line passing through the origin.
18. Log P-log V graph is a straight line making a negative intercept on log V axis.
19. In Boyles law experiment , the graph drawn between 'h' (excess pressure) and $1/l$ (where l is the length of the air column) is a straight line making a negative intercept on 'h' axis whose magnitude gives atmospheric pressure

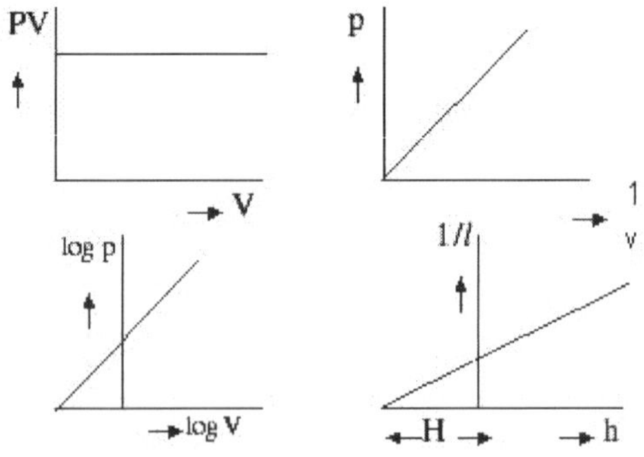

20. Real gases obey Boyles law only at high temperatures and low pressures.
21. At constant temperature, when the pressure of a gas is increased by x%, percentage decrease in the volume is $\frac{100x}{(100+x)}\%$
22. Two vessels of volumes V_1 and V_2 contain air at pressures P_1 and P_2 at same temperature. If both the vessels are interconnected then common pressure,

$$P = \frac{P_1V_1 + P_2V_2}{V_1 + V_2}$$

23. A quill tube contains air column of length l_1 with its open end vertically upwards , l_2 with its open end vertically downwards and l_3 when it is tilted through an angle θ from vertical open end up position .
therefore $P_1 = H + h$,
$P_2 = H - h$
and $P_2 = H + h\cos\theta$.
Here 'h' is the height of Hg column.
According to Boyles law,
$(H + h)l_1 = (H - h)l_2 = (H + h\cos\theta)l_3$
24. If an air bubble formed at the bottom of a liquid column raises to the top where its volume becomes n times that at the bottom then the depth of the liquid column is given by
$$h = \frac{76x13.6}{d}(n-1)\ \text{cm}$$
where d is the density of the liquid in CGS system.
If the radius of the bubble becomes 'n' times then h = $\frac{76x13.6}{d}(n^3-1)$
here atmospheric pressure is taken as 76cm of mercury.
25. In general , $h - \frac{Hd_b(n-1)}{d_l}$
here H = atmospheric pressure
d_b = density of liquid in barometer
d_l = density of liquid in lake
$n = \frac{V_2}{V_1}$,the ratio of final to initial volumes.
26. Charle's first law:- At constant pressure, the volume of a gas is directly proportional to its absolute temperature. ie $\frac{V_1}{T_1} = \frac{V_2}{T_2} = \cdots\cdots$ a constant
27. Charles' second law (or) Gay Lussac's Law :- At constant volume, for a given mass of gas $\frac{P_1}{T_1} = \frac{P_2}{T_2} = \cdots\cdots$ a constant
28. The slope of V-T graph of a given mass of gas increases with the decrease of pressure because $\frac{V}{T} \alpha \frac{1}{P}$
29. The slope of P-T graph of a given mass of gas increases with the decrease in the volume of the gas because $\frac{P}{T} \alpha \frac{1}{V}$.
30. At constant temperature and pressure, the volume of a gas is directly proportional to the mass of the gas.
31. At constant temperature and volume, the pressure of a gas is direclty proportional to the mass of the gas.
32. Gas equation :
a) When the mass of the gas is varying :
$$\frac{P_1V_1}{m_1T_1} = \frac{P_2V_2}{m_2T_2}$$
(or) $\frac{PV}{mT}$ = r (r = specific gas constant)

b) when the mass of the gas remains constant;
$$\frac{P_1V_1}{T_1} = \frac{P_2V_2}{T_2} = \ldots.$$
(or) $\frac{PV}{T}$ = constant
The value of the constant depends upon the mass of the gas.
c) when the mass of the gas is 1gm
$\frac{PV}{T}$ =r (r = specific gas constant)
value of r changes from gas to gas.

d) for 1gm - mole of a gas, $\frac{PV}{T} = R$
value of R is same for all gases and is called universal gas constant.
$R = 8.314\ J\ mol^{-1}\ K^{-1} = 2\ cal\ mol^{-1}\ {}^0C^{-1}$.
Dimensional formula of
$R = [ML^2\ T^{-2}\ K^{-1}mol^{-1}]$
Relation between R and r is, R=Mr where M is the molecular weight of the gas
e) for any amount of gas ,
In the following equations, m, N_A, N, n represent the mass of the gas, Avagadro's number, number of molecules and number of moles respectively.
i) interms of r : $\frac{PV}{T} = mr$
ii) interms of R : $\frac{PV}{T} = nR$
iii) interms of density $\rho : \frac{P}{\rho T} = \frac{R}{M}$ = constant
iv) interms of Boltzmann constant K,
$\frac{PV}{T} = n\,N_A\,K$: $[nN_A = N]$ & $[K = \frac{R}{N_A}]$
v) In terms of number of molecules per unit volume, $\mu : \frac{P}{\mu T} = K$ (here $\mu = \frac{N}{V}$)
33. When gases which are seperately in the states (P_1, V_1, T_1) . . . are mixed and if the resultant mixture is at (P,V,T) then $\frac{P_1V_1}{T_1} + \frac{P_2V_2}{T_2} + \ldots\ldots = \frac{PV}{T}$

34. In the above case, while mixing, if the external work done is zero and the energy of the system remain constant then $n_1T_1+n_2T_2 ..= (n_1+n_2+ . .)T$ where n_1, n_2 . . . are the number of moles in those gases respectively.

35. If two vessels of same capacity containing air at T_1 and T_2 K and at pressures P_1 and P_2 atmospheres then the common pressure is given by

$$P = \frac{P_1T_2 + P_2T_1}{T_1 + T_2}$$

EXERCISE-I

1. **A real gas deviates from perfect gas behaviour as its molecules**

   1) have definite size 2) attract each other
   3) are not spherical 4) both 1 and 2

2. **An ideal gas is that which**

   1) cannot be liquified
   2) can be easily liquified
   3) has strong inter molecular forces
   4) has a larger size of molecules

3. **It is decided to verify Boyle's law over a wide range of temperature and pressures. The most suitable gas to be selected for this purpose is [CPMT - 78]**

   1) Carbon dioxide 2) Helium
   3) Oxygen 4) Hydrogen

4. **Boyle's law is applicable in**

   1) Isobaric process 2) Isochoric process
   3) Isothermal process 4) Adiabatic process

5. **When a gas is forced in a smaller volume without change in temperature, its pressure increase because its molecules**

   1) strike unit area of the container walls more number of times
   2) strike the unit area of the container walls at higher speed
   3) strike the unit area of the container wall with greater area.
   4) have more energy

6. **The pressure of a gas does not depend up on [Alligarh 85 - 86]**

   1) Kinetic energy of its molecules
   2) The temperature of gas
   3) The density of the gas
   4) The volume of the gas when the density remains unchanged

7. **When the pressure of a gas is doubled under constant temperature the density becomes**

   1) double 2) halved
   3) does not change 4) decrease to 1/4.

8. **When the volume of a saturated vapour is decreased, its pressure [82 E]**

   1) increases according to Boyle's law
   2) decreases according to Boyle's law
   3) changes but not according to Boyle's law
   4) remains constant

9. **A closed vessel contains some gas at atmospheric pressure and room temperature. It is placed in a fast moving train. Then the temperature of the gas**

   1) increases 2) decreases
   3) remains same
   4) depends on the nature of the gas

10. **When the temperature of a gas is increased at constant pressure, the number of molecules per c.c.**

    1) decreases 2) increases
    3) remains same 4) zero

11. **Under which of the following conditions PV=RT is obeyed most closely by a real gas ?**

    1) high pressure and high temperature
    2) low pressure and low temperature
    3) low pressure and high temperature
    4) high pressure and low temperature

12. **Equation of gas in terms of pressure (p), absolute temperature (T) and density (d) is [2005 M]**

    1) $\frac{P_1}{T_1d_1} = \frac{P_2}{T_2d_2}$ 2) $\frac{P_1T_1}{d_1} = \frac{P_2T_2}{d_2}$

    3) $\frac{P_1d_2}{T_2} = \frac{P_2d_1}{T_1}$ 4) $\frac{P_1d_1}{T_1} = \frac{P_2d_2}{T_2}$

13. **A sample of an ideal gas occupies a volume 'V' at pressure 'P' and absoulte temperature 'T'. The mass of each molecule is m. The expression for the density of gas is [88 E&M]**

1) mKT 2) P/KT

3) P/KTV 4) Pm/KT

14. **The ratio of the volume expansivity of Helium to pressure expansivity of Hydrogen is**

1) 1 2) 2 3) 273 4) $\frac{1}{273}$

15. **The difference between volume and pressure coefficient of an ideal gas is [2005 E]**

1) 1/273 2) 273 3) 2/273 4) zero

16. **The law that leads to absolute scale of temperature is**

1) Kelvin's law 2) Charels' law

3) Boyles law 4) None

17. **The scale of temperature which does't have negative temperatures is**

1) Centigrade 2) Fahrenheit

3) absolute 4) None

18. **The temperature at which the volume of all the gases is zero is**

1) 274 K 2) -273°C

3) 273 K 4) -273K

19. **Universal gas constant has the units**

1) dynes /mole-°C 2) erg / mole-K

3) erg-cm/mole-K 4) watt/mole-K

20. **Which of the following methods will enable the volume of an ideal gas to be made 4 times**

1) Absolute temperature and pressure must be doubled

2) at constant pressure the absolute temperature must be increased by 4 times

3) at constant temperature the pressure must be increased by 4 times

4) quarter the absolute temperature at constant pressure.

21. **The graph between temperature in °C and pressure of a perfect gas is**

1) hyperbola

2) a straight line passing though the origin

3) A straight line parallel to pressure axis intercepting temperature axis as -273°C

4) a straight line with +ve intercept on pressure axis and intercepting the temperature axis as -273°C

22. **A real gas can be approximated to an ideal gas at [98E]**

1) low density 2) high pressure

3) high density 4) low temperature

23. **For Boyles law to hold good the gas should be**

1) Perfect and at constant temperature and mass

2) Real and at constant temperature and mass

3) Perfect and at constant temperature but variable mass.

4) Real and at constant temperature but variable mass

24. **The increase in volume of the bubble as it rises from the bottom of the beaker with water at constant temperature is due to [82 E]**

1) Charle's law 2) Boyle's law

3) Avogadro's law 4) Dalton's law

25. **We write the Boyles law as PV=C when the temperature remain constant. In this relation the magnitude of C depend upon**

1) The nature of the gas used in the experiment

2) The magnitude of g in the laboratory

3) The atmospheric pressure

4) The quantity of the gas enclosed

26. **Molar gas constant is the same for all gases because at the same temperature and pressure, equal volumes of all gases have the same**

1) number of molecules

2) K.E. 3) density

4) Ratio of specific heats

27. **A gas deviates maximum from the ideal gas laws at**

1) high temperature and high pressure

2) low temperature and low pressure

3) high temperature and low pressure

4) low temperature and high pressure

MULTIPLE ANSWERS & MATCHING TYPE

28. Real gas deviats from perfect gas behaviour as its molecules

a) Have definete size

b) Attract each other

c) Show brownian motion

d) Are not spherical

1) a,b & c are true 2) a & b are true

3) a, b & d are true 4) a & c are true

29. According to kinetic theory of gases, at zero kelvin

a) Pressure of ideal gas is zero

b) Volume of ideal gas is zero

c) Internal energy of ideal gas is zero

d) Matter exists in gaseous state only

1) a & d are true 2) a, b & d are true

3) a, b & c are true 4) all are true

30. The perameters that determine the physical state of a gas are

a) Pressure b) Volume

c) No.of moles d) Temperature

1) a & b 2) a, b & c

3) a. b & d 4) a, c & d

31. PV = nRT holds good for

a) Isobaric process b) Isochoric process

c) Isothermal process d) Adiabatic process

1) a & b 2) a, b & c 3) a, b & d 4) all

32. In the equation PV = constant, the numerical value of constant depends upon

a) Temperature b) Mass of the gas

c) System of units used d) Nature of the gas

1) a & b 2) b & c 3) c & d 4) all

33. Boyles law is applicable when

a) Temperature is constant

b) Gas is at high temperature & low pressure

c) The vessel enclosing the gas is a good conductor

d) The process is isothermal

1) a & b 2) b, c & d

3) a, b & c 4) a,b,c&d

34. Boyle's law can be verified by

a) Boyle's law apparatus b) Quill tube apparatus

c) Regneult's apparatus d) Jolly's bulb apparatus

1) a & b are true 2) a & c are true

3) b & d are true 4) b & c are true

35. Real gases aproaches ideal gas at high temperature & low pressures because

a) Interatomic separation is large

b) Size of the molecule is negligible when compared to inter atomic seperation.

1) a & b are true 2) Only a is true

3) Only b is true 4) a & b are false

36. The gas constant for a molecule is (terms have usual meaning)

a) K b) R/N c) R/NT d) R/m

1) Only 'a' 2) only C

3) Both a & b 4) only d

37. The units of universal gas constant is

a) J mol^{-1} K^{-1} b) J kg^{-1} k^{-1}

c) Litre atm $mol^{-1}k^{-1}$ d) cal mol^{-1} k^{-1}

1) a & b 2) a & c 3) a, c & d 4) all

38. LIST - I LIST - II

LIST - I	LIST - II
a) Barometer	e) Charles Law
b) Specific gas constant	f) J mol^{-1} k^{-1}
c) Gas thermometer	g) Boyle's law
d) Universal gas constant	h) $JKg^{-1}K^{-1}$

1) a - h, b - e, c - f, d - g

2) a - g, b - h, c - e, d - f

3) a - f, b - g, c - h, d - e

4) a - e, b - f, c - g, d - h

39. LIST - I LIST - II

LIST - I	LIST - II
a) Real Gas	e) Boltzmann constant
b) Gas constant for 1 molecule of a gas	f) Molecular forces are present
c) Gas const. for 1 mole of a gas	g) Molecular forces are absent
d) Ideal gas	h) Universal Gas constant

1) a - e, b - f, c - g, d - h

2) a - f, b - e, c - h, d - g

3) a - g, b - h, c - e, d -f

4) a - h, b - g, c - f, d - e

40. LIST - I / LIST - II

LIST - I	LIST - II
a) P - V graph (T is constant)	e) St.line cutting temp. axis at -273^0C
b) P - T graph (V is constant)	f) Rectangular Hyperbola
c) V - t graph (P is constant)	g) A.St line parallel to pressure axis
d) PV - P graph (T is constant)	h) st.line passing thorugh orgin

1) a - g , b - e, c - h, d - f

2) a - h, b - f, c - g, d - e

3) a - e, b - g, c - f, d - h

4) a - f, b - h, c - e, d - g

ASSERTION & REASON TYPE QUESTIONS

Instruction For Questions :

In each of the following questions, a statement of Assertion (A) is given followed by a corresponding statement of reason (R) just below it. Of the statement mark the correct answer.

1) Both 'A' and 'R' are true and 'R' is the correct explanation of 'A'

2) Both 'A' and 'R' are true and 'R' is not the correct explanation of 'A'

3)'A' is true and 'R' is false

4)'A' is false and 'R' is true

41. A) : The volume coefficient of any gas is $\frac{1}{273}/^0C$

R) : The pressure coefficient of any gas is $\frac{1}{273}/^0C$

42. A) : In Jolly's bulb apparatus,as reservoir is moved up, the mercury level raises into the bulb

R) : The pressure on the enclosed gas increases

43. A) : v-t graph for gas at constant pressure is a straight line passing through origin [t - temp in celsius]

R) : Volume is directly proportional to temperature (Kelvin) at constant pressure

44. A) : At constant temperature, when pressure of gas is increased by 10% , then volume decreases by less than 10%

R) : At constant temperature the volume of gas is inversely proportional to the pressure acting on it.

45. A) : Pressure of gas is same every where inside a closed container

R) : The gas molecules under go elastic collisions among themselves and with walls

46. A) : Ideal gasses do not undergo cooling during expansion.

R) : Molecules of ideal gas do not attract mutually

47. A) : With increase in temperature, the pressure of given gas increases

R) : Increase in temperature causes decrease in no.of collission of molecules with walls of container.

48. A) : Volume of ideal gas is taken as volume of its container

R) : Volume of real gas molecules is negligible when compared with that of container.

49. A) : P-V graph (at contant temperature) for idal gas is rectangular hyporbola

R) : Ideal gas obeys Charles' law.

50. A) : Ideal gas molecules posses potential energy.

R) : There is no force of attraction between molecules of ideal gas.

51. A) : PV/T = constant for 1 gram of gas. This constant varies from gas to gas.

R) : 1 gram of different gases at NTP occupy different volumes

52. A) : $\frac{PV}{T}$ = constant for 1 mole of gas. This constant is same for all gases

R) : 1 mole of different gases at NTP occupy same volume of 22.4 litres.

53. A) : When a solid iron ball is heated , the percentage change in its volume is greatest

R) : $\beta=2\alpha$ and $\gamma=3\alpha$ where the symbols have their usual meaning

54. A) : Water at 4^0C expands both on heating and cooling

R) : H_2O has least volume at 4^0C due to the molecular orientation

55. A) : Anamalous expansion of water is favorable for acquatic animals

R) : The lowest possible temperature at bottom of water lake is 4^0C

56. A) : In a bimetallic strip, the element with more linear expansion coefficient will occupy the outer side of curvature

R) : For all materials , the linear expansion coefficient is three times the cubical expansion coefficient.

ANSWERS

1) 4	2) 1	3) 4	4) 3	5) 1
6) 4	7) 1	8) 3	9) 3	10) 1
11) 3	12) 1	13) 4	14) 1	15) 4
16) 2	17) 3	18) 2	19) 2	20) 2
21) 4	22) 1	23) 1	24) 2	25) 4
26) 1	27) 4	28) 2	29) 3	30) 3
31) 4	32) 4	33) 4	34) 1	35) 1
36) 3	37) 3	38) 2	39) 2	40) 4
41) 2	42) 1	43) 4	44) 1	45) 1
46) 1	47) 3	48) 2	49) 2	50) 4
51) 1	52) 1	53) 2	54) 1	55) 1
56) 4				

EXERCISE-II(A)

1. The volume of agas at 20^0C is 100CC at normal pressure when it is heated to 100^0C, its volume is 125CC at the same pressure the volume coeficient of the gas is

1) $\frac{1}{200}/^0C$ 2) $\frac{1}{400}/^0C$

3) $\frac{1}{300}/^0C$ 4) $\frac{1}{273}/^0C$

2. If a given mass of a gas occupies a volume 100 cc at one atmospheric pressure and a temperature of 100^0C. What will be its volume at 4 atmospheric pressure, the temperature being the same? [NCERT - 77]

1) 100 cm^3 2) 400 cm^3

3) 25 cm^3 4) 200 cm^3

*3. A litre of air is heated from 27^0C to 177^0C at constant pressure. Find its volume.

1) 1.5 litres 2) 2 litres

3) 2.5 litres 4) 3.5 litres

*4. The pressure of a given mass of gas at 27^0C is 75 cm of mercury. Find the temperature in ^0C at which the pressure is doubled, the gas being heated at constant volume.

1) 227^0C 2) 327^0C 3) 420^0C 4) 627^0C

5. A gas at 27^0C and pressure of 30 atm is allowed to expand to atmosphere pressure and volume 15 times larger. The final temperature of the gas is..... [85 E]

1) -123^0C 2) 123^0C 3) -132^0C 4) 132^0C

6. If an air bubble rises from the bottom of a mercury tank to the top its volume becomes $1\frac{1}{2}$ times. When normal pressure is 76cm of Hg then the depth of the Hg tank is

1) 38 cm 2) 132 cm 3) 76 cm 4) 49 cm

7. What should be the percentage increase in the pressure so that the volume of a gas may decrease by 5% at constant temperature ?

1) 5% 2) 10% 3) 5.26% 4) 4.26%

8. If the presure of an ideal gas contains in a closed vessel is increased by 0.5%, the increase in temperature is 2^0C. The initial temperature of the gas is [2003 M]

1) 27^0C 2) 127^0C 3) 300^0C 4) 400^0C

9. A given amount of a gas is heated till the volume and pressure both increase by 2% each. The percentage change in temperature of the gas is nearly

1) 2% 2) 3% 3) 1% 4) 4%

10. At constant pressure, the ratio of increase in volume of an ideal gas per degree rise in Kelvin temperature to its original volume is [2004 M]

1) $\sqrt{3}$ 2) $\sqrt{273}$ 3) $\frac{1}{273}$ 4) $\frac{1}{\sqrt{3}}$

11. Two samples of Hydrogen and Oxygen of same mass possess same pressure and volume. The ratio of their temperatures is

1) 1 : 8 2) 1:16 3) 8:1 4) 16:1

12. The volume occupied by 8gm of oxygen at S.T.P is

1) 11.2 lt 2) 22.4 lt

3) 2.8 lt 4) 5.6 lt

13. 1gm of H_2 at S.T.P occupies 11.2 litre. The volume occupied by 3gm of H_2 at 300K and 0.7m of Hg will be

1) 40.1 lit 2) 46.7 lit

3) 46.6 lit 4) 47.9 lit

14. Two gases A and B having the same pressure, P, volume V and temperature T are mixed. If the mixture has volume and temperature as V and T respectively the pressure of the mixture is [98M] [2007E]

1) 2P 2) P 3) P/2 4) 4P

15. For an ideal gas V-T curves as constant pressures P_1 & P_2 are shown in figure - from the figure

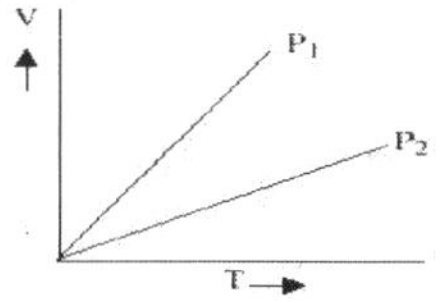

1) $P_1 > P_2$
2) $P_1 < P_2$
3) $P_1 = P_2$
4) $P_1 \geq P_2$

16. Two different curves at const temperature. The relationship between volume V and the pressure P at a given temp. of same ideal gas are shown for masses m_1 and m_2 of the gas respectively. Then

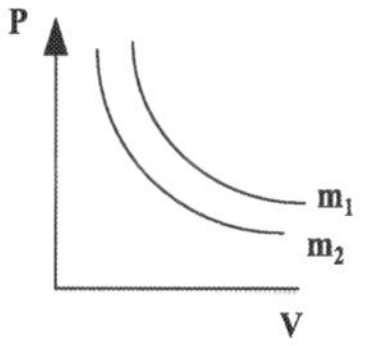

1) $m_1 > m_2$
2) $m_1 = m_2$
3) $m_1 < m_2$
4) $m_1 = 2\,m_2$

17. The equation of state crresponding to 8 gm of O_2 is

1) PV = RT 2) $PV = \frac{RT}{4}$
3) $\frac{RT}{3}$ 4) $PV = \frac{RT}{2}$

18. A vessel is filled with an ideal gas at a pressure of 20 atm and is at a temperature of 27°C. One half of the mass of the gas is removed from the vessel and the temperature of the remaining gas is increased to 87°C. At this temperature the pressure of the gas will be

1) 80 atm 2) 8.3atm
3) 12atm 4) 0.12 atm

19. The volume of a mass of gas at 37°C and a pressure of 75 cm of mercury is 620 c.c. Find the volume at N.T.P.

1) 500CC 2) 480CC
3) 326CC 4) 538.8CC

20. A Steel tank is filled with a gas at 150 atmosphere and at 20°C. If the pressure raise to 250 atmospheres, the tank explodes. Find the temperature at which the tank explodes.

1) 215.3°C 2) 273.3°C
3) 100.3°C 4) 373.3°C

21. A glass vessel contains air at 60°C. To what temperature must it be heated to expel one third of the air, the pressure remaining constant. (Neglect the expansion of the vessel)

1) 127°C 2) 226.5°C
3) 327°C 4) 427°C

22. A balloon of volume V contains a gas of mass m at a pressure P and temperature 15°C. Gas is pumped into the balloon so that its volume is doubled and the pressure is trebled. If the temperature increases 6°C in the process, find the ratio of the increase in mass to the original mass.

1) 34 : 5 2) 239 : 49
3) 199 : 21 4) 3 : 4

23. The density of a gas at N.T.P. is 1.3 kg / m³ the density of the gas at – 15°C and at a pressure of 84 cm of Hg ....

1) 1.52 kg/m³ 2) 0.52 kg/m³
3) 0.052 kg/m³ 4) 15.2 kg/m³

24. Calculate the value of the gas constant for one gram of hydrogen, given that the density of hydrogen at N.T.P is 0.00009 g/cm³

1) 4.12J/K-g 2) 2.12J/K-g
3) 5.12J/K-g 4) 3.12J/K-g

25. The mass of a litre of dry air at N.T.P is 1.293 g. Find the mass of 3 litres of air at 117°C and a pressure of 4 atmospheres ?

1) 10.86 g 2) 5.6 g
3) 6.4 g 4) 7.2 g

ANSWERS

1) 3	2) 3	3) 1	4) 2	5) 1
6) 1	7) 3	8) 2	9) 4	10) 3
11) 2	12) 4	13) 1	14) 1	15) 2
16) 1	17) 2	18) 3	19) 4	20) 1
21) 2	22) 2	23) 1	24) 1	25) 1

EXERCISE - II(B)

1. The pressure of a gas at 0°C is 50cm of Hg at constant volume it is heated to 80°C, then pressure is 64cm of Hg the pressure coefficient of gas

   1) 0.0035/ °C 2) 0.035/ °C

   3) 0.00035/ °C 4) 0.000035/ °C

2. A gas at a temperature 300 K and pressure 30 atm is allowed to expand to atmospheric pressure. If the volume becomes 10 times its initial volume, the final temperature beocmes [97 E]

   1) 100°C 2) 373K 3) 373°C 4) 100K

3. A given of gas occupies a volume of 100 c.c at one atmospheric and at 100°C. At the same temperature, how much volume the same gas occupies at 4 atmospheric pressure ?

   1) 25c.c 2) 15 c.c 3) 35 cc 4) 10 cc

4. The pressure of a gas ta 100°C is 2 atm. When the gas is heated by keeping volume constant, at what temperature the pressure raises to 3 atm ?

   1) 286.5°C 2) 168.5°C

   3) 186.5°C 4) 366.5°C

5. A gas at 27°C temperature and 30 atmospheric pressure is allowed to expand to the atmospheric pressure. If the volume becomes 10 times its initial volume, then the final temperature becomes [CBSE - 2001]

   1) 100°C 2) 173°C 3) 273°C 4) -173°C

6. A bubble rises from the bottom of a lake 90m deep on reaching the surface, its volume becomes (take atmospheric pressure as 10 m of water)

   1) 4 times 2) 8 times

   3) 10 times 4) 3 times

7. To decrease the pressure of the gas by 10% at constant temperature then change in volume should be

   1) 10% decrease 2) 10% increase

   3) 11.11% increase 4) 9.1% increase

8. A gas is heated through 1°C in a closed vessel. Its pressure is increased by 0.4%. The initial temperature of the gas is [2003E]

   1) $250^0 C$ 2) $100^0 C$

   3) $-75^0 C$ 4) $-23^0 C$

9. A given amount of a gas heated till the volume and pressure are each increased by 1%, then temperature increases by

   1) 0.5% 2) 1% 3) 2% 4) 4%

10. At constant volume the ratio of increase in pressure of an ideal gas per degree rise in kelvin temperature to its original pressure

    1) $\sqrt{3}$ 2) $\frac{1}{273}$ 3) $\sqrt{273}$ 4) $\frac{1}{\sqrt{3}}$

11. A sample of O_2 gas and a sample of H_2 gas both have the same mass, the same volume and the same pressure. Assuming them to be perfect gases, the ratio of the temperature of O_2 gas to the temperature of the H_2 gas is

    1) 16 2) 8 3) 32 4) 4

12. The density of air at N.T.P is 1.293×10^{-3} kg/litre. Calculate its gas constant.

    1) 0.287 J $gm^{-1} K^{-1}$ 2) 1.267 J $gm^{-1} K^{-1}$

    3) 3.87 J $gm^{-1} K^{-1}$ 4) 4.167 J $gm^{-1} K^{-1}$

13. Mass of 1litre of air at N.T.P is 1.293g. The mass of 10 litres of air at 273° C and 57cm of mercury pressure is

    1) 4.489gm 2) 4.849 gm

    3) 4.4 gm 4) 4.9 gm

14. A vessel containing 10 lit of air at atmospheric pressure (760 mm of Hg) is connected with an evacuated 9 litre vessel. The resultant air pressure will be

    1) 1440 mm of Hg 2) 760 mm of Hg

    3) 400 mm of Hg 4) 40 mm of Hg

15. The variation of pressure with volume for a given mass of a gas at two different temperatures T_1 and T_2 are represented as shown in the graph, then

    1) $T_1 > T_2$

    2) $T_2 > T_1$

    3) $T_1 = T_2$

    4) $T_1 \underset{<}{\overset{>}{=}} T_2$

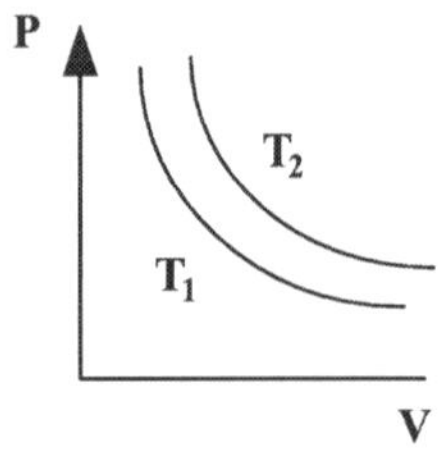

16. From the following P – T diagram, the inference drawn is

1) $V_2 > V_1$

2) $V_2 < V_1$

3) $V_1 = V_2$

4) none of these

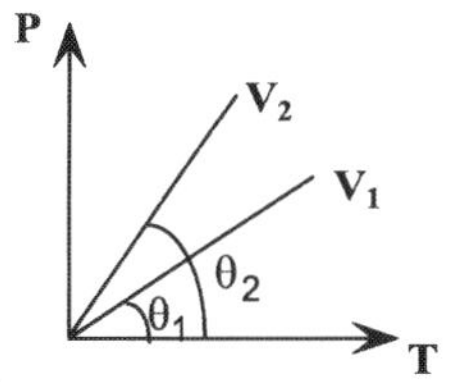

17. The equation of state corresponding to 2gm H_2 is

1) PV = RT 2) $PV = \frac{RT}{4}$

3) $PV = \frac{RT}{3}$ 4) $PV = \frac{RT}{2}$

18. A vessel contains 8gm of a gas at a pressure P and temperature 500K. The gas leaves through a small hole. The mass of the gas leaked out when the pressure is P/2 and temperature is 400K is

1) 5 gm 2) 3 gm 3) 4 gm 4) 6 gm

19. One litre of Helium under a pressure of 1 atm and at a temperature of 0°C is heated until the pressure becomes 1/3 and volume are doubled. The final temperature attained by the gas is

1) –123°C 2) –100°C 3) –13°C 4) -91°C

20. An ideal gas is enclosed in a glass bottle at 0^0c and 1 atmospheric pressure. The bottle bursts at a presssure just greater than 2 atmospheres. To which temperature the gas can be heated with out breakling the bottle ?

1) 273^0C 2) 373^0C 3) 473^0C 4) 173^0C

21. A vessel is filled with some gas at 76cm of Hg. If the mass of the gas in the vessel is increased by 50% at the same temperature then pressure of the gas in the vessel is

1) 114cm of Hg 2) 76cm of Hg

3) 38cm of Hg 4) 70cm of Hg

22. A flask is filled with 13g of an ideal gas at 27^0C and its temperature is raised to 52^0C. The mass of thc gas that has to be released to maintain the pressure constant is [2000 E]

1) 5gm 2) 6 gm 3) 7 gm 4) 1 gm

23. The density of a gas at NTP is 9×10^{-5} g/ cm^3 the density at 240^0C and at a pressure of 228 cm of Hg

1) $14.36\times10^{-5} g/cm^3$

2) $1.436\times10^{-5} g/cm^3$

3) $143.6\times10^{-5} g/cm^3$

4) $.1436\times10^{-5} g/cm^3$

24. The mass of 1 litre of hydrogen gas at N.T.P. is 0.9 g. What is the mass of the gas of same volume at 27^0C and 75 cm of Hg pressure ?

1) 0.81 g 2) 0.5g 3) 0.62g 4) 0.4g

25. The density of oxygen at N.T.P is 1.429g/ litre. A certain mass of the gas is enclosed in a cylinder, whose volume is 2.5 litres, under a pressure of 780 mm at a temperature of 27^0C. What is the mass of the gas in the cylinder ?

1) 3.336 g 2) 4.2 g 3) 5.12 g 4) 6.8 g

ANSWERS

1) 1	2) 4	3) 1	4) 1	5) 4
6) 3	7) 3	8) 4	9) 3	10) 2
11) 1	12) 1	13) 2	14) 3	15) 2
16) 2	17) 1	18) 2	19) 4	20) 1
21) 1	22) 4	23) 1	24) 1	25) 1

EXERCISE - III(A)

1. One litre of He gas at a pressume of 76cm of Hg and temperature 27^0C is heated till its pressure and volume are doubled. The final temperature attained by the gas is

1) 900^0C 2) 927^0C 3) 627^0C 4) 327^0C

2. When an air bubble of radius 'r' rises from the bottom to the surface of a lake, its radius becomes 5r/4 (the pressure of the atmosphere is equal to the 10m height of water column). If the temperature is constant and the surface tension is neglected,the depth of the lake is

1) 3.53 m 2) 6.53 m [2001 E]

3) 9.53 m 4) 12.53m

3. The pressure and temperature of an ideal gs in a closed vessel are 720 KPa and 40°C respectively. If $(1/4)^{th}$ of the gas is realeased from the vessel and the temperature of the reamining gas is raised to 353°C. the final pressure of the gas is [2000 M]

1) 1440 kPa 2) 1080 kPa
3) 720 kPa 4) 540 kPa

4. An ideal gas is trapped between mercuy thread of 12cm and the closed lower end of a narrow vertical tube of uniform cross section. Length of the air column is 20.5 cm, when the open end is kept upward. If the tube is - making 30° with the horizontal then the length of the air column is (assuming temperature to be constant and atmospheric pressure = 76cm of Hg)

1) 22 cm 2) 18 cm 3) 24 cm 4) 20.2 cm

5. 2gm of monoatomic gas occupies 2 litre volume at 8.3×10^5 pascal and at 127°C. then that gas is

1) Hydrogen 2) Neon
3) Helium 4) Argon

6. Two thermally insulated vessels 1 and 2 are fillcd with air at temperature (T_1, T_2), volume (V_1, V_2) and pressure (P_1, P_2) respectively. If the valve joining the two vessels is opened, the temperature inside the vessel at equilibrium will be [AIEEE - 2004]

1) $T_1 + T_2$

2) $T_1T_2 (P_1V_1 + P_2V_2) / (P_1V_1T_1 + P_2V_2T_2)$

3) $T_1T_2 (P_1V_1 + P_2V_2) / (P_1V_1T_2 + P_2V_2T_1)$

4) $(T_1+T_2) / 2$

7. During an experiment an ideal gas is found to obey an additional law VP^2 = constant. The gas is initially at a temperature 'T' and volume 'V'. When it expands to a volume 2V, the temperature becomes

1) T 2) 2T 3) $\sqrt{2}$ T 4) $\frac{T}{\sqrt{2}}$

8. An ideal gas expands in such a way that its pressure and volume comply with the condition PV^2 = constant. During this process, the gas is [CPMT - 2002]

1) Heated 2) Cooled
3) First '1' then '2' 4) Neither '1' nor '2'

9. One litre of oxygen at a pressure of 1 atm and two litres of nitrogen at a pressure of 0.5 atm are introduced into a vessel of volume 1litre. If there is no change in temperature, the final pressure of the gas in atm is [2008 E]

1) 1.5 2) 2.5 3) 2 4) 4

10. The mass of oxygen gas occupying a volume of 11.2 litres at a temperature 27°C and a pressure of 760mm of mercury in kilograms is (Molecular wieght of oxygen=32) [2002 E]

1) 0.001456 2) 0.01456 3) 0.1456 4) 1.1456

11. At the top of a mountain a thermometer read 7°C and barometer reads 70cm of Hg. At the bottom of the mountain the barometer reads 76cm of Hg and thermometer reads 27°C. The density of air at the top of the mountain is __times the density at the bottom.

1) 0.99 2) 0.9 3) 0.89 4) 0.95

12. Two identical vessels A and B with frictionless pistons contain the same ideal gas at the same temperature and the same volume V. The masses of gas in A and B are m_A and m_B respectively. The gases are allowed to expand isothermally to the same final volume 3 V. The change in pressures of the gas in A and B are found to be ΔP and 1.5 ΔP respectively. Then [2004 E]

1) $9m_A = 4m_B$ 2) $3m_A = 2m_B$
3) $2m_A = 3m_B$ 4) $4m_A = 9m_B$

13. A uniform tube with a piston in the middle and containing a gas at 0°C is heated to 100°C at one side. If the piston moves 5cm, find the length of the tube containing the gas at 100°C.

1) 37.5 cm 2) 75 cm 3) 16.75cm 4) 30 cm

14. A closed container of volume 0.02m³ contains a mixture of neon and argon gases, at a temperature of 27°C and pressure of $1 \times 10^{+5}$ Nm^{-2}. The total mass of the mixture is 28gm. If the gram molecular weights of neon and argon are 20 and 40 respectively. Find the masses of the individual gases in the container, assuming them to be ideal. (Universal gas constant R = 8.314 J/mol.k)

1) m_1 = 4gm, m_2 = 24 gm
2) m_1 = 8 gm, m_2 = 24 gm
3) m_1 = 4gm, m_2 = 12 gm
4) $m_1 = 6gm$, $m_2 = 12gm$

15. A column of Hg of 10 cm length is contained in the middle of a narrow horizontal 1m long tube which is closed at both ends. Both the halves of the tube contain air at a pressure 76cm of Hg. By what distance will the column of Hg be displaced if the tube is held vertical?

1) 3cm 2) 1cm 3) 4 cm 4) 5cm

16. A horizontal uniform glass tube of 100cm length sealed at both ends contains 10cm mercury column in the middle. The temperature and pressure of air on either side of mercury column are respectively 0^0C and 80cm of mercury. If the air column at one end is kept at 0^0C and the other end at 273^0C, the pressure of air which is 0^0C is (in cm of Hg) [2003 E]

1) 76 2) 88.2 3) 120 4) 132

ANSWERS

1) 2	2) 3	3) 2	4) 1	5) 3
6) 3	7) 2	8) 3	9) 3	10) 2
11) 1	12) 2	13) 1	14) 1	15) 1
16) 3				

EXERCISE - III(B)

1. One litre of helium under a pressure of 1 atmospheres and a temperature of 0°C is heated until its pressure and volume are doubled. The final temperature attained by helium gas is

1) 1092°C 2) 273°C

3) 927K 4) 819°C

2. An air bubble doubles its radius on raising from the bottom of water reservoir to the surface of water in it. If the atmospheric pressure is equal to 10m of water, the height of water in the reservoir is [99 M]

1) 10m 2) 20m 3) 70 m 4) 80 m

3. A bottle with an open mouth contains air at 47°C. The temperature to which the bottle must be heated so that 1/5th of the mass of air leaves the bottle is _______ °C

1) 172 2) 217 3) 127 4) 177

4. An ideal gas is trapped between a mercury column and the closed end of a narrow vertical tube of uniform base containing the column. The upper end of the tube is open to the atmosphere. The atmospheric pressure equals 76 cm of mercury. The lengths of the mercury column and the trapped air column are 20 cm and 43 cm respectively. What will be the length of the air column when the tube is tilted slowly in a vertical plane through an angle of 60^0 ? Assume the temperature to remain constant.

1) 48 cm 2) 38 cm 3) 24 cm 4) 12 cm

5. Given samples of 1 cm^3 of Hydrogen and 1 cm^3 of Oxygen both at N.T.P., which sample has larger number of molecules

1) Hydrogen 2) Oxygen

3) both have equal number of molecules

4) Both have zero molecules in 1 c.c.

6. Two vessels of volumes 5 and 3 litres contain air at 3 atm and x atm. When they are connected by a narrow pipe the resultant pressure was 4.5 atm. Then x =

1) 7 atm 2) 4 atm 3) 2 atm 4) 3.5 atm

7. A gas obeys PV^2 = constant in addition to PV =RT. If on heating, the temperature is doubled , the volume of the gas is

1) V/2 2) 2V 3) $V/\sqrt{2}$ 4) $\sqrt{2}\,V$

8. If V – T graph for a given mass of a ideal gas is shown in the figure. The variation of P with T will be

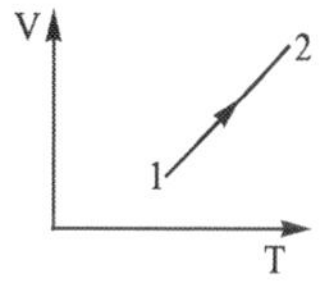

1) On heating thegas pressure increases

2) On heating thegas pressure decreases

3) On heating the gas pressure remains constant

4) Any one of the above

9. A closed cylinder has a freely moving piston separating the cylinder in two chambers A and B. Chamber A contains 25 gms of N_2 and chamber B contains 20 gm of H_2. When the equilibrium established, what will be the ratio of number of moles of N_2 to number of moles of H_2

1) 0.089 2) 0.0089 3) 0.89 4) 89

10. A closed copper vessel contains water equal to half of its volume. When the temperature of the vessel is raised to 447^0C, the pressure of the steam in the vessel is (treat stream as an ideal gas, R=8310JK^{-1}/kg mole, density of water = 1000kgm^{-3} molecular weight of water = 18) [2000 M]

1) $33.24\times10^7 Pa$ 2) $16.62\times10^7 Pa$

3) $10.31\times10^7 Pa$ 4) $8.31\times10^7 Pa$

11. A vessel of volume 2V contains 4 moles of a gas at certain temperature. Another vessel of volume 3V contains 3 moles of the same gas at same temperature then the ratio of the pressures is

1) 2 : 1 2) 1 : 2 3) 4 : 1 4) 1 : 1

12. Two identical containers A and B with frictionless pistons contain the same ideal gas at the same temperature and same volume V. The mass of the gas in A is m_A and that in B is m_B. The gas in each cylinder is now allowed to expand isothermally to the same final volume 2 V. The changes in the pressure in A and B are found to be $4\Delta P$ and $9\Delta P$ respectively. Then

1) $4\ m_A = 9\ m_B$ 2) $2m_A = 3m_B$

3) $3m_A = 2m_B$ 4) $9m_A = 4\ m_B$

*13. If the volume of nitrogen of mass 14 kg is 0.4 m^3 at 30^0C. Calculate the pressure.

1) $25 \times 10^5\ N/m^2$ 2) $42 \times 10^5\ N/m^2$

3) $34 \times 10^5\ N/m^2$ 4) $31.49 \times 10^5\ N/m^2$

14. A closed hollow insulated cylinder is filled with gas 0^0C and also contains an insulated piston of negligible weight and negligible thickness at the middle point. The gas on one side of the piston is heated to 100^0C. If the piston moves 5cm, the length of the hollow cylinder is [2001 E]

1) 13.65 cm 2) 27.3 cm

3) 38.6 cm 4) 64.6 cm

15. A vessel of volume V = 30 litres contains ideal gas at S.T.P. After a portion of the gas has been let out, the pressure in the vessel decreased by Δ P=0.78 atm (the temperture remaining constant).

Find the mass of the released gas. The gas density under the normal conditions is ρ = 1.3 gm/litre.

1) 30g 2) 40g 3) 50g 4) 20g

16. A glass tube, sealed at both ends, is 100 cm long. It lies horizontally with the middle 10 cm containing mercury. The two ends of the tube contain air at 27^0C and at a pressure 76 cm of mercury. The air column on one side is maintained at 0^0C and the other side is maintained at 127^0 C. Calculate the length of the air column on the cooler side. Neglect the changes in the volume of mercury and of the glass

1) 36.5 cm 2) 13.5 cm

3) 46.5 cm 4) 12.5 cm

ANSWERS

1) 4	2) 3	3) 3	4) 1	5) 3
6) 1	7) 1	8) 2	9) 1	10) 2
11) 1	12) 4	13) 4	14) 4	15) 1
16) 1				

COMPREHENSION QUESTIONS

Comprehension - I (1 - 4) :

The total number of independent ways in which the particles of a system can acquire energy are called the degrees of freedom. Each degree of freedom contributes some kinetic energy to the particles. Thus a molecule free to move in space has three translational degrees of freedom. Molecules of a monatomic gas have only translational degrees of freedom a diatomic molecule has also three translational degrees of freedom. In addition to this it rotates about its centre of mass. It can rotate about two independent axes of rotation perpendicular to the axis joining the atoms. Thus it has two rotational degrees of freedom. In addition at higher temperatures the molecule vibtates. This vibrational mode contributes the energy consisting of PE and KE. In equilibrium the total energy is equally distributed in all possible energy modes. The energy due to each degree of freedom is $\frac{1}{2}K_BT$. This is called the law of equipartition of energy.

1. The number of degrees of freedom for monatomic and diatomic molecules at ordinary temperatures.

1) 3,3 2) 3,5 3) 3,7 4) 5,3

2. The translational degrees of freedom in diatomic molecule

1) 3 2) 5 3) 7 4) 1

3. The temperature of a gas depends on

1) translational kinetic energy of molecules only

2) Translational and rotational kinetic energy

3) rotational kinetic energy only

4) the total energy of the gas

4. Pick up the correct statement

a) At absolute zero the degrees of freedom get frozen

b) According to classical physics degrees of freedom must remain unchanged

c) Quantum mechanics reuqires a minimum non zero amount of energy before a degree of freedom comes into play

1) a & b are true 2) b & c are true

3) c & a are true 4) a,b,c are true

ANSWERS

1) 2 2) 1 3) 1 4) 4

Chapter - 9

SIMPLE HARMONIC MOTION

SIMPLE HARMONIC MOTION

SYNOPSIS

1. The motion of a particle is said to be Simple Harmonic (i) If it moves to and fro about a fixed point (ii) its acceleration is always directed towards the fixed point and (iii) the acceleration is directly proportional to the displacement and opposite to it.
   $a \propto -y \qquad a = -\omega^2 y$
2. All vibratory motions are periodic. But all periodic motions need not be vibratory eg : Motion of earth around the sun is periodic but not vibratory.
3. All periodic motions need not be simple harmonic
4. **Example for Simple harmonic Motion :**
   1) Simple Pendulum
   2) Motion of needle of a sewing machine
   3) Motion of the prongs of a vibrating tuning fork
   4) Vertical oscillatory motion of a loaded spring
5. The condition for a body to be in simple harmonic motion is that the restoring force must be linearly dependent on displacement
   eg : 1) Simple pendulum → $F = -mg\sin\theta = -mg.\ \theta$
   2) Spring -mass system → $F = -k.\ x$
   3) Torsion pendulum → $C = -\tau\theta$
6. **Damped Oscillations :** Many oscillating bodies do not move back and forth between precisely fixed limits, because frictional forces dissipate the energy of motion. Such oscaillations are called damped oscillations.
   eg : Stretched violin string soon stops vibrating.
7. **Equilibrium Position :** The point at which no net force acts on the oscillating body is known as equilibrium position or **Mean position.**
   At the Mean position
   i) displacement of the body is Minimum.
   (ii) velocity of the body is Maximum.
   (iii) acceleration of the body is Minimum.
   (iv) P.E. of the body is Minimum.
   (v) K.E. of the body is Maximum
8. **Extreme Position :** The point at which maximum restoring force acts on the oscillating body is known as Extreme Position.
   At the extreme position
   (i) displacement of the body is maximum.
   (ii) velocity of the body is minimum.
   (iii) acceleration of the body is maximum.
   (iv) P.E. of the body is maximum
   (v) K.E. of the body is minimum.
9. **Representation of S.H.M. :** The motion of a particle executing periodic motion along a circular path is not simple harmonic motion. But the foot of the perpendicular dropped from the instantaneous position of a particle, executing periodic motion along any diameter is simple harmonic motion.
   In general the simple harmonic motion is represented as $Y = A\sin(\omega t + \phi)$
   Y = instantaneous displacement
   A = Amplitude
   $(\omega t + \phi)$ = phase ; ϕ is called initial phase.
   (i) if the motion starts from Mean Position $\phi = 0$
   (ii) If the motion starts from the extreme position $\phi = \pi/2$
10. A simple harmonic motion can be expressed by periodic functions like sinωt, cosωt or combination of these functions.
    $Y = A\sin\omega t + B\cos\omega t$
11. S.H.M. can also be represented in the following ways
    i) $F = -ky$ $F = ma = \dfrac{md^2y}{dt^2}$
    $F = m.\dfrac{d^2y}{dt^2} = -KY$ or $\dfrac{d^2y}{dt^2} + \left(\dfrac{k}{m}\right)y = 0$
    ii) $F = -ky$ $\dfrac{dp}{dt} + ky = 0$ P = momentum

CHARACTERISTICS OF S.H.M.

12. **Amplitude :** The maximum displacement of the vibrating particle in S.H.M from its mean position is called Amplitude
    The maximum displacement is same on either side of the mean position. S.I. Unit for Amplitude is metre.
13. **Instantaneous displacement :** The distance of the particle from mean position in a particular direction at any instant of time is known as instataneous displacement.
    It is given by $Y = A\sin(\omega t + \phi)$
    Displecement - time graph

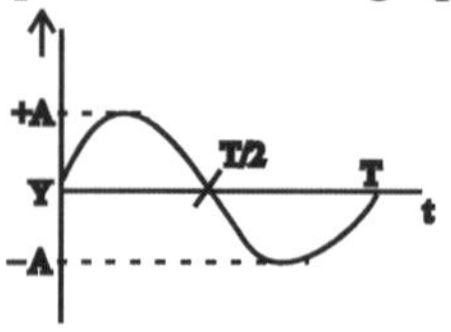

14. **Velocity :** The rate of displacement is called velocity.

$\therefore\ v = \dfrac{dy}{dt} = A\omega\cos(\omega t + \phi)$ Velocity - time graph

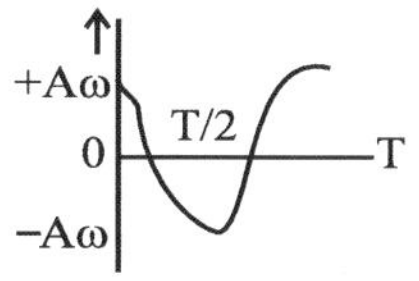

(as a function of t)

$v = \omega\sqrt{A^2 - Y^2}$ (as a function of 'y')

Displacement - velocity graph

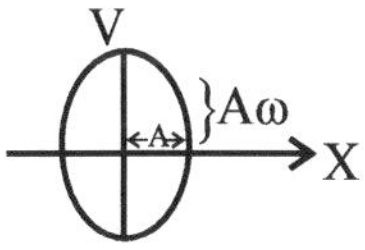

(iv) $V_{max} = \sqrt{\dfrac{v_1^2 Y_2^2 - v_2^2 Y_1^2}{Y_1^2 - Y_2^2}}$

15. **Acceleration :** The rate of change of velocity of a particle in SHM is called acceleration.

$\therefore\ a = \dfrac{dv}{dt} = -A\omega^2 \sin(\omega t + \phi) =$

Acceleration time-graph

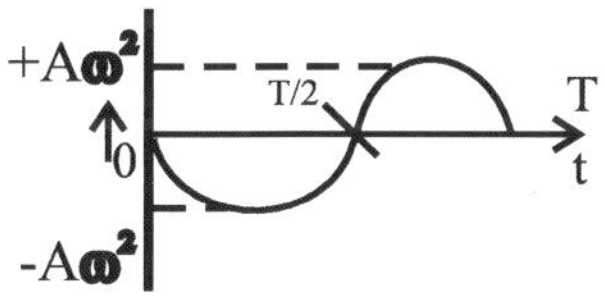

$a = -\omega^2 y \qquad = a\ \alpha - y$

Acceleration displacement - graph

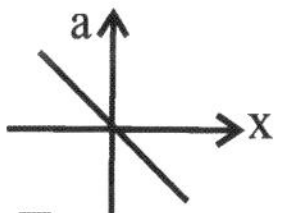

16. **Restoring Force :**

$F = ma \qquad F = -m\omega^2 A \sin(\omega t + \phi)$

Force - time graph

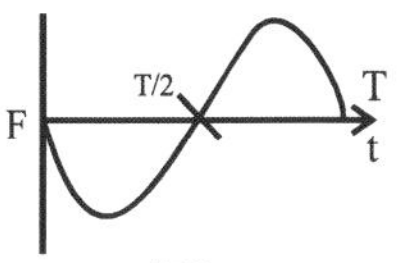

(i) $F = -\dfrac{dU}{dy}$ here U = P.E. (ii) $F = -m\omega^2 y$

Force - displacement

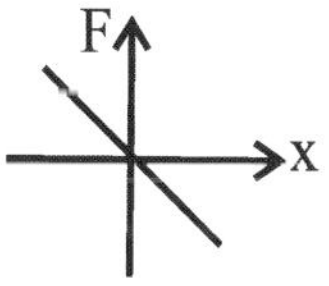

(iii) $F = -m\dfrac{d^2y}{dt^2}$

18. The displacement velocity, acceleration and force acting on the body vary with time, but their values repeat themselves at regular intervals of time (called time period) kinematical equations like s=ut, v=u+at ....etc, should not be used while solving problems on SHM.
19. Velocity is ahead in phase by 90° with displacement and acceleration leads the displacement by 180°
20. An oscillation corresponds to a phase of 2π
21. **Phase :** Phase at any instant is that which gives the state of the vibrating particle with respect to time in a specified direction with reference to a fixed point (Mean position)

    Ex : (1) If the phase is zero, ie., the particle is crossing the mean position.

    (2) If the phase is $\pi/2$ ie., the particle is at the extreme position.
22. The inital phase at t = 0 of a particle in S.H.M. is called phase constant or epoch.

ex : If $Y = A\sin(\omega t + \pi/4)$ at $t = o$ $\phi = \pi/4$

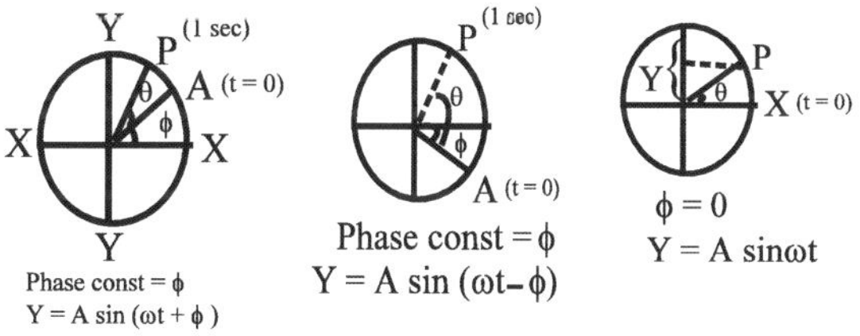

23 **Time Period :** Time taken by vibrating particle in S.H.M. to complete one vibration is called Time period or period of oscillation.

General formula : $T = 2\pi\sqrt{\dfrac{\text{displacement}}{\text{acceleration}}} =$

$= 2\pi\sqrt{\dfrac{\text{mass of the body}}{\text{restoring force per unit displacement}}}$

Time period of SHM independent of the amplitude. The oscillations are known as isochronus oscillations.

i) $\therefore\ T = 2\pi\sqrt{\dfrac{y}{a}}$

ii) Time period of a simple pendulum = $T = 2\pi\sqrt{\dfrac{l}{g}}$

l = length of the simple pendulum

g = acceleration due to gravity at a place.

iii) Time period of Torsion pendulum $T = 2\pi\sqrt{\dfrac{I}{C}}$

I = moment of Inertia about the suspension wire

C = couple per unit twist.

iv) Time period of a loaded spring

(a) $T = 2\pi\sqrt{\frac{m}{k}}$

K = Force constant or spring constant

m = mass attached to the spring.

(b) $T = 2\pi\sqrt{\frac{x}{g}}$

x = extension produced in the spring due to attachment of the load 'm'

g = acceleration due to gravity.

(v) when a hole is drilled along the diameter of the earth and if a body is dropped in it, it moves to and fro about the centre of the earth and is in S.H.M. with a time period of

$T = 2\pi\sqrt{\frac{R}{g}} = 84.6$ minutes or $T = \sqrt{\frac{3\pi}{GD}}$

D = Mean density of the earth

G = Gravitational constant

Note : 1) A horizontal board moves up and down in SHM of amplitude 'A'. If the object on that board does not loose contact with the board, then the smallest permissible time period is $T = 2\pi\sqrt{\frac{A}{g}}$

When the board is at the highest position, normal reaction on the object is

$N_1 = mg - m\omega^2 A$

When the board is at the lowest position, normal reaction on the object is

$N_2 = mg + m\omega^2 A$

So, the object may lose contact only at highest position if $N_1 = 0$, from which we can obtain 'T'

2) A horizontal board moves to and fro in the horizontal plane performing SHM of amplitude 'A'. If the object on the board does not slide relative to the board, then its time period

$T = 2\pi\sqrt{\frac{A}{\mu g}}$

(μ = coefficient of fricition between object and board)

17. Simple Pendulum : (i) The arrangement, in which a point mass is suspended from a massless flexible and inextensible thread clamped to a rigid support, is defined as ideal simple pendulum

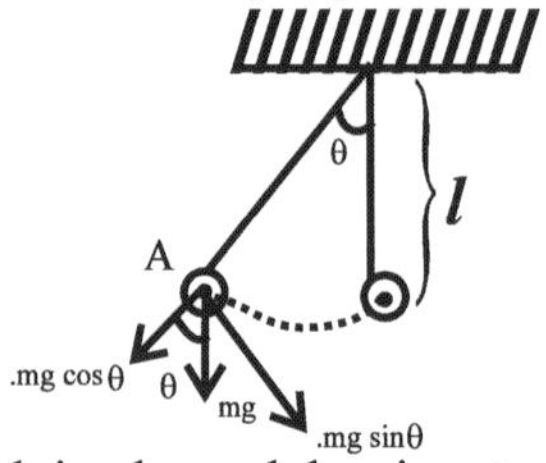

ii) An ideal simple pendulum is not possible hence a heavy bob suspended from a light inextensible string acts as a simple pendulum.

iii) The tension in the string at any position is equal to $T = mg\cos\theta + \frac{mv^2}{l}$

* When the bob is it extreme position, V = 0, then Tension is the string is minimum

$T_{min} = mg\cos\theta$

* When the bob is at mean position $\theta = 0$, then Tension is the string is maximim

$T_{max} = \frac{mv^2}{l} + mg$

$\left(V = \left(\sqrt{2gl(1-\cos\theta)}\right.\right.$ at mean position$\left.\right)$

If θ is the angular amplitude of a simple pendulum then the maximum tension in the thread is

$T_{max} = mg\left(1+\theta^2\right)$ A= linear amplitude

$= mg\left[1+\left(\frac{A}{\ell}\right)^2\right]$

iv) Equation of motion of a simple pendulum in differntial form is $\frac{d^2y}{dt^2} + \omega^2 y = 0$

$\omega^2 = g/l$

v) (a) Time period of a simple pendulum

$T = 2\pi\sqrt{\frac{l}{g}}$ for smaller amplitudes

vi) The period of oscillation of a simple pendulum of constant length is independent of the size, shape, mass and material of the bob provided it is not made of a very light substance like cork.

vii) The period of oscillation of a simple pendulum of constant length is independent of the amplitude, provided the amplitude is small.

viii) $T \,\alpha\, \sqrt{l}$ or $\frac{l}{T^2}$ = constant (at a place)

L - T graph for a simple pendulum

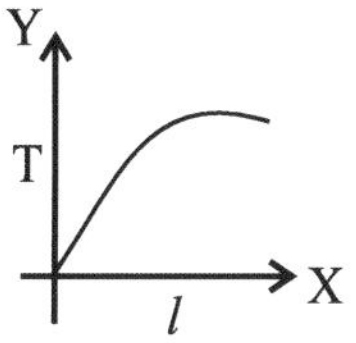

L-T^2 graph for a simple pendulum

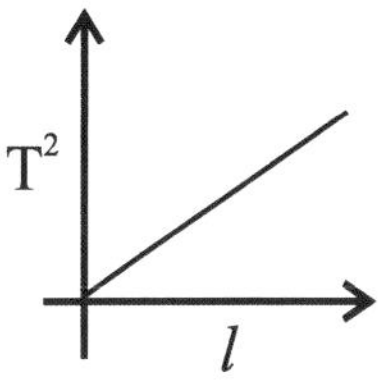

$$T = 2\pi\sqrt{\frac{l}{g}}$$

$$T^2 = \frac{4\pi^2}{g} l \quad y = mx \quad \text{Slope } m = \frac{4\pi^2}{g}$$

ix) $\frac{l_1}{l_2} = \frac{T_1^2}{T_2^2}$

x) for percentage calculation of Time period, when % increase in length is given

a) $\left(\frac{T_2}{T_1} - 1\right) = \left[\sqrt{\frac{l_2}{l_1}} - 1) \times 100\%\right]$

b) for smaller percentages, $\frac{\Delta T}{T} \times 100 = \frac{1}{2}\frac{\Delta l}{l} \times 100$

xi) $T \alpha \frac{1}{\sqrt{g}}$ or gT^2 = constant (if length of the pendulum is constant)

xii) $\frac{T_1}{T_2} = \sqrt{\frac{g_2}{g_1}}$

xiii) $g \alpha l$ or $\frac{g}{l}$ = constant (When T is constant)

xv) When a pendulum clock is taken from the equator to the poles the time period decreases. Hence it makes more oscillations and gains time and moves fast.

xvi) When a pendulum is taken from the earth to moon, the time period increases (as g is less on moon). Hence it makes less number of oscillations and looses time or moves slow.

xvii) When a pendulum clock is taken from the earth to the moon, to keep the time correct its length must be decreased.

xviii)If the pendulum of a clock is made of metal, it runs slow during summer and fast during winter due to thermal expansion or contraction.

xix) If a boy sitting in a swing stands up, as centre of Mass rised up, length of the pendulum decreases and hence the period of the swing decreases.

xx) If the bob of a pendulum is made hollow and filled with water, and the water is drained up as the water goes down, centre of mass shifts down, and then rises to its original position. Hence time period first increase and then decreases and reaches its original value.

xxi) When a simple pendulum is arranged in a lift moving up with an acceleration 'a', its time period decreases.

$$T = 2\pi\sqrt{\frac{l}{g+a}} \text{ and frequency } n = \frac{1}{2\pi}\sqrt{\frac{g+a}{l}}$$

xxii) When a simple pendulum is arranged in a lift moving down with an acceleration'a' its time period increases.

$$T = 2\pi\sqrt{\frac{l}{g-a}} \text{ and frequency } n = \frac{1}{2\pi}\sqrt{\frac{g-a}{l}}$$

In the freely falling lift $\quad a = g \; T = \infty \quad n = 0$

Therfore pendulum doesnot oscillate

xxiii) When the lift is stationary or moving up or down with constant velocity, its time period doesn't change. (acceleration $a = 0$)

$$T = 2\pi\sqrt{\frac{l}{g}} \text{ and frequency } n = \frac{1}{2\pi}\sqrt{\frac{g}{l}}$$

xxiv) When a simple pendulum is arranged to the roof of a car moving horizontally with an acceleration'a' its time period decreases.

$$T = 2\pi\sqrt{\frac{l}{\sqrt{g^2 + a^2}}} \; ; \; \tan\theta = \frac{a}{g}$$

$\theta =$ angle between the thread and the vehical of the mean position

xxv) When a simple pendulum is arranged to the roof of a car moving with a constant speed v in a circular path of radius 'r' its time period decreases.

$$T = 2\pi\sqrt{\frac{1}{\sqrt{g^2 + \left(\frac{v^2}{r}\right)^2}}}$$

xxvi) When a simple pendulum is arranged into the roof of a car moving down a smooth inclined plane of inclination θ its time period increases.

$$T = 2\pi\sqrt{\frac{l}{g\cos\theta}}$$

xxvii) If a simple pendulum is arranged in an artificial satellite its time period becomes infinity.

xxviii) If the length of the pendulum is considerably large $T = \frac{2\pi}{\sqrt{g\left(\frac{1}{l}+\frac{1}{R}\right)}}$

R = radius of the earth

l = length of the pendulum.

xxix) If the length of the pendulum is infinite i.e., $l = \alpha$ T = 84.6 minutes.

xxx) If the length of the pendulum is equal to radius of the earth, then its time period is

$$T = 2\pi\sqrt{\frac{R}{2g}} = 59\text{min } \& \text{ } 5 \text{ sec.}$$

xxxi) work done by the string of the simple pendulum during one complete vibration is equal to zero.

xxxii) A simple pendulum fitted with a metallic bob of density d_s has a time period T. when it is made to oscillate in a liquid of density d_L then its time period increases.

$$T = 2\pi\sqrt{\frac{l}{g\left(1-\frac{d_l}{d_s}\right)}}$$

xxxiii) When two simple pendulum of lengths l_1, and l_2, are get into vibration in the same direction at the same instant with same phase. Again they will be in same phase after the shorter pendulum has completed n oscillations. To find the value of n,

$n T_S = (n-1) T_L$ and $T \alpha \sqrt{l}$

$$\therefore \frac{n}{n-1} = \frac{T_L}{T_S} \quad \text{or} \quad \frac{n}{n-1} = \sqrt{\frac{l_L}{l_s}}$$

S - shorter L - Longer

24. Seconds pendulum : (i) The simple pendulum whose time period equal to 2 seconds is called seconds pendulum.

ii) Its length at place where g = 9.8 m/s² is 100 cm.

iii) Since T = 2 sec

$$L = \frac{g.T^2}{4\pi^2} \quad L = \frac{g}{4\pi^2}.4 \quad \therefore L = \frac{g}{\pi^2}$$

iv) The length of a pendulum at a place where $g = g_1$ is l_1 and its length at a place where $g = g_2$ is l_2

To keep the time period constant at T = 2 sec, its length has to be decreased or increased corresponding to the value of 'g' at that place

Decrease in length $= \frac{g_1 - g_2}{\pi^2}$ (if $g_1 > g_2$)

Increase in length $= \frac{g_2 - g_1}{\pi^2}$ (if $g_2 > g_1$)

25. Spring - Mass system :

i) Restoring force F = –kx

ii) When the load is pulled down and released, it makes vertical oscillations

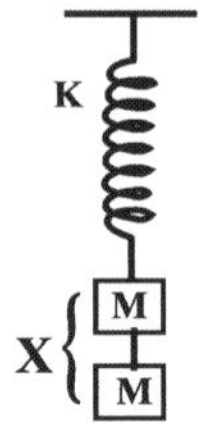

$T = 2\pi\sqrt{\frac{M}{K}}$ F = Kx Mg = Kx

$\frac{M}{K} = \frac{x}{g}$ $T = 2\pi\sqrt{\frac{x}{g}}$

iv) Time period of spring mass system is independent of acceleration due to gravity.

v) When a clock filled with spring mass system is taken to the surface of the moon, it's time period will remain constant.

vi) If mass of the spring is considered,

$$T = 2\pi\sqrt{\frac{M+\frac{m}{3}}{k}} \text{ (m = mass of the spring)}$$

Spring oscillates on its own weight then time period

$$T = 2\pi\sqrt{\frac{m}{3k}} \quad (M = 0)$$

vii) Spring constant 'K' is inversely proportional to its length.

viii) When a spring of force constant K and length l is cut into two parts of lengths l_1 and l_2 having force constants K_1 and K_2

$K \alpha \frac{1}{l}$ or $kl = k_1 l_1 = k_2 l_2$

ix) If a spring of spring constant 'K' is divided into 'n' equal points, then the spring constant of each part is 'nK'.

x) If a spring of spring constant 'K' and length 'l' is cuts in to two springs of lengths 'l_1' and 'l_2', then the spring constants of the two parts is $K_1 = \frac{K(l_1 + l_2)}{l_1}$

and $K_2 = \frac{K(l_1 + l_2)}{l_2}$

xi) When a spring of force constant K is cut into two equal parts each part will have a force constant 2k.

xii) When two springs of force constants K_1 and K_2 are connected in series. Then the effective force constant is

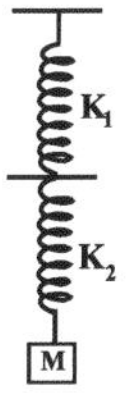

$$K = \frac{K_1 K_2}{K_1 + K_2}$$

xiii) When two springs of force constants k_1 and k_2 are connected in parallel. The effective spring constant is $K = k_1 + k_2$

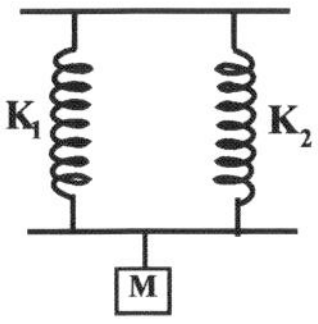

xiv) For the same mass attached, the time periods of two different springs T_1 and T_2

a) Whey they are joined in series to the same mass, time period of new arrangement T_s is given by

$$T_s - \sqrt{T_1^2 + T_2^2}$$

b) When they are joined in parallel to the same mass, time period of new arrangement T_p is given by

$$\frac{1}{T_p^2} = \frac{1}{T_1^2} + \frac{1}{T_2^2} \Rightarrow T_p = \frac{T_1 T_2}{\sqrt{T_1^2 + T_2^2}}$$

xv) Two masses m_1 and m_2 are suspended from a spring of spring constant 'k'. When the masses are in equilibrium, m_1 is gently removed, then

a) The angular frequency of oscillation of m_2 is

$$\sqrt{\frac{k}{m_2}}$$

b) The amplitude of oscillation of m_2 is $\frac{m_1 g}{k}$

26. **Remember:** Removed mass determines amplitude Remaining mass determines time period

27. **Potential Energy :** (i) The P.E. of particle in S.H.M. is given by

$$P.E = \frac{1}{2} m\omega^2 y^2 = \frac{1}{2} m\omega^2 A^2 \sin^2(\omega t \pm \phi)$$

Potential Energ - time graph

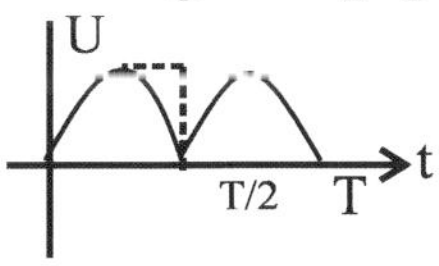

Potential energy - displacement graph

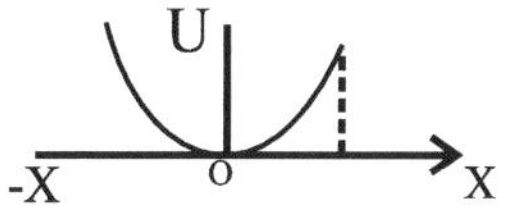

ii) At mean position (y = 0) P.E. is minimum.
iii) At extreme position (y = A) P.E. is maximum.
v) During one complete vibration average potential energy is given by

$$= \frac{1}{4} m\omega^2 A^2$$

28. **Kinetic Energy :** (i) The K.E. of a particle in S.H.M is given by

$$K.E. = \frac{1}{2} m\omega^2 (A^2 - y^2)$$

Kinetic energy- displacement graph

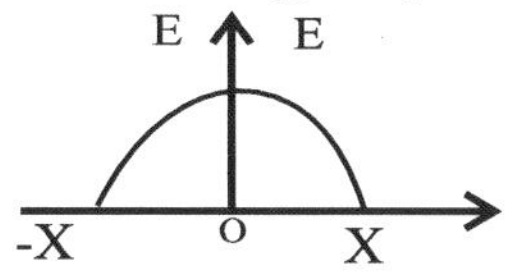

$$= \frac{1}{2} m\omega^2 A^2 \cos^2(\omega t \pm \phi)$$

Kinetic energy - time graph

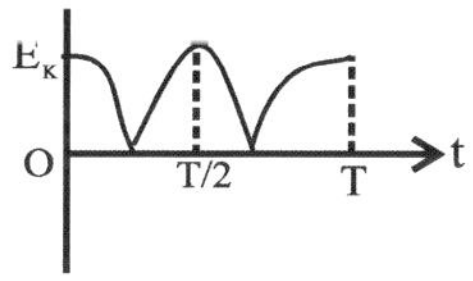

ii) At mean position (y = o) K.E is maximum

iii) At extreme position (y = A) K.E. is zero.

iv) During one complete vibration average kinetic Energy is $= \frac{1}{4} m\omega^2 A^2$

29. **Total Energy :**

(i) T.E. = P.E. + K.E. = $(1/2)m\omega^2 A^2$

Total energy - time graph

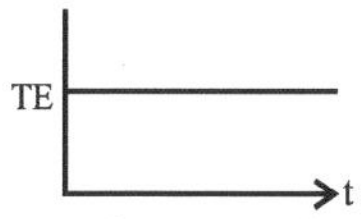

Total energy displcement graph

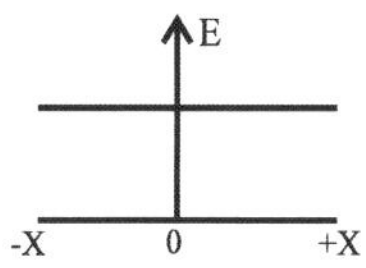

ii) When a particle is in S.H.M. At any position T.E. is constant.

30. The P.E of a particle in S.H.M. varies with double the frequency of the S.H.M.

31. The K.E. of a particle in S.H.M. varies with double the frequency of the S.H.M.

Note : Total energy of particle in SHM is independent of position

Time periods of various harmonic oscillatons

1. SHM of a liquid in U tube :

If a body of density ρ contained in a vertical U tube performs S.H.M. in its two limbs. Then time period

$$T = 2\pi\sqrt{\frac{L}{2g}} = 2\pi\sqrt{\frac{h}{g}}$$

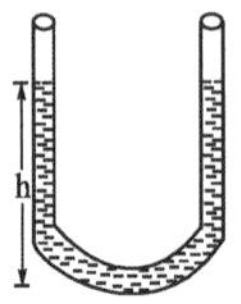

where L = Total length of liquid column, h = Height of undisturbed liquid in each limb (L = 2h)

2. SHM of a floating cylinder :

If a cylinder of mass 'm' and cross section 'A' is floating in a liquid of density 'σ' with length 'l' inside the liquid. is slightly depressed and released, if oscillation with a time period

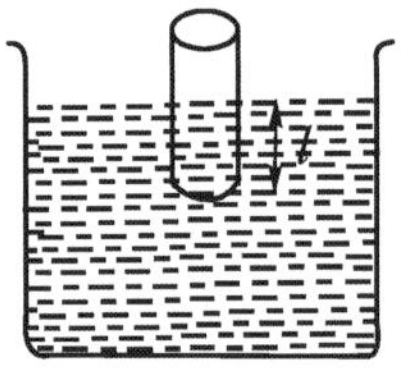

$$T = 2\pi\sqrt{\frac{m}{K}} = 2\pi\sqrt{\frac{m}{A\sigma g}}$$

$$T = 2\pi\sqrt{\frac{l}{g}}$$

3. SHM of a floating body

A piece of wood has dimensions a, b and c. Its relative density is 'd'. It is floating in water such that the side 'a' is vertical. If it is pushed down a little and then released, the time period of oscillation will be

$$T = 2\pi\sqrt{\frac{ad}{g}}$$

4. SHM of a small ball rolling down in hemi-spherical bowl

$$T = 2\pi\sqrt{\frac{R-r}{g}}$$

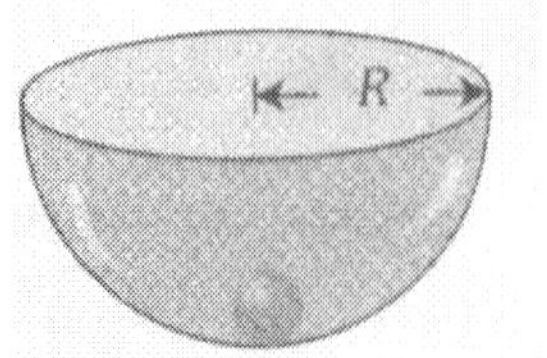

R = Radius of the bowl

r = Radius of the ball

5. SHM of a piston in a cylinder :

$$T = 2\pi\sqrt{\frac{Mh}{PA}}$$

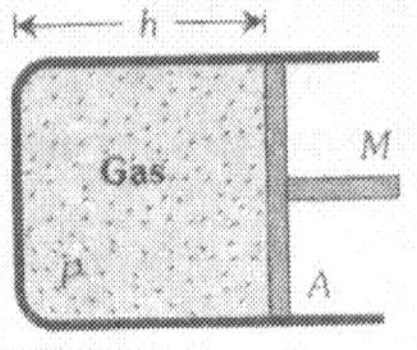

M = Mass of the pistion
A = area of crosssection
h = height of cylinder
P = pressure in a cylinder

6. Longitudinal oscillation of an elastic wire :

Wire or string pulled a distance Δl and left. It executes longitudinal oscillations.

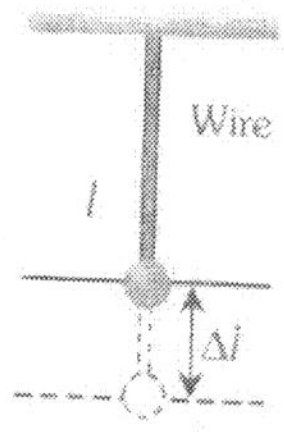

Restoring force

$$F = -YA\frac{\Delta l}{l}$$

$$F = -\left[\frac{YA}{l}\right]\Delta l = \Rightarrow F = -K\Delta l$$

force constant $K = \frac{YA}{l}$

A = Area of cross section of wire.

$$T = 2\pi\sqrt{\frac{m}{k}} = 2\pi\sqrt{\frac{ml}{YA}}$$

7. If the lower surface of a cube of side 'L' and of rigidity modulus 'n' is fixed while fixing a particle of mass 'm' on the upper face, a force parallel to upper force is applied to mass 'm' and with drawn. The period oscillation of mass 'm' is

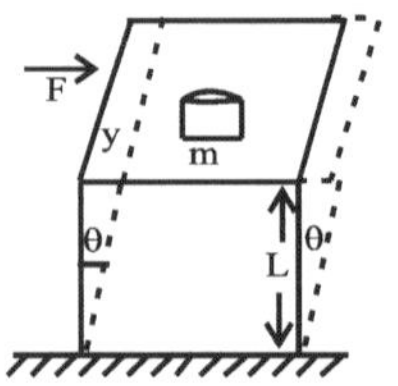

$$T = 2\pi\sqrt{\frac{m}{K}} = 2\pi\sqrt{\frac{m}{nL}}$$

EXERCISE - I

EQUATION OF S.H.M

1. A particle is moving in a circle with uniform speed. Its motion is

1) Periodic and simple harmonic
2) Periodic but not simple harmonic
3) a periodic
4) None of the above

2. In S.H.M, there is always a constant ratio between the displacement of the body and its

1) Velocity 2) Acceleration
3) Mass of the particle 4) All of the above

3. A body executing simple harmonic motion has maximum acceleration

1) At the mean positions
2) At the two extreme position
3) At any position 4) The question is irrelevant

4. Which of the following is a necessary and sufficient condition for simple harmonic motion

1) Constant period 2) Constant acceleration
3) Proportionality between acceleration and restoring force
4) Proportionality between restoring force and displacement from equilibrium position

5. Which of the following expression does not represent SHM

1) A cos wt 2) A sin 2wt
3) A sin wt + B coswt 4) $A \sin^2 wt$

6. A particle moves on the x-axis according to the equation $x = x_0 \sin 2\omega t$. The motion is simple harmonic.

1) With amplitude $x_0/2$
2) With amplitude $2x_0$
3) With time period π/w
4) With time period $\frac{2\pi}{\omega}$

7. A particle moves on the x-axis according to the equation x = A + B sinwt. The motion is simple harmonic with amplitude

1) A 2) B 3) A + B 4) $\sqrt{A^2+B^2}$

8. The motion of a particle in S.H.M. is represented by the equation y = A sin wt + B cos wt. The resultant amplitude is

1) A 2) B 3) A + B 4) $\sqrt{A^2+B^2}$

9. The ratio of maximum acceleration to maximum displacement of a Simple Harmonic Motion is equal to

1) Square of its angular velocity
2) Square of its time period
3) Its time period
4) Its angular velocity

10. A particle is executing S.H.M. Then the graph of acceleration as a function of displacement is

1) A sinecurve 2) A circle
3) A straight line 4) Parabola

11. The displacement of a particle in S.H.M is in opposite phase with

1) Velocity 2) Acceleration
3) Amplitude 4) Frequency

12. At the equilibrium position of a particle executing S.H.M

1) Restoring force acting on the particle is zero
2) The velocity of the particle is maximum
3) Potential energy is minimum
4) All the above

13. For a particle executing S.H.M along x-axis force is given by

1) – kx 2) A cos kx 3) Ae^{-kx} 4) Akx

14. The graph plotted between force and displacement of a particle executing S.H.M is

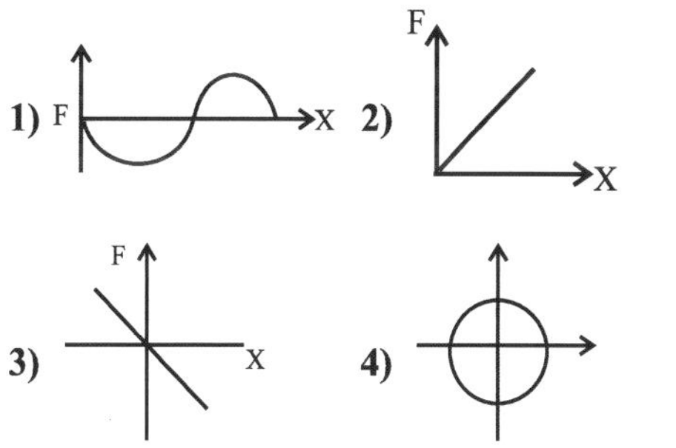

15. If the maximum acceleartion of a S.H.M. is a and the maximum velocity is b then amplitude of vibration is given by

1) b^2a 2) a^2b 3) $\frac{\beta^2}{\alpha}$ 4) $\frac{\alpha^2}{\beta}$

16. A particle is executing simple harmonic motion with an amplitude A and time period T. The displacement of the particles after 2T period from its initial position is (Eamcet 2008 E)

1) A 2) 4A 3) 8 A 4) Zero

17. The amplitude of a simple harmonic oscillator is doubled. Which of the following is also doubled ?

1) Its frequency 2) Its period

3) Its maximum velocity 4) Its total energy

18. In simple harmonic motion, when the displacement is minimum then

1) K.E is maximum and P.E. is minimum

2) P.E is maximum and K.E. is minimum

3) Both K.E and P.E are maximum

4) Both P.E and K.E are minimum

19. When the amplitude of oscillation of a particle in S.H.M is increased to two times, the time period

1) Is doubled 2) Is halved

3) Is unaltered 4) Increased to $\sqrt{2}$ times

20. For a boby in Simple Harmonic Motion

1) Speed in uniform

2) Acceleration is non uniform, positive while approaching and negative while moving away from mean position

3) Angular velocity is non uniform

4) Kinetic energy remains constant

21. For a particle executing S.H.M, which of the following statements is not correct ?

1) The total energy of a particle always remains the same

2) The restoring force is always directed towards a fixed point

3) The restoring force is maximum at the extreme positions

4) The acceleration of the particle is maximum at the equilibrium position

22. Which of the following four statements is false ?

1) A body can have zero velocity and still be accelerated

2) A body can have a constant velocity and still have a varying speed

3) A body can have a constant speed and still have a varying velocity

4) The direction of the velocity of a body can change when its acceleration is constant

23. Choose the correct statement. (ATB)

a) Any motion that repeats itself in equal intervals of time along the same path is called periodic motion.

b) The displacement of a particle in periodic motion can always be expressed in terms of sine and cosine functions of time.

c) A body in periodic motion moves back and forth over the same path is called oscillatory or vibrating motion

d) Simple harmonic motion is a particular case of periodic motion.

1) Only a, b, d are true 2) Only b , c , d are true

3) Only a, c, d are true 4) All are true

24. In a periodic motion when a body moves to and fro about a fixed mean position its acceleration.

1) Proportional to displacement of body from mean position and is always directed towards the mean position.

2) Inversely proportional to displacement of body from mean position and is always directed away from mean position.

3) Proportional to displacement of body from mean position and is always directed away from mean position

4) May be proportional to displacement but unspecified direction.

25. In S.H.M. (ATB)

1) The acceleration and displacement of a body are proportional to each other and opposite in direction.

2) The accelerations and displacement of body are proportional to each other and same in direction.

3) The acceleration and displacement of body are inversely proportional to each other and opposite in direction

4) The acceleration and displacement are inversely proportional to each other and same in direction.

26. The uniform circular motion in general can be described as a combination of two simple harmonic motions. (ATB)

1) Acting perpendicular to each other

2) Acting parallel to each other

3) Acting antiparallel to each other

4) Acting incline to each other with less than 90^0

27. Statement (a) : The velocity of simple harmonic oscillator is maximum at mean position

Statement (b) : At extreme position the acceleration of simple harmonic oscillator is maximum.

Statement (c) : The velocity of simple harmonic oscillator is minimum at extreme position

(ATB)

1) a , b are true 2) Only a is true

3) b , c are true 4) All are true

28. For a simple harmonic oscillator the frequency of oscillation is independent of (ATB)

1) Time period 2) Acceleration

3) Angular velocity 4) Amplitude

29. The phase difference between velocity and acceleration of simple harmonic oscillator.

1) π 2) $\frac{\pi}{2}$ 3) $\frac{\pi}{4}$ 4) $\frac{\pi}{3}$ **(ATB)**

30. The phase difference between acceleration and displacement of simple harmonic oscillator

1) π 2) $\frac{\pi}{2}$

3) $\frac{\pi}{4}$ 4) $\frac{\pi}{3}$ **(ATB)**

31. **Statement (a) :** During simple harmonic oscillation kinetic energy converted in potential energy and vice - versa

Statement (b) : Total mechanical energy of simple harmonic oscillator is directly proportional to square of the frequency of oscillation

Statement (c) : Simple harmonic oscillator obeys the law of conservation of energy.

Statement (d) : Total mechanical energy of oscillator is directly proportional to a square of the amplitude of the oscillation **(ATB)**

1) a , b are true 2) c , d are true

3) a , b , d are true 4) a , b , c , d are true

32. Total energy of particle performing S.H.M. depends on (ATB)

1) Amplitude , time period

2) Amplitude , Time period and displacement

3) Amplitude , displacement

4) Time period , displacement

33. The work done by the body which is in S.H.M, against the restoring force is stored in the form of (ATB)

1) K.E. 2) P.E.

3) Both P.E . & K.E. 4) Total energy

34. The phase of simple harmonic motion at t =0 is called (ATB)

1) Phase constant 2) Initial phase

3) Epoch 4) All the above

SIMPLE PENDULUM

35. The period of oscillation of a bob suspended by a string of length l is $2\pi\sqrt{\frac{l}{g}}$. This expression is valid only for

1) A light bob and no friction

2) Small amplitude of oscillation and a small bob

3) A bob of low density

4) A bob which is not ferromagnetic and large amplitude

36. An oscillating simple pendulum is arranged in a vacuum jar, then

1) It stops oscillating 2) It continues to oscillate

3) It oscillates with slightly more time period

4) It oscillates with slightly less time period

37. The bob of a simple pendulum is hollow. Now its time period is T. If it is filled with water the time period will be

1) More than T 2) Less than T

3) Equal to T 4) Zero

38. Pendulum clock shows correct time at O°C. At a higher temperature the clock

1) Loses time 2) Gains time

3) Neither gains nor loses time

4) All the above are possible depending on the co-efficient of linear expansion of the pendulum

39. A seconds pendulum is taken from the surface of the earth to that of the moon. In order to maintain the period constant

1) Length of the pendulum has to be decreased

2) Length of the pendulum has to be increased

3) Amplitude of the pendulum has to be increased

4) Amplitude of the pendulum has to be decreased

40. The period of a simple pendulum suspended from the ceiling of a car is T when the car is at rest. If the car moves with a constant acceleration the period of the pendulum

1) Unaltered 2) Decreases 3) Increases 4) None

41. If iron sphere is replaced by wooden sphere of same mass, time period

1) Increases 2) Decreases

3) Remains same 4) None of the above

42. A simple pendulum is a hollow sphere containing mercury suspended by means of a wire. If a little mercury is drained off, the period of the pendulum will

1) Increase 2) Decrease

3) Remains unchanged 4) Becomes erratic

43. For a simple pendulum graph between L and T^2 will be a

1) Rectangular hyperbola 2) Parabola
3) St line passing through origin
4) A st line having a + ve y - intercept

44. A hole is drilled along the diameter of the earth and pen is dropped into it. The time taken by it is reach other end of the earth is

1) 84 min 2) 42 min
3) 21 min 4) 102 hours

45. The work done by the string of a simple pendulum in one complete oscillation is equal to

1) Zero
2) total energy of the pendulum
3) P.E. of the pendulum
4) K.E. of the pendulum

46. The bob of a simple pendulum of period T is given a negative charge. If it is allowed to oscillate with an identical charge at the point of suspension. The new time period will be

1) equal T 2) more than T
3) less than T 4) infinite.

47. The time period of a pendulum in stationary lift is 'T', if lift starts accelerating in the downward direction, the time period will

1) Increase 2) Decrease
3) No change 4) Nothing certain

48. A rubber ball filled with water, having a small hole is used as the bob of a simple pendulum. The period of such a pendulum

1) Is a constant 2) Decreases with time
3) Increases with time
4) First increase and then decrease finally having same value as at the begining

49. Identify the correct statement among the following

1) The greater the mass of the pendulum bob, the shorter is its frequency of oscillation
2) A simple pendulum with a bob of mass'm' swing with an angular amplitude of 40°. When its angular amplitude is 20° the tension in the string is less than Mg cos 20°
3) As the length of a simple pendulum is increased, the maximum velocity of its bob during its oscillation will also increase
4) The fractional change in the time period of a pendulum on changing the temperature is in independent of the length of the pendulum

50. The restoring force which brings the bob towards the mean position is equal to

1) weight of bob **(ATB)**
2) Radial component
3) Tangential component of weight of bob
4) Resultant of radial and tangential component of weight of bob

51. A simple pendulum is vibrating in an evacuated chamber. It will

1) come to rest eventually
2) Oscillate forever with the same amptitude and frequency
3) oscillate with same frequency but amptitude will decrease with time
4) Oscillate with same amplitude but frequency will decrease with time.

52. To show that a simple pendulum is in a simple harmonic motin it is necessary to assume that [98 M]

1) The length of the pendulum is small
2) The mass of the pendulum is small
3) The amplitude of oscilation is small
4) The acceleration due to gravity is small

53. A simple pendulum hanging freely and at rest is vertical because in that position

1) K.E is zero 2) K.E minimum **[98 E]**
3) P.E. is zero 4) P.E. minimum

54. A seconds pendulum is taken from the surface of the earth to that of the moon in order to maintain the period constant [81]

1) the length of the pendulum has to be decreased
2) the length of the pendulum has to be increased
3) the amplitude of thependulum should be decreased
4) no change is requred

SPRINGS

55. A bob is hung from an elastic spring to a support and set in motion in a vertical plane in a way similar to the oscillations of a simple pendulum. Its length will be maximum

1) At its left extreme position
2) At its right extreme position
3) At its mean position
4) Will remain the same at all positions

56. A clock 'S' is based on oscillations of a spring and a clock P is based on pendulum motion. Both clocks run at the same rate on earth on a planet having the same density as earth but twice the radius

1) S will run faster than P

2) P will run faster than S

3) Both will run at same rate 4) None

57. A spring of force constant K is cut into two parts whose lengths in the ratio 2 : 3. The part having larger force constant is

1) Shorter spring 2) Longer spring

3) Both have equal force constants

4) Nothing can be said

58. The time period of stiffer spring as compared to that of a soft spring is

1) Less 2) More 3) Infinity

4) Depends upon the spring material

59. When a block is suspended by means of spring the restoring force per unit displacement of the block is called (ATB)

1) force constant 2) pressure

3) stress 4) strength of spting

60. One of the two clocks on earth is controlled by a pendulum and other by a spring. If both the clocks be taken to the moon, then which clock will show same time as that on earth [AIEEE2002]

1) Spring clok 2) pendulum clok

3) both '1' and '2' 4) neither '1' and '2'

MATCHINGS

61. When a body in SHM match the statements in column A with that in column B.

Column - I	Column II
a) Velocity is maximum	e) At half of the amplitude
b) Kinetic energy is 3/4th of total energy	f) At the mean position
c) P.E. is 3/4th of total energy	g) At extreme position
d) Acceleration is maximum	h) At $\frac{\sqrt{3}}{2}$ times amplitude

1) a - f, b - e, c - h, d - g 2) a - e, b - f, c - g, d - h

3) a - g, b - h, c - e, d - f 4) a - h, b -e, c - f, d - e

62. In SHM match the following

Column - I	Column II
a) Max. velocity	e) $(1/2)M\omega^2A^2$
b) Max. Acceleration	f) $(1/4)M\omega^2A^2$
c) Max. Force	g) $A\omega$
d) Max. total energy	h) $\omega^2 A$
	i) $m\omega^2 A$

1) a - g ; b- h ; c - i ; d - e 2) a - h; b - g ; c - i ; d - e

3) a - g ; b - h ; c - f ; d - e 4) a - g ; b - i ; c - h ; d - f

63. The time period of oscillation of the perticle in SHM is 'T'. Then match the following

Column - I	Column II
a) $\frac{3}{8}$th of oscillation from extreme position	e) $\frac{2T}{3}$
b) $\frac{3}{8}$th of oscillation from mean position	f) $\frac{T}{3}$
c) $\frac{5}{8}$th of oscillation from extreme position	g) $\frac{7T}{12}$
d) $\frac{5}{8}$th of oscillation from mean position	h) $\frac{5T}{12}$

1) a - e ; b- g ; c - f ; d - h
2) a - f; b - h ; c - e ; d - g
3) a - f ; b - e ; c - h ; d - g
4) a - e ; b - f ; c - g ; d - h

64. Match the following

Column - I	Column II
a) Displacement of body in S.H.M.	e) $r\omega\cos\omega t$
b) Velocity of body in S.H.M.	f) $r\sin\omega t$
c) Acceleration of body in S.H.M.	g) $r\omega$
d) Maximum velocity of body	h) $-r\omega^2\sin\omega t$

1) a - f , b - e , c - h , d - g
2) a - h , b - f , c - g , d - e
3) a - f , b - g , c - g , d - e
4) a - g , b - h , c - e , d - g

More than one answers Questions

65. If amplitude of particle executing SHM is doubled, which of the following quantities are doubled

a) Time period **b) Maximum velocity**
c) Maximum acceleration
d) Total energy

1) b and c
2) a, b and c
3) a and c
4) a, b, c and d

66. In S.H.M. at the equilibrium position

a) Amplitude is minimum
b) acceleration is zero
c) Velocity is maximum
d) P.E is maximum

1) All true
2) b, c, d true
3) b, c true
4) a, b, d true

67. If the length of the simple pendulum is equal to the radius of the earth, and it oscilates just above the surface of the earth then its time period is

a) $2\pi\sqrt{\frac{R}{2g}}$ **b)** $2\pi\sqrt{\frac{R}{g}}$

c) nearly 59.5 minutes **d) 84.6 minutes**

1) a is only correct
2) b is only correct
3) a and c are correct
4) a and d are correct

68. If a simple pendulum is arranged in an artificial satellite its

a) Time period becomes infinity
b) Frequency becomes infinity
c) Both time period and frequency become infinity
d) It does not oscillate

1) a and d are correct
2) a and b are correct
3) b, c correct
4) c, d are correct

69. When a hole is drilled along the diameter of the earth and if a body is dropped into it, it executes SHM with a period T equal to (R & D are radius and density of the earth)

a) $2\pi\sqrt{\frac{R}{g}}$ **b)** $\sqrt{\frac{3\pi}{GD}}$

c) $2\pi\sqrt{\frac{\pi}{GD}}$ **d)** $2\pi\sqrt{\frac{R}{2g}}$

1) a and b are true
2) a and c are true
3) b and c are true
4) c and d correct

70. Arrange the following simplependulum in ascending order of their periods of oscillation

a) of length 1 m at a place of g = 8m/s
b) of length 2 m at a place of g = 10 m/s
c) of length 4 m at a place of g = 4m/s

1) a, b, c 2) b, a, c 3) b, c, a 4) c, b, a

Assertion & Reason Questions

Instruction For Questions :

In each of the following questions, a statement of Assertion (A) is given followed by a corresponding statement of reason (R) just below it. Of the statement mark the correct answer.

1) Both 'A' and 'R' are true and 'R' is the correct explanation of 'A'
2) Both 'A' and 'R' are true and 'R' is not the correct explanation of 'A'
3)'A' is true and 'R' is false
4)'A' is false and 'R' is true

71. A) : A loaded spring oscillating with frequency 'f' continues to maintain the same frequency in an earth satellite

R) : Frequency of loaded spring is independent of acceleration due to gravity

72. A) : The displacement time graph for a particle in SHM is sine curve, when the motion begins from mean position.

R) : The diplacement of a particle in SHM is given by y = A sin wt

73. A) : When a spring is cut into two equal parts, spring constant of each piece is doubled

R) : Spring constant is inversly proportional to length of spring

74. A) : The pendulum clock made up of a metal losses time at high temperature

R) : When a metal rod is heated it contracts

75. A) : The length - time period graph for simple pendulum is parabola

R) : The length and time period at a given place are connected by the relation $l = KT^2$ (where 'K' is constant)

76. A) : In simple harmonic motion, the motion is to and fro and periodic

R) : Velocity of the particle $V = \omega\sqrt{A^2 - x^2}$ where 'x' is displacement as measured from extreme position.

77. A) : A wooden cube of side 'a' floats in a non viscous liquid of density 'ρ'. When it is slightly pressed and released it executes SHM

R) : The net force responsible for SHM is the resultant of buoyancy force and true weight of the body.

78. A) : The motion of sewing needle is an example for SHM

R) : A liquid is taken in U-tube. Liquid in one limb is pressed and released It executes SHM

79. A) : In damped vibrations, Amplitude of oscilation decreases.

R) : Damped vibrations indicate loss of energy due to air resistance

80. A) : SHM is an example of varying velocity and varying acceleration.

R) : For a particle performing SHM in nonviscous medium its total energy is constant

81. A) : A pendulum clock shows correct time at 0^0C, at a higher temp the clock loses time

R) : The period of oscillation increases due to increase in length so the clock loses time since it makes less no. of oscillation in a day

82. A) : When a clock fitted with a spring mass system is taken to the surface of the moon, its time period is same as that on the earth

R) : The time period of a spring mass system is not affected by the variation of g

83. A) : The work done by the tension in the string of a simple pendulum in one complete oscillation is zero

R) : No work is done by the tension in the string since tension is always at right angles to the motion of bob.

ANSWERS

1) 2	2) 2	3) 2	4) 4	5) 4
6) 3	7) 2	8) 4	9) 1	10) 3
11) 2	12) 4	13) 1	14) 3	15) 3
16) 4	17) 3	18) 1	19) 3	20) 2
21) 4	22) 2	23) 4	24) 1	25) 1
26) 1	27) 4	28) 4	29) 2	30) 1
31) 4	32) 1	33) 2	34) 4	35) 2
36) 2	37) 3	38) 4	39) 1	40) 2
41) 1	42) 1	43) 3	44) 2	45) 1
46) 1	47) 1	48) 4	49) 4	50) 3
51) 2	52) 3	53) 4	54) 1	55) 3
56) 1	57) 1	58) 1	59) 1	60) 1
61) 1	62) 1	63) 2	64) 1	65) 1
66) 3	67) 3	68) 1	69) 1	70) 1
71) 1	72) 1	73) 1	74) 3	75) 1
76) 3	77) 1	78) 2	79) 1	80) 2
81) 1	82) 1	83) 1		

EXERCISE - II(A)

Equations of SHM:

1. A particle is subjected to two S H Ms $x_1 = A_1 \sin wt$ and $x_2 = A_2 \sin(wt + \pi/4)$. The resultant S H M will have an amplitude of

1) $\frac{A_1 + A_2}{2}$ 2) $\sqrt{A_1^2 + A_2^2}$

3) $\sqrt{A_1^2 + A_2^2 + \sqrt{2}A_1A_2}$ 4) A_1A_2

2. The maximum displacement of an oscillating particle is 0.05m. If its time period is 1.57s (i) What is the velocity at the mean position ? (ii) What is its acceleration at the extreme position?

1) 0.2 ms^{-1}, 0.2ms^{-2} 2) 0.8 ms^{-1}, 0.8ms^{-2}

3) 0.2 ms^{-1}, 0.8ms^{-2} 4) 0.8 ms^{-1}, 0.2ms^{-2}

3. The time period of a paricle performing linear SHM is 12s. What is the time taken by it to make a displacement equal to half its amplitude?

1) 1sec 2) 2sec 3) 3sec 4) 4sec

4. The amplitude and time period of a particle of mass 0.1 kg executing simple harmonic motion are 1m and 6.28s, respectively. Find its (i) angular frequency, (ii) acceleration and (iii) velocity at a displacement of 0.5m.

1) 1rad/sec, –1 ms^{-2}, $\sqrt{3}ms^{-1}$

2) 0.5 rad/sec, –0.5 ms^{-2}, $\frac{\sqrt{3}}{2}ms^{-1}$

3) 1rad/sec, –0.5 ms^{-2}, $\frac{\sqrt{3}}{2}ms^{-1}$

4) 1rad/sec, –0.5 ms^{-2}, $\sqrt{3}ms^{-1}$

5. The equation motion of a particle in S.H.M is a + $16p^2x = 0$. In the equation 'a' is the linear acceleration (in m/sec^2) of the particle at a displacement 'x' in meter. The time period of S H M in seconds is (2004 M)

1) $\frac{1}{4}$ 2) $\frac{1}{2}$ 3) 1 4) 2

6. The angular velocity and amplitude of a simple pendulum are 'w' and A respectively. At a displalement 'x'. If its KE is 'T' and PE is 'U' then the ratio of 'T' to 'U' is

1) $\frac{x^2w^2}{A^2-w^2x^2}$ 2) $\frac{x^2}{A^2-x^2}$

3) $\frac{A^2-w^2x^2}{x^2w^2}$ 4) $\frac{A^2-x^2}{x^2}$

7. The amplitude of oscillation of a particle is 0.05m. If its time period is 1.57s (i) What is the velocity at the mean position ? (ii) What is its acceleration at the extreme position ?

1) 0.2 ms^{-1}, 0.2ms^{-2} 2) 0.8 ms^{-1}, 0.8ms^{-2}

3) 0.2 ms^{-1}, 0.8ms^{-2} 4) 0.8 ms^{-1}, 0.2ms^{-2}

8. Two simple harmonic oscillators with amplitudes in the ratio 1 : 2 are having the same total energy. The ratio of their frequencies is

1) 1:4 2) 1: 2 3) 2 : 1 4) 4 : 1

9. A particle is performing SHM. Its path length is 8 cm and time period 0.2s. If the particle is initially in the mean position find the time at which it will be at 1cm, and the phase angle at 1cm.

1) 0.008046sec, 14.48^0 2) 0.006046sec, 12.48^0

3) 0.004046sec, 16.48^0 4) 0.002046sec, 18.48^0

10. A simple harmonic oscillator of mass 0.5 kg, force constant 10N/m and amplitude 3cm. a) What is the maximum speed ? b) What is the speed when x = 2cm ? c) If the oscillator is released from rest at x = A when t = O determine the position and velocity of the oscillator at t = 2s

1) 0.114ms^{-1}, 0.2ms^{-1}, 0.087m, 0.052ms^{-1}

2) 0.134ms^{-1}, 0.1ms^{-1}, 0.027m, 0.062ms^{-1}

3) 0.144ms^{-1}, 0.4ms^{-1}, 0.037m, 0.032ms^{-1}

4) 0.124ms^{-1}, 0.6ms^{-1}, 0.017m, 0.042ms^{-1}

11. A simple harmonic oscillator is of mass 0.1kg. It is oscillating with a frequency of $5/\pi$ Hz. If its amplitude of vibration is 5cm find the force acting on the particle at its extreme position.

1) 0.5N 2) 2.5N 3) 3.5N 4) 4.5N

12. The displcement of a particle executing SHM is given by $Y = 10 \sin(3t + \pi/3)$m and 't' is in seconds. The initial displacement and maximum velocity of the particle are respectively (92 E)

1) $5\sqrt{3}$ m and 30m/sec

2) 15m and $15\sqrt{3}$ m/sec

3) $15\sqrt{3}$ m and 30 m/sec

4) $20\sqrt{3}$ m and 30 m/sec

13. A particle is vibrating in SHM with an amplitude of 4cm . At what displacement from the equilibrium position it has half potential and half kinetic

1) 1cm 2) $\sqrt{2}$ cm 3) 2 cm 4) $2\sqrt{2}$ cm

14. A particle moves according to the law $x = a\cos\frac{\pi t}{2}$. The distance covered by it in the time interval between t = 0 to t = 3 sec is

1) 2 a 2) 3a 3) 4 a 4) a

15. If a body of mass 36gm moves with S.H.M of amplitude A = 13 cm and period T = 12sec . At a time t = 0 the displacement is x = +13cm . The shortest time of passage from x = + 6.5 cm to x = – 6.5 is

1) 4sec 2) 2 sec 3) 6 sec 4) 3 sec

16. For a body in S.H.M the velocity is given by the relation $v = \sqrt{144-16x^2}$ m/sec . The maximum acceleration is

1) 12 m/sec^2 2) 16 m/sec^2

3) 36 m/sec^2 4) 48 m/sec^2

17. A 20g particle moves in SHM with a frequency of 3 oscillations per second and amplitude of 5 cm. Through what total distance does the particle move during one oscillation? What is its average speed ?

1) 20 cm, 20cm/sec 2) 40 cm, 40cm/sec

3) 40 cm, 60cm/sec 4) 20 cm, 60cm/sec

18. Acceleration displacement graph of a particle executing S.H.M is as shown in given figure. The time period of its oscillation is (in sec)

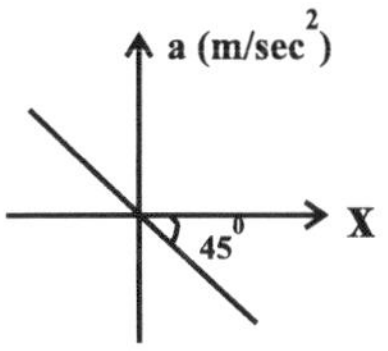

1) p/2 2) 2p 3) p 4) p/4

19. The displacement of a particle from its mean position (in meter) is given by Y = 0.2 sin $(10\pi t+1.5\pi)$ cos $(10\pi t+1.5\pi)$. The motion of the particle is

1) Periodic but not SHM 2) Non - Periodic

3) SHM with period 0.1sec

4) SHM with period 0.2 sec

20. Two SHMs are represented by the equations y_1= 10 sin (3pt+π/4) and y_2 = 5 [sin $3\pi t+\sqrt{3}$ cos $3\pi t$]. Their amplitudes are in the ratio

1) 1:2 2) 2:1 3) 1:3 4) 1 :1

21. A Particle of mass 4 gm is moving simple harmonically with a period 8 sec and amplitude 8cm. What is its velocity when the displacement is 8cm

1) 8 cm/sec 2) 6 cm/sec

3) 4 cm/sec 4) zero

22. Which of the following quatities are always positive in a S H M

1) $\overline{F}.\overline{a}$ 2) $\overline{a}.\overline{r}$ 3) $\overline{v}.\overline{r}$ 4) $\overline{F}.\overline{r}$

23. For a particle executing SHM. The KE 'K' is given by K = K_0 $\cos^2$ wt. The maximum value of PE is

1) K_0 2) zero 3) $\frac{K_0}{2}$ 4) $\frac{K_0}{4}$

24. A body executing SHM at a displacement 'x' its PE is E_1 , at a displacement 'Y' its PE is E_2. The P.E at a displacementt (x + y) is

1) $\sqrt{E}=\sqrt{E_1}-\sqrt{E_2}$ 2) $\sqrt{E}=\sqrt{E_1}+\sqrt{E_2}$

3) $E = E_1 + E_2$ 4) $E = E_1 - E_2$

25. An object is attached to the bottom of a light vertical spring and set vibrating. The maximum speed of the object is 15 cm / sec and the period is 628 m sec. The amplitude of the motion in centimeter is [EAMCET 2003 E]

1) 3 2) 2 3) 1.5 4) 1.0

26. At what displacement is the KE of a particle performing SHM of amplitude 10 cm, three times its PE ?

1) 10cm 2) 5 cm 3) 15 cm 4) 20cm

27. A particle execute S.H.M from the mean position. Its amplitude is A, its time period is 'T'. At what displacement, its speed is half of its maximum speed.

1) $\frac{\sqrt{3}A}{2}$ 2) $\frac{\sqrt{2}}{3}A$ 3) $\frac{2A}{\sqrt{3}}$ 4) $\frac{3A}{\sqrt{2}}$

28. An object executing SHM has an acceleration of 0.2m/s^2 at a distance of 0.05m from the mean position. Calculate the period of SHM.

1) 5.764sec 2) 4.654sec 3) 3.142sec 4) 2.764sec

29. The angular velocities of three bodies in SHM are w_1 w_2 w_3 with their respective amplitudes as A_1 A_2 A_3. If all three bodies have same mass and velocity then [BHU 2002]

1) $A_1 w_1 = A_2 w_2 = A_3 w_3$

2) $A_1 w_1^2 = A_2 w_2^2 = A_3 w_3^2$

3) $A_1^2 w_1 = A_2^2 w_2 = A_3^2 w_3$

4) $A_1^2 w_1^1 = A_2^2 w_2^2 = A_3^2 w_3^3$

30. The equation for the displacement of a particle executing SHM is y = {5 sin 2π t} cm. Find (i) the velocity at 3cm from the mean position, (ii) acceleration after 0.5s after leaving the mean position.

1) $8\pi cms^{-1}$, zero 2) $6\pi cms^{-1}$, zero

3) $4\pi cms^{-1}$, zero 4) $2\pi cms^{-1}$, zero

31. The velocity of a particle under going SHM at the mean position is 2ms^{-1}. Find the velocity of the particle at the point where the displacement from the mean position is equal to half the amplitude.

1) 1.234ms^{-1} 2) 1.732ms^{-1}

3) 2.732ms^{-1} 4) 1.414ms^{-1}

32. The frequency of a particle performing SHM is 12Hz. Its amplitude is 4cm. Its initial displacement is 2cm towards positive extreme position. Write its equation for displacement.

1) x=0.04 sin$\left(12\pi t+\frac{\pi}{6}\right)$ 2) x=0.04 sin$\left(48\pi t+\frac{\pi}{6}\right)$

3) x=0.04 sin$\left(24\pi t+\frac{\pi}{6}\right)$ 4) x=0.04 sin$\left(44\pi t+\frac{\pi}{6}\right)$

33. A block is placed on a horizontal platform. The system is making horizontal oscillations about a fixed point with a frequency of 0.5Hz. Find the maximum amplitude of oscillation if the block is not to slide on the horizontal oscillations about a fixed point with a frequency of 0.5Hz. Find the maximum amplitude of oscillation if the block is not to slide on the horizontal platform ? (Coefficient of friction between the block and the platform = 0.4)

1) 0.3971m 2) 0.4871m 3) 0.2971m 4) 0.5271m

SIMPLE PENDULUM

34. The periods of a pendulum on two planets are in the ratio 3 : 4. The acceleration due to gravity on them are in the ratio

1) 9 : 16 2) 3 : 4 3) 4 : 3 4) 16 : 9

35. The shape of l – T graph of simple pendulum is

1) Straight line 2) parabola
3) curve 4) Hyperbola

36. A simple pendulum is of length 50cm. Find its time period and frequency of oscillation. ($g = 9.8\ m/s^2$).

1) 2.419 sec, 0.8045Hz 2) 3.219 sec, 0.5045Hz
3) 5.619 sec, 0.6045Hz 4) 1.419 sec, 0.7045Hz

37. Pendulum beating seconds at the equator ($g = 978\ cm / sec^2$) is taken to Antartica ($g = 983\ cm / sec^2$), its length is to be increased by [83 M]

1) 0.1 cm 2) 0.2 cm 3) 0.4 cm 4) 0.5 cm

38. The mass and diameter of a planet are twice that of the earth. What will be the time period of oscillation of a pendulum on this planet, if it is a seconds pendulum on earth. [86E]

1) $\sqrt{2}$ sec 2) 2 sec 3) $\frac{1}{\sqrt{2}}$ sec 4) $2\sqrt{2}$ sec

39. If the value of 'g' on the surface of the moon is g / 6, the time period of a pendulum on the surface of the moon w.r.t time period on the earth will be

1) $\sqrt{6}$ times 2) 6 times
3) $\frac{1}{\sqrt{6}}$ times 4) $\frac{1}{6}$ times

40. A simple pendulum of length 1m is oscillated at a place where $g = 9.8 m/s^2$. Find the maximum velocity of the bob of the simple pendulum if the amplitude of oscillation is 3cm.

1) $5.291 cms^{-1}$ 2) $7.261 cms^{-1}$
3) $6.181 cms^{-1}$ 4) $9.391 cms^{-1}$

41. The period of a simple pendulum measured inside a stationary lift is 'T'. If the lift starts moving upwards with a acceleration g / 3. What will be the time period

1) T / 3 2) 3 T 3) $\frac{\sqrt{3}T}{2}$ 4) $\sqrt{\frac{3}{2}}T$

42. If a simple pendulum of length 'l' is about to start to make SHM, when it is brought a position so that its length has maximum angular displacement 'q'. If the mass of the bob is 'm", then the maximum K.E at its mean position is [AFMC 2002]

1) $\frac{1}{2}m\left(\frac{l}{g}\right)$ 2) $\frac{1}{2}$ mgl sin q
3) 2 mg l 4) mgl (1 – cos q)

43. Restoring force on the bob of a simple pendulum of mass 100 gm when its amplitude is 1^0 is

1) 0.017 N 2) 1.7 N
3) 0.17 N 4) 0.034 N

44. A child swiming on a swing in sitting position stands up, then the time period of swing will [EAMCET 2008 M, AIEEE 2002]

1) Increase 2) Decrease 3) Remains same
4) Increase if the child is long and decreases if thee child is short

45. On a planet, a body is let fall freely from a height 8m reaches the ground in 2 sec. If the length of a simple pendulum is 1m on that planet, its time period is

1) 3.14 sec 2) 6.24 sec
3) 1.57 sec 4) 4.71 sec

46. The time period of simple pendulum is 'T'. When the length increases by 10 cm, its period is T_1. When the length is decreased by 10 cm, itss period is T_2. Then the relation between T, T_1 and T_2 is

1) $\frac{2}{T^2} = \frac{1}{T_1^2} + \frac{1}{T_2^2}$ 2) $\frac{2}{T^2} = \frac{1}{T_1^2} - \frac{1}{T_2^2}$
3) $2T^2 = T_1^2 + T_2^2$ 4) $2T^2 = T_1^2 - T_2^2$

47. A simple pendulum has a time period 'T' in vaccum. Its time period when it is completely immerised in a liquid of density one – eighth of the density of the meterial of the bob is

1) $\sqrt{\frac{7}{8}}T$ 2) $\sqrt{\frac{5}{8}}T$ 3) $\sqrt{\frac{3}{8}}T$ 4) $\sqrt{\frac{8}{7}}T$

48. $T_1 T_2$ and T_3 the time periods of a given pendulum on the surface of the earth, at a depth 'h' in a mine and at an altitude 'h' above the earth's surface respectively, then

1) $T_1 = T_2 = T_3$ 2) $T_2 < T_1 < T_3$
3) $T_3 > T_2 > T_1$ 4) $T_1 > T_2 > T_3$

49. The period of oscillaion of a simple pendulum of length 'L' susspended from the ceiling of a vehicle which moves with out friction down in an inclined plane of inclination 'α' is given by

1) $2p\sqrt{\frac{L}{g\cos\alpha}}$ 2) $2p\sqrt{\frac{L}{g\sin\alpha}}$
3) $2p\sqrt{\frac{L}{g}}$ 4) $2p\sqrt{\frac{L}{g\tan\alpha}}$

*50. The bob of a simple pendulum of mass 100g is oscillating with a time period of 1.42s. If the bob is replaced by another bob of mass 150 g but of same radius what will be the new time period of oscillation ?

1) 2.24 sec 2) 1.42 sec
3) 1.22 sec 4) 3.26ec

*51. A pendulum that beats seconds on the surface of the earth were taken to a depth of (1/4)th the radius of the earth. What will be its time period of oscillation?

1) 5.107sec 2) 1.208sec 3) 3.401sec 4) 2.309sec

*52. The pendulum of a certain clock has time period 2.04s. How fast or slow does the clock run during 24 hour ?

1) 28.8minutes slow 2) 28.8minutes fast
3) 14.4minutes fast 4) 14.4 minutes slow

SPRINGS

53. A ball of mass 5kg hangs from a spring of force constant k N/m and oscillates with certain time period. If the ball is removed, the spring shortens by

1) $\left(\frac{K}{g}\right)$m 2) $\left(\frac{5g}{K}\right)$m 3) (2g) m 4) (g) m

54. A mass 'M' is suspended from a spring of negligible mass. The spring is pulled a little and then released so that the mass executes SHM of time period 'T, if the mass is increased by 'm', time period becomes $\frac{5T}{3}$, then the ratio of (m / M) is [AIEEE2003]

1) 3/2 2) 25/9 3) 16/9 4) 5/3

55. Two springs of force constans 1000 N/m and 2000 N/m are stretetched by same force. The ratio of their respective potential energies is

1) 2 : 1 2) 1 : 2 3) 4 : 1 4) 1 : 4 [2003 E]

56. A load 'm' is attached to a spring of force constant 'k' and stretched it through 'x' and released, it makes oscillation in a vertical plane with a time period 'T'. It is further pulled down through another 'x' and released. Now the time period of vertical oscillation will be

1) T / 2 2) 2 T 3) T 4) T / 4

57. K is the force constant of a spring. The work done is icreasing its extension from l_1 to l_2 will be

1) $K(l_2 - l_1)$ 2) $\frac{K}{2}(l_2 + l_1)$
3) $K(l_2^2 - l_1^2)$ 4) $\frac{K}{2}(l_2^2 - l_1^2)$

58. The elongation of a spring of length 'L' and of negligible mass due to a force is 'x'. The spring is cut into two pieces of length in ratio 1 : n. The ratio of the respective spring constants is [2002 M]

1) n : 1 2) 1 : n 3) $n^2 : 1$ 4) $1 : n^2$

59. Two springs with stiffness constants K_1 and K_2 ($K_2 > K_1$) are stretched by the same force. More work is done on the spring with spring constant [93 M]

1) K_1 2) K_2 3) Same for both 4) None

60. The natural angular frequency of a particle of mass 'm' attached to an ideal spring of force constant 'K' is [90 E]

1) $\sqrt{\frac{K}{m}}$ 2) $\sqrt{\frac{m}{K}}$ 3) $\left(\frac{K}{m}\right)^2$ 4) $\left(\frac{m}{K}\right)^2$

61. Springs of spring constants K, 2K, 4K, 8k, ... 2048 K are connected in series. A mass 'm' is attached to one end the system is allowed to oscillation. The time period is approximately

1) $2\pi\sqrt{\frac{m}{2K}}$ 2) $2\pi\sqrt{\frac{2m}{K}}$
3) $2\pi\sqrt{\frac{m}{4K}}$ 4) $2\pi\sqrt{\frac{4m}{K}}$

62. Two masses m and M are suspended together by a massless spring of force constant 'K'. When the masses are in equilibrium, M is removed with out distrubing the system. The amplitude of oscillation is

1) $\frac{Mg}{K}$ 2) $\frac{mg}{K}$ 3) $\frac{(M+m)g}{K}$ 4) $\frac{(M-m)g}{K}$

63. If a spring of force constant 'K' is cut into three equal parts, then the force constant of each part will be [AIEEE 2002]

1) K / 3 2) K 3) 3K 4) 6K

64. Three masses 0.1 kg, 0.3 kg and 0.4 kg are suspended at the end of the spring. When the 0.4 kg is removed, the system oscillates with a period of '2' sec. When 0.3 kg mass is also removed, the system will oscillate with a period

1) 1 sec 2) 2 sec 3) 3 sec 4) 4 sec

65. A spring of spring constant 5 ′ 10^3 N/ m is stretched initially by 5 cm from the unstretched position. Then the work required to stretch it further by 5 cm is [AIEEE 2003]

1) 12.50 Nm 2) 18.75 Nm

3) 25.00 Nm 4) 6.25 Nm

*66. A spring balance has a scale that reads 0 to 20 kg. The length of the scale is 10 cm. A body suspended from this balance, when displaced and released, oscillates with period of $\frac{\pi s}{10}$. What is the mass of the body

1) 4.9 kg 2) 9.8 kg 3) 19.6 kg 4) 14.7 kg

ANSWERS

1) 3	2) 3	3) 1	4) 3	5) 2
6) 4	7) 3	8) 3	9) 1	10) 2
11) 1	12) 1	13) 4	14) 2	15) 2
16) 4	17) 4	18) 2	19) 3	20) 4
21) 4	22) 1	23) 1	24) 2	25) 3
26) 2	27) 1	28) 3	29) 1	30) 1
31) 2	32) 3	33) 1	34) 4	35) 2
36) 4	37) 4	38) 4	39) 1	40) 4
41) 3	42) 4	43) 1	44) 2	45) 1
46) 3	47) 4	48) 3	49) 1	50) 2
51) 4	52) 1	53) 2	54) 3	55) 1
56) 3	57) 4	58) 1	59) 1	60) 1
61) 2	62) 1	63) 3	64) 1	65) 2
66) 1				

EXERCISE-II(B)

1. The displacment of a particle varies according to relation x = 4 [Cos pt + Sin pt]. The amplitude of the particle is

1) – 4 2) 4 3) $4\sqrt{2}$ 4) 8

2. The displacement equation of the particle in SHM is y = 10 sin(2pt +p/6)m. The initial displacement of the particle is

1) 5m $\frac{5}{}$m 3) 2.5 $\sqrt{3}$ m 4) 2.5 m

3. A particle encute SHM between x = –A and x = +A, The time taken for it to go from 0 to A/2 is T_1 and to go from A/2 to A is T_2, then [IIT Screening 2001]

1) $T_1 < T_2$ 2) $T_1 > T_2$

3) $T_1 = T_2$ 4) $T_1 = 2T_2$

4. A body exceuting S.H.M has maximum acceleration equal to 24m/sec and maximum velocity equal to 16m/sec. The amplitude of S.H.M is

1) $\frac{3}{32}$ m 2) $\frac{64}{9}$ m

3) $\frac{32}{3}$ m 4) $\frac{1024}{9}$ m

5. The motion of a particle is expressed by the equation a = –bx where 'a' is acceleration , 'x' is the displacement from the equilibrium position and 'b' is a constant. The periodic time will be

1) $\frac{2\pi}{b}$ 2) $\frac{2\pi}{\sqrt{b}}$

3) $2p\sqrt{b}$ 4) $2\sqrt{\frac{\pi}{b}}$

6. In a SHM, when the displacement is one- half the amplitude, what fraction of the total energy is kinetic

1) Zero 2) 1/4 3) 1/2 4) 3/4

7. A body of mass 100gm is performing SHM with a period of 4.5sec and amplitude7cm. Velocity of the body at equilibrium position in cm/sec is approximately

1) 10 2) 8 3) 6 4) 5

8. Ratio of average PE and average KE in SHM will be

1) 1 : 1 2) 2 : 1 3) 3 : 2 4) 2 : 3

9. A particle is performing SHM in a path of length 8 cm time period 4sec. Find the time taken and phase when particle is 2cm mean position

1) 1/3sec, 30^0 2) 2/3 sec, 60^0

3) 1/3 sec, 60^0 4) 2/3 sec, 30^0

10. The force constant of a particle of mass 1kg is SHM of amplitude 4cm is 100N/m when particle is at mean position at t = 0 then find maximum speed.

1) 0.4 m/s 2) 0.6 m/sec

3) 0.8 m/sec 4) 1 sec

11. **A body of mass 15g oscillates about a fixed point with a SHM of amplitude 8cm. If the body is attracted towards the fixed point, when at a distance of 4 cm from it with a force equal to the weight of 10g, find (a) the period of oscillation of the mass about the fixed point and (b) the maximum KE of the mass.**

1) 0.2916sec, $5.647\text{x}10^{-3}$ sec

2) 0.1917sec, $4.242\text{x}10^{-3}$ sec

3) 0.4917sec, $7.848\text{x}10^{-3}$ sec

4) 0.5913sec, $2.647\text{x}10^{-3}$ sec

12. **A simple harmonic oscillation is represented by the equation $y = 0.4 \sin\left(\frac{440t}{7} + 0.61x\right)$ Where 'Y' and 'X' are in m and time 't' in seconds respectively. The value of time period in seconds [91E]**

1) 0.1 2) 0.01 3) 1 4) 10

13. **In a S.H.M when the displacemt is one half of the amplitude, what fraction of the total energy is potential energy? [CPMT 99]**

1) 1/4 2) 1/ 2 3) 3/4 4) Zero

14. **The amplitude of SHM is 5cm, period 4sec and initial phase is p/6. Its displacement at t = 0 [initial displacement] is**

1) $2.5\sqrt{2}$ cm 2) 5 cm 3) 2.5 cm 4) $5\sqrt{2}$ cm

15. **Displacement time equation of a particle executing SHM is x = a sin (wt ı p/6). Time taken by the particle to go directly from x = –A/2 to x = + A/2 is**

1) $\frac{\pi}{3w}$ 2) $\frac{\pi}{2w}$ 3) $\frac{2\pi}{w}$ 4) $\frac{\pi}{w}$

16. **If the displacement (x) and velocity (v) of a particle executing S.H.M are related through the expression $4V^2 = 25 - x^2$, then its time period is [2002 E]**

1) p 2) 2p 3) 4p 4) 6p

17. **Maximum velocity in SHM is V_m. The average velocity during motion from one extreme point to the other extreme point will be**

1) $\frac{\pi Vm}{2}$ 2) $\frac{2}{\pi}Vm$

3) $\frac{4}{\pi}Vm$ 4) Zero

18. **A body of mass 0.01 kg executes S.H.M about x = 0 under the influence of a force shown below. The period of S.H.M is**

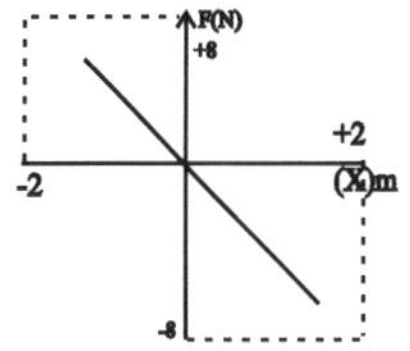

1) 1.05 sec 2) 3.14 sec

3) 0.25 sec 4) 0.314 sec

19. **The equation of motion of a particle is given by $\frac{dp}{dt} + mw^2x = 0$ whene 'p' is the momentum and 'x' is the position. Then the particle is [98E]**

1) Moves along a straight line

2) Moves along a parabola

3) Executes SHM

4) Falls freely under gravity

20. **Two SHM's are represented by the relations x = 4 sin (80t +p/2) and y = 2 cos(60t+p/2). The ratio of their time periods is**

1) 2 : 1 2) 1 : 2 3) 4 : 3 4) 3 : 4

21. **A particle of mass 10gm is describing an SHM along a straight line with a time period of 2sec and amplitude 10cm. Then its KE at 2 cm from the equilibrium position is**

1) 4.8 ′ 10^{-4} J 2) 9.6 ′ 10^{-4} J

3) 4.8 ′ 10^{3} J 4) 9.6 ′ 10^{3} J

22. **Which of the following quantities are always negative**

1) $\overline{F}.\overline{a}$ 2) $\overline{V}.\overline{r}$ 3) $\overline{a}.\overline{r}$ 4) $\overline{F}.\overline{V}$

23. **For a particle executing SHM. The kinetic energy 'K' is given by $k = k_0 \cos^2 wt$. The maximum value of KE is**

1) K_0 2) Zero 3) $\frac{K_0}{2}$ 4) $\frac{K_0}{4}$

24. **A body exceuting SHM at a displacement 'x' its PE is 16J, and at displacement 'y' its PE is 25J. The PE at a displacement (x+y) is**

1) 49 J 2) 64 J 3) 81J 4) 125 J

25. **A particle executes SHM has amplitude 6cm. Its acceleration at a distance from the mean position 2cm is 8cm/sec^2. The maximum speed of the particle is [2000E]**

1) 8cm/sec 2) 12cm/sec

3) 16cm/sec 4) 24cm/sec

26. If the displacement of a simple harmonic oscillators is half of its amplitude at an instant. PE of the oscillator is

1) 75% of total energy 2) 50% of total energy

3) 25% of total energy 4) 0% of total energy

27. A particle executes S.H.M with amplitude 'A'. If the displacement y = A/2 which of the following will be half of its maximum value

1) Kinetic energy 2) Speed

3) Restoring force 4) Potential energy

28. At a distance of 3cm from the mean postition, the acceleration of a particle performing S.H.M is 12cm/sec^2.

Theperiodic time of the particle is

1) 3.14sec 2) 1.57sec 3) 1.0sec 4) 2.0sec

29. Two particles P and Q start from origin and execute SHM along X - axis with same amplitude but the periods 3seconds and 6 seconds respectively. The ratio of the velocities P and Q when they meet is [2001E]

1) 1 : 2 2) 2 : 1 3) 2 : 3 4) 3 : 4

30. The equation for the displacement of a particle executing SHM is y = {5 sin π t} cm. Find (i) the velocity at 3cm from the mean position, (ii) acceleration after 0.5s after leaving the mean position.

1) $3\pi\,cm\,s^{-1}$, zero 2) $6\pi\,cm\,s^{-1}$, zero

3) $4\pi\,cm\,s^{-1}$, zero 4) $2\pi\,cm\,s^{-1}$, zero

31. The velocity of a particle under going SHM at the mean position is 4ms^{-1}. Find the velocity of the particle at the point where the displacement from the mean position is equal to half the amplitude.

1) $2\sqrt{3}ms^{-1}$ 2) $5\sqrt{3}ms^{-1}$

3) $4\sqrt{3}ms^{-1}$ 4) $3\sqrt{3}ms^{-1}$

32. A particle performing SHM with a frequency of 5 Hz and amplitude 2 cm is initially in the positive extreme position. What is the equation for its displacement ?

1) $x = 0.02\cos 10\pi t$ 2) $x = 0.02\cos 5\pi t$

3) $x = 0.02\cos 15\pi t$ 4) $x = 0.02\cos 20\pi t$

33. A block is placed on a horizontal platform. The system is making horizontal oscillations about a fixed point with a frequency of 0.5Hz. Find the maximum amplitude of oscillation if the block is not to slide on the horizontal oscillations about a fixed point with a frequency of 0.5Hz. Find the maximum amplitude of oscillation if the block is not to slide on the horizontal platform ? (Coefficient of friction between the block and the platform = 0.2) (g =10 ms^{-2})

1) $\frac{2}{\pi^2}$ 2) $\frac{5}{\pi^2}$ 3) $\frac{3}{\pi^2}$ 4) $\frac{4}{\pi^2}$

SIMPLE PENDULUM

34. A hollow sphere of simple pendulum is first filled with mercury and then with water. The time periods are in the ratio if their densities are in the ratio 13.6 :1

1) 1 : 1 2) 1 : $\sqrt{13.6}$

3) 1 : 13.6 4) 13.6 : 1

35. In a simple pendulum the period of oscillation (T) is related to the length of the pendulum (L) as [95M]

1) $\frac{L}{T}$ = constant 2) $\frac{L^2}{T^2}$ = constant

3) $\frac{L}{T^2}$ = constant 4) $\frac{L^2}{T}$ = constant

36. The time period of a pendulum of infinite length on earth'ssurface is

1) Infinity 2) Zero

3) 84.6 minutes 4) 42.3 $\sqrt{2}$ minutes

37. A pendulum beating seconds at one place (g = 985cm/sec^2) is taken to another place (g = 981cm/sec^2). If it is to beat seconds at another place, its length is to be decreases by

1) 0.1 cm 2) 0.2cm 3) 0.4cm 4) 0.5cm

38. The mass and radius of a planet are double that of the earth. The time period of simple pendulum on the planet, if T is the time period of simple pendulum on earth.

1) 2T 2) $\sqrt{2}$ T 3) T/2 4) T/ $\sqrt{2}$

39. The value of 'g'on moon is five times less than that on the earth. If the length of the seconds pendulum on the earth is 100cm, its length on the moon is [82M]

1) 20cm 2) 10cm 3) 100 cm 4) 25 cm

40. If the percentage increase in length of the pendulum is 3% and percentage increase in acceleration due to gravity is 1% then the percentage increase in time period of the pendulum is

1) 1% 2) 0.5% 3)1.5% 4) 2%

41. A seconds pendulum is arranged in a lift. If the lift is moving up with an acceleration g/4. Its new time period will be

1) $4\sqrt{2}$ sec 2) $\frac{1}{\sqrt{2}}$ sec 3) $\frac{4}{\sqrt{5}}$ sec 4) None

42. A simple pendulum consisting a ball of mass 'm' tied to a string of length 'l' is made to showing on a circular arc of angle 'q' in a vertical plane. At the end of this arc, another ball of mass 'm' is placed at rest. The momentum transfered to the ball at rest by swinging ball is

1) mql $\sqrt{\frac{l}{g}}$ 2) mql $\sqrt{g/l}$

3) 2qml $\sqrt{\frac{l}{g}}$ 4) Zero

43. Tangential acceleration on a bob of a simple - pendulum of mass 100gm when its amplitude is 1^0 is

1) 0.017 m/sec^2 2) 1.7 m/sec^2

3) 0.17 m/sec^2 4) 0.034 m/sec^2

44. A child swinging on a swing in standing position and she suddenly sits on it, then the time period of the swing will be

1) Increases 2) Decreases

3) Remains same 4) Can't say

45. On a planet a body is left fall freely from a height 2m reaches the ground $\sqrt{2}$ seconds. If the length of a simple pendulum is 2m on that planet, its time period is

1) 3.14 sec 2) 6.28sec 3) 1.57 sec 4) 4.71 sec

46. A simple pendulum of length L_1 has a time period 'T_1' and another pendulum of length L_2 has time period 'T_2' Then the time period of pendulum of length (L_1+L_2) is

1) T1+T2 2) $\sqrt{T_1+T_2}$

3) $\sqrt{T_1T_2}$ 4) $\sqrt{{T_1}^2+{T_2}^2}$

47. A simple pendulum has time period 3sec. The pendulum is completely immersed in a non - viscous liquid whose density is 1/10th of that of the material of the bob. Then time period of that pendulum is

1) $\sqrt{3}$ sec 2) 3sec 3) $\frac{3}{\sqrt{10}}$ sec 4) $\sqrt{10}$ sec

48. The period of simple pendulum whose bob is a hollow metallic sphere is 'T'. The period is T_1 when the bob is filled with sand , T_2 when thebob is filled with mercury and T_3 when it is half filled with sand then

1) $T = T_1 + T_2 > T_3$ 2) $T_1 = T_2 > T_3 > T$

3) $T_1 > T_3 > T_1 = T_2$ 4) $T = T_1 = T_2 < T_3$

49. A simple pendulum is fixed on a trolly which is sliding down on a frictionless plane at an angle 60^0. The ratio of time period when the trolly is on a horizontal plane to the time period when the trolly is sliding is

1) $\sqrt{3}$: 2 2) 1: 1 3) 1 : 2 4) 1 : $\sqrt{2}$

50. The bob of a simple pendulum of mass 200g is oscillating with a time 2sec. If the bob is replaced by another bob of mass 450g but of same radius what will be the new time period of oscillation ?

1) 1sec 2) 3sec 3) 4sec 4)2sec

51. A pendulum that beats seconds on the surface of the earth were taken to a depth of (1/3)th the radius of the earth. What will be its time period of oscillation ?

1) $\sqrt{6}$ sec 2) $\sqrt{5}$ sec 3) $\sqrt{2}$ sec 4) $\sqrt{3}$ sec

52. The pendulum of a certain clock has time period 2.04s. How fast or slow does the clock run during 12 hour ?

1) 11.4 minutes slow 2)12.4 minutes slow

3) 14.4 minutes slow 4)15.2 minutes slow

SPRINGS

53. A ball of mass 2kg hanging from a spring oscillates with a time period 2p seconds.Ball is removed when it is in equilibrium position, the spring shortens by

1) (g) m 2) $\left(\frac{g}{2}\right)$ 3) (2g)m 4) (2p)m

54. A mass 'm attached to a spring oscillates with a period 2seconds when the mass is increased by 2kg the period increases by 1second. The initial mass 'm' is

1) 4kg 2) 1.6kg 3) 0.8 kg 4) 2kg

55. Two springs of spring constants 1500N/m and 3000N/m respectively are stretched with the same force. They will have potential energy in the ratio

1) 4 : 1 2) 1 . 4 3) 2 . 1 4) 1 . 2

56. **A light vertical spring is stretched by 2cm when a weight of 10g is attached to its free end. The weight is further pulled down by 1.0 cm and released. Compute the frequency and amplitude of motion.**

1) 3.523 Hz, 2cm 2) 2.523 Hz, 5cm

3) 3.523 Hz, 1cm 4) 1.324 Hz, 1cm

57. **A long spring when stretched by 'x'cm has a potential energy 'V'. on increasing the stretching to 'nx' cm the potential energy stored in the spring will be [92M]**

1) v/n 2) nv 3) n^2V 4) v/n^2

58. **A spring of force constant 'K' is cut into two pieces such that one piece is double the length of the other. Then the long piece will have a force constant of [2000,IIT99]**

1) 2k/3 2) 3k/2 3) 3k 4) 6k

59. **Two springs having spring constants 25N/m and 16N/m are stretched by forces F_1 and F_2 have equal potential energy. The ratio of the forces F_1 and F_2 is [AIEEE2002]**

1) 4/5 2) 5/4 3) 16/25 4) 25/16

60. **Two bodies A and B of equal masses are suspended from two separate massless springs of force constants K_1 and K_2. If they oscillate vertically so that their maximum velocities are equal the ratio of their amplitudes is [AIEEE2003]**

1) $\frac{k_1}{k_2}$ 2) $\frac{k_2}{k_1}$ 3) $\sqrt{\frac{k_1}{k_2}}$ 4) $\sqrt{\frac{k_2}{k_1}}$

61. **Springs of constants k,2k,4k,8k,16k ...... are connected in series. The mass 'm' kg is attached to the lower end of the last spring and the system is allowed to vibrate. The frequency of oscillation is**

1) $\frac{1}{\pi}\sqrt{\frac{m}{2k}}$ 2) $\frac{1}{2\pi}\sqrt{\frac{k}{2m}}$ 3) $\frac{1}{2\pi}\sqrt{\frac{2m}{2k}}$ 4) $\frac{1}{2\pi}\sqrt{\frac{4m}{k}}$

62. **Two masses m_1 and m_2 are suspended from a spring of spring constant 'k'. When the masses are in equilibrium, m_1 is gently removed. The angular frequency and amplitude of oscillation of m_2 will be respectively**

1) $\sqrt{\frac{k}{m_2}}, \frac{m_1g}{k}$ 2) $\sqrt{\frac{m_2}{k}}, \frac{k}{m_1g}$

3) $\sqrt{\frac{k}{m_2}}, \frac{k}{m_1g}$ 4) $\sqrt{\frac{m_2}{k}}, \frac{m_1g}{k}$

63. **If a spring has time period 'T' and cut into 'n' equal parts, then the time period of each part wil be [AIEEE2002]**

1) $T\sqrt{n}$ 2) $T/\sqrt{n}$ 3) nT 4) T

64. **Three masses 700gm,500gm and 400gm are suspended at the end of the spring and they are in equilibrium. when the 700gm mass is removed, the system oscillates with a period of 3sec. when the 500gm mass is also removed, it will oscillate with a period of**

1)1 sec 2) 2 sec 3) 3 sec 4) $\sqrt{\frac{12}{5}}$ sec

65. **When some load is attached to the spring of spring constant then it elongates through 0.8m. If it is further stretches by 0.05m then the period of oscillation is**

1) $\frac{2\pi}{7}$ sec 2) $\frac{4\pi}{7}$ sec 3) $\frac{\pi}{7}$ sec 4) $\frac{\pi}{14}$ sec

66. **A spring balance has a scale that reads 0 to 10 kg. The length of the scale is 10 cm. A body suspended from this balance,when displaced and released, oscillates wih period of $\frac{\pi s}{10}$. What is the mass of the body ? ($g = 10ms^{-2}$)**

1) 2.5kg 2) 4.2 kg

3) 1.5 kg 4) 3.5 kg

ANSWERS

1) 3	2) 1	3) 1	4) 3	5) 2
6) 4	7) 1	8) 1	9) 1	10) 1
11) 3	12) 1	13) 1	14) 3	15) 1
16) 3	17) 2	18) 4	19) 3	20) 4
21) 1	22) 3	23) 2	24) 3	25) 2
26) 3	27) 3	28) 1	29) 2	30) 3
31) 1	32) 1	33) 1	34) 1	35) 3
36) 3	37) 3	38) 4	39) 1	40) 1
41) 3	42) 4	43) 3	44) 1	45) 2
46) 4	47) 4	48) 4	49) 4	50) 4
51) 1	52) 3	53) 1	54) 2	55) 3
56) 3	57) 3	58) 2	59) 2	60) 4
61) 2	62) 1	63) 2	64) 2	65) 2
66) 1				

EXERCISE-III(A)

Equations of SHM :

1. Two SHM are given by $Y_1 = A \sin\left(\frac{\pi}{2}t+\phi\right)$ and $Y_2 = B \sin\left(2\frac{\pi}{3}t+\phi\right)$ The phase difference between these two after '1' sec is

   1) p 2) p / 2 3) p / 4 4) p / 6

2. Four simple hormonic vibrations $x_1 = 8 \sin \omega t$, $x_2 = 6 \sin(\omega t + \pi / 2)$, $x_3 = 4 \sin(\omega t + \pi)$ and $x_4 = 2 \sin\left(\omega t+\frac{3\pi}{2}\right)$ are superimposed on each other. The resulting amplitude is

   1) 20 2) $8\sqrt{2}$ 3) $4\sqrt{2}$ 4) 4

3. A particle execute in S.H.M whose displacement equation is $Y = 5 \sin(10t - \pi/3)$. The minimum time after which the K.E of the particle will be maximum is

   1) $\frac{\pi}{30}$ sec 2) $\frac{\pi}{60}$ sec 3) $\frac{\pi}{90}$ sec 4) $\frac{\pi}{120}$ sec

4. The differential equation representing the S H M of a particle is $\frac{9d^2Y}{dt^2} + 4Y = 0$. The time period of the particle is given by

   1) $\frac{\pi}{3}$ sec 2) π sec 3) $\frac{2\pi}{3}$ sec 4) 3π sec

5. Two particles executive SHM of the same amplitude and frequency on parallel lines side by side. They cross one another when moving in opposite directions each time their displacement is $\frac{\sqrt{3}}{2}$ times their amplitude. What is the phase difference between them ?

   1) $\frac{2\pi}{3}$ 2) $\frac{\pi}{4}$ 3) $\frac{\pi}{6}$ 4) $\frac{\pi}{3}$

6. The displacement of a particle executing S.H.M from its mean porition is given by $x = 0.5 \sin(10\pi t + 1.5) \cos(10\pi t + 1.5)$. The ratio of the maximum velocity to the maximum acceleration of the body is given by [2000 E]

   1) 20π 2) $\frac{1}{20\pi}$ 3) $\frac{1}{10\pi}$ 4) 10π

7. A particle executing SHM has amplitude of 4cm, and its acceleration at a distance of 1 cm from the mean position is 3 cm s^{-2} . What will its velocity be when it is at a distance of 2 cm from its mean position ?

   1) 1 cm/ sec 2) 2 cm / sec
   3) 6 cm/sec 4) 4 cm/sec

8. The velocities of a particle performing linear SHM are 0.13m/s and 0.12 m/s when it is at 0.12m and 0.13m from mean position, respectively. Find the period and amplitude.

   1) 6.284 sec, 0.1769m
   2) 5.184 sec, 0.1863m
   3) 2.152 sec, 0.1261m
   4) 6.251 sec, 0.2753m

9. A body executes SHM under the influence to one force and has time period 'T_1' seconds and the same body executes SHM with period 'T_2' seconds when under the influence of another force. When both forces act simultaneously and in the same direction then the time period of the same body is

   1) $(T_1 + T_2)$ sec 2) $\sqrt{T_1^2+T_2^2}$ sec
   3) $\sqrt{\frac{T_1+T_2}{T_1T_2}}$ sec 4) $\sqrt{\frac{T_1^2T_2^2}{(T_1^2+T_2^2)}}$

10. The force on a body executing SHM is 2N when the displacement is 2cm. If the amplitude of oscillations is 5cm, what is the total energy associated with the SHM ?

    1) 1.25×10^{-2}J 2) 125×10^{-2}J
    3) 12.5×10^{-2}J 4) 0.125×10^{-2}J

11. A body executing S.H.M has a maximum acceleration equal to 48 m / sec^2 and maximum velocity equal to 12 m / sec. The amplitude of S.H.M is

    1) 3 m 2) 3 / 32 m
    3) 1024 / 9 m 4) 64 / 9 m

12. The displacement of an object attached to a spring and executing S.H.M is given by x = 2 ′ 10^{-2} cos pt m. The time at which the maximum speed first occurs is [AIEEE 2007]

    1) 0.5 sec 2) 0.75 sec
    3) 0.125 sec 4) 0.25 sec

13. **A particle executing SHM of period 12 seconds moves through a distance of 32 cm in one oscillation. Find the time it takes to travel 4cm from the positive extremity of its oscillation.**

1) 1 sec 2) 1.2 sec 3) 1.7 sec 4) 1.4 sec

14. **The total mechanical energy of a harmonic oscillator of amplitude 1m and force constant 200 N/m is 150J. Then [91 E]**

1) The minimum P E is Zero

2) The maximum P E is 100 J

3) The minimum P E is 50 J

4) The maximum P E is 50 J

15. **A particle is executing S.H.M with a frequency 'f', the frequency with K.E oscillates is**

1) f 2) f / 2 3) 2f 4) 4f

16. **A block of 0.5kg is placed on a horizontal platform. The system is making vertical oscillations about a fixed point with a frequency of 0.5Hz. Find the maximum amplitude of oscillation if the block is not lose contact with the horizontal platform ? Also find apparent weight of the block at the lower extreme position**

1) 0.6542m, $mg+m\omega^2A$ 2) 0.9927m, $mg+m\omega^2A$

3) 0.7428m, $mg-m\omega^2A$ 4) 0.852m, $mg-m\omega^2A$

17. **A particle of mass 10 gm is executing S.H.M with an amplitude of 0.5 m and angular frequency 10 rad/sec. The maximum value of force acting on the particle during the course of oscillation is [MP PMT 2000]**

1) 25 N 2) 5 N 2) 2.5 N 4) 0.5 N

18. **A small body of mass 10 gram is making simple harmonic oscillations along a straight line with a time period of $\pi/4$ and the maximum displacement is 10cm. Find the energy of oscillations. What is the average kinetic energy during one oscillation ?**

1) 32×10^{-2}J, 16×10^{-2}J

2) 32×10^{-2}J, 32×10^{-2}J

3) 16×10^{-2}J, 16×10^{-2}J

4) 16×10^{-2}J, 32×10^{-2}J

19. **A particle of mass 0.5kg is executing SHM along a straight line. Its path length is 10cm and time period is 8s. Calculate its KE, PE and TE when its phase angle is $\pi/6$ radian.**

1) 1.592×10^{-4}J, 0.562×10^{-4} J, 1.652×10^{-4}J

2) 2.371×10^{-4}J, 0.123×10^{-4} J, 2.154×10^{-4}J

3) 3.292×10^{-4}J, 0.462×10^{-4} J, 5.416×10^{-4}J

4) 2.892×10^{-4}J, 0.964×10^{-4} J, 3.856×10^{-4}J

20. **A particle is acted by a force F = kx, where 'K' is a positive constant. Its P.E at x = 0 is zero. Which curve correctly represents the variation of potential energy of the block with respect to 'x' [IIT Screening 2004]**

1) 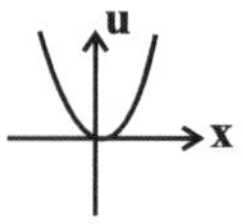

2) 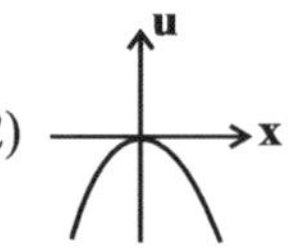

3) 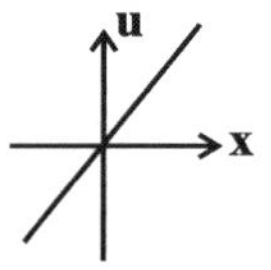

4) 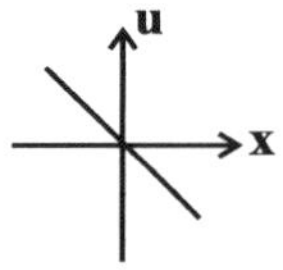

21. **A particle of mass 'm' is attached to a spring of spring constant ω_0. An external force F(t) proportional to cos ωt ($\omega \neq \omega_0$) is applied to the oscillator. The time displacement of the oscillator will be proportional to [AIEEE 2004]**

1) $\dfrac{m}{(\omega_0 - \omega^2)}$ 2) $\dfrac{m}{(\omega_0^2 + \omega^2)}$

3) $\dfrac{1}{m(\omega_0^2 + \omega^2)}$ 4) $\dfrac{1}{m(\omega_0^2 - \omega^2)}$

22. **At some instant velocity and acceleration of a particle of mass 'm' in SHM are 'a' and 'b' respectively. As angular frequency is w, its maximum K.E is**

1) $\frac{1}{2}m\left[a^2 + \frac{b^2}{w}\right]$ 2) $\frac{1}{2}m\left[a^2 - \frac{b^2}{w^2}\right]$

3) $\frac{1}{2}m\left[a^2 + \frac{b^2}{w^2}\right]$ 4) $\frac{1}{2}m\left[a^2 - \frac{b^2}{w}\right]$

23. **A piece of wood has dimensions a,b,c. Its relative density is 'd'. The wood is floating in water such that the side 'a' is vertical. It is pushed down a bit and released. The time period of SHM executed by it is**

1) $2\pi\sqrt{\frac{ab}{gc}}$ 2) $2\pi\sqrt{\frac{bc}{dg}}$ 3) $2\pi\sqrt{\frac{g}{da}}$ 4) $2\pi\sqrt{\frac{da}{g}}$

SIMPLE PENDULUM

24. A freely falling body takes 2 seconds to reach the ground on a plane, when it is dropped from a height of 8m. If the period of a simple pendulum is 'π' seconds on the planet, calculate the length of the pendulum.

1) 75cm 2) 25 cm 3) 50 cm 4) 100cm

25. A pendulum is taken 1 km inside the earth from sea level. Then the gain or loss of time of pendulum approximately

1) looses 13.5 sec per day 2) gains 13.5 sec per day
3) looses 7 sec per day 4) gain 7 sec per day

26. A clock regulated by a seconds pendulum keeps correct time. During summer the length of the pendulum increases to 1.02m. How much will the clock gain or lose in one day ?

1) 2280 sec 2) 1230 sec 3) 1180 sec 4) 1380 sec

27. A person measures the time period of a simple pendulum inside a stationary lift and finds it to be 'T'. If the lift starts accelerating down wards with an acceleration g/3, the time period of the pendulum will be

1) $\sqrt{3}\,T$ 2) $\sqrt{\frac{3}{2}}\,T$ 3) $\frac{T}{\sqrt{3}}$ 4) T / 3

28. A simple pendulum has a time period T_1 when on the earth's surface and T_2 when it taken a height 'R' above the earth's surface, where 'R' is the radius of the earth. The value of T_2 / T_1 is

1) 1 2) $\sqrt{2}$ 3) 4 4) 2

29. The period of a simple pendulum is found to increase by 50% when the length of the pendulum is increased by 0.6m. Calculate the initial length and the initial period of oscillation at a place where g = 9.8 m/s^2.

1) 0.28m, 1.491sec 2) 0.18m, 1.451sec
3) 0.68m, 1.141sec 4) 0.48m, 1.391sec

30. The angular amplitude of simple pendulum is q_0. The maximum tension in its string will be (assume q_0 is small)

1) mg $(1 + q_0)$ 2) mg$(1 - q_0)$
3) mg $(1 + q_0^2)$ 4) mg $(1 - q_0^2)$

31. A ball of radius 'a' is made to oscillate in a bowl of radius 'b' then time period of vibration of the ball is

1) $2p\sqrt{\frac{a}{g}}$ 2) $2p\sqrt{\frac{b-a}{g}}$
3) $2p\sqrt{\frac{b}{g}}$ 4) $2p\sqrt{\frac{b+a}{g}}$

32. The bob of a simple pendulum is displaced from its equilibrium position 'O' to a position 'Q' which is at a height 'h' above 'O' and the bob is then released. Assuming the mass of the bob to be 'm' and time period of oscillation to be 2.0 sec, the tension in the string when the bob passes through 'O' is

1) m (g + 2p^{2}h) 2) m (g + p^{2}h)
3) $m\left(g+\frac{\pi^2}{2}h\right)$ 4) $m\left(g+\frac{\pi^2}{3}h\right)$

33. A bob of a simple pendulum executes S.H.M in water with a period 't', while the period of oscillation of the bob is 'to' is air. Neglecting frictional force of water and given that the density of the bob is $\frac{4}{3}\times 1000$ kg / m^3. What relation ship between 't' and 'to' is true ? [AIEEE 2004]

1) $t = t_0$ 2) $t = 4\,t_0$ 3) $t = 2\,t_0$ 4) $t = \frac{t_0}{2}$

34. A simple pendulum have time periods 'T' and 5T/4. They start vibrating at the same instant from the mean position in the same phase. The phase difference between them when the pendulum with higher time period completes one oscillation is

1) 30^0 2) 45^0 3) 60^0 4) 90^0

35. Two pendulums of lengths 16l and l are in phase at the mean position at a certain time. If 'T' is the time period of shorter pendulum, the minimum time after which they will be again in phase is [98 E]

1) $\frac{1}{3}T$ 2) $\frac{2}{3}T$ 3) $\frac{3}{2}T$ 4) $\frac{4}{3}T$

36. Two simple pendulums of lengths 100m and 121 m start swinging together. They will swing together again after

1) the longer pendulum makes 10 oscillation
2) the shorter pendulum makes 10 oscillation
3) the longer pendulum makes 11 oscillation
4) the shorter pendulum makes 20 oscillaion

37. The maximum tension in the string of a pendulum is two times of the minimum tension. Let q_0 be the angular amplitude. The cos q_0 is

1) $\frac{1}{2}$ 2) $\frac{3}{4}$ 3) $\frac{2}{3}$ 4) $\frac{3}{5}$

SPRINGS

38. When a body is suspended from a spring its time period 'T'. If the spring is cut into two equal parts and the body is suspended from one part, its time period will be

1) $T/\sqrt{2}$ 2) $T/2$ 3) $\sqrt{2}T$ 4) $T/4$

39. If a body of mass 0.98 kg is made to oscillate on a spring of force constant 4.84 N/m, the angular frequency of the body is [CBSE 2001]

1) 1.22 rad / sec 2) 2.22 rad / sec
3) 3.22 rad / sec 4) 4.22 rad / sec

40. The time period of a body suspended from a spring is T. What will be the new time period, if the spring is cut into two equal parts and (i) the same mass is suspended from one part, (ii) The mass is suspended simultaneously from both the parts ?

1) $\frac{T}{2\sqrt{2}}, \frac{T}{2}$ 2) $\frac{T}{\sqrt{2}}, T$

3) $\sqrt{2}T, \frac{T}{2}$ 4) $\frac{T}{\sqrt{2}}, \frac{T}{2}$

41. Five identical springs are used in the following three configurations. The time period of vertical oscillations in configuration (i) (ii) and (iii) are in the ratio

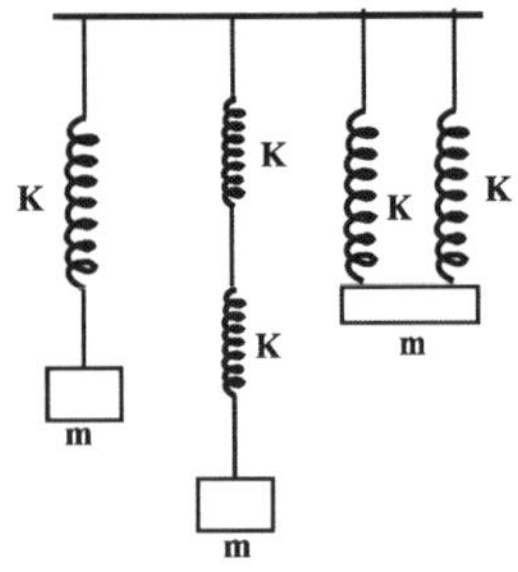

1) $1 : \sqrt{2} : \frac{1}{\sqrt{2}}$ 2) $2 : \sqrt{2} : \frac{1}{\sqrt{2}}$

3) $\frac{1}{\sqrt{2}} : \sqrt{2} : 1$ 4) $2 : \frac{1}{\sqrt{2}} : 1$

42. An ideal spring with a spring consant 'K' is hung from the ceiling and a block of mass 'm' is attached to its lower end. The mass is released with spring initially unstretched. Then the maximum extension in the spring is [IIT Screening 2002]

1) 4 mg / K 2) 2 mg / K
3) mg / K 4) mg / 2K

43. When a mass 'M' is attached to the spring of force constant 'K', then the spring stretches by 'l'. If the mass oscillates with amplitude 'l', what will be the maximum potential energy stored in the spring [BHU 2002]

1) $\frac{Kl}{2}$ 2) 2Kl 3) $\frac{1}{2}$ 4) mg l

44. A pan with certain weight is attached to a spring. The period of vertical oscillation is found to be 0.5 sec. If some weight is added in the pan, the time period becomes 0.6 sec. The additional distance through which the spring stretches due to additional load is

1) $\frac{11}{400}$ m 2) $\frac{9}{25}$ m
3) $\frac{11}{200}$ m 4) $\frac{9}{400}$ m

45. A particle at the end of the spring executes S.H.M with a period t_1, while the corresponding period for another spring is t_2. If the period of oscillation with two springs in series is 'T', then [AIEEE 2004]

1) $T = t_1 = t_2$ 2) $T^2 = t_1^2 + t_2^2$
3) $T^2 = t_1^{-1} + t_2^{-1}$ 4) $T^2 = t_1^{-2} + t_2^{-2}$

46. A horizontal spring block system of mass 'm' executes SHM of amplitude 'A' and frequency 'n'. When the block passing through its equilibrium position, an object of mass 'm' is put on it and the two move together. Find the new amplitude

1) $A\sqrt{\frac{M}{M+m}}$ 2) $A\sqrt{\frac{m}{M+m}}$

3) $A\sqrt{\frac{m+m}{M}}$ 4) $A\sqrt{\frac{M+m}{m}}$

47. A mass 'M' is suspended from a light spring An additional mass 'm' added displaces the spring further by a distance 'x' . Now the combined mass will oscillate the spring with a period

1) $T = 2\pi\sqrt{\frac{mg}{x(M+m)}}$ 2) $T = 2\pi\sqrt{\frac{(M+m)x}{mg}}$

3) $T = 2\pi\sqrt{\frac{M+m}{mgx}}$ 4) $T = 2\pi\sqrt{\frac{mgx}{M+m}}$

48. The variation of PE of a simple harmonic oscillator is as shown. Then force consant of the system is (PE 'U' is in joules, displacement 'x' is in mm)

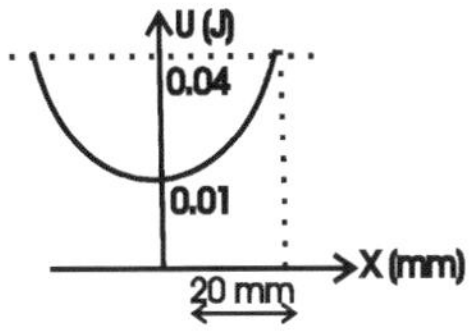

1) 100 N/m 2) 150 N/m

3) $\frac{200}{3}$ N/m 4) 300 N/m

49. The displacement of a particle of mass 3gm executing S.H.M is given by y = 3 sin (0.2t) in S.I units. The K.E of a particle at a point which is at a distance equal to 1/3 of its amplitude from its mean position (2007 M)

1) 12 x 10^3J 2) 25 x 10^{-3}J

3) 0.48 x 10^{-3}J 4) 0.24 x 10^{-3} J

50. The time period of a mass loaded spring is 'T'. If 20% of its turns are cut and removed, the same mass is attached to the remaining spring would oscillate with a time period of

1) $\frac{5T}{4}$ 2) $\frac{\sqrt{5}T}{2}$

3) $\frac{4T}{5}$ 4) $\frac{2T}{\sqrt{5}}$

ANSWERS

1) 4	2) 3	3) 1	4) 4	5) 4
6) 2	7) 3	8) 1	9) 4	10) 3
11) 1	12) 1	13) 3	14) 3	15) 3
16) 2	17) 4	18) 1	19) 4	20) 1
21) 4	22) 3	23) 4	24) 4	25) 3
26) 3	27) 2	28) 4	29) 4	30) 3
31) 2	32) 1	33) 3	34) 4	35) 4
36) 1	37) 2	38) 1	39) 2	40) 4
41) 1	42) 2	43) 4	44) 1	45) 2
46) 1	47) 2	48) 2	49) 3	50) 4

EXERCISE-III(B)

Eqation of SHM:

1. The displacement of two particles executing S.H.M are represented by y_1 = a sin(wt + ϕ) and y_2 = a coswt. The phase difference between the velocities of these particles is

1) – ϕ 2) + ϕ 3) ϕ +p/2 4) ϕ - p/2

2. Equations of S.H.M for 4 particles are x_1 = 8 sinwt, x_2 = 6 sin(wt+$\frac{\pi}{2}$), x_3 = 4 sin (wt+p)and x_4 = 3sin(wt+$\frac{3\pi}{2}$), If these four motions are superimposed simultaneously on a single particle and make it to oscillate, amplitude of oscillation of the particle is

1) 21 Units 2) 7 Units

3) 5 Units 4) 3 Units

3. The equation of motion of a particlestarted at t = 0, is given by y = 5sin(20t+$\frac{\pi}{3}$)cm.The least time affter which acceleration becomes zero is

1) $\frac{\pi}{90}$ sec 2) $\frac{\pi}{120}$ sec

3) $\frac{\pi}{60}$ sec 4) $\frac{\pi}{30}$ sec

4. The equation of S.H.M is $\frac{d^2x}{dt^2}$ = – 16x. The period of SHM is

1)8p/3 sec 2) 3sec 3) 1.57 sec 4) 4 sec

5. Two particles execute SHM of the same amplitude and frequency on parallel lines side by side. They cross each another when moving in opposite directions each time their displacement is half their amplitude. What is the difference between them ?

1) $\frac{\pi}{3}$ 2) $\frac{2\pi}{3}$ 3) $\frac{\pi}{6}$ 4) $\frac{\pi}{2}$

6. A body of mass 1kg is executing S.H.M. Its displacement 'y' cm and 't' in seconds is given by y = 6sin (100t+ p/4). Its maximum K.E is [2000E]

1) 6 J 2) 18 J 3) 24 J 4) 36 J

7. The maximum velocity of the particle executing S.H.M is 'V'. If the amplitude is doubled and the time period of oscillation decreased to $\frac{1}{3}$ of its original value, the maximum velocity becomes [99 E]

1) 18V 2) 12V 3) 6V 4) 3V

8. A particle makes SHM along a straight line and its velocity when passing through points 3cm and 4cm from the mean position is 16cm/sec and 12cm/sec respectively. Then its time period is

1) $\frac{\pi}{2}$ sec 2) π sec 3) $\frac{3\pi}{2}$ sec 4) 2π sec

9. A body executes SHM under the action of force 'F' with a time period 4/5 sec. If the force is changed to 'F_2' to execute SHM with time period (3/5)sec. If the both the forces F_1 and F_2 act simultaneously in the same direction on the body, its time period in seconds is (in see) . [2002 E]

1) $\frac{12}{25}$ 2) $\frac{12}{15}$

3) $\frac{25}{24}$ 4) $\frac{25}{12}$

10. Average energy in one time period of a simple harmonic oscillator whose amplitude is A, angular velocity 'w' and mass 'm' is

1) $\frac{1}{4}mw^2A^2$ 2) $\frac{1}{2}mw^2A^2$

3) mw^2A^2 4) Zero

11. A body executing SHM has a maximum velocity 1m/sec and maximum acceleration of 4m/sec^2. Its amplitude in meters is [2005M]

1) 1 2) 0.75 3) 0.5 4) 0.25

12. The S.H.M of a particle is represented by the equation $x = 4\cos(88t + \frac{\pi}{4})$.The frequency (in Hz) and initial displacement (in m) of the particle are [2006M]

1) 14, $2\sqrt{2}$ 2) 16, $2\sqrt{2}$

3) 14, $3\sqrt{2}$ 4) 16, $3\sqrt{2}$

13. The time period of a particle in SHM is 12 sec. At t = 0, it is at mean position. The ratio of the distance travelled in the second second and sixth second is

1) $(\sqrt{3}-1)$ 2) $(\sqrt{2}-1)$ 3) $\frac{1}{\sqrt{2}-1}$ 4) $\frac{1}{\sqrt{2}}$

14. A linear harmomic oscillator of force constant 2×10^6N/m and amplitude 0.01m has a total mechanical energy of 160 J. Its [IIT 89]

1) maximum PE is 100J

2) maximum KE is 100J

3) maximum PE is 120 J

4) maximum PE is zero

15. A particle is executing S H M with a frequency 'f' the frequency with which P.E oscillates is

1) f 2) f/2 3) 2f 4) 4f

16. A large horizontal surface moves up and down in SHM with an amlitude of 1cm. If a mass of 10kg (which is placed on this surface) is to remain continuously in contact with it. The frequency of S.H.M should not exceed ($g=10m/sec^2$)

1) 0.5 Hz 2) 1.5 Hz

3) 5.0 Hz 4) 10.0 Hz

17. When the P.E of the particle executing S.H.M is one fourth of its maximum value during the oscillation its displacement from the equilibrium in terms of its amplitude 'a' is

1) a/4 2) a/3 3) a/2 4) 2a/3

18. A particle executing SHM has total energy of 20 joules of the P.E at the mean position is 5J, the average K.E is

1) 10J 2) 5 J 3) 15 J 4) 7.5 J

19. A particle of mass 2 g is executing SHM of frequency 10 Hz. Find its KE, PE and TE when its displacement is 1 cm. Its amplitude is 4 cm.

1) 2.923×10^{-3}J, 2.949×10^{-4}J, 6.318×10^{-3}J

2) 5.923×10^{-3}J, 3.949×10^{-4}J, 6.318×10^{-3}J

3) 4.923×10^{-3}J, 1.949×10^{-4}J, 5.218×10^{-3}J

4) 1.923×10^{-3}J, 3.943×10^{-4}J, 4.118×10^{-3}J

20. **A body executes S H M . The potential energy (PE), kinetic energy (KE) and total energy (TE) are measured as function of displacement (x) . Which of the following statement is true (AIEEE 2003)**

1) K.E is maximum when x = 0

2) TE is zero at x = 0

3) KE maximum when 'x' is maximum

4) PE is maximum when x = 0

21. **In a forced oscillation of a particle the amplitude is maximum for a frequency w_1 of the force, while the energy is maximum for a frequency w_2 of the force, then (AIEEE -04)**

1) $w_1 = w_2$ 2) $w_1 < w_2$

3) $w_1 < w_2$ when the damping is small and $w_1 > w_2$ when the damping large

4) $w_1 > w_2$

22. **The velocites of a body executing SHM at displacement 'a' and 'b' are 'b' and 'a' respectively. The amplitude of S.H.M will be**

1) a + b 2) a-b

3) $\sqrt{a^2+b^2}$ 4) $\sqrt{a^2-b^2}$

23. **A cylindrical piston of mass 'm' and cross sectional area 'A' slides smoothly inside a long cylinder closed at one end enclosing a certain mass of the gas. The cylinder is kept with its axis horizontal. If the piston is disturbed from its equilibrium position, it oscillates simple harmonically. The period of oscillation will be (r = density of the enclosed gas)**

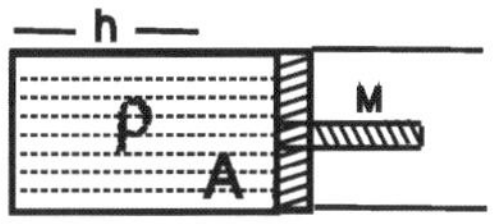

1) $2\pi\sqrt{\frac{Mh}{\rho a}}$

2) $2\pi\sqrt{\frac{MA}{\rho h}}$

3) $2\pi\sqrt{\frac{M}{\rho Ah}}$ 4) $2\pi\sqrt{Mph\rho}$

SIMPLE PENDULUM

24. **A simple pendulum from earth is shifted to moon (gm = ge/6). Forone minute on earth, the time on the moon as measured by the pendulum is (For same number of oscillations) [96 M]**

1) 24.5 sec 2) 2.45 sec

3) 245 sec 4) 147 sec

25. **A pendulum which gives correct time beats seconds on ground at certain place. If this is taken to the top of the tower of height 320m, the time lost by the pendulum wall clock in one day (R = 6400km) is approximately**

1) 8.6 sec 2) 2.1 sec 3) 4.3 sec 4) 5.2 sec

26. **The percentage change in the time period of a simple pendulum when its amplitude is reduced by 50 % is [99M]**

1) 75% 2) 25% 3) 0% 4) 50%

27. **A seconds pendulum is fixed on a lift which has upward acceleration 'g/4'. Its time period is**

1) $\sqrt{5}$ sec 2) $4/\sqrt{5}$ sec

3) $\frac{2}{\sqrt{5}}$ sec 4) $2\sqrt{5}$ sec

28. **If the length of the seconds pendulum is increased by 1% how many seconds will it loose in a day**

1) 864 sec 2) 432 sec 3) 216 sec 4) 108 sec

29. **Calculate the change in the length of a simple pendulum of length 1m, when its perid of oscillation changes from 2s to 1.5s.**

1) increases by 7/8m 2) decreases by 7/16m

3) decreases by 7/8m 4) increases by 7/16m

30. **A simple pendulum of length 'l' and mass of the bob 'm' is vibrating with a small amplitude 'a'. Then the maximum tension in the string is**

1) mg 2) $mg\left(1+\left(\frac{a}{\ell}\right)^2\right)$

3) $mg\left(1+\left(\frac{a}{\ell}\right)\right)^2$ 4) $mg\left(1+\left(\frac{a}{2\ell}\right)\right)^2$

31. **A simple pendulum executing S.H.M in a straight line has zero velocity at 'a' and 'b' whose distances from 'o' in the same line OAB are 'a' and 'b' . If the velocity half -way between them is 'v' then its time period is**

1) $\frac{\pi(b+a)}{v}$ 2) $\pi\left(\frac{b-a}{v}\right)$

3) $\left(\frac{b+a}{2v}\right)$ 4) $\left(\frac{b-a}{2v}\right)$

32. A simple pendulum with a bob of mass 'm' oscillates from A to C and back 'A' such that 'PB' is 'H'. If the acceleration due to gravity is 'g' then the velocity of the bob as it passes through 'B' is

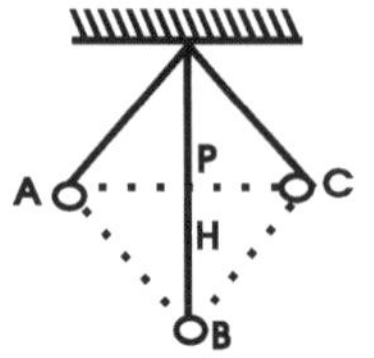

1) mg H 2) $\sqrt{2gH}$ 3) 2gH 4) zero

33. A simple pendulum with a metal bob has a time period 'T'. Now the bob is immersed in a liquid which is non viscous. Thus the time period is '3T' . Then the ratio of densities of metal bob and that of liquid is

1) 8 : 9 2) 9 : 8 3) 1 : 9 4) 9 : 1

34. Two harmomic oscillators starts oscillating simultaneously in the same direction. Their time periods are T and 5T/8 respectively. The phase difference between them, when second oscillator has completed one oscillation , will be

1) π 2) $\frac{3\pi}{2}$ 3) $\frac{3\pi}{4}$ 4) $\frac{5\pi}{6}$

35. Two simple pendulum have time periods T and $\frac{5T}{8}$.They start vibrating at the same instant from the mean position in the same phase. The minimum time after which they will be again in phase is

1) $\frac{3T}{2}$ 2) $\frac{4T}{3}$ 3) 4T 4) 5T

36. A simple pendulum of length 100cm and 144cm are swinging together. Initially they start from mean position in the same phase. They will swing together again after

1) The longer pendulum makes 5 oscillation

2) The shorter pendulum makes 5 oscillation

3) The longer pendulum makes 6 oscillations

4) None

37. A simple pendulum is oscillating with out damping when the displacement of the bob is less than maximum its acceleration vector'a' is correctly shown is [IIT SCREENING 2002]

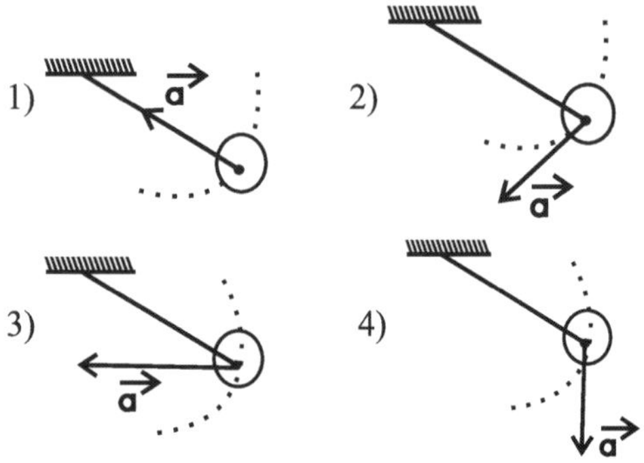

SPRINGS

38. The frequency of oscillation of mass 'm' suspended by a spring is n_1. If the spring is made half, the same mass oscillates with a frequency n_2. Then the value of n_2 is

1) $n_1/2$ 2) $2n_1$ 3) $\sqrt{2}\,n_1$ 4) $\frac{n_1}{\sqrt{2}}$

39. When a body of mass 1.0 kg is suspended from certain light spring hanging vertically, its length increases by 5 cm. By suspending 2.0kg block to the spring and if the block is pulled through 10cm and released, the maximum velocity in it in m/sec is [g =10m/sec^2] [2003 E]

1) 0.5 2) 1 3) 2 4) 4

40. The mass 'm' is attached to two springs according to the figure. If the frictional force is absent the period of horizontal oscillations of the body will be

1) $2\pi\sqrt{\frac{3k}{m}}$ 2) $2\pi\sqrt{\frac{k}{m}}$

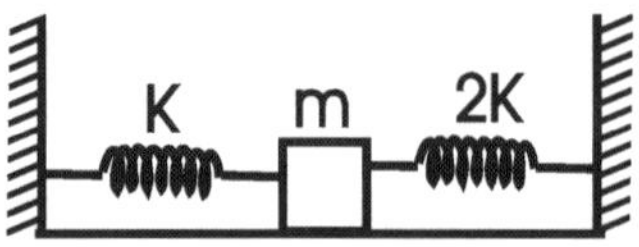

3) $2\pi\sqrt{\frac{m}{2k}}$

4) $2\pi\sqrt{\frac{m}{3k}}$

41. The total force constant of the springs shown in the figure will be

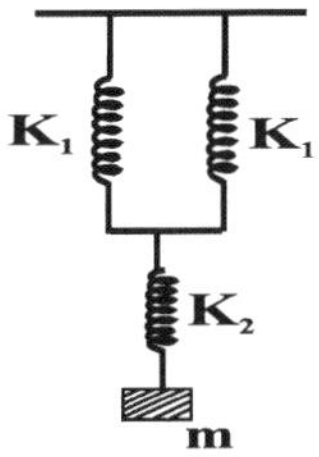

1) $\left(\frac{k_1}{2}+k_2\right)$ 2) $\left(\frac{1}{2k_1}+\frac{1}{k_2}\right)^{-1}$

3) $\left(\frac{1}{2k_1}+\frac{1}{k_2}\right)$ 4) $\left(2k_1+\frac{1}{k_2}\right)$

42. The free end of a spring hanging from the rigid support, a block of mass 'm' is hung and slowly allowed to come to its equilibrium position. Then the stretching in the spring is 'd'. If the same block is attached to same spring and allowed to fall suddenly the amount of stretching is [force constant K] [Eamcet 2006 E]

1) mg/k 2) 2d 3) $\frac{mg}{3k}$ 4) 4d

43. A body of mass 'm' is suspended to an ideal spring of spring constant 'k' . The expected change in position of a body due to an additional force 'F' acting vertically down wards is

1) $\frac{3F}{2k}$ 2) $\frac{F}{k}$ 3) $\frac{5F}{2k}$ 4) $\frac{4F}{k}$ **[2005 E]**

44. A pan with certain weight is attached to a spring. The period of vertical oscillations is found to be 0.2 sec. If some weight is added in the pan, the time period becomes 0.4sec. The additional distance through which the spring stretches due to additional load is

1) $\frac{5}{300}$m 2) $\frac{9}{75}$m 3) $\frac{3}{100}$m 4) $\frac{5}{100}$m

45. When a body is suspended from two light springs seperately, the periods of vertical oscillations are T_1 and T_2. If the same body is suspended from the two springs joined in parallel, the time period will be

1) $\sqrt{T_1T_2}$ 2) $\frac{T_1T_2}{\sqrt{T_1^2+T_2^2}}$

3) $\frac{\sqrt{T_1T_2}}{2}$ 4) $\frac{2T_1T_2}{\sqrt{T_1^2+T_2^2}}$

46. A block of mass 'm' collides perfectly inelastically with another idetical block to a spring of forceconstant 'K'. The amplitude of resulting S.H.M will be

1) $\sqrt{\frac{m}{K}}$ 2) $V\sqrt{\frac{m}{2K}}$

3) $V\sqrt{\frac{K}{m}}$

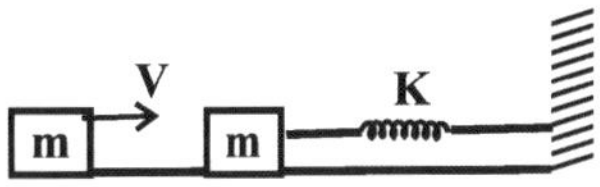

4) $V\sqrt{\frac{2K}{m}}$

47. A load of mass 'M is attached to the bottom of a spring of mass M/3 and spring consant 'K'. If the system is set into oscillation, the time period of oscillation is

1) $\frac{2\pi}{3}\sqrt{\frac{10M}{K}}$ 2) $4\pi\sqrt{\frac{M}{3K}}$

3) $4\pi\sqrt{\frac{M}{K}}$ 4) $2\pi\sqrt{\frac{M}{3K}}$

48. The graph between period oscillation (T) and mass attached (m) to a spring will be

1) 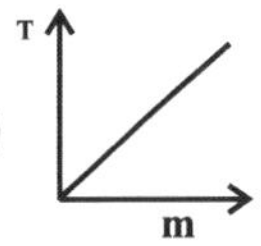

2) 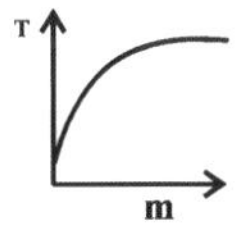

3) 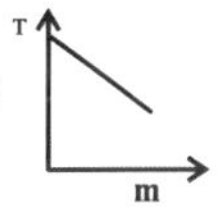

4) 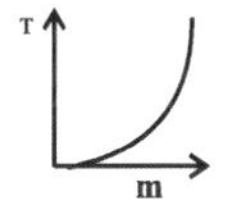

49. A point mass oscillation along the x–axis according to the law $x = x_0 \cos(wt - p/4)$. If the acceleration of the particle is written as $a = A\cos(wt+\delta)$ [AIEEE 2007]

1) $A=x_0,\ \delta=-p/4$ 2) $A=x_0w^2,\ \delta=-p/4$

3) $A=x_0p^2,\ \delta=p/4$ 4) $A=x_0w^2,\ \delta=3p/4$

50. Two identical springs are attached to a mass and the system is made to oscillate. T_1 is the time period when the springs are joined in parallel and T_2 is the time period when they are joined in series then

1) $T_1=2T_2$ 2) $T_1=\sqrt{2}\,T_2$

3) $T_2=2T_1$ 4) $T_2=\sqrt{2}\,T_1$

ANSWERS

1) 4	2) 3	3) 4	4) 3	5) 2
6) 2	7) 3	8) 1	9) 1	10) 2
11) 4	12) 1	13) 1	14) 2	15) 3
16) 3	17) 3	18) 4	19) 2	20) 1
21) 1	22) 3	23) 1	24) 4	25) 3
26) 3	27) 2	28) 2	29) 2	30) 2
31) 2	32) 2	33) 2	34) 3	35) 4
36) 1	37) 3	38) 3	39) 2	40) 4
41) 2	42) 2	43) 2	44) 3	45) 2
46) 2	47) 1	48) 2	49) 4	50) 3

COMPREHENSION QUESTIONS

Comprehension - I (1-3) :

A Simple harmonic oscillation has an amplitude 'A' and time period 'T'

1. The time required to travel from x = A to x = A/2 is

1) T/6 2) T/4 3) T/3 4) T/12

2. The time required to travel form x = -A/2 to x = +A/2 is

1) T/2 2) T/3 3) T/6 4) T/8

3. The time required to travel from x = -A to x = +A/2

1) T/2 2) T/3 3) T/6 4) T/8

Comprehension - II (4-5) :

A spring mass system whose spring constant is 'K' and carries mass 'm' is hanging from the ceilling of an elevator in equilibrium. The elevator suddenly starts accelerating upwards with an acceleration 'a', then

4. The extension in the spring in equilibrium is

1) $\frac{mg}{K}$ 2) $\frac{ma}{K}$ 3) $\frac{m(g+a)}{K}$ 4) $\frac{m(g-a)}{K}$

5. Amplitude of the oscillation of the system is

1) $\frac{mg}{K}$ 2) $\frac{ma}{K}$ 3) $\frac{m(g+a)}{K}$ 4) $\frac{m(g-a)}{K}$

Comprehension - III (6-9) :

Two masses m_1 and m_2 are suspended from the spring of constant 'K'. When the masses in equilibrium

6. Mass m_1 is gently removed, then the amplitude of oscillation of the system is

1) $\frac{m_1g}{K}$ 2) $\frac{m_2g}{K}$

3) $\frac{(m_1+m_2)g}{K}$ 4) $\frac{(m_1-m_2)g}{K}$

7. Mass m_2 is gently removed, then the amplitude of oscillation of the system is

1) $\frac{m_1g}{K}$ 2) $\frac{m_2g}{K}$

3) $\frac{(m_1+m_2)g}{K}$ 4) $\frac{(m_1-m_2)g}{K}$

8. Mass m_1 is gently removed, the angular frequency of the of the system is

1) $\sqrt{\frac{K}{m_1}}$ 2) $\sqrt{\frac{K}{m_2}}$

3) $\sqrt{\frac{k}{m_1+m_2}}$ 4) $\sqrt{\frac{k}{m_1-m_2}}$

9. Mass m_2 is gently removed, the angular frequency of the system is

1) $\sqrt{\frac{K}{m_1}}$ 2) $\sqrt{\frac{K}{m_2}}$

3) $\sqrt{\frac{K}{m_1+m_2}}$ 4) $\sqrt{\frac{K}{m_1-m_2}}$

Comprehension - IV (10-12):

In a simple hormonic motion, force acting on a particle is given as F = 4x, total mechanical energy of the particle is 10J and amplitude of oscillation is 2m. At time t = 0, acceleration of the particle is -16m/sec^2, mass of the particle is 0.5kg

10. Potential energy of the particle in mean position is

1) 10J 2) 8J 3) 6J 4) 2J

11. Displacement time equation of the particle is

1) $x = 2Sin(2t)$ 2) $x = 2\ Sin(4t)$

3) $x = 2\ Cos(t)$ 4) $x = 2\ Cos\ (2\sqrt{2}\ t)$

12. At x = +1m, potential energy and kinetic energy of the particle are

1) 2J and 5J 2) 8J and 2J

3) 6J and 4J 4) 4J and 6J

Comprehension - V (13-16) :

A linear hormonic oscillator of force constant is 2×10^6 N/m and amplitude is 0.01 m has a total mechanical energy of 160J. During the oscillation

13. Minimum kinetic energy of oscillator is

1) Zero 2) 60J 3) 100J 4) 160J

14. Maximum kinetic energy of oscillator is

1) Zero 2) 60J 3) 100J 4) 160J

15. Minimum potential energy of the oscillator is

1) Zero 2) 60J 3) 100J 4) 160J

16. Maximum potential energy of the oscillator is

1) Zero 2) 60J 3) 100J 4) 160J

Comprehension - VI (17-18) :

A particle execute simple hormonic motion. If it has velocities V_1 and V_2 at different displacements y_1 and y_2. Then

17. Angular frequency of oscillation of the particle is

1) $\dfrac{V_1^2 - V_2^2}{y_2^2 - y_1^2}$ 2) $\dfrac{y_2^2 - y_1^2}{V_1^2 - V_2^2}$

3) $\sqrt{\dfrac{V_1^2 - V_2^2}{y_2^2 - y_1^2}}$ 4) $\sqrt{\dfrac{y_2^2 - y_1^2}{V_1^2 - V_2^2}}$

18. Amplitude of oscillation of the particle is

1) $\dfrac{V_1^2 y_1^2 - V_2^2 y_2^2}{V_1^2 - V_2^2}$ 2) $\sqrt{\dfrac{V_1^2 y_1^2 - V_2^2 y_2^2}{V_1^2 - V_2^2}}$

3) $\dfrac{V_1^2 y_2^2 - V_2^2 y_1^2}{V_1^2 - V_2^2}$ 4) $\sqrt{\dfrac{V_1^2 y_2^2 - V_2^2 y_1^2}{V_1^2 - V_2^2}}$

ANSWERS

1) 1	2) 3	3) 2	4) 3	5) 3
6) 1	7) 2	8) 2	9) 1	10) 4
11) 4	12) 4	13) 1	14) 3	15) 2
16) 4	17) 3	18) 4		

Chapter - 10

SYSTEM OF PARTICLES

CENTRE OF MASS

SYNOPSIS

1. Centre of mass of a body or a system of particles is the point at which the whole mass of the body or system is supposed to be concentrated and moves as if the whole external force is applied at that point.
2. The motion of centre of mass of a body represents the motion of the whole body.

   Two particles of masses m_1 and m_2 are separated by a distance 'd'. If x_1 and x_2 are the distances of their centre of mass from m_1 and m_2, then

   $$m_1x_1 = m_2x_2$$

   $$x_1 = \frac{m_2\, d}{m_1 + m_2} \quad \text{and} \quad x_2 = \frac{m_1\, d}{m_1 + m_2}$$

   b) The centre of mass of heavier and lighter mass system lies nearer to the heavier mass .

   c) Number of particles lying along the x-axis.

   Particles of masses m_1, m_2, m_3 ------- are at distances x_1 x_2 x_3 ------ from the origin, the distance of centre of mass from the origin.

   $$x_{cm} = \frac{m_1x_1 + m_2x_2 + m_3x_3 + ...+ m_nx_n}{m_1 + m_2 + m_3 + ...+ m_n}$$

   d) Number of particles lying in a plane.

   Particles of masses m_1, m_2, m_3, --------- are lying in xy plane at positions (x_1,y_1), (x_2,y_2), (x_3,y_3) ------- , then the position co-ordinates of their centre of mass.

   $$x_{cm} = \frac{m_1x_1 + m_2x_2 + ...+ m_nx_n}{m_1 + m_2 + ...+ m_n}$$

   $$y_{cm} = \frac{m_1y_1 + m_2y_2 + ...+ m_ny_n}{m_1 + m_2 + ...+ m_n}$$

   e)Particles distributed in space.

   If $(x_1y_1z_1)$, $(x_2y_2z_2)$...... are the position co-ordinates of particles of masses m_1, m_2 ...... the position co-ordinates of their centre of mass are

   $$x_{cm} = \frac{m_1x_1 + m_2x_2 + ...+ m_nx_n}{m_1 + m_2 + ...+ m_n}$$

   $$y_{cm} = \frac{m_1y_1 + m_2y_2 + ...+ m_ny_n}{m_1 + m_2 + ...+ m_n}$$

   $$z_{cm} = \frac{m_1z_1 + m_2z_2 + ...+ m_nz_n}{m_1 + m_2 + ...+ m_n}$$

   f) In Vector notation.

   If $\vec{r}_1, \vec{r}_2, \vec{r}_3$ . . . are the position vectors of particles of masses m_1, m_2, m_3 ...... then the position vector of their centre of mass is

   $$\vec{r}_{cm} = \frac{m_1\vec{r}_1 + m_2\vec{r}_2 + ...+ m_n\vec{r}_n}{m_1 + m_2 + ...+ m_n}$$

   g) Relative to centre of mass :

   $$m_1\vec{r}_1 + m_2\vec{r}_2 + ...+ m_r\vec{r}_n = 0,$$

   where m_1, m_2 .... m_n have position vectors $\vec{r}_1, \vec{r}_2 .... \vec{r}_n$ relative to centre of mass i.e. Algebraic sum of the moments of masses of a system about its centre of mass is always zero.

   h) In case of continuous distribution of mass to locate the centre of mass we use.

   $$\therefore X_{cm} = \frac{1}{M}\int x\, dm, Y_{cm} = \frac{1}{M}\int y\, dm,\ Z_{CM} = \frac{1}{M}\int z\, dm$$

4. Locating centre of mass can be done in following ways.

   a) Method of symmetry

   Ex: I) For a circular hoop the centre of mass lies at its centre.

   II) For a homogeneous sphere, centre of mass lies at its geometric centre, because of its symmetry the sum of moments of mass about its centre is zero i.e., $\sum r dm = 0$

   III) For an equilateral traingle the centre of mass is at its centroid.

   IV) If two circular discs of radii r_1 and r_2 of same material are kept in contact then the distance of centre of mass of system from centre of a disc of radius r_1 is given by

   $$x_{cm} = \frac{r_2^2(r_1 + r_2)}{r_1^2 + r_2^2}$$

   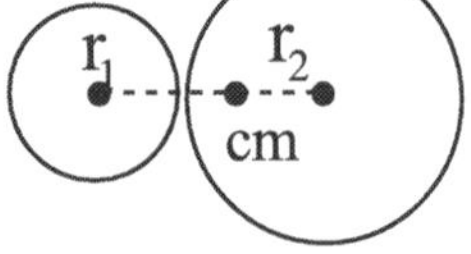

   V) If two spheres of radii r_1 and r_2 of same material are kept in contact, then the distance of centre of mass of the system from centre of a sphere of radius 'r_1' is given by

   $$x_{cm} = \frac{r_2^3(r_1 + r_2)}{r_1^3 + r_2^3}$$

   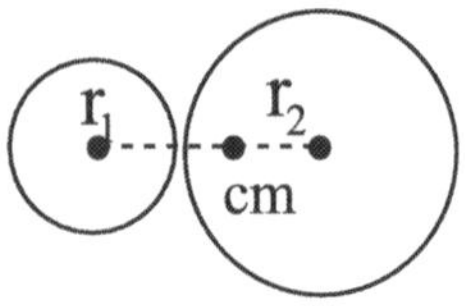

VI) If two thin uniform rods of lengths L_1 and L_2 of same material are joined to form 'T' shape as shown in the figure, then the distance of centre of mass of the system from centre of mass of first rod of length L_1 is given by

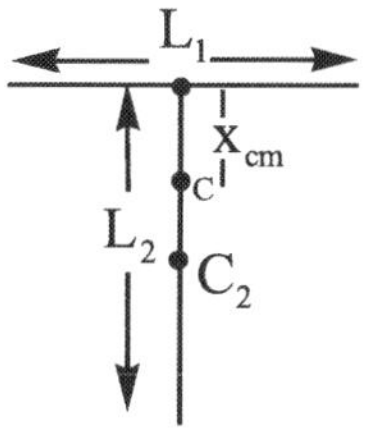

$$x_{cm} = \frac{L_2^2}{2(L_1 + L_2)}$$

VII) If two cylinders of lengths L_1 and L_2, radii r_1 and r_2 made up of same material are kept in contact as shown in the figure, then distance of centre of mass of the system from the centre of first cylinder is given by

$$x_{cm} = \frac{r_2^2 L_2}{\left(r_1^2 L_1 + r_2^2 L_2\right)}\left(\frac{L_1 + L_2}{2}\right)$$

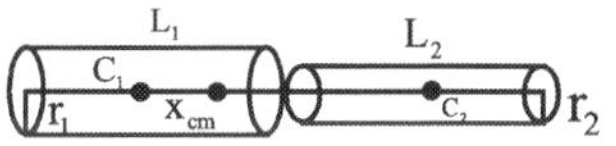

VIII) If a wire of length 'ℓ' is bent in the form a circular ring then the shift in center of mass is $x = \dfrac{\ell}{2\pi}$.

IX) If a uniform rod of length L is bent at the mid point so that the two halves are inclined by an angle θ with each other then the shift in center of mass, $x = \dfrac{L}{4}\cos\left(\dfrac{\theta}{2}\right)$.

X) Distance of centre of mass of a uniform cone of height 'h' and base radius R, from the vertex on the line of symmetry is $\dfrac{3h}{4}$.

b) Method of decomposition

Some of the surfaces can be divided into singular parts whose centre of mass can be found by the method of symmetry.

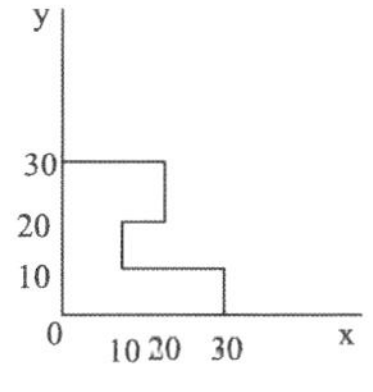

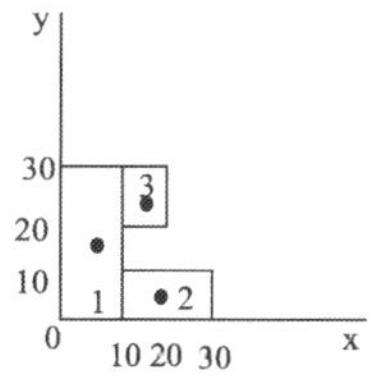

C. Pappus theorems

1st law

The volume 'V' generated as a result of the revolution of a closed plane area about an axis (such that every point moves perpendicular to the plane) equals to the area S of the palne times the circumference $(2\pi\, x_c)$ of the circle described by the centre of mass of the plane.i.e., $V = S\ 2\pi x_c$

Ex: Distance of centre of mass of a semi circular plate from its center is $\boxed{x_c = \dfrac{4r}{3\pi}}$

2nd law :

The surface area S generated by revolution of a plane curved line about an axis such that every point moves perpendicular to the line of the curve is equal to the product of length of the line and the distance through which the centre of mass moves.

Ex : The distance of centre of mass of a wire bent into the form of a half circle of radius R from its center is $x_c = \dfrac{2R}{\pi}$

5. Position of centre of mass of a body
   a) depends on shape of the body.
   b) depends on distribution of mass for a given shape of the body.
   c) coincides with centre of gravity of the body if the body is in uniform gravitational field.
6. There may or may not be any mass at centre of mass. Ex uniform ring

   Centre of mass may be within or outside the material of body.

   a) For a circular disc the centre of mass is at its geometric centre where there is mass.
   b) For a triangular plane lamina, the centre of mass is the point of intersection of the medians of the triangle.
7. If $\Delta x_1, \Delta x_2$ are shift in position of 1st and 2nd particle then shift in centre of mass along x - axis is given by

$$\Delta x_{cm} = \frac{\pm m_1 \Delta x_1 \pm m_2 \Delta x_2}{m_1 + m_2}$$

+ when shifted along + ve x – direction and
– when shifted along – ve x - direction
simillary along y - axis

$$\Delta y_{cm} = \frac{\pm m_1 \Delta y_1 \pm m_2 \Delta y_2}{m_1 + m_2}$$

If $\vec{v}_1, \vec{v}_2, \vec{v}_3$ .... $\vec{v}_n$ are the velocities of particles of masses ,m_1, m_2, m_3, ..., m_n then the velocity of their centre of mass.

$$\vec{v}_{cm} = \frac{m_1\vec{v}_1 + m_2\vec{v}_2 + m_3\vec{v}_3 + ... + m_n\vec{v}_n}{m_1 + m_2 + m_3 + ... + m_n}$$

9. $M\vec{V}_{cm} = \vec{P}_1 + \vec{P}_2 + ... + \vec{P}_n$ i.e total momentum of the system is the product of mass of the whole system and the velocity of the centre of mass.

10. If two particles of masses m_1 and m_2 are moving with velocities $\vec{v}_1$ and $\vec{v}_2$ at right angles to each other, then the magnitude of velocity of their centre of mass is given by

$$V_{cm} = \sqrt{\frac{m_1^2v_1^2 + m_2^2v_2^2}{(m_1 + m_2)^2}}$$

If $\vec{a}_1, \vec{a}_2, \vec{a}_3$, ... $\vec{a}_n$ are the acccelerations of particles of masses m_1, m_2, m_3 ....m_n then the accelaration of their centre of mass is

$$\vec{a}_{cm} = \frac{m_1\vec{a}_1 + m_2\vec{a}_2 + ... + m_n\vec{a}_n}{m_1 + m_2 + ... + m_n}$$

$m_1 + m_2 + m_3 + ...+m_n = M$, total mass of the system, then $m_1\vec{a}_1 + m_2\vec{a}_2 + ... m_n\vec{a}_n = \vec{F}_1 + \vec{F}_2 + ... + \vec{F}_n = \vec{F}_{ext}$

$M\vec{a}_{cm} = \vec{F}_{ext}$

12. Centre of mass can be accelerated only by a net external force.

13. Internal forces cannot accelerate the centre of mass or change the state of centre of mass.

14. In the absence of external forces,

a) the centre of mass of a system is at rest if the centre of mass is initially at rest.

b) if the centre of mass of a system is moving with constant velocity, it continues to move with the same velocity.

15. Shift in centre of mass when a small portion of mass is removed from a uniform body is

$$\boxed{x_{shift} = \frac{m_{removed}d}{(M_{total} - m_{removed})}}$$

16. Out of a uniform circular disc of radius R, if a circular sheet of radius 'r' is removed then the centre of mass of remaining part shifts by a distance $\frac{r^2d}{R^2 - r^2}$ where d is the distance of the centre of the smaller part from the centre of the original disc.

(i) If circular portion is removed from the edge then, shift is maximum and $\boxed{x_{shift} = \frac{r^2}{R+r}}$

(ii) If a circular portion of diameter R is removed from a uniform circular plate of radius R, from one edge then shift in the center of mass $x_{shift} = \frac{R}{6}$.

(iii) If the centre of removed plate and the original plate coincide then shift is zero.

17. Out of a uniform solid sphere of radius R, if a sphere of radius r is removed then the centre of mass of the remaining part shifts by $\frac{r^3d}{(R^3 - r^3)}$ where d is the distance of the smaller sphere from the centre of the original sphere.

(i) If spherical portion is removed from the edge then, shift is maximum and $\boxed{x_{shift} = \frac{r^3}{R^2 + r^2 + Rr}}$

(ii) If a spherical portion of diameter R is removed from a uniform sphere of radius R, from one edge then shift in the center of mass $x_{shift} = \frac{R}{14}$.

18. When a triangular portion is removed from one edge of a square plate of side 'a' as shown in the figure then shift in centre of mass is given by $x_{shift} = \frac{a}{9}$

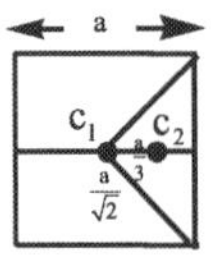

19. When a triangular plate is removed from one edge of a rectangular plate of length a and breadth b as shown in the figure then shift in centre of mass is given by $x_{shift} = \frac{a}{9}$

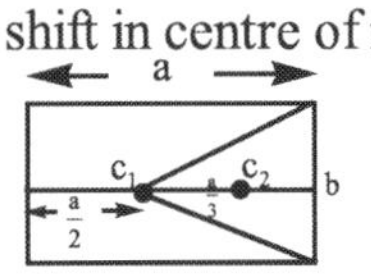

20. When a shell in flight explodes,

a) The acceleration of centre of mass before and immediately after explosion is $a_{cm} = g$ downward.

b) The centre of mass of all the fragments will continue to move along the same trajectory as long as all the fragments are still in space.

c) If all the fragments reach the ground simultaneously, the centre of mass will complete the original trajectory.

d) If some of the fragments reach the ground earlier than the other fragments, the acceleration of centre of mass changes.

21. When a person initially at rest in a boat at rest in still water, walks on a boat. Centre of mass of person + boat system is not displaced.

a) If the man walks a distance L on the boat, the boat is displaced in the opposite direction relative to shore or water by a distance

$$x = \frac{mL}{M+m}$$

(m = mass of man, M = mass of boat)

b) distance walked by the man relative to shore or water is (L-x)

22. If two masses starting from rest move under mutual force of attraction towards each other, then they meet at their centre of mass.

a) in the above case, $\vec{v}_{cm} = 0$ and $\vec{a}_{cm} = 0$

b) If the two particles are m_1 and m_2 and their velocities are $\vec{v_1}$ and $\vec{v_2}$, then $m_1\vec{v_1} = -m_2\vec{v_2}$

c) If the two particles have accelerations $\vec{a_1}$ and $\vec{a_2}$, then $a_{cm} = 0$

So, $m_1 \vec{a_1} = -m_2 \vec{a_2}$

d) If s_1 and s_2 are the distances travelled before they meet $m_1 s_1 = m_2 s_2$

EXERCISE - I

1. **When no external force is acting on a system of particles, the centre of mass of the system**
   1) remains at rest only
   2) moves with constant velocity only
   3) moves with constant velocity or will be at rest.
   4) moves with variable velocity

2. **Two particles, move towards each other from rest under a mutual force of attraction. At the instant, when the speed of A is V, the speed of B is V/2. The speed of centre of mass of the system is**

   1) zero 2) V/2 3) 3/2V 4) 3/4V

3. **Three identical spheres each of radius R are placed touching each other on a horizontal table. The centre of mass of the system is located**
   1) at one of the centres of the spheres
   2) at the mid point joining the centres of any two spheres.
   3) at the point of intersection of the medians of the triangle formed by the centres of the three spheres.
   4) at the mid point of a median of the triangle formed by the centres of the three spheres.

4. **Two balls are thrown at the same time in vacuum. While they are in vacuum, the accelaration of their centre of mass**
   1) depends on masses of the balls
   2) depends on direction of motion of the balls
   3) depends on speeds of the balls
   4) is equal to accelaration due to gravity

5. **Centre of mass of a body (ATB)**
   1) always lies inside the body
   2) always lies outside the body
   3) always lies on the surface of the body
   4) may lie inside or outside the body

6. **Three identical masses are kept at the corners of an equilateral triangle ABC. A moves towards B with a velocity V, B moves towards C with velocity V, and C moves towards A with same velocity V. Then the velocity of centre of mass of the system of particles is**

   1) V 2) zero 3) 3V 4) $\frac{V}{3}$

7. **A shell moving in a parabolic path explodes. The centre of mass of the fragments move**
   1) vertically down wards 2) vertically upwards
   3) horizontally 4) in the same parabolic path

8. **A bomb at rest explodes. The centre of mass of the system (ATB)**
   1) describes a parabola 2) vertically upwards
   3) horizontally 4) is at rest

9. **When an external force is applied at the centre of mass of a system of particles, then it undergoes (ATB)**
   1) Only translatory motion
   2) Only rotatory motion
   3) both translatory and rotatory motion
   4) an oscillatory motion

10. **A bomb moving in a parabolic path explodes into two fragments of equal masses. The accelaration of the centre of mass of the fragments when both are in air is equal to**
    1) g/2 2) 2g 3) g 4) zero

11. **A cylinder is completely filled with water. If 1/4 of the volume of water leaks out, its centre of mass,**
    1) moves up 2) moves down
    3) does not change
    4) moves towards vertical surface

12. Two bodies of masses m_1 and m_2 are at distances x_1 and x_2 from their centre of mass. Then, the correct statement of the following is (ATB)

1) $\frac{m_1}{m_2} = \frac{x_1}{x_2}$ 2) $\frac{m_1}{m_2} = \sqrt{\frac{x_1}{x_2}}$

3) $\frac{m_1}{m_2} = \frac{x_2}{x_1}$ 4) $\frac{m_1}{m_2} = \sqrt{\frac{x_2}{x_1}}$

13. A uniform metre stick is placed vertically on a horizontal frictionless surface and released. As the stick is in motion, the centre of mass moves

1) vertically up 2) vertically down
3) in a parabolic path 4) horizontally

14. A shell is thrown vertically up. The shell at the highest point explodes into two equal fragments. The centre of mass of the two fragments

1) goes further up and then falls
2) falls down with an initial speed
3) falls down with zero inital velocity
4) comes to rest

15. Planets moving around sun and each planet having satellites moving around them is an example of

1) rigid system 2) Non– rigid system
3) Both rigid and non rigid systems
4) Neither rigid nor non – risid systems.

16. Choose the wrong statement :

1) In the case of small bodies [Uniform gravitional field] centre of mass and center of gravity concide with each other.
2) In non – uniform gravitational field centre of mass & centre of gravity will not coincide with each other.
3) Centre of mass is independent of aceeleration due to gravity, but center of gravity depends on acelleration due to gravity.
4) Center of mass describes the stability of the body where as centre of gravity describes the nature of motion of the body.

17. A thin uniform wire is bent in the form of a semi circle of radius R. Distance of its centre of mass from the geometric centre is

1) $\frac{2R}{\pi}$ 2) $\frac{R}{\pi}$ 3) $\frac{R}{2}$ 4) $\frac{\pi R}{2}$

18. Choose the correct statement (ATB)

1) Centre of mass of two particles will be nearer to lighter particle.
2) Centre of mass of the rigidbody depends on reference frame used.
3) Centre of mass of the system of particles depends on the masses of the particles.
4) Centre mass must lie with in the body.

19. Distance of centre of mass of a thin uniform semi circular disc of radius R from its centre is

1) $\frac{R}{\pi}$ 2) $\frac{2R}{\pi}$ 3) $\frac{4R}{3\pi}$ 4) $\frac{3R}{4\pi}$

20. Explosion is due to

1) Internal forces 2) External forces
3) Both 1 and 2 4) Neither 1 and 2

21. The path of the centre of mass of earth - moon system around the sun is

1) Circular path 2) Helical path
3) Parabolic path 4) Elliptical path

22. Choose the wrong statement (ATB)

1) In the process of explosion some changes may occur in momentum of individual fragments due to internal forces but the motion of the centre of mass is unaltered.
2) Motion of centre of mass depends upon the external force.
3) The location of centre of mass depends on the reference frame used locate it
4) The position of centre of mass depends upon the shape of the body and the distribution of mass

23. A bomb travelling in a parabolic path under the effect of gravity explodes in mid air. The centre of mass of fragments will (1993 E)

1) Move vertically upwards and then vertically downwards
2) Move vertically upwards
3) Move in an irregular path
4) Move in the parabolic path the unexploded bomb would have travelled.

QUESTIONS ON MORE THAN ONE OPTION

24. a) Algebraic sum of moments of masses about centre of mass is zero
b) For small bodies centre of mass coincides with centre of gravity
c) Position of centre of mass depends on co-ordinate system
d) Position of centre of mass is independent of mass distribution

1) a and b are correct 2) b and c are correct
3) a, b and c are correct 4) a, b, c and da are correct

25. A trolley filled with sand moves on a smooth horizontal surface with a velocity V_0. A small hole is made at the base of it from which sand is leaking out vertically down at constant rate. As the sand leaks out

a) the velocity of the trolley remains constant

b) the velocity of the trolley increases

c) the velocity of trolley decreases

d) the momentum of trolley + leaked out sand is conserved

1) a & b are correct 2) c & d are correct

3) a & c are correct 4) b & d are correct

26. P is the centre of mass of a system of four point masses A, B, C and D, which are coplanar but not collinear

a) P may or may not coincide with one of the point masses

b) P must lie within or on the edge of at least one of the triangles formed by taking A, B, C and D three at a time

c) P must lie on a line joining two of the points A, B, C, D

d) P lies out side the quadrangle ABCD

1) a & b are correct 2) c & d are correct

3) a & c are correct 4) b & d are correct

27. Identify the correct statement. In the absence of external force,

a) the kinetic energy associated with the motion of centre of mass is conserved

b) Centre of mass is not accelerated

c) The centre of mass always move in curved line

d) Centre of mass must be at rest

1) a & b are correct 2) all are true

3) all are false 4) a & c are correct

28. Mass of a ring is non-uniformly distributed around its geometric centre. If R is radius of the ring, then

a) Centre of mass does not coincide with geometric centre

b) Position of centre of mass from the geometric centre will be x $(0 < x < R)$

c) Centre of mass will be nearer to the greater mass distribution

d) Centre of mass may lie out side the periphery

1) only a and b are correct 2) only b and c are correct

3) a, b and c are correct 4) a, b, c and d are correct

29. A disc of radius 'r' is removed from the disc of radius 'R' then

a) The minimum shift in centre of mass is zero

b) the maximum shift in centre of mass cannot be greater than $\frac{r^2}{(R+r)}$

c) Centre of mass must lie where mass exists

d) the shift in centre of mass is $\frac{r^2}{(R+r)}$

1) only a and b are correct 2) only a and c are correct

3) only a, b and d are correct 4) all are correct

30. In pure rolling motion of a ring

a) it rotates about instantaneous point of contact of ring and ground

b) its centre of mass moves in translatory motion only

c) its centre of mass will have translatory as well as rotatory motion

1) only a is correct 2) a and c are correct

3) a and b are correct 4) a, b and c are correct

31. A tank filled with water is connected to another empty tank lying to the right side of the first tank. Untill the level of water in the tanks become same

a) centre of mass of water shifts down at constant rate

b) centre of mass of water shifts down and rate of change of shift in centre of mass decreases gradually and becomes zero

c) centre of mass shifts towards right as well

1) only a and c are correct

2) only b and c are correct

3) only a and b are correct 4) all are correct

32. Statement A : When an iron sphere on the ground is heated its centre of mass rises slightly

Statement B : Centre of mass of system of particles depends on mass of the particles and their relative positions.

1) only A is correct 2) only B is correct

3) A and B are correct 4) Both A and B are wrong

33. Statement A : When no external force acts on a body its centre of mass will be at rest or under uniform motion.

Statement B : When a force acts on a body at its centre of mass then will have only translation motion

1) only A is correct 2) only B is correct

3) both A and B are correct

4) Both A and B are wrong

34. ***Statements :***

a) In translatory motion each point of the body experiences the same displacement as any other point as time goes on, so that the motion of any particle represents the motion of the whole body.

b) In rotatory motion, all the particles in the rigid body moves in concentric circles about the axis of rotation.

1) a is true, b is false 2) a is false, b is true

3) both a and b are true 4) both a and b are false

35. **Statements :**

a) Centre of mass is that fixed point of a system of particles or a rigid body with in the boundaries of the system where the entire mass is concentrated.

b) Center of gravity of a body is the point through which the whole weight of the body acts.

1) Both a and b are true 2) Both a and b are false

3) a is true, b is false 4) a is false, b is true

36. **Statements :**

a) Position of C.m depends on acceleration due to gavity where as position of C.g independent on aceeleration due to gravity.

b) Centre of gravity of a body is defined to know the amount of stability of the body and the center of mass of the body is defined to describe the nature of motion of the body.

1) a is true, b is false 2) a is false, b is true

3) both a and b are true 4) both a and b are false

37. **Statements : a) Ratio of the distances of the centre of mass of a two particle system are in the inverse ratio of their masses :**

b) The location of the centre of mass is independent of the reference frame used locate it

1) Only a is true 2) Only b is true

3) Both a and b are true 4) Both a and b are false

38. **The centre of mass of the system of particles depends on**

a) masses of the particles

b) Relative position of the particles

1) Only a is true 2) Only b is true

3) Both a and b are true 4) Both a and b are false

39. **Statements : a) Allgebric sum of moments of mass about centre of mass is equal to zero**

b) x - co ordinate of centre of mass of system of particles in a plane is represented by

$$x_{cm} = \frac{1}{M}\sum m_i x_i$$

C) x - co . ordinate of a rigidbody of continuous mass distribution represented by

$$x_{cm} = \frac{1}{M}\int x.dm$$

1) a and b are true 2) b and c are true

3) a and c are true 4) All a, b, c are true

40. **If no external force acts on the system**

a) acleleration of centre of mass is zero

b) Velocity of centre of mass is zero

c) Centre of mass may be at rest (or) moving with constant velocity

d) Velocity and momentum of centre of mass remains constant

1) a,b and c are true 2) a, c and d are true

3) b,c and d are true 4) a,b,c and d all are true

41. **Statements : a) The motion of the centre of mass can be studied using kepler's laws of motion.**

b) The internal forces will not effect the motion of the centre of mass

1) Both a and b are true 2) Both a and b are false

3) a is true b is false 4) a is false b is true

42. **Statements : a) Motion of centre of mass depends on external forces applied on it .**

b) The motion of the centre of mass of the body is called the translational motion of the body.

1) Both a and b are true 2) Both a and b are false

3) a is true b is false 4) a is false b is true

43. **The moon moves around the earth in a circular orbit and the earth moves around the sun in elliptical orbit then.**

a) For the earth - moon system , the gravitational attractions between them are the internal forces

b) The sun's attraction on the earth and moon are external forces acting on the centre of mass of the system.

1) a is true b is false 2) a is false b is true

3) both a and b are true 4) Both a and b are false

ASSERTION & REASON

These Questions consist of two statements each printed as Assertion and Reason. While answering these questions you are required to choose any one of the following four responses.

1) Both (A) and (R) are true and (R) is the correct explanation of (A)

2) Both (A) and (R) are true and (R) is not the correct

explanation of (A)

3) (A) is true but (R) is false

4) (A) is false but (R) is true

44. **(A) : Position of centre of mass of a body (w.r.t body frame)is independent of co-ordinate system**

(R) : Internal forces do not effect the motion of centre of mass

45. **(A) : For the body shown in the figure , the centre of mass lies outside the body**

(R) : Centre of mass of a body may or may not lie within the body

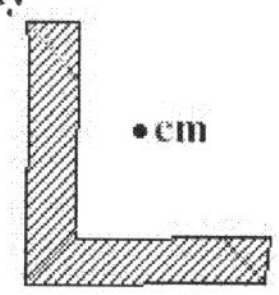

46. **(A) : Two particles (starting from rest) move towards each other under a mutual force of attraction. The velocity of their c.m is zero**

(R) : Internal forces do not alter the state of motion of c.m

47. **(A) : When an external force is applied at the centre of mass of a system of particles then it undergoes translatory motion only**

(R) : The torque of the force acting at c.m, about that c.m is zero

48. **(A) : When a person walks on a stationary boat in still water, centre of mass of person and boat system is not displaced**

(R) : Internal forces can not alter the position of c.m.

49. **(A) : A shell moving in a parabolic path explodes in mid air. The centre of mass of the fragments will follow the same parabolic path**

(R) : Explosion is due to internal forces, which can not alter the state of motion of centre of mass

50. **(A) : A vehicle is stopped due to friction between the wheels and the road**

(R) : Only external forces can cause the centre of mass of system to accelerate or decelerate

51. **(A) : Stationary bomb explodes into two fragments. They move with different velocities in opposite directions**

(R) : Explosion does not violate the law of conservation of linear momentum

52. **(A) : A block of mass 'm' rests on a wedge of mass M, which , in turn , is at rest on a horizontal table. All the surfaces are friction less. The block starts from rest. The position of c.m of the system will change only in vertical direction**

(R) : The system is free from external force along horizontal direction

53. **(A) : There are some passengers in a stationary compartment. A physical fight takes place between them because of different opinions. The position of c.m of system will not change**

(R) : The fight between passengers is internal force

54. **(A) : Heavy boxes are loaded along with empty boxes on a cart. Heavy boxes should be kept at the centre of gravity of the cart, for more stability**

(R) : The vertical line passing through the centre of gravity should fall within the base of cart for stability.

55. **Assertion :(A) In the uniform gravitational field centre of mass and centre of gravity coincide with each other. Where as in the case of non Unifrom gravitational field they will not coincide with each other.**

Reason : (R) Centre of mass is independent of acceleration due to gravity but centre of gravity depends on acceleration due to gravity.

56. **Assertion : (A) Internal forces do not effect the path and position of centre of mass.**

Reason : (R) According to newtons third law of motion internal forces will occur in equal and opposite pairs . so they cancel each other.

MATCHING TYPE QUESTIONS

57. List–1 | List–2

a) Centre of mass of large uniform ring — e) mass is present

b) Centre of mass of non uniform rod — f) Coincides with centre of gravity

c) Centre of mass of small funnel — g) on the heavier side

d) Centre of mass of a uniform disc — h) at Geometric centre

The correct match is

1) a – f; b–e; c – g; d – h
2) a – g; b– h; c–e; d – f
3) a – h; b – g; c – f; d – e
4) a – e; b – f; c – h; d – g

58. List–1 | List–2

a) Position of c.m of a triangle — e)at one-forth of maximum height from the base

b) Position of c.m of a square — f) at the point of intersection of medians

c) Position of c.m of a sphere — g) at the point of intersection of diogonals

d) Position of c.m of a conc — h) at the centre

The correct match is

1) a –f, b–g, c–h; d–e
2) a – g, b – h, c – e, d – f
3) a – g, b – h, c – f, d – e
4) a – g, b – f, c – e, d – h

59. List–1 | List–2

a) Position of centre of mass — e) is zero

b) The algebraic sum of moments of all the masses about centre of mass — f) in non. uniform gravitational field

c) Centre of mass and centre of gravity coincide — g) is independent of frame of reference

d) Centre of mass and centre of gravity do not coincide — h) in uniform gravitational field

The correct match is

1) a –e, b–g, c – f; d – h
2) a – g, b – e, c – f, d – h
3) a – g, b – e, c – h, d – f
4) a – h, b – e, c – f, d – g

60*. A square plate and a circular plate made up of same meterial are placed touching each other on a horizontal table. If the side length of square plate is equal to diam eter of the circular plate then the centre of mass of the system will be

1) inside the square plate
2) inside the circular plate
3) at the point of contact
4) outside the system

ANSWERS

1) 3	2) 1	3) 3	4) 4	5) 4
6) 2	7) 4	8) 4	9) 1	10) 3
11) 2	12) 3	13) 2	14) 3	15) 2
16) 4	17) 1	18) 3	19) 3	20) 1
21) 4	22) 3	23) 4	24) 1	25) 4
26) 1	27) 1	28) 3	29) 1	30) 3
31) 2	32) 3	33) 3	34) 3	35) 1
36) 2	37) 3	38) 3	39) 4	40) 2
41) 4	42) 1	43) 3	44) 2	45) 1
46) 1	47) 2	48) 1	49) 1	50) 1
51) 1	52) 1	53) 1	54) 1	55) 1
56) 1	57) 3	58) 1	59) 3	60) 1

EXERCISE - II(A)

1. Centre of mass of two particles with masses 2 kg and 1 kg located at (1, 0, 1) and (2, 2,0) has the co-ordinates of

1) $\frac{4}{3},\frac{2}{3},\frac{2}{3}$ 2) $\frac{2}{3},\frac{4}{3},\frac{2}{3}$ 3) $\frac{2}{3},\frac{2}{3},\frac{4}{3}$ 4) $\frac{1}{3},\frac{2}{3},\frac{2}{3}$

2. Particles of masses 100 and 300 gram have position vectors $(2\hat{i}+5\hat{j}+13\hat{k})$ and $(-6\hat{i}+4\hat{j}+2\hat{k})$. Position vector of their centre of mass is

1) $-\frac{16}{4}\hat{i}+\frac{17}{4}\hat{j}+\frac{7}{4}\hat{k}$ 2) $\frac{20}{4}\hat{i}+\frac{17}{4}\hat{j}+\frac{7}{4}\hat{k}$

3) $-\frac{16}{4}\hat{i}+\frac{17}{4}\hat{j}+\frac{19}{4}\hat{k}$ 4) $-\frac{16}{4}\hat{i}+\frac{13}{4}\hat{j}+\frac{19}{4}\hat{k}$

3. The distance between the centres of carbon and oxygen atoms in the carbon monoxide gas molecule is 1.13×10^{-10}m. The distance of centre of mass of the molecule relative to carbon atom is

1) 0.48×10^{-10}m 2) 0.64×10^{-10}m
3) 0.56×10^{-10}m 4) 0.36×10^{-10}m

4. The co-ordinates of centre of mass of particles of mass 10, 20 and 30 gm are (1, 1, 1) cm. The position co-ordinates of mass 40 gm which when added to the system, the position of combined centre of mass be at (0, 0, 0) are,

1) (3/2, 3/2, 3/2) 2) (-3/2, -3/2, -3/2)
3) (3/4, 3/4, 3/4) 4) (-3/4, -3/4, -3/4)

5. Two particles of masses 4 kg and 6 kg are at rest separated by 20 m. If they move towards each other under mutual force of attraction, the position of the point where they meet is

1) 12m from 4kg body
2) 12m from 6kg body
3) 8m from 4kg body
4) 10m from 4kg body

6. One end of a thin uniform rod of length L and mass M_1 is rivetted to the centre of a uniform circular disc of radius 'r' and mass M_2 so that both are coplanar. The centre of mass of the combination from the centre of the disc is (Assume that the point of attachment is at the origin) (2003 M)

1) $\frac{L(M_1+M_2)}{2M_1}$ 2) $\frac{LM_1}{2(M_1+M_2)}$
3) $\frac{2(M_1+M_2)}{LM_1}$ 4) $\frac{2LM_1}{(M_1+M_2)}$

7. A ball of mass 3M rolls from rest on a smooth flat table towards another ball of mass 2 M at rest with a constant velocity 5 m/s. When the two balls are separated by a distance 6 m, the position of centre of mass from the mass 3 M is

1) 3 m 2) 2.4m 3) 2.0 m 4) 1 m

*8. Three particles of masses 1 kg, 2 kg and 3 kg are placed at the vertices A, B and C of an equilateral triangle ABC. If A and B lie at (0,0) and (1,0) m, the co-ordinates of their centre of mass are

1) $\left(\frac{\sqrt{3}}{2}m \text{ and } \frac{7}{6}m\right)$ 2) $\left(\frac{7}{6}m \text{ and } \frac{\sqrt{3}}{4}m\right)$

3) $\left(\frac{7}{12}m \text{ and } \frac{\sqrt{3}}{4}m\right)$

4) $\left(\frac{7}{12}m \text{ and } \frac{7}{12}m\right)$

9. Three particle each of 1kg mass are placed at corners of a right angled triangle AOB, 'O' being the origin of coordinate system (OA and OB along the x-direction and +Ve y-direction). If OA = OB =1m, the position vector of centre of mass (in metres) is (2005 M)

1) $\frac{\hat{i}+\hat{j}}{3}$ 2) $\frac{\hat{i}-\hat{j}}{3}$ 3) $\frac{2(\hat{i}+\hat{j})}{3}$ 4) $\hat{i}-\hat{j}$

10. Four particles of masses 2,2,4,4 kg are arranged at the corners A,B,C,D of a square ABCD of side 2m as shown in the figure. The perpendicular distance of their centre of mass from the side AB is,

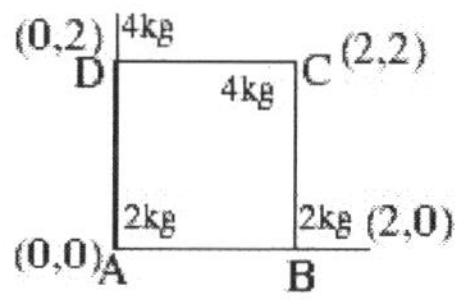

1) 1.33 m 2) 1 m 3) 1.5 m 4) 0.5 m

11. Two uniform rods A and B of lengths 5 m and 3 m are placed end to end. If their linear densities are 3 kg/m and 2 kg/m, the position of their centre of mass from their interface is

1) 19/14 m on the side of heavier rod
2) 8/7 m on the side of lighter rod
3) 2 m on the side of heavier rod
4) 2 m on the side of lighter rod.

12. **A uniform disc of radius R is put over another uniform disc of radius 2R of same thickness and density. The peripheres of the two discs touch each other. The position of their centre of mass is**
1) at R/3 from the centre of the bigger disc towards the centre of the smaller disc
2) at R/5 from the centre of the bigger towards the centre of the smaller disc
3) at 2R/5 from the centre of the bigger disc towards the centre of the smaller disc
4) none of the above

13. **A system consists of two identical particles one particle is at rest and the other particle has an acceleration 'a'. The centre of mass of the system has an acceleration of (2002 M)**
1) 2a 2) a 3) a/2 4) a/4

14. **Three identical spheres each of radius 'R' are placed touching each other so that their centres A,B and C lie on a straight line. the position of their centre of mass from A is**
1) $\frac{2R}{3}$ 2) 2 R 3) $\frac{5R}{3}$ 4) $\frac{4R}{3}$

15. **The velocities of three particles of masses 20g, 30g and 50g are 10i, 10 j and 10k respectively. The velocity of the centre of mass of the three particles is (2001 E)**
1) $2\vec{i}+3\vec{j}+5\vec{k}$ 2) $10(\vec{i}+\vec{j}+\vec{k})$
3) $20\vec{i}+30\vec{j}+5\vec{k}$ 4) $20\vec{i}+30\vec{j}+50\vec{k}$

16. **A thin uniform rod of length L is bent at its mid point as shown in the figure. The distance of centre of mass form the point 'O' is**

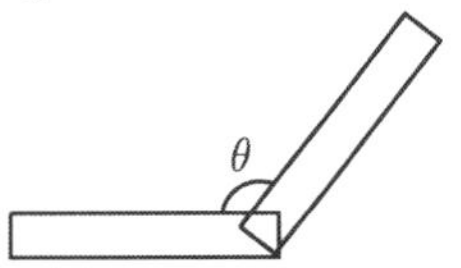

1) $\frac{L}{2}\cos\frac{\theta}{2}$ 2) $\frac{L}{4}\sin\frac{\theta}{2}$
3) $\frac{L}{4}\cos\frac{\theta}{2}$ 4) $\frac{L}{2}\sin\frac{\theta}{2}$

17. **Particles of masses m, 2m, 3m,...... nm grams are placed on the same line at distances *l*, 2*l*, 3*l*, ... n*l* cm from a fixed point. The distance of centre of mass of the particles from the fixed point in centimetre is (2002 E)**
1) $\frac{(2n+1)l}{3}$ 2) $\frac{l}{n+1}$
3) $\frac{n(n^2+1)l}{2}$ 4) $\frac{2l}{n(n^2+1)}$

18. **Four particles, each of mass 1kg, are placed at the corners of a square OABC of side 1m. 'O' is at the origin of the coordinate system. OA and OC are aligned along positive x-axis and positive y-axis respectively. The position vector of the centre of mass is (in 'm') (2006 M)**
1) $\hat{i}+\hat{j}$ 2) $\frac{1}{2}(\hat{i}+\hat{j})$
3) $(\hat{i}-\hat{j})$ 4) $\frac{1}{2}(\hat{i}-\hat{j})$

19. **Three particles of masses 8 kg, 4 kg and 4 kg situated at (4,1), (-2,2) and (1,-3) are acted upon by external forces $6\hat{j}$ N, $-6\hat{i}$ N and $14\hat{i}$ N. The accelaration of centre of mass of the system is**
1) 0.625 ms^{-2} 2) 6.25 ms^{-2}
3) 2.2 ms^{-2} 4) 22 ms^{-2}

20. **The mass of a uniform ladder of length 5 m is 20 kg. A person of mass 60 kg stands on the ladder at a height of 2 m from the bottom. The position of centre of mass of the ladder and man from the bottom nearly is**
1) 1 m 2) 2.5 m 3) 3.5 m 4) 2.125 m

21. **From a sphere of radius 1 m, a sphere of radius 0.5 m is removed from the edge. The shift in the C.M. is**
1) 13/16 m 2) 16/13m
3) 14/13 m 4) 1/14 m

22. **A circular plate has a uniform thickness and has a diameter 56 cm. A circular disc of diameter 42 cm is removed from one edge of the plate. The distance of centre of mass of the remaining portion from the centre of the circular plate is**
1) 18 cm 2) 9 cm
3) 27 cm 4) 4.5 cm

23. **A man of 50kg is standing at one end on a boat of length 25m and mass 200kg. If he starts running and when he reaches the other end, he has a velocity $2ms^{-1}$ with respect to the boat. The final velocity of the boat is (in ms^{-1}) (2006 E)**
1) 2/5 2) 2/3 3) 8/5 4) 8/3

24. **Two particles having masses m and 2m are travelling along x-axis on a smooth surface with velocities u_1 and u_2, collide. If their velocities after collision are v_1 and v_2, then the ratio of velocities of their centre of mass before and after impact is**
1) 2 : 1 2) 2 : 3 3) 1 : 1 4) 1 : 2

25. Two particles A and B initially at rest move towards each other under a mutual force of attraction. At the instant when the speed of A is V and the speed of B is 2v, the speed of the centre of mass of the system is

1) 0 2) V 3) 1.5V 4) 3V

26. A baloon of mass M is stationary in air. It has a ladder on which a man of mass m is standing. If the man starts climbing up the ladder with a velocity v relative to ladder, the velocity of balloon is

1) $\frac{m}{M}v$ upwards 2) $\frac{mv}{(M+m)}$ downwards

3) $\frac{mv}{(M+m)}$ upwards 4) $\frac{Mv}{(M+m)}$ downwards

27. A wooden log of mass 120kg is floating on still water perpendicular to the shore. A man of mass 80kg is standing at the centre of mass of the log and he is at a distance of 30m from the shore. When he walks through a distance of 10m towards the shore and halts, now his distance from the shore is

1) 20m 2) 24m 3) 28 4) 30m

28. A uniform wire of length L is bent in the form of a circle. The shift in its centre of mass is

1) $\frac{L}{\pi}$ 2) $\frac{2L}{\pi}$ 3) $\frac{L}{2\pi}$ 4) $\frac{L}{3\pi}$

29. Masses 1 kg, 1.5 kg, 2 kg and M kg are situated at (2,1,1), (1,2,1), (2,–2,1) and (–1,4,3). If their centre of mass is situated at (1,1,3/2), the value of M is

1) 1 kg 2) 2 kg 3) 1.5 kg 4) 3 kg

30. Three identical spheres each of radius 'R' are placed on a horizontal surface touching one another. If one of the spheres is removed, the shift in the centre of mass of the system is

1) $R/2\sqrt{3}$ 2) $R/2$ 3) $\sqrt{3}R/2$ 4) $R/\sqrt{3}$

31. Two bodies of masses 5kg and 3kg are moving towards each other with $2ms^{-1}$ and $4ms^{-1}$ respectively. Then velocity of centre of mass is

1) $0.25ms^{-1}$ towards 3kg 2) $0.5ms^{-1}$ towards 5kg
3) $0.25ms^{-1}$ towards 5kg 4) $0.5ms^{-1}$ towards 3kg

32. Two particles each of the same mass move due north and due east respectively with the same velocity 'V'. The magnitude and direction of the velocity of centre of mass is

1) $\frac{V}{\sqrt{2}}$ NE 2) $\sqrt{2}$ V NE 3) 2 V SW 4) $\frac{V}{2}$ SW

33*. If the distance between the centres of the atoms of potassium and bromine in KBr molecule is 0.282×10^{-9} m, the centre of mass of this two particle system from potassium is (mass of bromine = 80 u and of potassium = 39 u).

1) 9.3×10^{-2} nm 2) 18.96×10^{-2} nm
3) 27.9×10^{-2} nm 4) 3.3×10^{-2} nm

34. The co-ordinates of the centre of mass of the system as shown in figure are

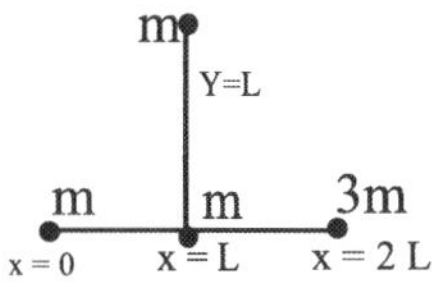

1) (0, 0)

2) (1.33L, 0.167 L)

3) (L, L)

4) (0.33L, 1.67 L)

35. Two bodies of masses 10 kg and 20 kg are located in x - y plane at (0, 1) and (1, 0). The position of their centre of mass is

1) (2/3, 1/3) 2) (1/3, 2/3) 3) (2, 1) 4) (1/3, 4/3)

36. A uniform disc of diameter R/2 is put over another uniform disc of radius R of same thickness and density. The peripheries of the two discs touch each other. The centre of mass of the system from centre of big disc is

1) R/10 2) R/20 3) 9R/10 4) 11R/10

37. Particles of masses 1kg and 3kg are at (2i + 5j + 13k) m and (–6i + 4j – 2k)m then instantaneous position of their centre of mass is

1) $\frac{1}{4}(-16i+17j+7k)m$

2) $\frac{1}{4}(-8i+17j+7k)m$

3) $\frac{1}{4}(-6i+17j+7k)m$

4) $\frac{1}{4}(-6i+17j+5k)m$

38. Two particles of same mass are projected simultaneously with same speed 20 ms^{-1} from the top of a tower of height 20m. One is projected vertically upwards and other projected horizontally. The maximum height attained by centre of mass from the ground will be
($g = 10\ ms^{-2}$)

1) $10\sqrt{2}$m 2) 25 m

3) $25\sqrt{2}$m 4) 5 m

39. Six identical particles each of mass m are arranged at the corners of a regular hexagon of side length a. If the mass of one of the particle is doubled, the shift in the centre of mass is

1) a 2) $\frac{6a}{7}$ 3) $\frac{a}{7}$ 4) $\frac{a}{\sqrt{3}}$

40. Three identical rods each of length L and mass m are joined to form a U-shape as shown in figure. The distance of centre of mass of the combined system from the corner 'O' is

1) $\frac{L}{2}$ 2) $\frac{L}{3}$

3) $\frac{\sqrt{8}L}{6}$ 4) $\frac{\sqrt{13}L}{6}$

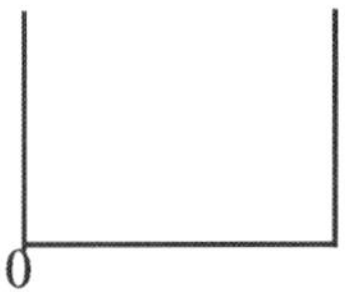
O

41. Four identical particles each of mass 1 kg are arranged at the corners of a square of side length $2\sqrt{2}$ m. If one of the particles is removed, the shift in the centre of mass is

1) 8/3m 2) 4/3m 3) 2/3m 4) 2m

ANSWERS

1) 1	2) 3	3) 2	4) 2	5) 1
6) 2	7) 2	8) 3	9) 1	10) 1
11) 1	12) 2	13) 3	14) 2	15) 1
16) 3	17) 1	18) 2	19) 1	20) 4
21) 4	22) 2	23) 1	24) 3	25) 1
26) 2	27) 2	28) 3	29) 3	30) 4
31) 3	32) 1	33) 2	34) 2	35) 1
36) 1	37) 1	38) 2	39) 3	40) 4
41) 3				

EXERCISE - II(B)

1. The centre of mass of two particles with masses 2kg and 6kg located at (1, 1,1), (2, 2, 1) respectively has the coordinates

1) $\left(\frac{7}{4}, \frac{7}{4}, \frac{7}{4}\right)$ 2) $\left(\frac{4}{3}, \frac{4}{3}, \frac{2}{3}\right)$

3) $\left(\frac{7}{4}, 4, 1\right)$ 4) $\left(\frac{7}{4}, \frac{7}{4}, 1\right)$

2. Two particles of masses 1kg and 3kg have position vectors $2\hat{i}+3\hat{j}+4\hat{k}$ and $-2\hat{i}+3\hat{j}-4\hat{k}$ respectively. The centre of mass has a position vector **(1999 E)**

1) $\hat{i}+3\hat{j}-2\hat{k}$ 2) $-\hat{i}-3\hat{j}-2\hat{k}$

3) $-\hat{i}+3\hat{j}+2\hat{k}$ 4) $-\hat{i}+3\hat{j}-2\hat{k}$

3. In a molecule of sodium chloride, the masses of sodium and chlorine atoms are in the ratio 23 : 36. If the separation between the two atoms is 'd', distance of centre of mass of the molecule from sodium atom

1) 36d/59 2) 23d/59 3) 59d/23 4) 59d/36

* **4. The centre of mass of three particles of masses 1kg, 2kg, 3kg is at (2, 2, 2). The position of fourth mass of 4kg to be placed in the system so that new centre of mass is at (0, 0, 0) is**

1) (–3, –3, –3) 2) (–3, 3, –3) **(2005 E)**

3) (2, 3, –3) 4) (2, –2, 3)

5. Two skaters A and B weighing 40 kg and 60 kg respectively stand facing each other 5 m apart. If they pull a light rope stretched between them, then they meet

1) 1.5 m from A 2) 2 m from A

3) 2.5 m from A 4) 3.0 m from A

6. A uniform bar 8 m long has a mass of 6 kg. A 10 kg mass is tied to one end. The position of centre of mass of the system is

1) 1.5m from 10 kg mass

2) 1.5 m from the centre of the rod

3) 2.5 m from 10 kg mass

4) 4m from the free end of the rod

7. Two persons of masses m and 2m are standing on a horizontal smooth surface of ice. They are initially seperated by a distance d. They are pulling each other holding a rope. They meet at a distance

1) 2d/3 from heavier person

2) d/3 from heavier person

3) 2d/4 from lighter person

4) d/3 from lighter person

8. Three particles of masses 1 kg, 2 kg and 2 kg are placed at the corners of an equilateral triangle. The co-ordinates of A,B,C are (0,0) (2,0) and (x,y), the co-ordinates of their centre of mass are

1) $\frac{6}{5}, \frac{3}{5}$ 2) $\frac{6}{5}, \frac{2\sqrt{3}}{5}$ 3) $\frac{6}{5}, \frac{1}{5}$ 4) $\frac{6}{5}, \frac{2}{5}$

9. Masses each 1 kg are placed at the verticies of an isosceles triangle ABC in which AC=BC=5 cm and AB = 8 cm. The distance of centre of mass of the system from the vertex C is

1) 2 cm 2) 1 cm 3) 1.5 cm 4) 3 cm

10. Four particles P,Q,R,S of masses 1 kg, 2 kg, 3kg and 4 kg are placed at the corners of a square of side 1 m as shown below. the co-ordinates of the centre of mass of the system are

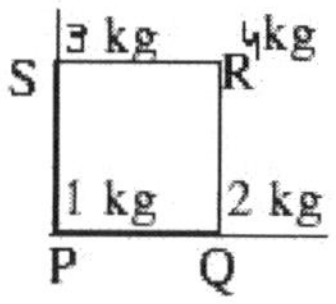

1) 0.5m, 0.7 m 2) 0.6 m, 0.7 m
3) 0.9 m, 0.8m 4) 0.4m, 0.5 m

11. Two uniform rods A and B of same diameter having length 2m and 3m and having mass per unit 4kg m^{-1} and 6kgm^{-1} respectively are joined end to end. The position of centre of mass of combined rod from the free end of A is

1) $\frac{71}{26}$m 2) $\frac{26}{71}$m 3) $\frac{41}{36}$m 4) $\frac{36}{41}$m

12. Two discs of radii 2 cm and 1 cm and made up of same material respectively are placed as shown below. The distance of centre of mass of the system from O_1 is

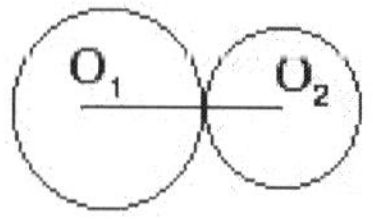

1) 0.4 cm 2) 1.5 cm 3) 0.6 cm 4) 0.8 cm

13. Two particles of masses 2 m and 3 m are at a distance 'd' apart. Under their mutual gravitational force, they start moving towards each other. The accelaration of their centre of mass when they are d/2 apart is

1) $\frac{12\,G\,m}{d^2}$ 2) $\frac{8\,G\,m}{d^2}$ 3) $\frac{24\,G\,m}{d^2}$ 4) zero

14. Three identical particles each of same mass are placed touching each other with their centres on a straight line. Their centres are at A, B and C respectively. Then distance of centre of mass of the system from A is

1) $\frac{AB+AC+BC}{3}$ 2) $\frac{AB+AC}{3}$
3) $\frac{AB+BC}{3}$ 4) $\frac{AC+BC}{3}$

15. Two objects of masses 200g and 500g possess velocities $10\vec{i}$ ms^{-1} and $3\vec{i}+5\vec{j}$ ms^{-1} respectiv ely. The velocity of their centre of mass in ms^{-1} is

1) $5\vec{i}-25\vec{j}$ 2) $\frac{5}{7}\vec{i}-25\vec{j}$ (2003 E)
3) $5\vec{i}+\frac{25}{7}\vec{j}$ 4) $25\vec{i}-\frac{5}{7}\vec{j}$

16. A uniform metre rod is bent into L shape with the bent arms at 90°to each other. The distance of the centre of mass from the bent point is

1) $\frac{1}{4\sqrt{2}}$m 2) $\frac{1}{2\sqrt{2}}$m 3) $\frac{1}{\sqrt{2}}$m 4) $\frac{1}{8\sqrt{2}}$m

17. Particles of masses 1gm, 2gm, 3gm and 4 gm are placed at x = 1 cm x = 2cm, x = 3 cm, x = 4cm respectively. Then X_{cm} =

1) 1cm 2) 2cm 3) 3cm 4) 4cm

18. Four particles, each of mass 1kg, are placed at the corners of a square of side one meter in the X – Y plane. If the point of intersection of the diagonals of the square is taken as the origin, the coordinates of the centre of mass are (2004 M)

1) (1, 1) 2) (–1, 1) 3) (1, –1) 4) (0, 0)

19. Three particles of masses 1kg, 2kg, 3kg are acted upon by forces $(\vec{i}+2\vec{j}),(2\vec{j}+3\vec{k})$ and $(\vec{i}-\vec{k})$ newton respectively. The magnitude of acceleration of centre of mass to ms^{-2} is

1) $\sqrt{3}$ 2) $2\sqrt{3}$ 3) $\sqrt{\frac{2}{3}}$ 4) 1

20. The mass of a uniform ladder of length 6 m is 20 kg. A mass 60 kg is placed on the ladder at a height of 2 m from the bottom. The position of centre of mass from the bottom is

1) 1.5 m 2) 2.25 m 3) 2.25 m 4) 1 m

21. A uniform sphere has radius R. A sphere of diameter R is cut from its edge as shown. Then the distance of centre of mass of remaining portion from the centre of mass of the original sphere is

1) R/7
2) R/14
3) 2R/7
4) R/18

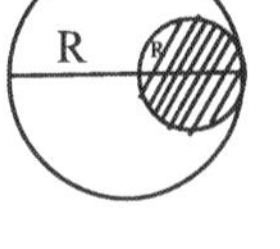

22. A uniform circular disc has a radius of 20cm. A circular disc of radius 10 cm is cut concentrically from the original disc. The shift in centre of mass is

1) 1/200 cm 2) 7/300 cm 3) 1/400 cm 4) zero

23. A man of mass 40kg in standing on a uniform plank of mass 60 kg lying on horizontal frictionless ice. The man walks from one end to the other end of the plank. The distance walked by the man relative to ice is (given length of plank = 5m)

1) 2m 2) 3m 3) 5 m 4) 4m

24. Two particles of masses 1kg and 3kg move towards each other under their mutual force of attraction. No other force acts on them. When the relative velocity of approach of the two particles is $2ms^{-1}$, the velocity of centre of mass is $0.5ms^{-1}$. When the velocity of approach becomes $3ms^{-1}$, the velocity of centre of mass is

1) 0.75 ms^{-1} 2) 0.5 ms^{-1}
3) 2.5 ms^{-2} 4) 0 ms^{-2}

25. Two identical particles move towards each other with velocity 2V and V respectively. The velocity of centre of mass is

1) V 2) V/3 3) V/2 4) Zero

26. Two spheres of masses 2M and M are initially at rest at a distance R apart. Due to mutual force of attraction, they approach each other. When they are at reparation R/2, the acceleration of the centre of mass of spheres would be

1) Zero 2) g ms^{-2} 3) 3g ms^{-2} 4) 12g ms^{-2}

27. A boy of mass 36 kg moves from one end to other end of a boat of mass 72 kg on still water. If the length of the boat is 2.7m, the displacement of the boat is

1) 2.7 m 2) 1.35m 3) 5.4m 4) 0.9m

28. A uniform wire of length 6.28 cm is bent in the form of a circle. The shift in its centre of mass is

1) 1cm 2) 2cm 3) 4cm 4) 3.14 cm

29. A block of mass M is placed on the top of a bigger block of mass 10M as shown in fig. All the surfaces are frictionless. The system is releated from rest. The distance moved by the bigger block by the time the smaller block reaches the ground is

1) 0.2m
2) 0.4m
3) 0.3m
4) 0.5m

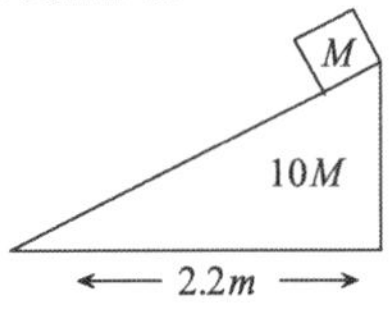

30. Three identical spheres each of radius 3cm are placed on a horizontal surface touching one another. The distance of centre of mass of system from any centre of the sphere is

1) 1/3cm 2) 3 cm 3) $2\sqrt{3}$ cm 4) $1/\sqrt{3}$ cm

31. Two bodies of masses 4kg and 2kg are moving towards each other with $3ms^{-1}$ and $2ms^{-1}$ respectively. Then velocity of centre of mass is

1) $0.5ms^{-1}$ towards 4kg 2) $0.25ms^{-1}$ towards 2kg
3) 4/3 ms^{-1} towards 2kg 4) 4/3 ms^{-1} towards 4kg

32. Two particles of masses 3kg and 2kg move due north and due east respectively with the velocities $2ms^{-1}$ and $3ms^{-1}$. The magnitude and direction of the velocity of centre of mass is

1) 1.2NE 2) $\sqrt{2}$ NE 3) 2 SW 4) $1.2\sqrt{2}$ NE

33. The distance between the two nucleii of HCl is 1.27A. If the mass of chlorine atom is 37 times that of Hydrogen, the centre of mass of the system from the Hydrogen nucleus is

1) 1.1A 2) 1.24A 3) 1.3A 4) 1.4A

34. A body of mass 4 kg is attached to another body of mass 2 kg with a massless rod. If the 4 kg mass is at (2i+5j)m and 2 kg mass is at (4i+2j)m, the centre of mass of the system is at (in m)

1) 8i + 12j 2) $\frac{1}{2}(8i+12j)$

3) $\frac{1}{3}(8i+12j)$ 4) $\frac{1}{4}(8i+12j)$

35. Two spheres of masses 4 kg and 8 kg are moving with velocities 2 ms^{-1} and 3 ms^{-1} away from each other along the same line. The velocity of centre of mass is

1) $\frac{8}{3}ms^{-1}$ towards second sphere

2) $\frac{8}{3}ms^{-1}$ towards first sphere

3) $\frac{4}{3}ms^{-1}$ towards second sphere

4) $\frac{4}{3}ms^{-1}$ towards first sphere

36. A disc of radius R is cut from a larger disc of radius 2R in such a way that the edge of the hole touches the edge of the disc. The centre of mass of the remaining portion is at

1) R/2 from the centre of big disc away from the centre of the hole

2) R/3 from the centre of big disc away from the centre of the hole

3) 3R/2 from the centre of big disc away from the centre of the hole

4) 5R/2 from the centre of big disc away from the centre of the hole

37. **Particals of masses 1kg and 3kg are at (10i – 7j – 3k) ms^{-1} and (7i + 9j + 6k)ms^{-1} then a instantaneous velociti of their centre of masses**

1) $\frac{1}{4}(31i - 34j + 15k)ms^{-1}$

2) $\frac{1}{4}(30i - 32j + 15k)ms^{-1}$

3) $\frac{1}{4}(25i - 28j + 15k)ms^{-1}$

4) $\frac{1}{4}(31i - 34j + 20k)ms^{-1}$

38. **Two particles of masses m_1 and m_2 ($m_1 > m_2$) are separated by a distance 'd'. The shift in the centre of mass when the two particles are interchanged.**

1) $\frac{(m_1 + m_2)d}{m_1 - m_2}$ 2) $\frac{(m_1 - m_2)d}{m_1 + m_2}$

3) $\frac{m_1 d}{m_1 - m_2}$ 4) $\frac{m_2 d}{m_1 \quad m_2}$

39. **Four identical particles each of mass m are arranged at the corners of a square of side 'a'. If mass of one of the particle is doubled, the shift in the centre of mass of the system is**

1) $\frac{a}{2\sqrt{2}}$ 2) $\frac{a}{3\sqrt{2}}$

3) $\frac{a}{4\sqrt{2}}$ 4) $\frac{a}{5\sqrt{2}}$

40. **If three rods of same mass are placed as shown in the figure, the co-ordinates of the centre of mass of the system are**

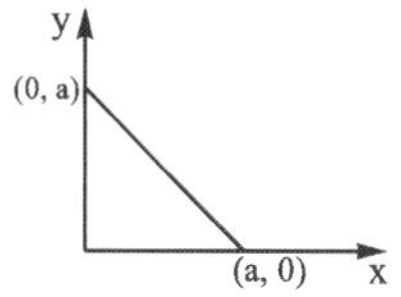

1) $\left(\frac{a}{2}, \frac{a}{2}\right)$ 2) $\left(\frac{a}{\sqrt{2}}, \frac{a}{\sqrt{2}}\right)$

3) $\left(\frac{a}{3}, \frac{a}{3}\right)$ 4) $\left(\frac{a}{\sqrt{3}}, \frac{a}{\sqrt{3}}\right)$

41. **Six identicles particles each of mass 0.5 kg are arranged at the corners of a regular hexagon of side length 0.5 m. If one of the particle is removed, the shift in the centre of mass is**

1) $\frac{1}{8}$m 2) $\frac{1}{10}$m 3) $\frac{1}{12}$m 4) $\frac{1}{14}$m

ANSWERS

1) 4	2) 4	3) 1	4) 1	5) 4
6) 1	7) 2	8) 2	9) 1	10) 2
11) 1	12) 3	13) 4	14) 2	15) 3
16) 1	17) 3	18) 4	19) 3	20) 2
21) 2	22) 4	23) 2	24) 2	25) 3
26) 1	27) 4	28) 1	29) 1	30) 3
31) 3	32) 4	33) 2	34) 1	35) 1
36) 2	37) 1	38) 2	39) 2	40) 3
41) 2				

EXERCISE - IIIA

1. **The C.M of a uniform card board cut in 'T' shape as shown in figure is**

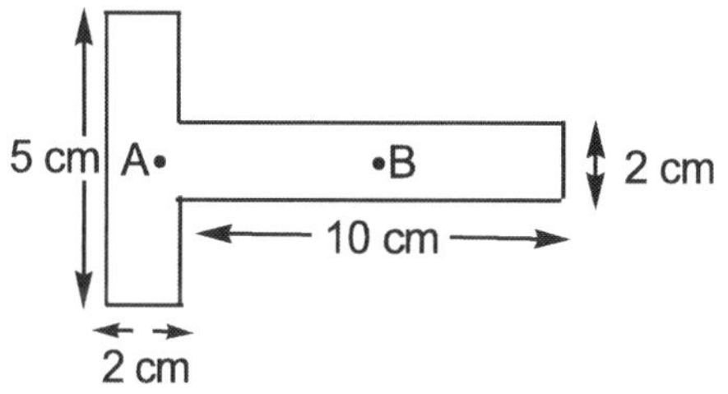

1) 4 cm from A towards B
2) 4 cm from B towards A
3) 3 cm from B towards A
4) 3 cm from A towards B

2. **A metal rod of length 1 m is suspended from a rigid support vertically at 0°C. It is heated to 20°C. The linear coefficient of the material of the rod is $\frac{18}{10^6}/K$. The shift in the centre of mass is**

1) $\frac{18}{10^5}$m 2) $\frac{9}{10^5}$m

3) $\frac{2}{10^5}$m 4) $\frac{3}{10^5}$m

3. **On a large tray of mass M an ice cube of mass m, edge L is kept. If the ice melts completely, the centre of mass of the system come down by**

1) $\frac{mL}{2(M+m)}$ 2) $\frac{(M-m)L}{2(M+m)}$

3) $\frac{(M+2m)L}{2(M+m)}$ 4) $\frac{2ML}{(M+m)}$

4. Two steel spheres of radius r and 2r are made to touch each other. The distance of their centre of mass from their point of contact is

1) at a distance $\frac{8r}{3}$ in the bigger sphere

2) at a distance r/3in the smaller sphere

3) at a distance $\frac{5r}{3}$ in the bigger sphere

4) at a distance $\frac{r}{3}$ in the smaller sphere

5. Two particles of equal mass have velocities $\overline{V}_1 = 4\bar{i}$ and $\overline{V}_2 = 4\bar{j}\,ms^{-1}$. First particle has an acceleration $\bar{a}_1 = (5\bar{i} + 5\bar{j})\,ms^{-2}$ while the acceleration of the other particle is zero. The centre of mass of the two particles moves in a path of (2004 E)

1) Straight line 2) Parabola

3) Circle 4) Ellipse

6. Two blocks of masses 10 kg and 20 kg are placed on the x-axis. 10 kg mass is pushed on the x-axis, by a distance of 2 cm towards the +ve x-axis. The shift in the centre of mass is

1) 2/3 cm 2) 2 cm 3) 1 cm 4) 6 cm

7. A ladder AP of length 5m inclined to a vertical wall is slipping over a horizontal surface with velocity of 2m/s, when A is at a distance 3m from O, the velocity of CM at this moment is

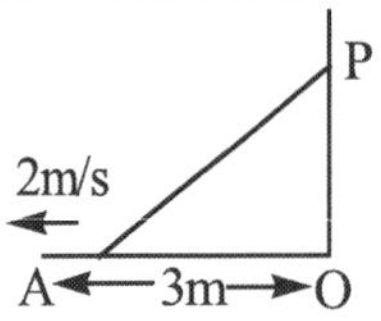

1) 1.5 ms^{-1} 2) 2.5 ms^{-1}

3) 12.5ms^{-1} 4) 1.25ms^{-1}

8. A shell in flight explodes into n equal fragments, k of the fragments reach the ground earlier than the other fragments. The acceleration of their centre of mass subsequently will be

1) g 2) (n-k)g

3) $\frac{(n-k)g}{k}$ 4) $\frac{(n-k)}{n}g$

9. Four identical blocks of length L are arranged one over the other as shown. The maximum distance of the uppermost block from the edge of the lowermost block is x such that no block tumbles. Then x is

1) $\frac{4L}{3}$ 2) $\frac{3L}{4}$

3) $\frac{11L}{12}$ 4) 2L

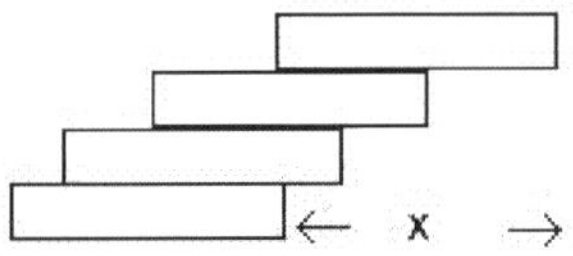

10. Four particles of masses m, 2m, 3m and 4m are at the verticies of a parallelogram in x – y plane with one of the adjacent angle 60^0 and smaller side 'a' and larger side 2a. The mass 'm' is at the origin and mass 4m x-axis. The centre of mas of the system is

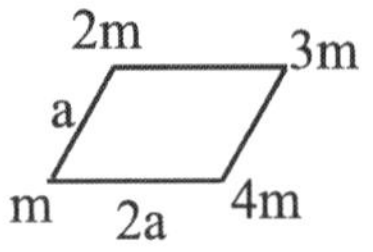

1) $\left(\frac{3a}{2};\frac{a}{2}\right)$

2) $\left(\frac{a}{2};\frac{3a}{2}\right)$

3) $\left(1.65a;\frac{\sqrt{3}a}{4}\right)$ 4) $\left(\frac{\sqrt{3}a}{4};0.82a\right)$

11. Two blocks of masses 10 kg and 30 kg are placed along a vertical line. The first block is raised through a height of 7 cm. By what distance should the second mass be moved to raise the centre of mass by 1 cm ?

1) 1 cm upward 2) 1 cm downward

3) 2 cm downward 4) 2 cm upward

12. A uniform square sheet has a side length of 2R. If one of the quadrants is removed, the shift in the centre of mass

1) $\frac{R}{3\sqrt{2}}$ 2) $\frac{R}{\sqrt{2}}$ 3) $\frac{R}{6\sqrt{2}}$ 4) $\frac{\sqrt{3}R}{2}$

13. A uniform square sheet has a side length of 2R. A circular sheet of maximum possible area is removed from one of the quadrants of the square sheet. The distance of centre of mass of the remaining portion from the centre of the original sheet is

1) $\frac{\pi R}{\sqrt{2}[16-\pi]}$ 2) $\frac{R}{[16-\pi]}$

3) $\frac{R}{\pi[16-\pi]}$ 4) $\frac{R\pi}{16-\pi}$

14. From a uniform circular disc of radius 2cm (its centre of mass is at O). A circular portion of radius 1cm is removed such that shift in centre of mass is maximum. The disc is now rotated about 'O' perpendicular to the plane through θ then the magnitude of displacement of new centre of mass is $\frac{1}{\sqrt{3}}$ cm, then θ is

1)30^0 2) 45^0 3) 60^0 4) 120^0

15. Two balls of masses 5m and m have radii 2R and R. Their centre of masses are separated by 12R. They move towards each other under their gravitational force. The distance moved by the centre of smaller sphere when the spheres touch each other is

1) 2. 5 R 2) 5R 3) 7.5 R 4) 10 R

16. A trolley of mass 360 kg is lying on a horizontal frictionless surface. A man of mass 40 kg runs with uniform speed on the trolley by a distance of 2m in 0.5 second. What is the velocity of the trolley relative to earth?

1) 0.8 m/s 2) 0.4 m/s

3) 1.6m/s 4) 2.2m/s

17. A boat of mass 80 kg is floating on still water. A dog of mass 20 kg on the boat is at a distance of 10m from the shore. The dog moves on the boat by a distance of 2m towards the shore. The distance of the dog from the shore is

1) 11.6 m 2) 8.4 m 3) 9.6 m 4) 10.4 m

18. A shell projected from a level ground has a range R, if it did not explode. At the highest point, the shell explodes into two fragments having masses in the ratio 1:3; with each fragment moving horizontally immediately after the explosion. If the lighter fragment falls at a distance R from the point of projection, behind the point of projection, the distance at which the other fragments falls from the point of projection is

1) 2R 2) 5R/3 4) 4R/3 4) 2R/3

19. A shell is projected from a level ground with a velocity of 20m/s at 45° to the horizontal. When the shell is at the highest point it breaks into two equal fragments. One of the fragments whose initial velocity after the explosion is zero falls vertically downward. At what distance from the point of projection does the other fragment fall? ($g = 10\ ms^{-2}$)

1) 30 m 2) 60 m 3) 90 m 4) 40 m

20. Two blocks of masses 2 kg and 1 kg respectively are tied to the ends of a string which passes over a light frictionless pulley. The masses are held at rest at the same horizontal level and then released. The distance transversed by centre of mass in 2 seconds is ($g = 10\ m/s^2$)

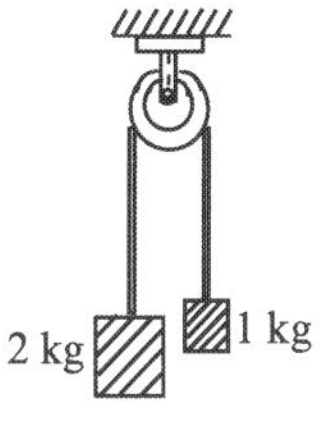

1) 1.42m

2) 2.22m

3) 3.12m

4) 3.33m

21. The linear density of a rod of length L varies as $\rho = A + Bx$ where x is the distance from the left end. The distance of centre of mass from O is

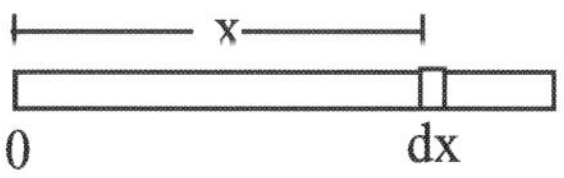

1) $\frac{3AL+2BL^2}{3(2A+BL)}$ 2) $\frac{2AL+2BL^2}{3(2A+BL)}$

3) $\frac{2AL+2BL^2}{4(2A-BL)}$ 4) $\frac{2AL-2BL^2}{4(2A-BL)}$

22. The co-ordinates of centre of mass of letter F which cut from a uniform metal sheet are (take orgin at bottom left corner)

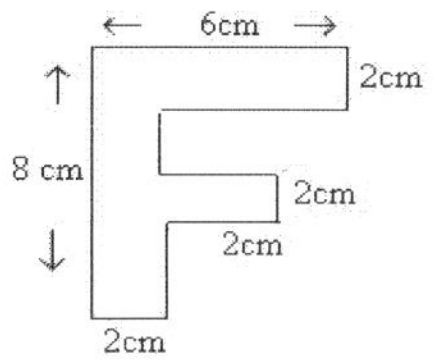

1) $\left(\frac{15}{7},\frac{33}{7}\right)$

2) $\left(\frac{15}{6},\frac{33}{6}\right)$

3) $\left(\frac{15}{4},\frac{33}{4}\right)$ 4) $\left(\frac{7}{15},\frac{7}{33}\right)$

23. A one meter long rod having a constant cross sectional area is made of four materials.The first 0.2m made of iron, the next 0.3 m of lead the following 0.2 m of aluminium and remaining part is made of copper. If the densities of iron lead, aluminium and copper are approximately $8 \times 10^3\ kg/m^3$, $11 \times 10^3\ kg/m^3$, $3 \times 10^3\ kg/m^3$, and $9 \times 10^3\ kg/m^3$ respectively, the centre fo mass of the rod from the left end of iron rod is

1) 0.38 m 2) 0.48 m

3) 0.58 m 4) 0.68 m

24. A circular disc of diameter d and a square plate of side d are placed as shown in fig. The centre of mass of this combination from centre of disc is (both the object are having same mass per unit area)

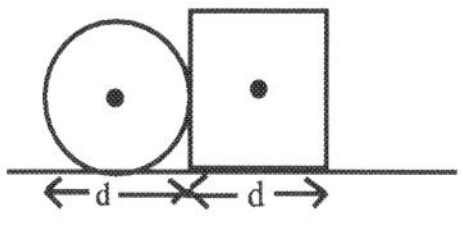

1) $\frac{4d}{4+\pi}$

2) $\frac{4d}{3+\pi}$

3) $\frac{2d+3\pi}{4+\pi}$ 4) $\frac{3d+7\pi}{4+\pi}$

25. If a cube of side R/2 is removed from a uniform sphere of radius R as shown in the figure, the shift in centre of mass is

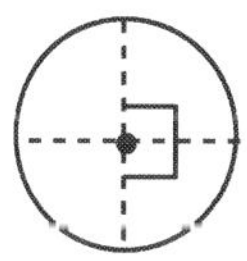

1) $\frac{3R}{2(8\pi-1)}$ 2) $\frac{2R}{3(8\pi-1)}$

3) $\frac{R}{3(8\pi-1)}$ 4) $\frac{R}{4(8\pi-1)}$

ANSWERS

1) 1	2) 1	3) 1	4) 3	5) 1
6) 1	7) 4	8) 4	9) 3	10) 3
11) 2	12) 1	13) 1	14) 4	15) 3
16) 2	17) 2	18) 2	19) 2	20) 2
21) 1	22) 1	23) 2	24) 1	25) 2

EXERCISE - III(B)

1. **A rod consists of two parts made up of a homogeneous material with square type cross – section as shown. The position of C.M from the end 'O' is**

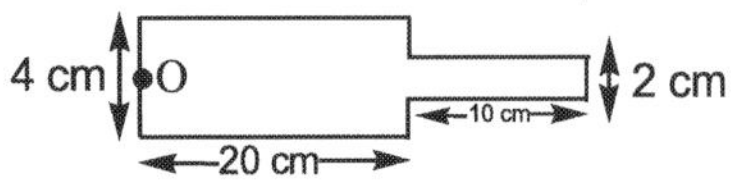

1) $\frac{35}{3}$ cm 2) $\frac{65}{3}$ cm 3) $\frac{55}{3}$ cm 4) 15 cm

2. **A uniform rectangular plate is heated from 0^0 to 100^0C. Initial area of plate is 10 cm^2. The linear coefficient of the material of the plate is $18\times10^{-6}K$. What is the shift of the centre of mass ?**

1) Zero 2) 1cm 3) 2cm 4) 3cm

3. **Half of the uniform rectangular plate of length 'L' is made up of material of density d_1 and the other half with density d_2. The perpendicular distance of centre of mass from AB is**

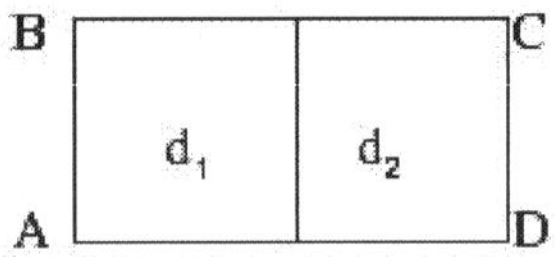

1) $\frac{2d_1+3d_2}{d_1+d_2} \times \frac{L}{4}$ 2) $\frac{d_1+3d_2}{d_1+d_2} \times \frac{L}{4}$

3) $\frac{3d_1}{d_1+d_2} \times \frac{L}{4}$ 4) $\frac{3d_2}{d_1+d_2} \times \frac{L}{4}$

4. **Two steel spheres of radius 2 cm and 5 cm are made to touch each other. The distance of their centre of mass from their point of contact is**

1) at a distance 4 cm in the bigger sphere

2) at a distance 1.5 cm in the smaller sphere

3) at a distance 4.57 cm in the bigger sphere

4) at a distance 0.56 cm in the smaller sphere

5. **Two particles of equal mass have velocities $\bar{V}_1 = 3\bar{i}$ and $\bar{V}_2 = 2\bar{j}\,ms^{-1}$. First particle has an acceleration $\bar{a}_1 = (5\bar{i})\,ms^{-2}$ while the acceleration of the other particle is zero. The centre of mass of the two particles moves in a path of**

1) Straightline 2) Parabola

3) Circle 4) Ellipse

6. **Two particles of masses 10kg and 30kg are lying on a straight line. The 10kg mass is shifted towards the 30kg mass by a distance of 2cm. By what distance should the 30kg mass be shifted so that the position of their centre of mass does not change**

1) 2/3 cm towards 10kg 2) 2/3 cm away from 10 kg

3) 3/2 cm towards 10 kg 4) 3/2 cm away from 10 kg

7. **Two blocks of masses 10kg and 4kg are connected by a spring of negligible mass and placed on a frictionless horizontal surface. An impulse gives a velocity of $14ms^{-1}$ to the heavier block in the direction of the lighter block. The velocity of the centre of mass is**

1) $30ms^{-1}$ 2) 20 ms^{-1} 3) 10 ms^{-1} 4) 5 ms^{-1}

8. **A shell of mass m in flight explodes into three equal fragments. One of the fragment reaches the ground earlier than the other fragments. The acceleration of the centre of mass of the shell system just after first one reaches ground is**

1) g 2) 2g/3 3) 3g/4 4) g/3

9. **Seven homogneous blocks each of length L are arranged as shown in th figure. Each block is displaced with respect to the one in contact by L/10. The x-coordinate of the centre of mass of the system relative to the origin O is**

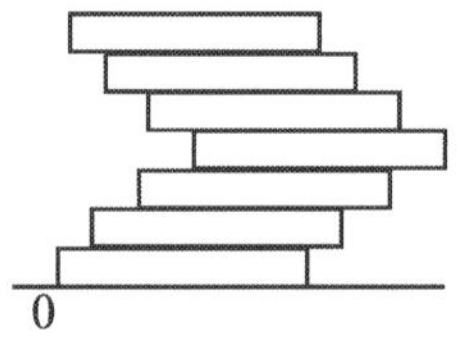

1) $\frac{11L}{12}$ 2) $\frac{22L}{25}$

3) $\frac{22L}{35}$ 4) $\frac{11L}{35}$

10. **Masses 8, 2, 4, 2 kg are placed at the corners A, B, C,D respectively of a square ABCD of diagonal 80cm. The distance of centre of mass from A is**

1) 20cm 2) 30cm 3) 40cm 4) 60cm

11. Consider a two-particle system with the particles having masses m_1 and m_2. If the first particle is pushed towards the centre of mass through a distance d, by what distance should the second particle be moved so as to keep the centre of mass at the same position ? (AIEEE 2004)

1) $d' = \frac{m_1}{m_2}d$ 2) $d' = \frac{m_2}{m_1}d$

3) $d' = \frac{m_1+m_2}{m_1}d$ 4) $d' = \frac{m_1-m_2}{m_1}d$

12. A uniform square sheet has a side length of 12 cm. If one of the quadrants is removed, the shift in the centre of mass

1) 1 cm 2) $\sqrt{3}$ cm 3) $\sqrt{2}$ cm 4) 2 cm

13. A uniform circular sheet has a radius r. A square sheet whose diagonal length is equal to radius of the plate is removed. The maximum distance of the centre of mass of the remaining plate from the centre of this original sheet is

1) $\frac{r}{2(2\pi-1)}$ 2) $\frac{r}{(\pi-2)}$

3) $\frac{2r}{\pi-3}$ 4) $\frac{r}{2(\pi-1)}$

14. The centre of mass coordinates of a block of shape shown in fig. is

1) $\left(\frac{L}{2},\frac{L}{2}\right)$

2) $\left(\frac{5}{12}L,\frac{5}{12}L\right)$

3) $\left(\frac{2}{3}L,\frac{2}{3}L\right)$

4) $\left(\frac{L}{4},\frac{L}{4}\right)$

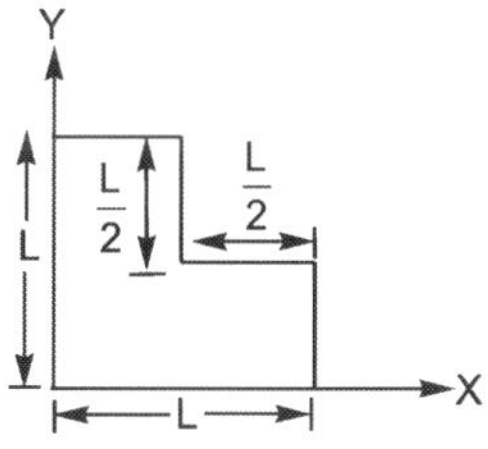

15. Two balls of masses 2M and 6M have radii 2R and 3R. Their centre of masses are separated by 10R. They move towards each other under their gravitational force. The distance moved by the centre of smaller sphere when the spheres touch each other is

1) 5/4R 2) 7/2R 3) 10/3R 4) 15/4R

16. A man of mass 80 kg is riding on a small cart of mass 40 kg which is rolling along a level floor at a speed of 2 m/s. He is running on the cart so that his velocity relative to the cart is 3 m/s in the direction opposite to the motion of cart. What is the speed of the centre of mass of the system.

1) 1.5 m/s 2) 1 m/s 3) 3 m/s 4) Zero

17. A boat of mass 40 kg is floating on water. A boy of mass 10 kg on the boat moves by 8 m towards the shore. The distance by which the boat moves away from the shore is

1) 4 m 2) 3 m 3) 2 m 4) 1.6 m

18. A shell projected from a level ground has a range R, if it did not explode. At the highest point, the shell explodes into two fragments having masses in the ratio 1:2 ; with each fragment moving horizontally immediately after the explosion. If the lighter fragment falls at a distance R/2 from the point of projection, behind the point of projection, the distance at which the other fragments falls from the point of projection is

1) R/2 2) 2R/3 4) 4R/3 4) 7R/4

19. A shell is projected from a level ground with a velocity u at angle q to the horizontal. When the shell is at the highest point it explodes into two equal fragments. One of the fragments retraces its path to the point of projection. At what distance from the point of projection does the other fragment fall.

1) $\frac{2u^2 Sin2\theta}{g}$ 2) $\frac{u^2 Sin2\theta}{g}$

3) $\frac{u^2 Sin2\theta}{2g}$ 4) $\frac{2u^2 Sin2\theta}{3g}$

20. Two blocks of equal mass are tied with a light string which passes over a massless pulley as shown in figure. The magnitude of acceleration of centre of mass of both the blocks is (neglect friction everywhere)

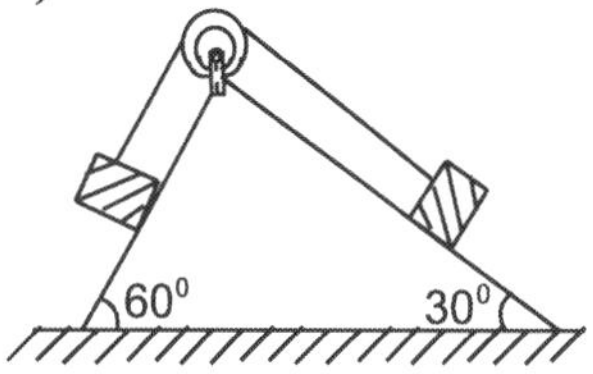

1) $\frac{\sqrt{3}-1}{4\sqrt{2}}g$

2) $\left(\sqrt{3}-1\right)g$

3) $\frac{g}{2}$ 4) $\left(\frac{\sqrt{3}-1}{\sqrt{2}}\right)g$

21. The centre of mass of a non uniform rod of length L whose mass per unit length $\lambda = \frac{Kx^2}{L}$, where K is a constant and x is the distance from one end is

1) $\frac{3L}{4}$ 2) $\frac{L}{8}$ 3) $\frac{K}{L}$ 4) $\frac{3K}{L}$

22. The co-ordinates of centre of mass of letter E which is cut from a uniform metal sheet are (Take origin at bottom left corner and width of letter 2 cm every where)

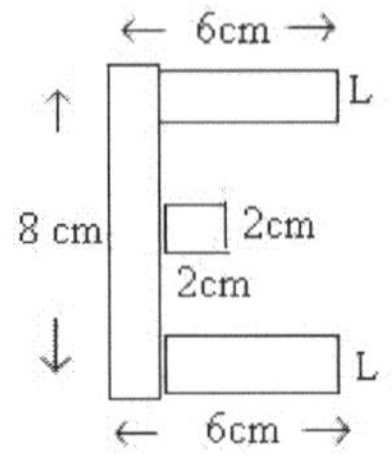

1) (2cm, 4cm)
2) (2.4cm, 5cm)
3) (3cm, 5cm)
4) 3.3cm, 5cm)

23. A cylindrical rod of length 1.2 m is made of three materials. The first part 0.5 m is made of iron, the second part 0.3 m is of copper and the last part 0.4 m is of aluminium. If the densities of iron, copper and aluminium are 8 x 10^3 kg/m^3, 9 x 10^3 kg/m^3, respecitively the centre of mass of the rod from right end of the aluminium rod is

1) 0.5 2) 0.67 m 3) 0.9 m 4) 1.2 m

24. If a triangular sheet is removed from one edge of a uniform rectangular sheet as shown in the figure, then the shift in the centre of mass of the system is

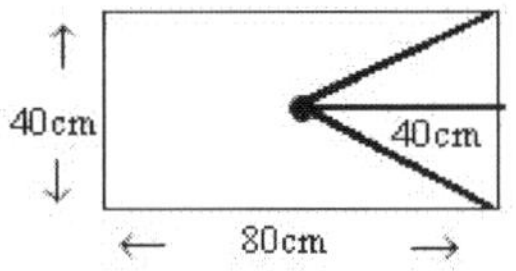

1) 3 cm
2) 6 cm
3) 9 cm
4) 12 cm

25. If a square of side R/2 is removed from a uniform circular disc of radius R as shown in the figure, the shift in centre of mass is

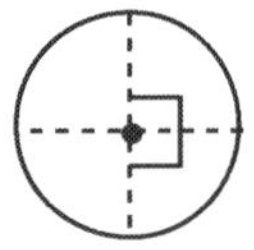

1) $\frac{R}{4\pi - 1}$ 2) $\frac{R}{2(4\pi - 1)}$

3) $\frac{R}{3(4\pi - 1)}$ 4) $\frac{R}{4(4\pi - 1)}$

ANSWERS

1) 1	2) 1	3) 2	4) 3	5) 2
6) 1	7) 3	8) 2	9) 3	10) 2
11) 1	12) 3	13) 1	14) 2	15) 4
16) 4	17) 4	18) 4	19) 1	20) 1
21) 1	22) 2	23) 2	24) 3	25) 4

COMPREHENSION QUESTIONS

Comprehension - I (1 - 3) :

Two particles of masses 3 kg and 2 kg are at (x_1, 0) and (x_2, 0) where $x_1 = 2t+3t^2$ and $x_2 = 5t-2t^2$, where x_1, x_2 are in metre and t is in seconds. Then

1. Position of centre of mass at t = 2s is

1) (0, 0) 2) (10.4, 0) 3) (52, 0) 4) (4.8, 0)

2. Velocity of centre of mass at t = 2s is

1) zero 2) 4.8 ms^{-1} 3) 7.2 ms^{-1} 4) 9.6 ms^{-1}

3. Acceleration of centre of mass at t = 2s is

1) zero 2) 2 ms^{-2} 3) 5.2 ms^{-2} 4) 10 ms^{-2}

Comprehension - II (4 - 6):

Two blocks of masses 4 kg and 1 kg respectively are tied to the ends of a string which passes over a light frictionless pully. The masses are held at rest at the same horizontal level and then released. Then (g=10ms^{-2})

4. Acceleration of 4 kg block is

1) 6 ms^{-2} 2) 5 ms^{-2} 3) 6.6 ms^{-2} 4) 4 ms^{-2}

5. Acceleration of centre of mass of the blcoks is

1) 10 ms^{-2} 2) 2.4 ms^{-2} 3) 3.6 ms^{-2} 4) 6 ms^{-2}

6. Displacement of centre of mass of the blocks in 2 sencods is

1) 3.6 m 2) 7.2 m 3) 14.4 m 4) 12 m

Comprehension - III (7 - 8) :

A Circular plate of diameter d is kept in contact with a square plate of edge d as shown in figure. The desnsity of the material and the thickness are same every where.

7. The centre of mass of the composite system will be

1) inside the circular plate 2) inside the square plate
3) at the point of contact 4) outside the system

8. Distance of centre of mass of the system from the centre of the circular disc is

1) $\frac{4d}{4+\pi}$ 2) $\frac{\pi d}{4+\pi}$ 3) $\frac{d}{1+\pi}$ 4) $\frac{\pi d}{1+\pi}$

Comprehension - IV (9 - 11):

A particle of mass m moving horizontally with v_0 strikes a smooth wedge of mass M, as shown in figure. After collision, the ball starts moving up the inclined face of the wedge and rises to a height h.

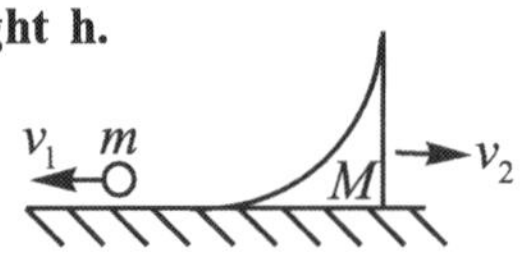

9. When the particle has risen to a height h on the wedge, then choose the correct alternative.

1) The particle is stationary with respect to ground

2) Both are stationary with respect to the centre of mass.

3) The kinetic energy of the centre of mass remains constant.

4) None of these

10. Identify the correct statement related to the situation when the particle starts moving downward.

1) The centre of mass of the system remains stationary.

2) Both the particle and the wedge remain stationary with respect to centre of mass.

3) When the particle reaches the horizontal surface its velocity relative to the wedge is v_0.

4) None of these

11. Choose the wrong statement related to the wedge M. (v_2 is final velocity of wedge)

1) Its kinetic energy is $K_f = \left(\dfrac{4m^2}{m+M}\right)gh$

2) $v_2 = \left(\dfrac{2m}{m+M}\right)v_0$

3) Its gain in kinetic energy is

$$\Delta K = \left(\frac{4mM}{(m+M^2)}\right)\left(\frac{1}{2}mv_0^2\right)$$

4) Its velocity is less than the velocity of centre of mass.

Comprehension - V (12 - 14) :

Two men Siva of mass m_1 and Krishna of mass m_2 are standing at the ends A and B of the trolley, respectively. The mass of the trolley is M and its length is L. Siva and Krishna can exchange their positions in three different ways :

Case–I : **m_1 moves towards B with U_{rel} and m_2 remains stationary until m_1 reaches its position ; and then m_2 starts moving and reaches the end A.**

Case–II : **m_2 moves towards A with U_{rel} and m_1 remains stationary until m_2 reaches its position, and then m_1 starts moving and reaches the end B.**

Case–III : **Both moves with U_{rel} with respect to trolley towards each other and reach then opposite ends.**

12. Choose the wrong statement related to *Case–I.*

1) As Siva moves, the trolley moves toward left and its velocity becomes maximum when he reaches the end B.

2) When Siva reaches the end B, then distance moved by the trolley is $\dfrac{m_1 L}{m_1 + m_2 + M}$

3) When Siva and Krishna has exchanged their positions, then displacement of the centre of mass of the system is zero.

4) When the Siva and Krishna have exchanged their positions, then final velocity of the trolley is zero.

13. Choose the correct statement related to *Case–III.*

1) As both the men above moves simultaneously, the velocity of the trolley at any instant is zero.

2) Both men reach their opposite ends simultaneously.

3) The distance travelled by both the men with respect to ground is same.

4) All the above.

14. Choose the correct statement related to the *three cases.*

1) The displacement of the trolley cannot exceed L

2) The displacement of the trolley is independent of the velocity of each man

3) The displacement of the trolley in all three cases is same

4) All the above

ANSWERS

1) 2	2) 3	3) 2	4) 1	5) 3
6) 2	7) 2	8) 1	9) 2	10) 3
11) 4	12) 1	13) 2	14) 4	

Chapter - 11

KINETIC THEORY OF GASES

KINETIC THEORY OF GASES

SYNOPSIS

1 Kinetic Theory of Gases relates macroscopic properties of gas such as pressure, temperature etc., to the microscopic properties of a gas molecule such as speed momentum, KE of molecules etc.

2. A gas consisting of large number of identical, tiny, spherical, neutral and elastic particles called molecules.

3. In a gas molecules are moving in all possible directions with all possible speeds.

4. The space occupied by the molecule is much smaller than the volume of the gas

5. There is no force of attraction among the molecules

6. The pressure of the gas is due to elastic collisions of the gas molecule with the walls of the container.

7. The time of contact of a moving molecule with the walls of container is negligible as compared to the intervals between two successive collisions on the same walls of container.

A. Different - Types of speeds of Gas molecules:

The motion of molecules in a gas is characterised by any of the following three speeds:

1) Root mean square speed(r.m.s):

It is defined as the square root of mean of squares of the speed of different - molecules ie.

$$V_{rms} = \sqrt{\left(\frac{V_1^2 + V_2^2 + ... + V_N^2}{N}\right)} = \sqrt{\left(\overline{V}\right)^2}$$

Hear $\left(\overline{V}\right)^2$ is the mean square velocity according to the kinetic theory of gases

$$V_{rms} = \sqrt{\frac{3PV}{mass\ gas}} = \sqrt{\frac{3P}{\rho}} = \sqrt{\frac{3RT}{M}} = \sqrt{\frac{3KT}{m}}$$

Where M is the molecular weight while m the mass of a single molecule.

2) Most probaable speed:

It is the speed which maximum number of molecules in a gas have at constant temperaature and is given by

$$V_{mp} = \sqrt{\frac{2RT}{M}} = \sqrt{\frac{2}{3}}\,V_{rms} = 0.816V_{rms}$$

3) Average speed:

It is the arithemetic mean of the speeds of molecules in a gas at a given temperature,

i.e. $V_{av} = \dfrac{(V_1 + V_2 + ... + V_n)}{N}$

according to kenetic theory of gases

$$V_{av} = \sqrt{\frac{8RT}{\pi M}} = \left(\sqrt{\frac{8}{3\pi}}\right)V_{rms} = 0.92V_{rams}$$

4) Relation between above three speeds

$V_{rms} > V_{av} > V_{mp}$

So a planet (or) satellite will have atmosphere only and only if $V_{rms} < V_e$. (V_e is the escape velocity)

B. Some important points concerning kinetic theory of gases

1) Pressure P exerted by an ideal gas is given by

$$p = \frac{1}{3}\frac{mN}{V}\left(\overline{V}\right)^2 = \frac{2}{3}E$$

hear, $\left(\overline{V}\right)^2$ is mean square velocity

hear E is the kinetic energy per unit volume

$$\therefore E = \frac{1}{2}\frac{mN}{V}\left(\overline{V}\right)^2 = \frac{1}{2}\rho\left(\overline{V}\right)^2$$

Density of the gas $\rho = \left(\dfrac{mN}{V}\right)$

Formula For Translational Kinetic Energy :

* The total Kinetic Energy of all the molecules of a gas is given by $E_T = \frac{3}{2}kTN$ (Where N are the total no. of gas molecules)
* If n moles of gas are there in a container then total Kinetic Energy $E_T = \frac{3}{2}nRT$
* Average translation K.E of a gas molecules depends only on its temperature and is independent of its nature.
* The average distance travelled by a molecule between two successive collisions is called mean free path and is given by

$$\lambda = \frac{1}{\sqrt{2}\pi d^2 n}$$

Where 'd' is the diameter of molecules and n is the number of molecules per unit volume.

Degrees of Freedom :

* The total number of independent modes or ways in which a particle can possess energy is called Degrees of freedom. (f)
* For mono atomic gas f = 3
* For di atomic gas f = 5

Law of equipartition of energy :

* According to it the energy of gas molecule is equally distributed among its various degrees of freedom. Each degree of freedom is associated with energy of $\frac{1}{2}KT$.
* For a monoatomic ideal gas, the energy for molecule is $\frac{3}{2}KT$
* For diatomic ideal gas the average energy for molecule is $\frac{5}{2}KT$
* For Polyatomic gas molecule, such as CO_2, H_2O which has 6 degrees of freedom, the energy for molecule is $\frac{6}{2}KT$
* The internal energy of 'n' moles of a gas in which each molecule has 'f' degrees of freedom will have internal energy $U = \frac{f}{2}(nRT)$
* For monoatomic gas molecule of 'n' moles $U = \frac{3}{2}(nRT)$
* For diatomic gas of 'n' moles $U = \frac{5}{2}(nRT)$

EXERCISE - I

1. **The moleculer weights of two gases are M_1 and M_2, then at a temperature the ratio of root mean square velocity v_1 and v_2 will be**

   1) $\sqrt{\frac{M_1}{M_2}}$ 2) $\sqrt{\frac{M_2}{M_1}}$

   3) $\sqrt{\frac{M_1+M_2}{M_1-M_2}}$ 4) $\sqrt{\frac{M_1-M_2}{M_1+M_2}}$

2. **At a given temperature the r.m.s velocity will be greater for**

   1) Hydrogen 2) Oxygen

   3) Equal for both 4) Nothing is definite

3. **In thermal equilibrium, the average velocity of gas molecules is**

   1) Proportional to $\sqrt{T}$ 2) Proportional to T^2

   3) Proportional to T^3 4) zero

4. **A gas is allowed to expand isothermally. The root mean square velocity of the molecules**

   1) Will increase 2) Will decrease

   3) Will remain unchanged

   4) Depends on the other factors

5. **Moon has no atmosphere because**

   1) The r.m.s. velocity of all gases is more than the escape velocity from moon's surface

   2) Its surface is not smooth

   3) It is quite far away from the earth

   4) It does not have population and plants

6. **Which of the following statement is true**

   1) Absolute zero degree temperature is not zero energy temperature

   2) Two different gases at the same temperature, pressure have equal root mean square velocities

   3) The root mean square speed of the molecules of different ideal gases, maintained at the same temperature are the same

   4) Given sample of 1 CC of hydrogen and 1 CC of oxygen both at NTP, oxygen sample has a large number of molecules.

7. **According to the kinetic theory of gases the r.ms. velocity of gas molecules is directly proportional to**

   1) T 2) $\sqrt{T}$ 3) T^2 4) $\frac{1}{\sqrt{T}}$

8. **The root mean square speed of the molecules of a gas is**

   1) Independent of its pressure but directly proportional to its Kelvin temperature

   2) Directly proportional to the square roots of both its pressure and its Kelvin temperature

   3) Independent of its pressure but directly proportional to the square root of its Kelvin temperature

   4) Directly proportional to both its pressure and its Kelvin temperature.

9. **Cooking gas co1ntainers are kept in a lorry moving with uniform speed. The temperature of the gas molecules inside will [AIEEE 2002]**
   1) Increase 2) Decrease
   3) Remain same
   4) Decrease for some, while increase for others

10. **At a given temperature if V_{rms} is the root mean square velocity of the molecules of a gas, V_s is the velocity of sound in it, then these are related as** $\left(\gamma = \frac{C_p}{C_v}\right)$
   1) $V_{rms} = V_s$ 2) $V_{rms} = \sqrt{\frac{3}{\gamma}} x V_s$
   3) $V_{rms} = \sqrt{\frac{\gamma}{3}} x V_s$ 4) $V_{rms} = \left(\frac{3}{\gamma}\right) x V_s$

11. **For a gas at a temperature T the root mean square velocity v_{rms}, the most probable speed v_{mp}, and the average speed a_{av} obey the relationship**
   1) $v_{av} > v_{rms} > v_{mp}$ 2) $v_{rms} > v_{av} > v_{mp}$
   3) $v_{mp} > v_{av} > v_{rms}$ 4) $v_{mp} > v_{rms} > v_{av}$

12. **If V_H, V_N and V_O denote the root mean square velocities of molecules of hydrogen, nitrogen and oxygen respectively at a given temperature, then**
   1) $V_N > V_O > V_H$ 2) $V_H > V_N > V_O$
   3) $V_O = V_N = V_H$ 4) $V_O > V_H > V_N$

13. **If the pressure in a closed vessel is reduced by drawing out some gas, the mean free path of the molecules**
   1) Is decreased 2) Is increased
   3) Remains unchanged
   4) Increases or decreases according to the nature of the gas

14. **At constant volume, temperature is increased. Then**
   1) Collission on walls will be less
   2) Number of collisions per unit time will increase
   3) Collisions will be in straight lines
   4) Collisions will not change

15. **A monoatomic gas molecule has**
   1) Three degrees of freedom
   2) Four degrees of freedom
   3) Five degrees of freedom
   4) Six degree of freedom

16. **A diatomic molecule has how many degrees of freedom**
   1) 3 2) 4 3) 5 4) 6

17. **If the degree of freedom of a gas are f, then the ratio of two specific heats C_p / C_v is given by**
   1) $\frac{2}{f}+1$ 2) $1-\frac{2}{f}$ 3) $1+\frac{1}{f}$ 4) $1-\frac{1}{f}$

18. **The specific heat of an ideal gas is**
   1) Proportional to T 2) Proportional to T^2
   3) Proportional to T^3 4) Independent of T

19. **Molar specific heat at constant volume is C_v for a monoatomic gas is**
   1) $\frac{3}{2}R$ 2) $\frac{5}{2}R$ 3) 3R 4) 2R

20. **The pressure exerted on the walls of the container by a gas is due to the fact that gas molecules are [CPMT 1972; NCERT 1980]**
   1) Losing their kinetic walls
   2) Sticking to the energy
   3) Changing their momenta due to collision with the walls
   4) Getting accelarated towards the wall

21. **The relation between the gas pressure P and average kinetic energy per unit volume E is**
   1) $P = \frac{1}{2}E$ 2) $P = E$ 3) $P = \frac{3}{2}E$ 4) $P = \frac{2}{3}E$

22. **Relationship between P, V and E for a gas is**
   1) $P = \frac{3}{2}EV$ 2) $V = \frac{2}{3}EP$
   3) $PV = \frac{3}{2}E$ 4) $PV = \frac{2}{3}E$

23. **Mean kinetic energy (or average energy) per gm molecule of a monoatomic gas is given by**
   1) $\frac{3}{2}RT$ 2) $\frac{1}{2}KT$ 3) $\frac{1}{2}RT$ 4) $\frac{3}{2}KT$

24. **The kinetic energy per gm mol for a diatomic gas at room temperature is**
   1) $3RT$ 2) $\frac{5}{2}RT$ 3) $\frac{3}{2}RT$ 4) $\frac{1}{2}RT$

25. **The translational kinetic energy of gas molecule for one mole of the gas is equal to**
   1) $\frac{3}{2}RT$ 2) $\frac{2}{3}RT$ 3) $\frac{1}{2}RT$ 4) $\frac{2}{3}KT$

26. At O K which of the following properties of a gas will be zero

1) Kinetic energy 2) Potential energy
3) Vibrational energy 4) Density

27. On absolute temperature, the kinetic energy of the molecules

1) Becomes zero 2) Becomes maximum
3) Becomes minimum 4) Remains constant

28. In the absence of intermolecular force of attraction, the observed pressure P will be

1) P 2) < P 3) > P 4) Zero

29. The molecules of an ideal gas at a certain temperature have

1) Only potential energy
2) Only kinetic energy
3) Potential and kinetic energy both
4) None of the above

30. In kinetic theory of gases, which of the following statements regarding elastic collisions of the molecules is wrong

1) Kinetic energy is lost in collisions
2) Kinetic energy remains constant in collision
3) Momentum is conserved in collision
4) Pressure of the gas remains constant in collisions

31. The rms speed of a gas molecule is

1) $\sqrt{(M/3RT)}$ 2) (M/3RT)
3) $\sqrt{(3RT/M)}$ 4) $(3RT/M)^2$

32. The average velocity of the molecules in a gas in equilibrium is [NCERT 1974;CPMT 1983]

1) $\propto \sqrt{T}$ 2) $\propto T$ 3) $\propto T^2$ 4) Equal to zero

33. The root mean square speed of molecules of ideal gases at the same tempersture are

1) The same **[CPMT 1981]**
2) Inversely proportional to the square root of the molecular weight
3) Directly proportional to the molecular weight
4) Inversely proportional to the molecular weight.

34. At a given volume and temperature the pressure of a gas will vary

1) Directly proportional to the mass of the gas
2) Inversely as the mass of the gas
3) Linearly as the square of the mass of the gas
4) Inversely as the square of the mass of the gas

35. When a gas is forced in a smaller volume without change in temperature, its pressure increases because its molecules.

1) Strike the unit area of the container walls more often
2) Strike the unit area of the container walls at higher speed
3) Strike the unit area of the container wall with greater force
4) Have more energy

36. When the temperature of a gas in a rigid container is raised, the pressure exerted by the gas on the walls of the container increases because [CBSE1993]

1) The molecules have higher average speed, so strike the walls more often
2) The molecules lose more KE each time they strike the walls
3) The molecules are now in contact with the walls for small intervals
4) The molecules collide with each other more often

37. Sometimes a cycle with well inflated tyres left in the sun has its tube burst open because:

1) The volume of air inside the tube increases
2) The pressure of air inside the tube increases
3) The tube containing air melts
4) The volume and pressure both increase

38. The kinetic energy of a given sample of an ideal gas depends only on its

1) Volume 2) Pressure 3) Density 4) Temperature

39. The mean translational kinetic energy of a molecule of an ideal gas at temperature T K is

1) 1/2KT 2) KT 3) 3/2KT 4) 2KT

40. A polyatomic gas with n degrees of freedom has a mean energy per molecule given by

1) nkT/N_A 2) $nkT/2N_A$ 3) nkT/2 4) 3kT/2

41. In a jar having a mixture of H_2 and He

1) Hydrogen has more mean KE
2) Helium has more mean KE
3) Both have same mean KE
4) Boh have same internal energy

42. There is no atmosphee on the moon because :

1) Moon has no gravity
2) The escape velocity from the surface of moon is very small
3) The r.m.s velocity of gas molecules is grater than escape velocity
4) The above statement is wrong;moon has full atmosphere this is why the american astronauts could land on it and could stay for sometime

43. **The Brownian motion refers to [MNR 1978]**
   1) The motion of molecules in fluid medium
   2) The motion of microscopic particles under the impact of molecules in a fluid medium
   3) The motion of partcles as recorded in a cloud chamber
   4) All the above

ANSWERS

1) 2	2) 1	3) 1	4) 3	5) 1
6) 1	7) 2	8) 3	9) 3	10) 2
11) 2	12) 2	13) 2	14) 2	15) 1
16) 3	17) 1	18) 4	19) 1	20) 3
21) 4	22) 4	23) 1	24) 2	25) 1
26) 1	27) 1	28) 3	29) 2	30) 1
31) 3	32) 1	33) 2	34) 1	35) 1
36) 1	37) 2	38) 4	39) 3	40) 3
41) 1	42) 3	43) 2		

EXERCISE - II (A)

1. **At a certain temperature, the r.m.s. velocity for O_2 is 400 m/sec. At the same temperature, the r.m.s veloctiy for H_2 molecules will be**
   1) 100 m/sec 2) 25 m/sec 3) 1600 m/sec 4) 6400 m/sec
2. **To what temperature should the hydrogen at 327°C be cooled at constant pressure, so that the root mean square velocity of its molecules become half of its previous value**
   1) –123°C 2) 23°C 3) –100°C 4) 0°C
3. **At a pressure of 24 x 10^5 dyne/cm², the volume of O_2 is 10litre and mass is 20 gm. The r.m.s velocity will be**
   1) 800 m/sec 2) 400 m/sec
   3) 600 m/sec 4) 200 m/sec
4. **The molecules of a given mass of a gas have a r.m.s. velocity of 200 m/sec at 27°C and 1.0 x 10^5 N/m² pressure. When the temperature is 127°C and pressure is 0.5 x 10^5 N/m², the r.m.s velocity in m/sec will be**
   1) $\frac{100\sqrt{2}}{3}$ 2) $100\sqrt{2}$ 3) $\frac{400}{\sqrt{3}}$ 4) 400
5. **The respective speeds of five molecules are 2, 1.5, 1.6, 1.6 and 1.2 km/sec. The most probable speed in km/sec will be**
   1) 2 2) 1.58 3) 1.6 4) 1.31
6. **Gas at a pressure P_0 is contained in a vessel.If the masses of all the molecules are halved and their speeds are doubled, the resulting pressure P will be equal to**
   1) $4P_0$ 2) $2P_0$ 3) P_0 4) $\frac{P_0}{2}$
7. **At a given temperature the root mean square velocities of oxygen and hydrogen molecules are in the ratio**
   1) 16:1 2) 1:16 3) 4:1 4) 1:4
8. **In the two vessels of same volume, atomic hydrogen and helium at pressure 1 atm and 2 atm are filled. If temperature of both the sample is same then average speed of hydrogen atoms $<C_H>$ will be related to that of helium $<C_{He}>$ as**
   1) $<C_H> = \sqrt{2}<C_{He}>$ 2) $<C_H> = <C_{He}>$
   3) $<C_H> = 2<C_{He}>$ 4) $<C_H> = \frac{<C_{He}>}{2}$
9. **What is the velocity of monoatomic gas having pressure 1 kilopasal and density 2.6kg/m³**
   1) 3.6 m/s 2) 8.9 m/s 3) zero 4) 25 m/s
10. **A cylinder of capacity 20 litres is filled with H_2 gas. The total average kinetic energy of translatory motion of its molecules is 1.5x10^5J. The pressure of hydrogen in the cylinder is**
    1) 2x10^6N/m² 2) 3x10^6N/m²
    3) 4x10^6N/m² 4) 5x10^6N/m²
11. **The value of universal gas constant is 8.3 J/mole/K, the mean kinetic energy of 32gm of oxygen at -73°C will be**
    1) 480J 2) 4980J 3) 2490J 4) 100 J
12. **If number of molecules of H_2 are double than that of O_2, then ratio of kinetic energy of hydrogen and that of oxygen at 300 K is**
    1) 1:1 2) 1:2 3) 2:1 4) 1:16
13. **At which of the following temperature would the molecules of a gas have twice the average kinetic energy they have at 20°C**
    1) 40°C 2) 80°C 3) 313°C 4) 586°C

ANSWERS

1) 3	2) 1	3) 3	4) 3	5) 3
6) 2	7) 4	8) 3	9) 4	10) 4
11) 3	12) 1	13) 3		

EXERCISE - II (B)

1. **At what temperature is the root mean square velocity of gaseous hydrogen molecules is equal to that of oxygen molecules at 47°C**
   1) 20K 2) 80K 3) –73K 4) 3K
2. **A sample of gas is at 0°C. To what temperature it must be raised in order to double the r.m.s speed of the molecule**
   1) 270°C 2) 819°C 3) 1090°C 4) 100°C

3. **At what temperature will the oxygen molceules have the same root mean squre speed as hydrogen molecules at 200K**
1) 800K 2) 1600K 3) 2400K 4) 3200K

4. **Root mean square velocity of a particle is υ at pressure P. If pressure is increased two times, then the r.m.s velocity becomes**
1) 2υ 2) 3υ 3) 0.5υ 4) υ

5. **The respective speeds of the molecules are 1, 2, 3, 4 and 5 km/sec. The ratio of their r.m.s velocity and the average velocity will be**
1) $\sqrt{11}:3$ 2) $3:\sqrt{11}$ 3) 1: 2 4) 3: 4

6. **Let A and B the two gases and given $\frac{T_A}{M_A}=4\frac{T_B}{M_B}$, where T is the temperature and M is molecular mass. If v_A and v_B are the r.m.s speed then the ration $\frac{v_A}{v_B}$ will be equal to**
1) 2 2) 4 3) 1 4) 0.5

7. **At a given temperature the ratio of r.m.s velocities of hydrogen molcecule and helium atom will be**
1) $\sqrt{2}:1$ 2) $1:\sqrt{2}$ 3) 1: 2 4) 2: 1

8. **If mass of He atom is 4 times that of hydrogen atom then mean velocity of He is**
1) 2 times of H-mean value
2) 1/2 times of H-mean value
3) 4 times of H-mean value
4) Same as H- mean value

9. **The r.m.s speed of the molecules of a gas in a vessel is 400 ms^{-1}. If half of the gas leaks out, at constant temperature, the r.m.s speed of the remaining molecules will be [Kerala (E) 2002]**
1) 800 ms^{-1} 2) $400\sqrt{2}ms^{-1}$ 3) 400 ms^{-1} 4) 200 ms^{-1}

10. **The root mean square speed of hydrogen molecules of an ideal hydrogen gas kept in a gas chamber at 0° C is 3180 meters/second. The pressure on the hydrogen gas is (Density of hydrogen gas is $8.99x10^{-2}$kg /m^3 , 1 atmosphere = $1.01x10^5N/m^2$)**
1) 1.0atm 2) 1.5 atm 3) 2.0 atm 4) 3.0atm

11. **The kinetic energy of translation of 20gm of oxygen at 47°C is (molecular wt. of oxygen is 32gm/mol and R=8.3 J/mol/K)**
1) 2490joules 2) 2490ergs
3) 830 joules 4) 124.5 joules

12. **The average kinetic energy of a gas at –23° C and 75cm pressure is $5x10^{-14}$erg for H_2.The mean kinetic energy of the O_2 at 227°C and 150 cm pressure will be**
1) $80x10^{-14}$erg 2) $20x10^{-14}$erg
3) $40x10^{-14}$erg 4) $10x10^{-14}$erg

13. **At what temperature is the kinetic energy of a gas molecule double that of its value of 27°C**
1) 54°C 2) 300K 3) 327°C 4) 108°C

ANSWERS

1) 1	2) 2	3) 4	4) 4	5) 1
6) 1	7) 1	8) 2	9) 3	10) 4
11) 1	12) 4	13) 3		

EXERCISE - III (A)

1. **Four particles have speeds 2, 3, 4 and 5 cm/s respetively. Their rms speed is [CPMT-93]**
1) 3.5 cm/s 2) (27/2)cm/s
3) $\sqrt{54}\ cm/s$ 4) $\sqrt{54}/2\ cm/s$

2. **At room temperature the rms speed of the molecules of a certain diatomic gas is found to be 1930 m/s. The gas is [IIT 1983]**
1) H_2 2) F_2 3) O_2 4) Cl_2

3. **The absolute temperature of a gas increases 3 times. The root mean square velocity of the molecules will become [CPMT 1989]**
1) 3 times 2) 9 times
3) (1/3) times 4) $\sqrt{3}$ *times*

4. **The temperature of an ideal gas is increased from 27°C to 927°C. The rms speed of its molecules will becomes. [NCERT-83; CBSE-94]**
1) Twice 2) Half 3) Four times 4) One fourth

5. **If the rms velocity of oxygen molecule at certain temperature is 0.5 km/s, the rms velocity for hydrogen molecule at the same temperature will be [CPMT 1984; EAMCET 1993]**
1) 2 km/s 2) 4 km/s 3) 9 km/s 4) 16 km/s

6. **The average translational energy and the rms speed of molecules in a sample of oxygen gas at 300 K are 6.21 x 10^{-21} J and 484 m/s respectively. The corresponding values at 600 K are nearly (assuing ideal gas behaviour)**
1) 12.42 x 10^{-21} J, 968 m/s
2) 8.78 x 10^{-21} J, 684 m/s
3) 6.21 x 10^{-21} J, 968 m/s
4) 12.42 x 10^{-21} J, 684 m/s

7. The temperature of an ideal gas is increased from 120 K to 480K. If at 120 K the root mean square velocity of the gas molecules is ν, at 480 K it becomes [IIT 1996]

1) 4ν 2) 2ν 3) $\nu/2$ 4) $\nu/4$

8. At what temperature is the rms speed o gaseous hydrogen molecules equal to that of oxygen molecules at 47^0C [CPMT 1983]

1) 20 K 2) 80 K 3) –73 K 4) 3 K

ANSWERS

1) 4 2) 1 3) 3 4) 1 5) 1
6) 4 7) 2 8) 1

EXERCISE - III (B)

1. The root mean square velocity of molecules of a gas whose density is 1.4 kg m^{-3} at a pressure of 76 cm of mercury (specific gravity of mercury = 13.6 and g = 9.81 m s^{-2}) is

1) 4.66 x 10^2 ms^{-1} 2) 2.5 x 10^2 ms^{-1}
3) 2.33 x 10^2 ms^{-1} 4) 6.66 x 10^2 ms^{-1}

2. The rms velocity of a gas at a given temperature is 300 m/s. What will be the rms velocity of a gas having twice the molecular weight and half the temperature in K. [CPMT-82]

1) 300 m/s 2) 600 m/s
3) 75 m/s 4) 150 m/s

3. Two vessels having equal volume contain molecular hydrogen at one atmosphere and helium at two atmospheres respectively; if both samples are at the same temperature, the mean velocity of hydrogen molecules is [NCRET 1979;CPMT 1985]

1) Equal to that of helium
2) Twice that of helium
3) Half that of helium
4) $\sqrt{2}$ times that of helium

4. The molecular weight of a gas is 2. Find the root mean square velocity of its moleculars at 100^0C (R = 8.3 J mol^{-1} K^{-1})

1) 2.155 x 10^3 m/s 2) 1.5 x 10^3 m/s
3) 3.155 x 10^3 m/s 4) 4.155 x 10^3 m/s

5. Three closed vessels A, B and C are at the same temperature T and contain gases which obey Maxwellian distribution of velocities. Vessel A contain O_2, B only N_2 and C mixure of equal quantities of O_2 and N_2. If the average speed of the O_2 molecules in the vessel A is V_1, that of N_2 molecules in the vessel B is v_2, the average speed of the O_2 molecules in vessel C is [IIT 1992]

1) $\frac{(v_1+v_2)}{2}$ 2) v_1 3) $\sqrt{4v_2}$ 4) $\sqrt{3kT/m}$

6. The average translational kinetic energy of O_2 (molar mass 32) molecules at a particular temperature is 0.048 eV. The translational kinetic energy of N_2 (molar mass 28) molecules in eV at the same temperature is [IIT 1997(II)]

1) 0.0015 2) 0.003
3) 0.048 4) 0.768

7. The ratio of the mean KE of chlorine and oxygen molecules at the same temperature is [NCRET 1981]

1) 16 : 35.5 2) 35.5 : 16
3) $(35.5)^2 : (16)^2$ 4) 1 : 1

8. A vessel contains a mixture of one mole of oxygen and two moles of nitrogen at 300 K. The ratio of the average rotational kinetic energy per O_2 molecule to that per N_2 molecule is [IIT 1998]

1) 1 : 1 2) 1 : 2 3) 2 : 1
4) depends on the moment of inertia of the two molecules

ANSWERS

1) 1 2) 4 3) 4 4) 1 5) 2
6) 3 7) 4 8) 1

Made in the USA
Monee, IL
05 November 2026

67811010R10168